Mixed Severity Fire
Nature's Phoenix

Mixed Severity Fires
Nature's Phoenix

Second Edition

Edited by

Dominick A. DellaSala
Chief Scientist, Wild Heritage, a project of the Earth Island Institute, Berkeley, CA, United States

Chad T. Hanson
Research Ecologist and Director, John Muir Project of Earth Island Institute, Berkeley, CA, United States

ELSEVIER

Elsevier
Radarweg 29, PO Box 211, 1000 AE Amsterdam, Netherlands
125 London Wall, London EC2Y 5AS, United Kingdom
50 Hampshire Street, 5th Floor, Cambridge, MA 02139, United States

Copyright © 2024 Elsevier Inc. All rights are reserved, including those for text and data mining, AI training, and similar technologies.

Publisher's note: Elsevier takes a neutral position with respect to territorial disputes or jurisdictional claims in its published content, including in maps and institutional affiliations.

No part of this publication may be reproduced or transmitted in any form or by any means, electronic or mechanical, including photocopying, recording, or any information storage and retrieval system, without permission in writing from the publisher. Details on how to seek permission, further information about the Publisher's permissions policies and our arrangements with organizations such as the Copyright Clearance Center and the Copyright Licensing Agency, can be found at our website: www.elsevier.com/permissions.

This book and the individual contributions contained in it are protected under copyright by the Publisher (other than as may be noted herein).

Notices

Knowledge and best practice in this field are constantly changing. As new research and experience broaden our understanding, changes in research methods, professional practices, or medical treatment may become necessary.

Practitioners and researchers must always rely on their own experience and knowledge in evaluating and using any information, methods, compounds, or experiments described herein. In using such information or methods they should be mindful of their own safety and the safety of others, including parties for whom they have a professional responsibility.

To the fullest extent of the law, neither the Publisher nor the authors, contributors, or editors, assume any liability for any injury and/or damage to persons or property as a matter of products liability, negligence or otherwise, or from any use or operation of any methods, products, instructions, or ideas contained in the material herein.

ISBN: 978-0-443-13790-7

For information on all Elsevier publications visit our website at
https://www.elsevier.com/books-and-journals

Publisher: Candice Janco
Acquisitions Editor: Maria Elekidou
Editorial Project Manager: Rupinder Heron
Production Project Manager: Paul Prasad Chandramohan
Cover designer: Matthew Limbert

Typeset by TNQ Technologies

Contents

Contributors   xiii
Preface   xv
Acknowledgments   xxxiii

Section I
Biodiversity of Mixed- and High-Severity Fires

1. Setting the Stage for Mixed- and High-Severity Fire

Chad T. Hanson, Dominick A. DellaSala, Rosemary L. Sherriff, Richard L. Hutto, Thomas T. Veblen and William L. Baker

1.1.	Earlier Hypotheses and Current Research	3
	Do Open and Park-like Structures Provide an Accurate Historical Baseline for Dry Forest Types in Western US Forests?	6
	Does Time Since Fire Influence Fire Severity?	6
	What is the Evidence for Mixed- and High-Severity Fire?	8
1.2.	Ecosystem Resilience and Mixed- and High-Severity Fire	14
1.3.	Mixed- and High-Severity Fires Increases Are Equivocal	15
1.4.	Conclusions	17
	References	18

2. Ecosystem Benefits of Megafires

Dominick A. DellaSala and Chad T. Hanson

2.1.	Just What are Megafires?	27
2.2.	Megafires as Global Change Agents	30
2.3.	Megafires, Large Severe Fire Patches, and Complex Early Seral Forests	31
2.4.	Historical Evidence of Megafires	34
	Rocky Mountain Region	35
	Eastern Cascades and Southern Cascades	36
	Oregon Coast Range and Klamath Region	36
	Sierra Nevada	38
	Southwestern United States and Pacific Southwest	38
	Black Hills	38

2.5.	Megafires and Landscape Heterogeneity	39
2.6.	Are Megafires Increasing?	43
2.7.	Language Matters	45
2.8.	Conclusions	46

Appendix 2.1. Some Fires of Historical Significance From Records Compiled by the National Interagency Fire Center (http://www.nifc.gov/fireInfo/fireInfo_stats_histSigFires.html) From 1825 to 2013   49
References   55
Further reading   62

3. Using Bird Ecology to Learn About the Benefits of Severe Fire

Richard L. Hutto, Monica L. Bond and Dominick A. DellaSala

3.1.	Introduction	63
3.2.	Insights From Bird Studies	64
	Time Since Fire	65
	Old-Growth Forests	66
	Postfire Vegetation Conditions	68
	Bird Species in Other Regions That Seem to Require Severe Fire (Lesson 3 Continued)	79
3.3.	Postfire Management Implications	84
3.4.	Fire Risk Reduction Should Be Focused on Human Population Centers	84
3.5.	Fire Suppression Should be Focused on Human Communities	84
3.6.	High-Severity Fires Beget Mixed-Severity Results	85
3.7.	Mitigate Fire Severity Through Thinning Only Where Ecologically Appropriate	85
3.8.	Postfire "Salvage" Logging in the Name of Restoration or Rehabilitation is Always Ecologically Inappropriate and Misdirected	86
3.9.	We can do More Harm Than Good Trying to "Mimic" Nature	87
3.10.	Concluding Remarks	89
References		89

4. Mammals and Mixed- and High-Severity Fire

Monica L. Bond

4.1.	Introduction	99
4.2.	Bats	101
4.3.	Small Mammals	104
	Chaparral and Coastal Sage Scrub	106
	Forests and Woodlands	106
	Deserts	109
	Deer Mice	110

4.4.	Carnivores	111
	Mesocarnivores and Large Cats	111
	Bears	113
4.5.	Ungulates	116
4.6.	Management and Conservation Relevance	120
4.7.	Conclusions	121

Appendix 4.1 The Number of Studies by Taxa Showing Directional Response (Negative, Neutral, or Positive) to Severe Wildfire Over Three Time Periods Following Fire. Studies Cited Include Unburned Areas compared to Severely Burned Areas With no Postfire Logging, and Excluded Prescribed Burns. For Small Mammals, Only Species With Enough Detections to Determine Directional Response Were Reported   121
References   125

Section II
Global Perspectives on Mixed- and High-Severity Fires

5. Bark Beetles and High-Severity Fires in Rocky Mountain Subalpine Forests

Dominik Kulakowski and Thomas T. Veblen

5.1.	Fire, Beetles, and their Interactions	133
5.2.	How Do Outbreaks Affect Subsequent High-Severity Fires?	137
	Methodological Considerations	137
5.3.	How Do High-Severity Fires Affect Subsequent Outbreaks?	148
	Lodgepole Pine Forests	148
5.4.	How Are Interacting Fires and Bark Beetles Affecting Forest Resilience in the Context of Climate Change?	150
5.5.	Conclusions	153
Acknowledgments		154
References		154

6. High-Severity Fire in Chaparral: Cognitive Dissonance in the Shrublands

Richard W. Halsey and Alexandra D. Syphard

6.1.	Chaparral and the Fire Suppression Paradigm	163
6.2.	The Facts About Chaparral Fires: They Burn Intensely and Severely	166
6.3.	Fire Misconceptions are Pervasive	168
	Confusing Fire Regimes	169
	Native American Burning	171
	Succession Rather Than Destruction	171
	Decadence, Productivity, and Old-Growth Chaparral	172
	Allelopathy	174

		Fire Suppression Myth	174
		Too Much Fire Degrades Chaparral	175
		Type Conversion and Prescribed Fire	179
		Combustible Resins and Hydrophobia	180
	6.4.	Reducing Cognitive Dissonance	182
		Local Agency	183
		State Agency	184
		Media	185
	6.5.	Paradigm Change Revisited	186
	6.6.	Conclusion: Making the Paradigm Shift	189
	References		190

7. Regional Case Studies: Southeast Australia, Sub-Saharan Africa, Central Europe, and Boreal Canada

Case Study: The Ecology of Mixed-severity Fire in Mountain Ash Forests — 197

Laurence E. Berry and Holly Sitters

7.1.	The Setting	197
7.2.	Mountain Ash Life Cycle	199
7.3.	Influence of Stand Age on Fire Severity	201
7.4.	Distribution of Old-growth Forests	202
7.5.	Mixed-severity Fire and Fauna of Mountain Ash Forests	202
7.6.	Fauna and Fire-affected Habitat Structures	203
7.7.	Faunal Response to the Spatial Outcomes of Fire	205
7.8.	Conservation Challenges and Future Fire	205
Acknowledgments		207
References		207

Case Study: The Importance of Mixed- and High-severity Fires in Sub-Saharan Africa — 211

Ronald W. Abrams

7.9.	The Big Picture	211
7.10.	Where Is Fire Important in Sub-Saharan Africa?	212
7.11.	What About People and Fire?	213
7.12.	Coevolution of Savannah, Herbivores, and Fire	214
7.13.	Herbivores and Fire	215
7.14.	Beyond Africa's Savannah Habitat	218
7.15.	Habitat Changes Forest Loss/Gain and Other Considerations	218
7.16.	Habitat Management Through Controlled Burns	220
7.17.	Southwestern Cape Renosterveld Management	223
7.18.	Conclusion	225
References		226

Case Study: Response of Invertebrates to Mixed- and High-severity Fires in Central Europe 231
Petr Heneberg

7.19.	The Setting	231
7.20.	Aeolian Sands Specialists Alongside the Railway Track Near Bzenec-Přívoz	232
7.21.	Postfire Succession Near Jetřichovice: A Chance for Dead Wood Specialists	233
7.22.	Conclusions	236
References		237

The Role of Large Fires in the Canadian Boreal Ecosystem 238
André Arsenault

7.23.	The Green Halo	238
7.24.	Land of Extremes	239
7.25.	Vegetation	240
7.26.	Plants Coping With Fire	242
7.27.	Fire Regime of the Canadian Boreal Forest	242
7.28.	Temporal Patterns of Fire and Other Changes in the Boreal	246
7.29.	Biodiversity	248
7.30.	Conclusion	250
References		252

8. What's Driving the Recent Increases in Wildfires?

Dominick A. DellaSala and Chad T. Hanson

8.1.	Understanding the Past, Present, and Future of Wildfires	257
8.2.	Looking Back Over the Paleo-Record (Back Casting)	258
	Sedimentary Charcoal Analysis	260
	Fire History Across a Moisture Gradient	262
8.3.	Western USA Fire History Case Studies	263
8.4.	Historical Range of Variation	267
	Command and Control versus Conservation Science Approaches	268
8.5.	Linking Wildfire to Anthropogenic Climate Change	269
	How Might the Climate Tipping Point Affect Wildfires and People?	269
	Is There a Rigorous Methodology for Attributing Wildfires to ACC?	272
	How do Extreme Wildfires Impact the Built Environment?	275
8.6.	Conclusions	281
Acknowledgement		283
References		283
Further reading		291

Section III
Managing Mixed- and High-Severity Fires

9. Postfire Logging Disrupts Nature's Phoenix

   Dominick A. DellaSala, David B. Lindenmayer, Chad T. Hanson and Jim Furnish

9.1.	Postfire Logging and the Knee-Jerk Response to Fire	295
9.2.	Cumulative Effects of Postfire Logging and Related Activities	297
9.3.	Case Study Postfire Logging Lessons	301
	Biscuit Fire of 2002, Southwest Oregon	301
	Rim Fire of 2013, Sierra Nevada, California	308
	Jasper Fire of 2000, Black Hills, South Dakota	315
	2009 Wildfires, Victoria, Australia	318
9.4.	Conclusions	321
	Appendix 9.1 Effects of Postfire Management Across Regions Where Most Studies Have Been Conducted	323
	References	328
	Further reading	333

10. Forest Managers Play the Backcountry Logging Fiddle as Towns Burn down

    Dominick A. DellaSala

10.1.	The Day Climate Change Came Knocking at my Door	335
10.2.	The Fiddle Players	342
	All Fire Management is Politics	342
	TNC and Collaboratives as Backcountry Fiddlers	344
10.3.	Countering Fire Hyperbole and Doublespeak	354
10.4.	Fire, Fire, Homes on Fire, Again!	355
10.5.	What is "Active Management" and Will it Work?	356
10.6.	Has Active Management Become a Religion of Sorts? (Concluding Thoughts)	358
	Acknowledgments	360
	References	360
	Further reading	362

11. Misinformation About Historical and Contemporary Forests Leads to Policy Failures: A Critical Assessment of the "Overgrown Forests" Narrative

    Chad T. Hanson and Bryant C. Baker

11.1.	The Popular Narrative of "Overgrown Forests"	363
11.2.	Are Contemporary Western US Dry Forests "Overgrown?"	364

	11.3.	Do Denser, Mature, and Old Forests Burn More Severely?	364
	11.4.	Does "Thinning" Reduce Overall Severity in Wildfires?	365
	11.5.	Is High-Severity Fire Converting Dense Dry Forests to Nonforest?	367
	11.6.	Implications of Prologging Misinformation	372
	11.7.	Conclusions	376
	References		376

12. Out of the Ashes, Nature's Phoenix Rises

Dominick A. DellaSala and Chad T. Hanson

	12.1.	Pyrodiversity Begets Biodiversity Reaffirmed	379
	12.2.	A Nature-based Correction Is Needed in Attitudes and Approaches	381
	12.3.	Respond to the Root Causes and Not Just the Effects	384
	12.4.	Does Active Management Work?	385
	12.5.	Public Attitudes Are Shifting but Nature's Phoenix Remains Undervalued	389
	12.6.	The Disconnect Between Independent Research and Wildfire Attitudes	391
	12.7.	Lessons Learned and Closing Remarks	392
	References		395

Index   399

Contributors

Ronald W. Abrams, Dru Associates, Inc., Glen Cove, NY, United States; ADA Pty. Ltd, Chintsa, South Africa

André Arsenault, Memorial University at Grenfell, Corner Brook, NL, Canada; Université du Québec en Abitibi-Témiscamingue, Rouyn-Noranda, QC, Canada; Natural Resources Canada, Canadian Forest Service—Atlantic Forestry Centre, Corner Brook, NL, Canada

Bryant C. Baker, Wildland Mapping Institute, Ventura, CA, United States

William L. Baker, Program in Ecology and Evolution, University of Wyoming, Laramie, WY, United States

Laurence E. Berry, Department of Energy, Environment and Climate Action, Melbourne, VIC, Australia

Monica L. Bond, Wild Nature Institute, Concord, NH, United States

Dominick A. DellaSala, Wild Heritage, A Project of Earth Island Institute, Berkeley, CA, United States

Jim Furnish, Consulting Forester, Gila, NM, United States

Richard W. Halsey, The California Chaparral Institute, Escondido, CA, United States

Chad T. Hanson, John Muir Project of Earth Island Institute, Berkeley, CA, United States

Petr Heneberg, Charles University, Prague, Czech Republic

Richard L. Hutto, Division of Biological Sciences, University of Montana, Missoula, MT, United States

Dominik Kulakowski, Graduate School of Geography, Clark University, Worcester, MA, United States

David B. Lindenmayer, Fenner School of Environment and Society, The Australian National University, Canberra, ACT, Australia

Rosemary L. Sherriff, Department of Geography, Environment & Spatial Analysis, California State Polytechnic University, Humboldt, Arcata, CA, United States

Holly Sitters, Australian Wildlife Conservancy, Subiaco East, WA, Australia; School of Ecosystem and Forest Sciences, University of Melbourne, Creswick, VIC, Australia

Alexandra D. Syphard, The California Chaparral Institute, Escondido, CA, United States; Conservation Biology Institute, Corvallis, OR, United States

Thomas T. Veblen, Department of Geography, University of Colorado-Boulder, Boulder, CO, United States

Preface

Dominick A. DellaSala and Chad T. Hanson

WHY A SECOND EDITION?

When the first edition of this book was published in 2015, some wildfire scientists scoffed at the idea of large fire complexes producing mixed-severity vegetation effects that necessarily include a significant high-severity fire component (where most trees are fire-killed). They were largely stuck in a dated paradigm that envisioned fires as frequent low to moderate severity events in dry low-mid elevation forests that historically lacked large high-severity burn patches. This paradigm viewed contemporary fires as unprecedented, based on the assumption that there is too much high-severity fire. We exposed the limitations of this thinking, based on newer and more robust evidence, and challenged the now-discredited notion that only low- and moderate-severity effects represent so called "good fires" (Baker et al., 2023). The previous fire-paradigm is a house-of-cards concept based on an incomplete historical baseline that was shifted by proponents to a more recent time period that omits evidence of more active fire periods prior to fire suppression and logging. Under this past paradigm, land managers have doubled-down on failed tactics—fire suppression, and logging under the guise of "fuel reduction" that is not stopping or substantially curbing climate-driven fires and, in fact, is exacerbating climate change contributions to extreme wildfires through logging impacts and related emissions. Meanwhile, the public and decision makers are lured into a false sense of security to cavalierly expand into fire-prone areas so long as the government (e.g., U.S.) pours billions of dollars into fire suppression and so called "active management."

The fire-human problem cannot be solved by even more of the same tactics that are failing to change fire behavior at scale. More suppression and logging cannot put the fire genie back in the bottle nor tame it. In particular, at large spatial scales, logging causes more emissions than wildfires (Bartowitz et al., 2022). This, in turn, contributes to a dangerous fire-climate feedback loop (Lindenmayer et al., 2011). The proponents of more logging and suppression are dropping a "climate bomb" on the fire problem that in the long-run will only make fires more catastrophic for people at the expense of ecologically beneficial mixed-severity fire complexes. More logging, regardless of the euphemism being used ("thinning," "fuel reduction," "active management," "restoration," "forest health") is not the solution. In addition to exacerbating the climate crisis, it changes the forest microclimate in ways that can increase the rate of wildfire spread (Stephens et al., 2021a), often toward towns (Downing et al., 2022), while also tending to increase overall severity and tree mortality, according to a growing body of science (Bradley et al., 2016; Baker and Hanson, 2022; Bartowitz et al., 2022; DellaSala et al., 2022). Logging is acting in concert with climate change to create the imperfect storm now destroying towns, sometimes more than once at the same location.

We started the first edition of this book with definitions, as any thematic science book would, to be clear on what we meant by vernacular routinely thrown around in fire circles. Most notably, the difference between wildfire intensity and wildfire severity is fundamental to any discussion about wildfires. By fire intensity, we mean the relative heat energy produced during a fire (e.g., low to high heat intensity). In turn, fire severity refers to the localized or landscape-scale effects of wildfire on ecosystems once the fires are out (e.g., low- to high-severity effects on tree mortality). The most severe burn patches in large fire complexes, for instance, are dominated by the death of most of the overstory trees that for the untrained eye may appear like an environmental disaster. Instead, we showed how when trees die in small or large burn patches that is not the end, rather, it is a sort of rebirthing of the forest experiencing the circularity of succession from young to old and back again that is intimately tied to the miracle of nature's self-renewal processes (i.e., "Nature's Phoenix"). Fires of low (some trees killed) to moderate severity (many trees killed but not most mature/old trees) are ecologically important components of mixed-severity fires that benefit certain groups of native species (Hutto and Patterson, 2016). They do not, however, produce the postfire ecological benefits of high-severity patches (most trees killed) that create biodiverse "snag forest habitat" that many species need. Lower burn severity patches lack the habitat elements—especially abundant snags, downed logs, and montane chaparral patches—that are created by severe burn patches, essential to jumpstarting Nature's Phoenix (DellaSala, 2021). Some like it hot, as our friend and colleague, Dr. Richard Hutto, likes to say (Hutto, 2008), and the many such species have every bit as much intrinsic value and right to exist as those adapted to lower severity fire patches.

Natural disturbances like wildfires are nature's spice-of-life as they generate complexity, which, in turn, is associated with diversity, that is, the pyrodiversity begets biodiversity hypothesis (Tingley et al., 2016; DellaSala et al., 2017; Stephens et al., 2021b; Hutto et al., 2020). Ecosystems are uniquely adapted to wildfire and many thrive in the postdisturbance landscape as long as wildfire is on nature's fire-return schedule (DellaSala et al., 2022). But in a world increasingly dominated by humanity's ecological footprint that includes unprecedented fire suppression coupled with widespread logging and a radically changing climate, wildfires and the landscapes they shape—ecologically and socially—are changing faster than what we can possibly handle. Even though change is a necessary process in disturbance-adapted ecosystems, too much of it, too fast can flip ecosystems to novel states especially when anthropogenic disturbances have piled on spatially and temporally in the same area (cumulative effects).

A case in point is chaparral ecosystems. Left alone, they have an infrequent but beneficial association with fire. However, these diverse shrublands are ever more prone to too much fire caused by anthropogenic ignitions in many communities experiencing expansive valley to montane development footprints. The built environment has penetrated deeper into these fire-unsafe areas, raising collateral damages to chaparral and society as fire's expansive reach grows. Too many anthropogenic fires (Balch et al., 2017) are now causing ecosystem type conversions from biodiverse chaparral to weed-infested, flammable grasslands—all of which are exacerbated by climate chaos.

Since the book's first edition, the demonization of high-severity fire patches in forests as "bad fire" has taken on even greater hyperbole by those that promote false solutions, cloaked in seemingly benign fuels reduction double-speak ("thinning," "salvage," "restoration," "resiliency") with little attention to associated ecosystem damages and climate impacts (DellaSala et al., 2022). Specifically, decision makers, land managers, and researchers with ties to federal agency "fuels reduction" funding, and organizations involved in logging for fire risk reduction like The Nature Conservancy are grossly downplaying "active management" damages while overstating their efficacy in dampening down wildfire effects. It simply will not work in a rapidly changing climate and will cause much more harm in the long run as ecosystems and the climate are transformed more by the logging than the fires themselves. Not admitting limitations and overselling active management is a form of confirmation bias that allows decision makers to throw unprecedented amounts of money at backcountry "thinning" and "fuel reduction" logging operations that do not stop wildland fires driven mainly by extreme fire weather and climate, and leave vulnerable human communities unprotected.

As the years since the first edition progressed, so did the wildfire debate. Despite initial opposition to mixed-severity fire as the predominant wildfire regime in fire-adapted forests (Odion et al., 2014), most scientists now recognize fire regimes in dry pine and mixed conifer forests (low-mid elevations) when viewed at the appropriate landscape scale are indeed a mixture of severities (Odion et al., 2016). The debate has since shifted from denial to qualified acceptance by what is typically overgeneralized by opponents as we need more "good fires" and fewer "bad fires" on the landscape and that we can somehow tilt that balance via command-and-control massive suppression and logging (DellaSala et al., 2022). So-called good fire advocates push for creating low—moderate severity effects via prescribed burning and mechanical thinning (logging) recognizing that "some" high-severity burn patches are inevitable so long as they are small patches. They advocate for more commercial logging as if it could magically put the severe-fire genie back in the bottle to avoid too much of that "bad" fire. While there are situations where prescribed fire by itself (also cultural burning practices) can indeed reduce fire severity, commercial logging under the guises of "thinning" and "fuel reduction" are not necessary precursors to such burning, even in dense, long-unburned forests in places (Knapp and Keeley, 2006; Knapp et al., 2007; van Mantgem et al., 2013), and logging too much of the overstory can increase rates of wildfire spread (Stephens et al., 2021a) and overall fire severity (Hanson, 2022a; Baker and Hanson, 2022; DellaSala et al., 2022) while exacerbating carbon emissions relative to fire alone (Bartowitz et al., 2022). Land managers taking victory laps over the effectiveness of thinning are generally cherry-picking site-specific thinned forests that happened to burn at low severity, while ignoring the many counter-examples nearby and the effects of fire weather that can change hourly (Baker and Hanson, 2022; DellaSala et al., 2022).

It is our view in this book that fire in dry forests is neither "good" nor "bad," and any dichotomous labeling as such is deeply rooted in a hubris-filled notion that we can somehow tame fire at will if only we can do even more of the same. This type of 20th century pseudo-moralizing led to the near extinction of apex predators like wolves and grizzly bears, whose removals triggered cascading ecological effects on food-web dynamics. Likewise, wildfire is a keystone process with predatory-like prowess that balances the push–pull between forest renewal and forest aging, creating essential landscape heterogeneity, including complex early seral forest habitat upon which much of the forest biodiversity depends (Swanson et al., 2011; DellaSala et al., 2014; DellaSala et al., 2017; Hutto et al., 2020). The magnitude and severity of wildfires have risen and fallen sharply over millennia based on fires' tight association with the climate—drought years produce many more and intense fires and wet ones dampen down fire occurrence and intensity. The fire-climate dance is against the back drop of culturally important Indigenous fire ignitions that also have been influential in localized places but suppressed overall through cultural subjugation and manifest destiny. In sum, mixed-severity fires, including high-severity fire patches, are intrinsic in conifer forest ecosystems, providing numerous ecological benefits for species that evolved to associate with, and even depend upon, postfire habitat long before any humans even walked the Earth.

This second edition tells an inconvenient truth aimed at exposing wildfire hyperbole and misinformation, debunks the false narrative of some (Jones et al., 2020), and responds to character assassinations typically aimed at those who legitimately take issue with challenging claims about fire and active management by seeking much deeper scientific dialogue. We have entered a new climate-fire era where fire is increasingly governed by the cumulative emissions we put in the atmosphere from both the unprecedented burning of fossil fuels and the destruction of Nature's Phoenix via logging and related actions. At the same time, character attacks aimed at those who question the dominant fire-logging paradigm are attempts to shut off honest and well-intended debate in the scientific literature. We want to be clear that we are not opposed to active management when done responsibly by working with nature instead of against it. We have recommended numerous "no regrets" management actions that would benefit forest ecosystems, mainly be restoring natural processes and habitat structures, rather than by removing them (Hanson et al., 2010). We do believe there is far too much emphasis on active management via logging (e.g., commercial thinning that removes larger trees and postfire logging) and there are limitations (e.g., legitimate prescribed fire and culturally burning can work with nature at local scales, but ultimately the climate signal will decide how fires increasingly behave).

This second edition updates the preface, nearly all chapters (except for the original Chapter 5, 10, and 12 where we could not get a commitment from the authors), and adds three new chapters (Chapters 10−12) plus more current literature materials for helping scientists and advocates position for a better way forward.

FIRE IN THE BACKCOUNTRY!

As a global change agent, fire has been around since the dawn of terrestrial plants in the Paleozoic Era some 400 million years ago. That is a long time for plants and fire to work things out evolutionarily. Over millennia, fire acted in concert with the regional and global climate in shaping fire-adapted ecosystems aided at local scales in recent millennia by Indigenous burning practices. Charcoal sediments long buried in peat deposits and the fossil record have left forensic clues about fire's indelible mark over the ages that by far predates humanity.

Witness, some 66 million years ago, a mountain-sized asteroid on a collision course with our primeval blue planet, striking what is now the Yucatan Peninsula in Mexico and triggering massive fires that may have contributed to trophic cascades that pushed dinosaurs further into oblivion and facilitated the evolution of our mammalian ancestors on the tree-of-life (Balcerak, 2013). Millions of years later, the "Big Burn of 1910" (Egan, 2009) affected far more acres in northern Montana and Idaho compared to any recent fire yet contemporary fires are typically referred to as "unprecedented" by decision makers and the media. There is some truth and myth in this too, and this book seeks to separate the two by demonstrating where fire is misbehaving, how, why, and what can be done about it.

Today, nearly every terrestrial biome and continent (except for polar regions) on the planet is influenced by fire to a degree and, in any given year, fires are active across a significant proportion of the planet (Fig. 1). NASA's Moderate Resolution Imaging Spectroradiometer (MODIS) on board the Terra satellite has been tracking wildfire pulses for over 2 decades, displaying startling images of a world periodically on fire as a moving window of natural and human-caused fire activity marches on (https://earthobservatory.nasa.gov/global-maps/MOD14A1_M_FIRE).

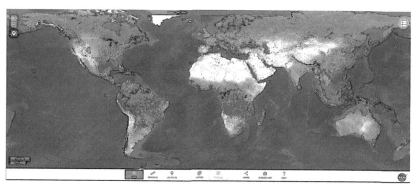

FIGURE 1 MODIS Rapid Response System Global Fire Maps for September 8, 2022, showing vast regions burning in red at the time (https://firms.modaps.eosdis.nasa.gov/map/#d:24hrs;@0.0,0.0,3z).

Clearly, fire is hardly uniform in its frequency, intensity, or spatial extent (Fig. 2) and as discussed throughout this book it ebbs and flows based largely on climatic factors and secondarily by vegetation and topography, with climate change increasingly becoming the top dog regulating fire behavior (Bradley et al., 2016; Evers et al., 2022).

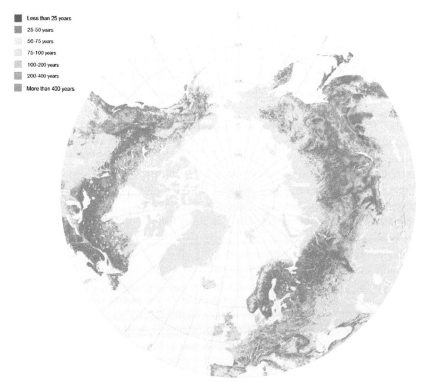

FIGURE 2 Variability in fire-return intervals of northern (high latitude) forests at the global scale
Courtesy of Alexey Yaroshenko, Greenpeace International.

On a philosophical note, wildfire can be viewed as a "self-willed" force of nature that will often do its thing regardless of our actions and sometimes in direct opposition to what we would like to see happen. We have an auspicious history with fire from the earliest humans that used it for heating and cooking, to Indigenous Peoples' use of fire to forge a biocultural relationship with culturally important plants and animals. In recent centuries, manifest destiny, industrialized fire suppression, and "active management" with chainsaws and bulldozers have sought in vain to turn the fire valve off as if it were some spicket handed to us from the gods to manage!

Each year, the Forest Service attempts large-scale fire suppression wherever they can, despite the fact that their "National Cohesive Wildland Fire Management Strategy" allows the agency to refrain from suppression when fires are burning safely in the backcountry or in wilderness areas far removed from towns (https://forestsandrangelands.gov/strategy/index.shtml; accessed September 8, 2022). Commercial logging, including removal of large fire-resistant trees to pay for "fuels reduction," is most often championed as "restoration" despite evidence to the contrary that it does nothing to save homes, has a spotty track record in reducing fire severity, and can be ecologically damaging (DellaSala et al., 2022). Meanwhile, land developers build homes, directly in the high-risk path of fire, that are joyfully purchased by millions of unsuspecting homeowners seeking a more natural connection with nature, thus upping the ante to employ even more fire suppression or else face political, libelous, and other consequences.

We are also changing the global climate in a way that propagates even more fire over large regions. For instance, one study showed how anthropogenic increases in temperature and vapor pressure deficit have enhanced vegetation aridity across western US forests, contributing to 75% more forested area burning and extending the number of fire days per year (Abatzoglou and Williams, 2016).

For the large fires burning in extreme fire weather (e.g., droughts, heat domes, high winds), attempting to stop or slow them over vast areas has proven risky to fire fighters, costly economically and ecologically, and ineffective (Schoennagel et al., 2017; DellaSala et al., 2022). Furthermore, in some years, wildfire activity is so intense that it can spin off pyro-tornadoes that devastate entire towns that no "fuels reduction" project has a prayer of altering. This is the new fire-climate era that we have created. It is the product of what many call the "Anthropocene" (DellaSala and Goldstein, 2017), the age of humanity's transformative and unprecedented ecological footprint.

FROM FIRE COEXISTENCE TO COMMAND-AND-CONTROL FIRE DISRUPTIONS

Over millennial time scales, Indigenous Peoples developed a deep respect for wildland fire use, conducting burns in culturally important locations within a broader landscape context of fire regimes driven by lightning ignitions and climate fluctuations (Vachula et al., 2019). Native Peoples purposefully set fire in cocreating fire-adapted ecosystems, often closest to villages and hunting areas to prepare the conditions for culturally important fire-loving plants and animals for food. With the genocide of Indigenous populations globally, the beneficial biocultural bond was severely damaged.

Today, fire-phobia and command-and-control tactics are engrained in modern societies as we witness shocking yearly accounts of walls of flames racing toward towns and homes unwittingly carved out in fire's path. Consequently, we spend billions of dollars every year attempting to suppress fires all over the world by sending thousands of firefighters into the backcountry in military style fire-attack strategies often with high risk to them. While effective in putting out nearly all fire starts in low-moderate fire weather, this seldom tames any of the small percentage of fires that escape containment during extreme fire weather conditions (high winds, drought, heat domes) and comprise nearly all of the annual area burned. In short, the beneficial links between humans and fire were deeply damaged during European colonization as respect for fire shifted to fear of it through mechanized suppression at all costs, and widespread "fuels reduction" logging unmindful of the true consequences and collateral damages.

Command-and-control attitudes on fire are pervasive worldwide and while they are necessary to save lives and homes when fire is dangerously too close, such attempts have harmed fire-dependent plants and wildlife over vast fire-dependent regions. The popular press and even some fire ecologists sensationalize the severe element of fire as a "catastrophic" event that "destroys" the green forest, terraforming living landscapes into "wastelands" or "moonscapes." Others have argued that forests cannot recover on their own without logging and tree planting over large landscapes, especially as the climate-fire era reshuffles plant communities and fire associations, despite abundant evidence to the contrary (Owen et al., 2017; Hanson and Chi, 2021). And still some well-intentioned conservationists would much rather see an intensively thinned and "fire-proofed" forest neatly manicured with European-forestry precision to resist fire rather than the tapestry of the biologically rich and complex postfire landscapes produced by a mixed-severity fire event.

In the United States, the fire-phobic media has shaped public opinion so much so that when land managers need to work with fires for ecosystem benefits under safe conditions there is little public support (Kauffman, 2004). Attitudes about fires in the United States can be traced to at least Smokey Bear, a fictional character created in 1944 to symbolize fire suppression and whose cartoon mantra is "Remember, ONLY YOU can prevent forest fires." Smokey was immensely popular with the public and led to an important safety awareness about proper campfire etiquette but it also unfortunately became the rallying cry of the US Forest Service and its quest to put out every fire start by 10 a.m. the next day. In the era of Smokey Bear, the Forest Service would become the de facto "Fire Service" as the agency's fire suppression budgets skyrocketed while fires got larger and more frequent in extreme climate years regardless of suppression forces (DellaSala et al., 2022). And, while the agency has moderated its fire suppression policies at times via the National Cohesive Wildland Fire Management Strategy, it has a tendency to flip-flop when the going gets tough. For instance, in August 2021, US Forest Service Chief Randy Moore announced that because of a "national fire crisis" the agency would temporarily suspend its policy to "manage fires for resource benefit." In doing so, the agency returned to its backwards 20th century fire suppression policies even though they cannot possibly achieve that objective during high to extreme fire conditions (Hanson, 2021).

FIRES' UNDER-APPRECIATED ROLE AS NATURE'S ARCHITECT

Contrary to what many think, decades of fire suppression have contributed to a *fire deficit* in forests of the Western USA (Marlona et al., 2012; Odion et al., 2014; Parks et al., 2015) rather than a surplus as often claimed. The big fires ("megafires"), which occur infrequently, but affect proportionately large areas when they do occur, have gotten almost all the attention from the press and politicians who have responded with unprecedented pre- and postfire logging and even more suppression demands.

Despite the fire deficit in forests, nature's fire-balance sheet is about to close the gap (e.g., the boreal and some dry forested regions) as global climate change primes the fire pump for more active fire seasons (Littell et al., 2009; Pechony and Shindell, 2010; Abatzoglou and Williams, 2016). However, that is even more the case for chaparral, which already has too much human-caused fire ignitions. Consequently, given all the attention about fires (perhaps hundreds of stories each fire season in the popular press globally) and the increasing likelihood that more of them will occur in the fire-climate era, it is imperative that we understand the ecological benefits of fire and not just the destruction of the built environment. Aside from legitimate public-safety concerns, ecologically speaking, fire is an essential natural force greatly underappreciated for its role as one of nature's chief architects. In nature, there are short-term winners and losers in any natural disturbance event and fire provides no exception—some species thrive (pyrogenic) in the immediate postfire environment, others avoid fire or move on after (fire avoiders) (Taylor et al., 2012; Kelly et al., 2014), but most *all* species benefit at some point in postfire succession before the development of older forest conditions (Hutto and Patterson, 2016). Thus, fire is nature's way of rejuvenating itself via "pyrodiversity," the mixture of postfire tree age classes uniquely within fire mosaics (Moritz et al., 2014; DellaSala et al., 2017).

This book describes how plants and wildlife have not only evolved to cope with fires but to thrive in the rich postfire environment. The geographic scope remains mainly focused on western U.S. and Canada and the second edition carries forward the prior materials published for other global ecosystems to show broad patterns in fire behavior, ecological effects of fires, and conservation importance of postfire landscapes. We cover expansive geographies from sub-Sahara Africa, southeast Australia, Europe, Canadian boreal, and other areas where very large fires have influenced vegetation and wildlife dynamics for millennia. And we did not restrict our treatise solely to forests because chaparral, Mediterranean shrublands, fynbos, and grasslands also exhibit codependencies with fire's beneficial role when the natural fire regime has not been altered by human development, too many anthropogenic ignitions, and invasive species.

In sum, while severe fire has been treated by society as a comic book-like harbinger of forest death, in reality it soon results in a floral phoenix, an explosion of seed banks long dormant before the first flame but that soon colonize fire's footprint with the changes in soil nutrients that prepare the way for new growth. Fire's perpetuality means plants and wildlife have had time to coevolve with it as a persistent agent of change. In return, some plants, particularly chaparral species, produce waxy flammable leaves that encourage severe fire to return. These fire-dependent communities are quite resilient, and many plant species actually depend on intense fire to germinate successfully (Odion et al., 2009). Other species, like certain conifers, have produced unique adaptations to resist fires of moderate to high intensity such as thick fire-resistant tree bark acquired over eons of natural selection. Witness giant sequoia (*Sequoiadendron giganteum*), arguably the most fire-resistant conifer in the world, capable of surviving most intense fires that scorches almost the entire tree crown but also depends on higher-severity fire for effective reproduction (Hanson, 2022b). Although concerns have been raised that severe fire are killing far too many giant sequoia groves, the desire to protect sequoia is complicated by an ecosystem that is largely failing to reproduce due to the absence of fire from a history of suppression.

As in the first edition, we clearly documented how numerous birds, small mammals, big game, invertebrates, and pollinators prosper in postfire landscapes because the renewed plant growth brings a pulse of biological activity. Keystone members of this pulse are the biological legacies consisting of fire-killed trees that provide nesting sites, foraging sites, and hiding cover, "nurse logs" as wildlife habitat and substrate for seedlings, and patches of fire-following montane chaparral, which provide habitat for pollinators and shrub-nesting birds. Resprouting native flowering shrubs also provide habitat for birds and small mammals that nest and den in montane chaparral, and for insect pollinators that are attracted to the flowers of the native shrubs that, in turn, provide food for bats and insectivorous birds. Woody debris created by intense burns also assist in development of stream channel morphology and hiding cover for aquatic organisms. The ecology of these process-driven biotic interactions with fire builds on the first edition. While there are numerous fire ecology text books, most researchers have skipped over the ecological importance of mixed- and high-severity fires, especially large ones, believing that these underappreciated postfire landscapes are catastrophes of nature and not worth attention. That assumption has contributed to the perception that wildfire is worse than logging, which we show is not the case in both editions.

WHAT WE COVER

No book can cover all of the necessary topics of fire ecology (there are thousands of articles and books on different aspects of fire). This second edition updates prior sections and provides an even more comprehensive global fire ecology edition on the benefits of mixed- and high-severity fires produced by large fire events. It is organized in three mostly updated sections and 12 chapters (3 new ones were added). By fire ecology, we mean the interdependencies among fire and fire-adapted plants and animals, along with the factors that govern fire behavior (particularly climatic processes), the role of fire in ecological and evolutionary processes (much like that of large floods, wind-storms, and other large natural disturbances), and effects of fire on past, present, and future ecosystem dynamics. Both Sections I and II describe the postfire environment where nature rejuvenates fire-adapted systems (Nature's Phoenix), bridging the past with the future in a pulse of postfire productivity that lasts far beyond the initiating fire event. The global Section II demonstrates the prevalence and broad ecological importance of large fire events.

The last section, Section III, includes chapters on the ecological use of fires over large landscapes. Fire-phobic and inappropriate land management policies are breaking the ecological bonds among fire, nature, and people via damaging treatments before (via inappropriate thinning as a fire suppression tool), during (via suppression activities), and after (via postfire logging, planting, and seeding) fire. Notably, postfire logging and related land uses are especially disruptive to postfire landscapes compared to the ecologically beneficial role that fire itself plays (Thorn et al., 2018; Georgiev et al., 2020). It is our hope that with a better understanding of the ecology of large mixed- and high-severity fires, a deeper ecological appreciation for fire's rejuvenating powers will begin to emerge with new policies that recognize the importance of large fires as restorative rather than destructive. We must begin to acknowledge that burned forests are not in need of "restoration" or "recovery" actions by land managers, and promote ways to live in fire-sheds where fires are inevitable but where we are better prepared for them (coexistence).

Chapter authors of this book's second edition contributed their unique perspectives on the ecological importance of fires based on an extensive research pedigree acquired on the ground from many places around the world. In putting the pieces together, we recognize that this new edition could possibly be even more controversial, but truth-telling, given the information presented cuts against the grain of many prevailing fire management policies and public attitudes, including policies that have resulted in billions of dollars of fire suppression and many misdirected pre- and postfire management actions (e.g., Hessburg et al., 2021; Jones et al., 2020 vs. DellaSala et al. 2022). During largely the past decade, however, the scientific literature has been making a cogent and increasingly powerful case for the importance of mixed- and high-severity fires. Lags in ecological fire management responses are mostly due to economic drivers associated with a rush to log postfire landscapes by devaluing them in an "all-bets-are-off" and financially motivated suspension of environmental safeguards to get the cut out by public land management agencies such as the US Forest Service and the Bureau of Land Management (BLM). Additionally, some researchers and conservation groups (like The Nature Conservancy) obtain government research grants to conduct studies on how to make "fuels reduction" more effective without addressing the climate and ecosystem consequences of greatly expanding their logging proposals (DellaSala et al., 2022). In essence, anyone who makes legitimate scientific arguments against the policies of logging as "fuels reduction" is branded as a heretic (DellaSala, 2021; Lee et al., 2021). Attempting to shut off scientific debate (Hessburg et al., 2021; Jones et al., 2020) on controversial topics has been going on since the days of pioneering scientists like Galileo and Copernicus, despite the need for debate to advance the entire field of inquiry that brings us closer to the truth (DellaSala, 2021; Lee et al., 2021).

In contrast, some may view this book update as a timely need for spotlighting the shortcomings of the dominant postfire management responses that are even more rampant today as the active management crowd lashes out unprofessionally (see Lee et al., 2021; Hanson, 2022b for responses) when the consequences and collateral damages of their actions are exposed. In any case, legitimate scientific controversy is often needed so new discoveries about nature spotlight false narratives, hopefully change perceptions, and counter destructive policies. This book is bound to get the juices flowing whether you are inquisitive about the ecological importance of fire in general or disagree with us about its ecosystem importance and management. We hope to shine an ecological spotlight on why fires are restorative agents of landscape change rather than the harbinger of eco-death. Notably, we agree with author Dr. Richard Hutto and fire activist George Wuerthner who have stated that recovery is not about a charred forest becoming a green forest eventually, but rather a charred forest is actually the recovery event from a green forest; although we celebrate the ecological complexity of both green and charred as interconnected stages in the ongoing cycle of life and death in nature.

NEVER JUDGE A POSTFIRE LANDSCAPE BY THE INITIATING EVENT

We also prepared this book because most people view a burned area as a single event in time—that is, right after a fire has "destroyed" their favorite hiking spot, campsite, or, unfortunately, home. Few take the time to actually go back into the burn area 1, 3, or even 10 years later to witness the remarkable rebirthing event that has taken place unseen, where no postfire logging or other vegetation manipulation has occurred. Consequently, the ecological story of fire has yet to attract sufficient or equal media attention needed to infuse a bit of reality into the fire debate especially when homes or lives are not at risk.

In 2003, editor Dominick DellaSala took a film crew from CNN into the Quartz Creek fire area of southwest Oregon, one of the few times the media were actually willing to explore a large burn area after the flames were extinguished. This was repeated in 2012 in a separate tour, 10 years after the historic Biscuit Fire in southwest Oregon to see nature's rebirth at work. In both instances, members of the press were surprised by Nature's Phoenix—even in the most intensely burned patches life was superabundant shortly after the fires and continued to prosper a decade or so later except for areas that were postfire logged. Similarly, Richard Hutto's field trip into two severely burned forests in Montana, which was conducted in association with the Large Wildland Fire Conference of 2014, is still available via a video documentary provided by the Northern Rockies Fire Science Network (http://nrfirescience.org/event/fire-effects-field-trip-dick-hutto; accessed September 21, 2022). However, there is no substitute for first-hand experience in a severely burned forest born of such fires. A new ground-break documentary, Elemental (https://www.elementalfilm.com/; accessed September 21, 2022) does that remarkably well with exceptional video coverage and interviews with scientists, Indigenous Peoples, and those that have unfortunately lost their homes and ways to avoid that from happening again.

The media's hesitancy in covering postfire succession is nothing new. For instance, in 1988, news crews raced to cover the Yellowstone fires, reporting that fires had "destroyed" American's iconic park. There were calls to remove the park superintendent and even the Director of the National Park Service was implicated. Clearly, someone was at fault for this massively "destructive" fire! A year or two later scientists were busy counting the proliferation of lodgepole pine seedlings carpeting the ground while hungry elk took advantage of an explosion of flowering plants, and birds not seen in the park for years began to colonize the fire mosaic. No film crews were present in Yellowstone that year, but the Park Service gives routine fire ecology tours to photo-snapping tourists because of the living tapestry of wildflowers, regenerating forests with abundant snags, and prolific wildlife habitat created by the "destructive" 1988 fire—especially in the areas that burned hottest. Fire ecology tutorials like these are vital in changing public attitudes by showing how there is beauty and ecological value in postfire forests that may appear lifeless soon after fire but that are actually rejuvenated the moment the fire ceased. That beauty, like anything else, if properly appreciated, might lead to a sense of understanding, an intrinsic value of fire as a self-willed force, and not a wasteland that needs to be logged because it is devalued by the misinformed or those with economic interests.

Unfortunately, we have a long way to go because postfire logging creates a destructive feedback loop that begins with a large fire burning beneficially through an area, including large patches of high severity. This is soon followed by massive postfire logging project right smack in the middle of the most ecologically important burn patches that are then transformed by logging practices into more fire-prone landscapes due to flammable logging slash and crop-like uniformity from artificial tree replanting. We have seen this knee-jerk response over and over again and it is accelerating as fire begins to fill the deficit created by decades of fire suppression forces.

THE LONG-VIEW OF FIRE AND A NEW POST-FIRE FRAMEWORK

In this book, we take the long-view of fires by examining their history using various back-casting techniques to reconstruct the past environment and fires' ecosystem effects. By long timelines, we mean 300–1000 years or more because fire cannot be fully understood by what happened last year or even the last few decades or century. Examining long timelines is especially relevant at uncovering the cyclical nature of fire and its linkages to global climatic processes that are now poised to shift fire regimes in ways that are increasingly climate-dominated and far less responsive to "fuel" treatments (Willis et al., 2007; Littell et al., 2009; Abagatzlou and Williams, 2016). In ecology, knowledge of the past is important for understanding present conservation and restoration needs (Willis et al., 2007) and for determining whether a current or future event is characteristic or not. With multiple or continuous timelines, rather than a single point in time, scientists and managers can reconstruct proper historical baselines, and avoid shifting the baseline inappropriately to a more recent timeline. They can also more effectively forecast the future of fires in a changing climate and begin to interact with fire purposely to perform its ecological functions while reducing the risks to human communities (Chapters 10 and 12).

We also provide the basis for a new and much needed communications framework on fires so that we can embrace it ecologically while reducing risks to people through effective land and fire-zoning approaches. It is especially crucial that ecologists and conservation groups develop more objective language when it comes to describing fire's ecological effects. We suggest that they abandon terms such as catastrophic, destroyed, damaged, or consumed by fire, and replace those with restored, rejuvenated, recovered by fire. Even the term fire severity implies negative connotations and needs a new ecologically based lexicon that is less value laden. Importantly, there is no ecological justification for postfire "salvage" logging and the term itself needs to be replaced with what it truly is—postfire logging—as nothing ecologically is being "salvaged" in the aftermath of fire (DellaSala et al., 2022). Rather, the complex early seral forest, rich in legacies and fire-dependent species (Swanson et al., 2011; DellaSala et al., 2017), is being damaged or eliminated for myopic interests in "salvaging" economic value of postfire forests viewed primarily as commodities. A change in terminology is the first step in translating fire science to the public, decision makers, and the press. Thus, we urge the profession of fire ecologists and conservationists to lead by example and start the discussion around a new lexicon of fire coexistence (Moritz et al., 2014, the film Elemental as well).

Importantly, there needs to be substantial investments not only on the ecology associated with fire but also in communications aimed at replacing Smokey Bear's fire-phobic messaging with ways to coexist in "firesheds," a term we use in this book to describe the interface where human-built and natural systems come together. As you will see throughout this book, we present compelling evidence of fire's ecological role that is likely to challenge many land managers, decision makers, and even some scientists and conservation groups. But if we are going to coexist with fire, an inevitable force of nature, then opening up a new dialogue that recognizes its rarity, beauty, and magic (in the words of Dr. Richard Hutto) of postfire landscapes is an important first step.

Wildfires will continue to be part of dry regions through the next millennium and we need to ensure, above all, that communities are protected from wildland fires. But instead of blaming forest protections on fires, as many politicians and land managers do (Bradley et al., 2016), we should be working together, as many communities and conservation groups already are, in reducing flammable vegetation closest to where people live, supporting home hardening with new policies and funding, and encouraging land managers to move away from longstanding commercial logging in forests distant from homes under outdated policies that divert precious resources and attention away from community safety and often make fires move faster toward towns.

We encourage readers to take a hard look at what is actually happening to forests and shrublands after a fire. Go out and see the effects of "megafires" for yourself and come back to the area periodically to witness the progression of fire's rejuvenating magic. See what these areas look like years after fire has passed through and created the fire mosaic, as we hope you will agree that they are not wastelands but are Nature's Phoenix personified. In sum, fire is to fire-dependent ecosystems as rain is to rainforests, inseparable and necessary.

Here are some award-winning videos that complement this book and serve as an introduction to fire's ecological benefits for inquiring open minds:

International Wildlife Film Festival Award—"Disturbance:" http://vimeo.com/groups/future/videos/8627070; accessed September 20, 2022

Photos capturing the ecological magic of severely burned forests: https://vimeo.com/75533376; accessed September 20, 2022

Fire video from PBS: http://www.youtube.com/watch?v=iTl-naywNyY&list=PL7F70F134E853F520&index=15; accessed September 20, 2022

"Forests Born of Fire" by Wild Nature Institute: http://www.youtube.com/watch?v=1BmTq8vGAVo&feature=youtu.be; accessed September 20, 2022
"Blacked-backed woodpeckers and fire:" http://www.fs.usda.gov/detail/r5/news-events/audiovisual/?cid=stelprdb5431394; accessed September 20, 2022

Fire field trip: http://nrfirescience.org/event/fire-effects-field-trip-dick-hutto; accessed September 20, 2022.

REFERENCES

Abatzoglou, J.T., Williams, A.P., 2016. Impact of anthropogenic climate change on wildfire across western US forests. Proc. Natl. Acad. Sci. 113, 11770–11775. https://doi.org/10.1073/pnas.1607171113.

Baker, B.C., Hanson, C.T., 2022. Cumulative tree mortality from commercial thinning and a large wildfire in the Sierra Nevada, California. Land 11, 995.

Baker, W.L., Hanson, C.T., Williams, M.A., DellaSala, D.A., 2023. Countering omitted evidence of variable historical forests and fire regime in western USDA dry forests: the low-severity-fire model rejected. Fire 6, 146. https://doi.org/10.3390/fire6040146.

Balch, J.K., Bradley, B.A., Abatzoglou, J.T., Nagy, R.C., Fusco, E.J., Mahood, A.L., 2017. Human-started wildfires expand the fire niche across the United States. PNAS 114, 2946–2951. https://www.pnas.org/content/114/11/2946.

Balcerak, E., 2013. Global fires after asteroid impact probably caused mass extinction, Eos Transactions. Am. Geophys. Union 94, 188.

Bartowitz, K.J., Walsh, E.S., Stenzel, J.E., Kolden, C.A., Hudiburg, T.W., 2022. Forest carbon emission sources are not equal: putting fire, harvest, and fossil fuel emissions in context. Front. For. Glob. Change 5, 867112.

Bradley, C., Hanson, M.C.T., DellaSala, D.A., 2016. Does increased forest protection correspond to higher fire severity in frequent-fire forests of the western USA? Ecosphere 7, e01492.

DellaSala, D.A., Bond, M.L., Hanson, C.T., Hutto, R.L., Odion, D.C., 2014. Complex early seral forests of the Sierra Nevada: what are they and how can they be managed for ecological integrity? Nat. Areas J. 34, 310–324.

DellaSala, D.A., Hutto, R.L., Hanson, C.T., Bond, M.L., Ingalsbee, T., Odion, D., Baker, W.L., 2017. Accommodating mixed-severity fire to restore and maintain ecosystem integrity with a focus on the Sierra Nevada of California, USA. Fire Ecol. 13, 148–171.

DellaSala, D.A., Goldstein, M.I., 2017. The Anthropocene (275 multi-authored articles). Elsevier, Oxford.

DellaSala, D.A. (Ed.), 2021. Conservation science and advocacy for a planet in crisis: speaking truth to power. Elsevier, Oxford. https://doi.org/10.1016/C2016-0-03650-1.

DellaSala, D.A., Baker, B.C., Hanson, C.T., Ruediger, L., Baker, W.L., 2022. Have western USA fire suppression and megafire active management approaches become a contemporary Sisyphus? Biol. Conserv. 268, 109499.

Downing, W.M., Dunn, C.J., Thompson, M.P., Caggiano, M.D., Short, K.C., 2022. Human ignitions on private lands drive USFS cross-boundary wildfire transmission and community impacts in the western US. Sci. Rep. 12, 2624. https://doi.org/10.1038/s41598-022-06002-3.

Egan, T., 2009. The big burn. Teddy Roosevelt and the fire that save America. Marinera Books, Boston, MA.

Evers, C., et al., 2022. Extreme winds alter influence of fuels and topography on megafire burn severity in seasonal temperate rainforests under record fuel aridity. Fire 5, 41.

Georgiev, K.B., et al., 2020. Salvage logging changes the taxonomic, phylogentic and functional successional trajectories of forest bird communities. J. Appl. Ecol. 57, 1103−1112.

Hanson, C.T., 2021. Smokescreen: Debunking Wildfire Myths to Save Our Forests and Our Climate. University Press Kentucky, Lexington, KY.

Hanson, C.T., 2022a. Cumulative severity of thinned and unthinned forests in a large California wildfire. Land 11, 373.

Hanson, C.T., 2022b. A Decade of Change in the Sierra Nevada: Conservation Implications. In: DellaSala, D.A. (Ed.), Imperiled: The Encyclopedia of Conservation. Elsevier Inc., Waltham, MA, USA.

Hanson, C.T., Chi, T.Y., 2021. Impacts of postfire management are unjustified in spotted owl habitat. Front. Ecol. Evol. 9, 596282.

Hanson, C.T., Odion, D.C., DellaSala, D.A., Baker, W.L., 2010. More-comprehensive recovery actions for Northern Spotted Owls in dry forests: reply to Spies et al. Conserv. Biol. 24, 334−337.

Hessburg, P.F., Prichard, S.J., Hagmann, R.K., Povak, N.A., Lake, F.K., 2021. Wildfire and climate change adaptation for intentional management. Ecol. Appl., e02432 https://doi.org/10.1002/eap.2432.

Hutto, R.L., 2008. The ecological importance of severe wildfires: some like it hot. Ecol. Appl. 18, 1827−1834.

Hutto, R.L., Hutto, R.R., Hutto, P.L., 2020. Patterns of bird species occurrence in relation to anthropogenic and wildfire disturbance: management implications. For. Ecol. Manag. 461, 117942.

Hutto, R.L., Patterson, D.A., 2016. Positive effects of fire on birds may appear only under narrow combinations of fire severity and time-since-fire. Int. J. Wildland Fire 25, 1074−1085.

Jones, G., et al., 2020. Spotted owls and forest fire: comment. Ecosphere 11 (12). https://doi.org/10.1002/ecs2.3312.

Kauffman, J.B., 2004. Death rides the forest: perceptions of fire, land use, and ecological restoration of western forests. Conserv. Biol. 18, 878−882.

Kelly, L.T., Bennett, A.F., Clarke, M.F., McCarthy, M.A., 2014. Optimal fire histories for biodiversity conservation. Conserv. Biol. 29, 473−481. https://doi.org/10.1111/cobi.12384.

Knapp, E.E., Keeley, J.E., 2006. Heterogeneity in fire severity within early season and late season prescribed burns in a mixed-conifer forest. Int. J. Wildland Fire 15, 37−45.

Knapp, E.E., Schwilk, D.W., Kane, J.M., Keeley, J.E., 2007. Role of burning on initial understory vegetation response to prescribed fire in a mixed conifer forest. Can. J. For. Res. 37, 11−22.

Lee, D.E., Bond, M.L., Hanson, C.T., 2021. When scientists are attacked: strategies for dissident scientists and whistleblowers. In: DellaSala, D.A. (Ed.), Conservation Science and Policy.

Lindenmayer, D.B., Hobbs, R.J., Likens, G.E., Krebs, C.J., Banks, S.C., 2011. Newly discovered landscape traps produce regime shifts in wet forests. PNAS 108, 15887−15891. www.pnas.org/cgi/doi/10.1073/pnas.1110245108.

Littell, J.S., McKenzie, D., Peterson, D.L., Westerling, A.L., 2009. Climate and wildfire area burned in western U.S. ecoprovinces, 1916−2003. Ecol. Appl. 19, 1003−1021.

Marlona, J.R., Bartlein, P.J., Gavin, D.G., Long, C.J., Anderson, R.S., Brilese, C.E., Brown, K.J., Hallett, D.J., Power, M.J., Scharf, E.A., Walsh, M.K., 2012. Long-term perspective on wildfires in the western USA. PNAS 109, E535−E543. www.pnas.org/lookup/suppl/.

Moritz, M.A., Batllori1, E., Bradstock, R.A., Gill, A.M., Handmer, J., Hessburg, P.F., Leonard, J., McCaffrey, S., Odion, D.C., Schoennagel, T., Syphard, A.D., 2014. Learning to coexist with fire. Nature 515, 58−66. https://doi.org/10.1038/nature13946.

Odion, D.C., Hanson, C.T., Arsenault, A., Baker, W.L., DellaSala, D.A., Hutto, R.L., Klenner, W., Moritz, M.A., Sherriff, R.L., Veblen, T.T., Williams, M.A., 2014. Examining historical and current mixed-severity fire regimes in ponderosa pine and mixed-conifer forests of western North America. PLoS ONE 9, 1–14.

Odion, D.C., Hanson, C.T., Baker, W.L., DellaSala, D.A., Williams, M.A., 2016. Areas of agreement and disagreement regarding ponderosa pine and mixed conifer forest fire regimes: a dialogue with Stevens et al. PLoS ONE 11, e0154579.

Odion, D.C., Moritz, M.A., DellaSala, D.A., 2009. Alternative community states maintained by fire in the Klamath Mountains, USA. J. Ecol. https://doi.org/10.1111/j.1365-2745.2009.01597.x.

Owen, S.M., Sieg, C.H., Sánchez Meador, A.J., Fulé, P.Z., Inigueza, J.M., Baggett, L.S., Fornwalt, P.J., Battaglia, M.A., 2017. Ponderosa pine regeneration in high-severity burn patches. For. Ecol. Manag. 405, 134–149.

Parks, S.A., Miller, C., Parisien, M.-A., Holsinger, L.M., Dobrowski, S.Z., Abatzoglou, J., 2015. Wildland fire deficit and surplus in the western United States 1984-2012. Ecosphere. https://doi.org/10.1890/ES15-00294.1.

Pechony, O., Shindell, D.T., 2010. Driving forces of global wildfires over the past millennium and the forthcoming century. PNAS 107, 19167–19170. https://doi.org/10.1073/pnas.1003669107.

Schoennagel, T., et al., 2017. Adapt to more wildfire in western north american forests as climate changes. Proc. Natl. Acad. Sci. 114, 4582–4590. https://doi.org/10.1073/pnas.1617464114.

Stephens, S.L., et al., 2021a. Forest restoration and fuels reduction: convergent or divergent? BioScience 71, 85–101.

Stephens, S.L., et al., 2021b. Fire, water, and biodiversity in the Sierra Nevada: a possible triple win. Environ. Res. Commun. 3, 081004.

Swanson, M.E., Franklin, J.F., Beschta, R.L., Crisafulli, C.M., DellaSala, D.A., Hutto, R.L., Lindenmayer, D.B., Swanson, F.J., 2011. The forgotten stage of forest succession: early-successional ecosystems on forested sites. Front. Ecol. Environ. 9, 117–125. https://doi.org/10.1890/090157.

Taylor, R.S., Watson, S.J., Nimmo, D.G., Kelly, L.T., Bennett, A.F., Clarke, M.F., 2012. Landscape-scale effects of fire on bird assemblages: does pyrodiversity beget biodiversity? Divers. Distrib. 18, 519–529.

Thorn, S., et al., 2018. Impacts of salvage logging on biodiversity – a meta-analysis. J. Appl. Ecol. 55, 279–289.

Tingley, M.W., Ruiz-Gutiérrez, V., Wilkerson, R.L., Howell, C.A., Siegel, R.B., 2016. Pyrodiversity promotes avian diversity over the decade following forest fire. Proc. R. Soc. B 283, 20161703.

Vachula, R.S., Russell, J.M., Huang, Y., 2019. Climate exceeded human management as the dominant control of fire at the regional scale in California's Sierra Nevada. Environ. Res. Lett. 14, 104011.

van Mantgem, P.J., Nesmith, J.C.B., Keifer, M., Brooks, M., 2013. Tree mortality patterns following prescribed fire for Pinus and Abies across the southwestern United States. For. Ecol. Manag. 289, 463–469.

Willis, K.J., Araújo, M.B., Bennett, K.D., Figueroa-Rangel, B., Froyd, C.A., Myers, N., 2007. How can knowledge of the past help conserve the future: biodiversity conservation and the relevance of long-term ecological studies. Philos. Trans. R. Soc. Biol. Sci. 362 (1478), 175–187.

FURTHER READING

Krawchuk, M.A., Moritz, M.A., Parisien, M., Van Dorn, J., Hayhoe, K., 2009. Global pyrogeography: the current and future distribution of wildfire. PLoS ONE 4, e5102.

Acknowledgments

The second edition of this book is dedicated to the many scientists and activists willing to speak on behalf of nature's phoenix. You are not "minority opinions" as some might say, nor are you alone in taking a stand for nature in all its forms, functions, and beauty whether it was burned or not. Editor and author Dr. Dominick DellaSala was inspired to begin working with Dr. Chad Hanson on this project in 2015 when he took a hike with his then 8-year-old daughter, Ariela Fay DellaSala, in a severely burned forest. She spied the amazing cacophony of sounds from pollinators buzzing about, woodpeckers drilling snags, and songbirds marking their territories in what was a living ecosystem, not destroyed by fire, but rejuvenated by it. It is through her eyes, I learned the meaning of a complex early seral ecosystem that would inspire the many publications that followed. This book is also dedicated to his other amazing daughter Janelle, for love and acceptance into her family, and his joyful grandchildren Michael, Stella, Becky, and Nathan. Author and editor, Dr. Chad Hanson also dedicates this book to his wife, Rachel.

We acknowledge the influence of deceased Dr. Dennis Odion for his understanding of the pyrodiversity begets biodiversity hypothesis. Dr. William Baker is appreciated for publishing studies that address the blind spots in agency thinking. We acknowledge all the conservation groups involved in *Nature's Phoenix* that continue to speak truth to power and especially Doug Bevington for his tireless advocacy for the conservation of fire-needing ecosystems and Dr. Richard Hutto for teaching us that restoration of a forest begins for the black-backed woodpecker and an entire community of fire-loving species at the moment the flames go out. We appreciate the beauty that unfolds in *Nature's Phoenix* that emerges from the ashes like the mythical bird.

Section I

Biodiversity of Mixed- and High-Severity Fires

Chapter 1

Setting the Stage for Mixed- and High-Severity Fire

Chad T. Hanson[1], Dominick A. DellaSala[2], Rosemary L. Sherriff[3], Richard L. Hutto[4], Thomas T. Veblen[5] and William L. Baker[6]
[1]*John Muir Project of Earth Island Institute, Berkeley, CA, United States;* [2]*Wild Heritage, A Project of Earth Island Institute, Berkeley, CA, United States;* [3]*Department of Geography, Environment & Spatial Analysis, California State Polytechnic University, Humboldt, Arcata, CA, United States;* [4]*Division of Biological Sciences, University of Montana, Missoula, MT, United States;* [5]*Department of Geography, University of Colorado-Boulder, Boulder, CO, United States;* [6]*Program in Ecology and Evolution, University of Wyoming, Laramie, WY, United States*

1.1 EARLIER HYPOTHESES AND CURRENT RESEARCH

In the late 19th century to early 20th century, fire—especially patches of high severity wherein most or all of the dominant vegetation is killed—was generally considered to be a categorically destructive force. Clements (1936) hypothesized that the old-growth stage of succession would result in a stable "climax" condition and described natural disturbance forces such as fire as a threat to this state, characterizing old forests that experienced high-severity fire as a "disclimax" state. One early report opined that there is no excuse or justification for allowing fires to continue to occur at all in chaparral and forest ecosystems (Kinney, 1900). After a series of large fires in North America in 1910 ("the Big Burn"), land managers established a policy goal of the complete elimination of fire from all North American forests (a "one-size-fits-all" policy) through unsuccessful attempts to achieve 100% fire suppression (Pyne, 1982; Egan, 2010). Through the mid-20th century, and in recent decades (Odion et al., 2014, 2016; McLauchlan et al., 2020), views have shifted to broadly acknowledge the importance of low- and low/moderate-severity fire, but less so for moderate- to high-severity fire in conifer forest systems. In this chapter, we focus on drier montane forests of western North America as a case study of how diverse, competing, and rather complex sets of evidence have converged on a new story that embraces not just low-severity fire but also mixed- and high-severity fire in these ecosystems.

A commonly articulated hypothesis is that dry forests at low elevations in western North America were historically open and park-like, and heavily dominated by low-severity and low/moderate-severity fire (Weaver, 1943; Cooper, 1962; Covington, 2000; Agee and Skinner, 2005; Stephens and Ruth, 2005; Hessburg et al., 2021). Under this hypothesis, high-severity fire patches were rare, or at least were believed to be small to moderate in size, and larger patches (generally hundreds of hectares or larger) often are considered to be unnatural and ecologically harmful (Hessburg et al., 2021; Prichard et al., 2021). While this model is a relatively closer fit in some low-elevation, xeric forest systems (Perry et al., 2011; Williams and Baker, 2012a, 2013; Odion et al., 2014), it has been extrapolated far beyond where it seems to apply best. Instead, most or all fire-dependent vegetation types of western North America experienced higher fire severities historically, albeit at a wide variety of spatial and temporal scales (Veblen and Lorenz, 1986; Mast et al., 1998; Taylor and Skinner, 1998; Brown et al., 1999; Kaufmann et al., 2000; Heyerdahl et al., 2001, 2012; Wright and Agee, 2004; Sherriff and Veblen, 2006, 2007; Baker et al., 2007, 2018, 2023; Hessburg et al., 2007; Klenner et al., 2008; Amoroso et al., 2011; Perry et al., 2011; Schoennagel et al., 2011; Williams and Baker, 2012a; Marcoux et al., 2013; Odion et al., 2014; Hanson and Odion, 2015a; Parks et al., 2015; Law and Waring, 2015; Baker and Hanson, 2017; Baker and Williams, 2018).

A key extension of the concept of historical forests characterized by open structure coupled with a low- or low/moderate-severity fire regime is that current areas of dense forest structure—and larger, higher-severity fire patches in such areas—are the result of unnatural fuel accumulation from decades of fire suppression policies, leading to higher-severity fire effects outside the natural range of variability (Hessburg et al., 2021). The most fire-suppressed forests (i.e., those that have gone without fire for periods that exceed their "average" natural fire cycles) are, therefore, expected to experience unnaturally high proportions of higher-severity fire when they burn (Covington and Moore, 1994; Covington, 2000; Agee, 2002; Agee and Skinner, 2005; Stephens and Ruth, 2005; Roos and Swetnam, 2012; Williams, 2012; Stephens et al., 2013; Steel et al., 2015; Hessburg et al., 2021; Prichard et al., 2021).

We recognize that the historical low-severity fire regime described above has never been applied to all forest types in western North America (e.g., Romme and Despain, 1989; Agee, 1993). The idea has, however, been widely applied to most forest types, and widespread acceptance of the low- and low/moderate-severity fire regime has been the primary basis driving fire management policy in an overwhelmingly large proportion of montane forests in the western United States by agencies such as the USDA Forest Service, which is now investing billions of dollars in "fuel reduction" projects involving logging and burning (DellaSala et al., 2022). Thus, many management plans explicitly adopt a low-severity fire regime model without rigorously examining evidence of its applicability to the ecosystem type under consideration (Baker et al., 2023). A key research need has been to determine the particular ecosystem types to which the low-severity fire regime applies. Scientists recently rigorously investigated the hypothesis that forests are burning in a largely unnatural fashion and found that historical forest structure and fire regimes were far more variable than previously believed, and that ecosystem responses to large, intense fires often differ from past assumptions (Figure 1.1; see also Chapters 2–4). We discuss these notions in greater depth throughout this book.

FIGURE 1.1 Natural regeneration of native vegetation—including conifers, deciduous trees, and shrubs—in large high-severity fire patches. Top: Star Fire of 2001; bottom: Storrie Fire of 2000. *Photo by Hanson (2007, 2013), see also Chapter 2.*

Do Open and Park-like Structures Provide an Accurate Historical Baseline for Dry Forest Types in Western US Forests?

Using multiple lines of evidence, including spatially extensive tree ring field data, historical landscape photographs from the late 19th and early 20th centuries, early aerial photography from the 1930s through 1950s, and direct records from late 19th-century land surveyors, many recent studies have been able to reconstruct the historical structure of conifer forests in the western United States (see Odion et al., 2014, 2016; Baker and Hanson, 2017, Baker et al., 2023 for evidence reviews). A portion of the historical montane forest landscape in any given region undoubtedly comprised open forest dominated by low-severity fire (e.g., Brown et al., 1999; Fulé et al., 2009; Iniguez et al., 2009; Perry et al., 2011; Williams and Baker, 2012a; Hagmann et al., 2013; Baker, 2014; Baker and Williams, 2018), and some forest types (e.g., ponderosa pine [*Pinus ponderosa*]) often had a preponderance of low-severity fire in low-elevation or xeric-type forest environments throughout western North America. Nevertheless, landscape-level evidence indicates that vast areas of ponderosa pine and mixed-conifer forests also comprised moderate to very dense forests characterized by a mixed-severity fire regime, wherein higher-severity fire patches of varying sizes occurred within a burn mosaic of low- and moderate-severity fire patches (Veblen and Lorenz, 1986, 1991; Baker et al., 2007, 2018; Sherriff and Veblen, 2007; Hessburg et al., 2007; Perry et al., 2011; Baker, 2012, 2014; Williams and Baker, 2012a, 2012b; Baker and Williams, 2015, 2018; Hanson and Odion, 2015a; Baker and Hanson, 2017). In general, in historical ponderosa pine and mixed-conifer forests of the western United States, local variability was indeed characteristic (Brown et al., 1999; Fulé et al., 2009, Iniguez et al., 2009; Hessburg et al., 2007; Perry et al., 2011; Baker, 2012, 2014; Williams and Baker, 2012a, 2012b, 2013; Baker and Williams, 2015, 2018; Hanson and Odion, 2015a). In sum, these and other studies indicate that historically there was high variability in fire effects (low to high severity) and fire-severity patch sizes and configurations depending on the regional climate and biophysical setting (DellaSala and Hanson, 2019).

Does Time Since Fire Influence Fire Severity?

The predominant view in North American fire science has been that as woody ecosystems age, they steadily increase in their potential for higher-severity fire. Thus, in the fire exclusion/fuels buildup model applied to relatively dry conifer forests and woodlands (e.g., Covington and Moore, 1994), long fire-free intervals caused by effective fire suppression result in fuel accumulation and changes in fuel arrangements (e.g., vertical fuel continuity) that lead to substantially increased fire severity. Likewise, even for forest ecosystems known to burn primarily in severe stand-replacing fires, many classical models of fire potential (in this case the instantaneous chance of fire occurrence) assume that fire severity increases with time since the last fire as a result of fuel load accumulation (Johnson and Gutsell, 1994); some research supports this (Steel et al., 2015). Nevertheless, empirical and modeling studies have demonstrated that in many ecosystem types, including temperate forests, flammability is relatively stable with regard to time since fire (Kitzberger et al., 2012; Perry et al., 2012; Paritsis et al., 2014). We suggest that the predominance of the viewpoint in the western United States that flammability and potential fire severity inexorably increase with time since fire has been an important contributor to the expectation that 20th-century fire suppression—if assumed to have effectively reduced fire frequency—should result in increased, and unnaturally high, fire severity in the modern landscape. This relationship does not seem to hold in at least some conifer forest types for a wide variety of reasons (Veblen, 2003; Odion et al., 2004; Baker, 2009; van Wagtendonk et al., 2012; Lesmeister et al., 2021). Even if it held everywhere, this preoccupation with changes from historical proportions of higher-severity fire skirts the key management questions of whether a change in the proportion of higher-severity patches renders a forest incapable of "recovery" after such fires (precious few papers deal with this key management question) or whether the overall spatiotemporal extent of higher-severity fire (i.e., rotation intervals) exceeds historical levels.

A second assumption about fire regimes in the western United States is implied by language commonly used to describe modern fire regimes in terms of "missed fire cycles." While fire cycle may be a useful descriptor of fire regimes, the assertion that a particular place or patch has missed one or more fire cycles implies a regularity to fire return intervals that are more varied; i.e., there is always variation around a mean. Even in dry forests characterized by relatively frequent fires, the historical fire frequency is typically characterized by such a high degree of variance that descriptors such as means or cycles are often misleading. Using the term *missed fire cycle* in mixed-severity fire regimes, among which the frequency and severity of fires are inherently diverse, is particularly problematic. Usage of *missed fire cycles* connotes a consistency and degree of equilibrium in the historical fire regime that is not supported by actual fire history evidence, which shows large variations in fire intervals (e.g., Baker and Ehle, 2001; Baker, 2012, 2017). Though it seems to make intuitive sense that, with increasing time since fire, fuels would accumulate to create a higher probability of higher-severity fire effects, numerous countervailing factors modulate fire severity as stands mature since the previous fire.

Notably, many studies of this issue have found that, in some areas, the most long-unburned forests are burning mostly at low/moderate severity and are not experiencing higher levels of high-severity fire than forests that have experienced less fire exclusion (Odion et al., 2004, 2010; Odion and Hanson, 2006, 2008; Miller et al., 2012; van Wagtendonk et al., 2012; Lesmeister et al., 2019, 2021). Further, forests with the largest amounts of surface fuels (based on prefire measurements) and small trees do not necessarily always experience more severe fire (Azuma et al., 2004). Debate about this issue remains, however. For example, Steel et al. (2015: Table 7 in particular) modeled time since fire and fire severity in California's forests and predicted that, in mixed-conifer forests, high-severity fire would range from 12% a decade after fire to 20% 75 years after fire. However, the modeling for mixed-conifer forests seems to have been based on what appears to be very limited data for forests that experienced fire <75 years earlier (Steel et al., 2015: Figure 4), weakening inferences about a time since fire/severity relationship. Regardless, the high-severity fire values reported by Steel et al.—even for forests that had not previously burned for 75−100 years—remain well within the range of natural variation of high-severity fire proportions in these forests found by most studies (Beaty and Taylor, 2001; Bekker and Taylor, 2001; Baker, 2014; Odion et al., 2014; Hanson and Odion, 2015a; Baker and Hanson, 2017; Baker et al., 2018).

Although the notion that fire severity would not necessarily increase with time since fire is seemingly counterintuitive, a number of factors help explain it. For example, as forests mature with increasing time since the last fire, canopy cover increases, creating more cooling shade, facilitating moister surface conditions, and slowing wind speeds and thus rates of fire spread. Also, increasing shade in the forest understory can cause a reduction in sun-dependent shrubs and understory trees, making it more difficult to initiate or sustain crown fire (Odion et al., 2004, 2010; Odion and Hanson, 2006). Much more important, however, is that severe fire events are largely driven by extreme-fire weather (Finney et al., 2003) and have less to do with the amount of fuel available (Azuma et al., 2004). Given weather and the climate crisis largely explain much of the variation in wildfire behavior (e.g., see Dennison et al., 2014; Westerling, 2006; Abatzoglou and Williams, 2016; Holden et al., 2018; Hawkins et al., 2022; Turco et al., 2023; Dahl et al., 2023), forest thinning programs will do little to stop the recent increases in fire size—that is a climate change issue, not a fuels issue per se.

An analogous assumption about the role of fuels was previously made regarding chaparral, one of the most fire-dependent plant communities in the world—i.e., that, historically, there was less fuel and more moderate fire effects. This idea is also inconsistent with the scientific evidence (Keeley and Zedler, 2009); see also Chapter 6.

What is the Evidence for Mixed- and High-Severity Fire?

In recent decades, a growing number of studies have investigated historical fire regimes using a variety of methods to determine the extent and frequency of mixed- and high-severity fire, particularly in the ponderosa pine and mixed-conifer forests of western North America (Table 1.1). Regardless of the method used, most landscape-level studies of dry forest types, for example, tend to find evidence for mixed-severity fire regimes that included low-,

TABLE 1.1 Summary of Historical Higher-Severity Fire Proportions Found in Various Reconstruction Study Areas at Least 1000 ha Within Mixed-Conifer and Ponderosa Pine Forests of Western North America

Region	Study	Study Area (ha)	Higher-Severity Fire (%)
Baja California	Minnich et al. (2000)	~75,000	16
Sierra Nevada	Baker (2014)	330,000	31–39
	Hanson and Odion (2015a)	65,296	26
Klamath	Taylor and Skinner (1998)	1570	12–31
Southern Cascades	Beaty and Taylor (2001)	1587	18–70
		2042	52–63
	Bekker and Taylor (2001)	400,000	26
	Baker (2012)		
Northern/Central Cascades	Hessburg et al. (2007)	303,156	37
Blue Mountains (Oregon)	Williams and Baker (2012a)	301,709	17
Front Range (Colorado)	Williams and Baker (2012a)	65,525	65
	Sherriff et al. (2014: Figure 6)	564,413	~72[a]
Southwestern United States (Arizona)	Williams and Baker (2012a)	556,294	15–55
British Columbia, Canada	Heyerdahl et al. (2012)	1105	10

[a]Includes mixed- and high-severity fire.

moderate-, and high-severity fire (both small and large patches) in most forest types and regions across the western United States (Odion et al., 2014; DellaSala and Hanson, 2019), with few exceptions. Here, we describe some of the more common methods that researchers have used to determine historical fire regimes, mostly in western North America.

Aerial Photos

Many researchers have used early aerial photos of montane forests to determine the historical occurrence of high-severity fire. Specifically, researchers have used such photos to determine (1) the number of emergent trees that survived previous high-severity fire (Beaty and Taylor, 2001, 2008; Bekker and Taylor, 2001, 2010); (2) broad stand-structure categories consistent with past low-, moderate-, and high-severity fire (Hessburg et al., 2007); and (3) levels of forest canopy mortality consistent with low-, moderate-, and high-severity fire. Such studies concluded that mixed- and high-severity fire effects were generally dominant in both lower- and middle-montane forests, including mixed-conifer forests, as well as some upper montane forests. Comparisons of modern and historical aerial photographs have revealed important variability in forest changes along environmental gradients in areas experiencing similar land use and fire exclusion histories. For example, in the Colorado Front Range, comparison of aerial photographs showed no significant increase in tree densities in the upper montane zone of ponderosa pine and mixed-conifer forests from 1938 to 1999 (Platt and Schoennagel, 2009) in an area with mixed- and high-severity fire (Schoennagel et al., 2011), whereas low-elevation areas near the ecotone with the plains grasslands exhibited a moderate degree of invasion of grasslands by trees during the same period (Mast et al., 1997) in an area known to have had predominantly lower-severity fire (Veblen et al., 2000; Sherriff and Veblen, 2007; Sherriff et al., 2014).

Historical Reports

Scientists have reviewed evidence in early historical US government forest reports, finding widespread occurrence of small and large higher-severity fire patches in all forest types, including ponderosa pine and mixed-conifer forests (Shinneman and Baker, 1997; Baker et al., 2007; Williams and Baker, 2012a [Appendix S1], 2014; Baker, 2014; Hanson and Odion, 2015a; Baker and Hanson, 2017; Baker et al., 2018). Evidence in these reports includes detailed descriptions of low-, mixed-, and high-severity fires; maps of slightly to severely burned forests; estimates of total area burned at mixed and high severity; and photographs of the landscapes after these fires.

Direct Records and Reconstructions From Early Land Surveys

Field data from unlogged forests collected by the US General Land Office in the 19th century before fire suppression has been extensively analyzed across large landscapes, and historical stand structure has been correlated to fire severities that facilitated or stimulated those forest structures. Based on these analyses, large areas of ponderosa pine and mixed-conifer forests across the western United States were dominated by a mixed-severity fire regime that includes evidence of substantial occurrence of high-severity fire (Baker, 2012, 2014; Williams and Baker, 2012a,b, 2013; Odion et al., 2014, 2016; Baker and Williams, 2018, Baker et al., 2023), typically intermixed with areas of predominantly low/moderate-severity fire. Importantly, note that nearly all tree ring reconstructions that found open, park-like historical forests in some areas have been supported by these land survey reconstructions for those same areas, but the land surveys show definitively that these park-like forests grew only in portions of most dry forests in the western United States (Williams and Baker, 2014). Historical mixed- and high-severity fires shown by the land surveys led to diverse landscapes at scales of a few townships (e.g., 25,000 ha) within each region. These landscapes contained intermixed patches of open forests, dense forests, complex early seral forests, old-growth forests, and dense shrub fields important to wildlife. Such landscape diversity, which is rooted in pyrodiversity concepts, was missed by tree ring reconstructions because using tree ring methods without abundant extant large, old trees is unreliable, and thus more heavily burned historical forests were avoided or missed (Baker and Ehle, 2001; Williams and Baker, 2012a; Baker, 2017).

The land survey records also show (Baker and Williams, 2015) that historical dry forests were numerically dominated by small trees (e.g., <40 cm in diameter) but also included abundant large trees, which together provided "bet-hedging" resilience against a variety of forest disturbances that produce high levels of tree mortality (e.g., insect outbreaks, severe droughts, mixed- and high-severity fires). Large surviving trees are particularly important after severe fires, but smaller trees can differentially survive insect outbreaks and droughts (Baker and Williams, 2015).

Though some (e.g., Fule et al., 2014) have recently questioned some findings of Williams and Baker (2012a), an in-depth analysis of the critique found that most of its points were founded on mistakes, misunderstandings, and omission and misuse of evidence by critics (Williams and Baker, 2014). The accuracy of land survey methods has undergone extensive checking and cross-validation methods, and the findings are strongly corroborated by other published sources, including historical US government fire-severity mapping and reports (Williams and Baker, 2010, 2011, 2012a, 2014; Baker, 2012, 2014; Baker and Hanson, 2017; Baker et al., 2018), and 1908−09 forest atlases that mapped moderate- to high-severity fires of the late-1800s (Baker 2018). These checks show that land survey reconstructions can achieve accuracies almost as high as those from tree ring reconstructions and can do so across very large land areas (e.g., ≥400,000 ha).

Tree Ring Reconstructions of Stand Densities and Fire History

Many scientists have used stand-age data from unlogged forests, often in combination with fire-scar dating of past fires, to reconstruct historical fire regimes and changes in the rate of new stand initiation from mixed- to high-severity fire. In mixed-conifer and ponderosa pine forests of western North America, researchers have found regional stand-age distributions consistent with a mixed-severity fire regime that maintained a mix of age classes and successional stages (e.g., Taylor and Skinner, 1998; Heyerdahl et al., 2012; Odion et al., 2014; Baker et al., 2023). Reconstructions of stand structures and fire history are most effective when supported by diverse evidence, gathered independently, that converges to the same overall interpretations. For example, in the Colorado Front Range, tree ring evidence, historical landscape photographs, and General Land Office surveys converge to the same conclusions. They demonstrate that the historical (i.e., before 1920) fire regime of ponderosa pine and mixed-conifer forests included low-severity fires (i.e., not lethal to large, fire-resistant trees) as well as patches of high-severity fires (i.e., killing >70% of canopy trees) (Veblen and Lorenz, 1986, 1991; Mast et al., 1998; Schoennagel et al., 2011; Williams and Baker, 2012b). The conclusion that most of the montane zone forests dominated by ponderosa pine in the Front Range were characterized by a mixed-severity fire regime is further supported by independently conducted studies by researchers based on tree ring evidence of past fires and their ecological effects (Brown et al., 1999; Kaufmann et al., 2000; Huckaby et al., 2001).

Clear delineation of the spatial extent of past fire regimes is a major concern of ecosystem managers in the context of ecological restoration and management of wildfire. Fire histories and stand structures reconstructed using tree ring data are most useful for guiding management decisions where data sets are sufficiently robust to produce high-resolution spatial layers to compare historical and modern landscape conditions. As an example, in the Colorado Front Range a data set consisting of 7680 tree cores and 1262 fire-scarred tree samples collected at 232 field sites allowed for a spatially explicit comparison of historical fire severity (before fire exclusion in 1920) with observed modern fire severity and modeled potential fire behavior across 564,413 ha of montane forests (Sherriff et al., 2014). Forest structure and tree ring fire history were used to characterize fire severity at the 232 sites. Then, historical fire severity was spatially modeled across the entire study area using biophysical variables that had successfully predicted (retrodicted) fire severity at the 232 sampled sites. Only 16% of the study area recorded a shift from historically low-severity fire to a higher potential of contemporary crown fire. A historical fire regime of more frequent, low-severity fires at elevations below 2260 m is consistent with the view among land managers that these forests be managed both to restore historical structure and to reduce "fuels" in this area of widespread exurban development. By contrast, at mid-to-high elevations and in the upper montane zone (i.e., 2260–3000 m), mixed-severity fires were predominant historically as they are today. Thus, fuels reduction at mid-to-high elevations of the montane zone is inappropriate if the management goal is ecological restoration via fire maintenance. Comparison of the severity of nine large fires that occurred between 2000 and 2012 with the severity of fires before the 20th century revealed no significant increase in fire severity from the historical to the modern period except for a few fires that occurred within the lowest elevations (16%) of the montane study area (Sherriff et al., 2014). This spatially extensive tree ring–based reconstruction is strongly corroborated by land survey records of higher-severity fire patches across the same area (Williams and Baker, 2012b).

Charcoal and Sediment Reconstructions

Paleoecologists have explored fire-induced sediment layers in alluvial fans (e.g., Pierce et al., 2004) and charcoal sediments (e.g., Whitlock et al., 2008; Colombaroli and Gavin, 2010; Jenkins et al., 2011; Marlon et al., 2012) to reconstruct historical fire occurrence. They found numerous periods of large and severe fire activity over the past several centuries in North American mixed-conifer and ponderosa pine forests (see Chapter 8 for many additional citations). Thus, paleoecological methods and other evidence further corroborate findings based upon methods, discussed above, regarding historical mixed- and high-severity fire in these forests.

Plant and Animal Fire Adaptations

Living organisms evolve traits that enhance their reproductive and survival success within a particular environment. These evolutionary results or "adaptations" are embedded in the distributions and biological characteristics of plants and animals that live in a place that experiences a certain kind of fire. For example, many plant characteristics (e.g., serotinous cones, stump sprouting, needle flushing) can be understood only when placed in the context of the kinds of fires (intense, large, infrequent, and duff-consuming) that burn where trees harboring those kinds of adaptations live. The most striking plant and animal fire adaptations (e.g., thick bark, adventitious buds, seed release, seed germination, seed dispersal, extreme habitat specialization by mushrooms, certain beetles and birds) are clearly associated with severe fire, not low-severity understory fire (see discussions of this method in e.g., Hutto et al., 2008; Paula et al., 2009; Pausas and Keeley, 2009; Keeley et al., 2011; Hutto et al., 2015; Pausas, 2015; Hutto et al., 2016; Pausas, 2017; Pausas et al., 2017; Hutto et al., 2020). This method of uncovering historical fire regimes is powerful but underappreciated by most.

Despite the concern expressed about the unnaturalness of large high-severity fire patches and, in particular, interior areas of such patches, hundreds of meters from live-tree edges (Hessburg et al., 2021; Prichard et al., 2021; Steel et al., 2022), some encouraging data are emerging from recent research. For example, one analysis found that interior areas of large (>400 ha) high-severity fire patches, >100 m and >300 m from live-tree edges, comprised only 0.6% and 0.13%, respectively, of ponderosa pine and mixed-conifer forests across the western United States over a three-decade period, 1984–2015 (DellaSala and Hanson, 2019: Tables 3 and 4). Notably, while only a fraction of these forests are comprised of interior spaces of larger high-severity fire patches, 13% of bird species were found to select high-severity fire patch interiors (Steel et al., 2022), indicating that such habitat has an important ecological role to play. Moreover, some studies have reported surprisingly vigorous natural postfire forest and conifer regeneration deep in the interior of large high-severity fire patches in these forest types (Owen et al., 2017; Hanson and Chi, 2021).

1.2 ECOSYSTEM RESILIENCE AND MIXED- AND HIGH-SEVERITY FIRE

Along with the surge in scientific investigation into historical fire regimes over the past 10—15 years has come enhanced understanding of the naturalness and ecological importance of mixed- and high-severity fire in many fire-adapted forest and shrub ecosystems. Contrary to the historical assumption that higher-severity fire is inherently unnatural and ecologically damaging, evidence suggests otherwise. Ecologists now conclude that in vegetation types with mixed- and high-severity fire regimes, fire-mediated age-class diversity is essential to the full complement of native biodiversity and fosters ecological resilience and integrity in montane forests of North America (Hutto, 1995, 2008; Swanson et al., 2011; Bond et al., 2012; Williams and Baker, 2012a; DellaSala et al., 2014, 2017; also see Chapters 3 and 4). Ecological resilience is essentially the opposite of "engineering resilience," which pertains to the suppression of natural disturbance to achieve stasis and command and control of nature (DellaSala et al., 2022). Ecological resilience is the ability to ultimately return to predisturbance vegetation types after a natural disturbance, including high-severity fire. This sort of dynamic equilibrium, where a varied spectrum of succession stages is present across the larger landscape, tends to maintain the full complement of native biodiversity on the landscape (aka— the pyrodiversity hypothesis). Forests that are purported to be burning at unprecedented levels of high-severity fire are generally responding well in terms of the forest succession process and native biodiversity in most places (see Chapters 2—4), so the widespread fear of too much severe fire seems to be unfounded in the vast majority of cases (see, e.g., Kotliar et al., 2002; Bond et al., 2009; Donato et al., 2009; Burnett et al., 2010; Malison and Baxter, 2010; Williams and Baker, 2012a, 2013; Buchalski et al., 2013; Baker, 2014; Odion et al., 2014; Sherriff et al., 2014; Hanson and Odion, 2015a; DellaSala and Hanson, 2019). We acknowledge that more research is needed for some forest regions, such as some areas of the southwestern United States experiencing increasing fire severity (Dillon et al., 2011), to determine the effects of climate change on forest resilience over time.

As discussed above, in mixed-severity fire regimes, higher-severity fire occurs as patches in a mosaic of fire effects (Williams and Baker, 2012a; Baker, 2014). In conifer forests of North America, higher-severity fire patches create a habitat type, known as complex early seral forest (DellaSala et al., 2014), that supports levels of native biodiversity, species richness, and wildlife abundance that are generally comparable to, or even higher than, those in unburned old forest (Raphael et al., 1987; Hutto, 1995; Schieck and Song, 2006; Haney et al., 2008; Donato et al., 2009; Burnett et al., 2010; Malison and Baxter, 2010; Sestrich et al., 2011; Swanson et al., 2011; DellaSala et al., 2014, 2017). Many rare, imperiled, and declining wildlife species depend on this habitat (Hutto, 1995, 2008; Kotliar et al., 2002; Conway and Kirkpatrick, 2007; Hanson and North, 2008; Bond et al., 2009; Buchalski et al., 2013; Hanson, 2013, 2014; Rota, 2013; Siegel et al., 2013; DellaSala et al., 2014, 2017; Baker, 2015a; see also Chapters 3 and 4). The scientific literature reveals the naturalness and ecological importance of multiple age classes and successional stages following higher-severity fire, as well as the common and typical occurrence of natural forest regeneration after such fire (Shatford et al., 2007; Donato et al., 2009; Crotteau et al., 2013; Cocking et al., 2014; Odion et al., 2014; Hanson and Chi, 2021). These and other studies suggest that mixed-severity fire, including higher-severity fire patches, is part of the intrinsic ecology of these forests and has been shaping fire-dependent biodiversity and diverse landscapes for millennia (Figure 1.2; also see DellaSala and Hanson, 2019).

1.3 MIXED- AND HIGH-SEVERITY FIRES INCREASES ARE EQUIVOCAL

Fire history studies show that for many montane forests, including mixed conifer and ponderosa pine forests, fire frequencies in most forested regions were substantially less during the 20th century (and the early 21st century) compared with the previous few centuries (e.g., Odion et al., 2014; Parks et al., 2015). Nonetheless, factors responsible for this fire decline vary from region to region and include fire suppression, changes in forest structure as a result of logging, and removal of fine fuels by livestock grazing. The result is that all fire types, including high-severity fire, have been reduced substantially in broad regions since the early 20th century (Veblen et al., 2000; Odion and Hanson, 2013; Odion et al., 2014; Hanson and Odion, 2015a; Parks et al., 2015; DellaSala and Hanson, 2019). Nevertheless, some forest types or local areas within regions may have more high-severity fire than they did historically (e.g., some low-elevation or other particular environments of xeric montane forests; Perry et al., 2011; Sherriff et al., 2014). While some chaparral/shrub ecosystems (and some forests) are in close proximity to large human populations and associated unplanned human-caused ignitions resulting in an excess of fire relative to historical rates (see Chapter 6), these are the exception, not the rule—at least for conifer forests with mixed-severity fire regimes.

Recent climate-induced increases in fire frequency driven largely by climate change (Kasischke and Turetsky, 2006; Westerling et al., 2006; Dennison et al., 2014; Westerling et al., 2006; Abatzoglou and Williams, 2016; Holden et al., 2018; Hawkins et al., 2022; Turco et al., 2023; Dahl et al., 2023)

FIGURE 1.2 Though high-severity fire patches in montane forests may initially seem to be relatively lifeless landscapes, within the first weeks and months after fire, by the first spring after fire, and for many springs thereafter, native shrubs, conifers, and deciduous trees naturally regenerate, creating an ecologically rich habitat for numerous wildlife species. (a) Star Fire of 2001, Eldorado National Forest, Sierra Nevada. (b) McNally Fire of 2002, Sequoia National Forest, Sierra Nevada. *(a) Photo by Hanson (2013). (b) Photo by Hanson (2014).*

have led to increases in total area burned in most regions of western North America, but this has not necessarily resulted in similar trends in high-severity fire, particularly with regard to the proportion of the total burned area comprised of high-severity fire and high-severity patch sizes (e.g., see Hanson et al., 2009; Dillon et al., 2011; Miller et al., 2012; Hanson and Odion, 2014; Hanson and Odion, 2015b; DellaSala and Hanson, 2019; also see Chapter 8).

Some notable exceptions include the southwestern United States, where overall high-severity fire area (but not proportion of high severity) has increased (Dillon et al., 2011). Additionally, in the southern Rocky Mountains, both high-severity fire area and proportion have increased (Dillon et al., 2011). However, in both of these regions high-severity fire rates are not outside of the natural range of historical variation and generally remain in deficit (Baker, 2015b). Not only is the habitat created by higher-severity fire incredibly biodiverse and—in most forest regions—highly degraded by postfire logging (see Chapter 9), it also is often severely threatened by the inertia of historical misconceptions about the effects of high-severity fire and the responses of ecosystems and biodiversity to such fire (Bond et al., 2012; DellaSala et al., 2014; Hanson, 2014; DellaSala et al., 2022; also see Chapter 11). This results in forest management policies that continue to focus on aggressive fire suppression, pre- and postfire logging, postfire shrub eradication and plantation establishment, and homogenous low-severity prescribed burning designed to prevent mixed- and high-severity fire that are implemented across landscapes to further degrade complex early seral forest habitat (Lindenmayer et al., 2004; Hutto, 2006; Hanson and North, 2008; Bond et al., 2009; DellaSala et al., 2014; Hanson, 2014; DellaSala et al., 2022; also see Chapters 10–12).

1.4 CONCLUSIONS

Historical forest structure and fire regimes in mixed-conifer and ponderosa pine forests of western North America were far more variable than reflected in current management regimes. Mixed- and high-severity fires are a natural and ecologically beneficial part of many forests and shrublands as we demonstrate throughout this book. Yet the unique and ecologically rich habitat created by such fire (Stephens et al., 2021) remains demonized and, in nearly all places, is threatened by fire suppression, and pre- and postfire logging (Chapter 9) that is degrading to complex post fire habitat. Ecologists are increasingly urging a shift in policies that would allow more mixed- and high-severity fire in the wildlands away from homes, while focusing on fuel reduction and fire suppression activities adjacent to homes to provide for public safety (Gibbons et al., 2012; Calkin et al., 2014, 2023; Moritz et al., 2014; Abatzoglou and Williams, 2016; Schoennagel et al., 2017; Balch et al., 2017; Law et al., 2023; see Chapter 10). A paradigm shift in land management policies is needed to restore mixed-severity fire by allowing wildland fires to burn safely in the backcountry while protecting postfire habitat from the ecologically destructive practices of postfire logging, shrub removal, and artificial plantation establishment (Lindenmayer et al., 2004; Bond et al., 2012; DellaSala et al., 2014; Hanson, 2014). Such a management shift would help forests draw down more of the dangerous excess of CO_2 in our atmosphere to mitigate the climate crisis (e.g., Depro et al., 2008; Law et al., 2018).

REFERENCES

Abatzoglou, J.T., Williams, A.P., 2016. Impact of anthropogenic climate change on wildfire across western US forests. Proc. Natl. Acad. Sci. USA. www.pnas.org/cgi/doi/10.1073/pnas.1607171113.

Agee, J.K., 1993. Fire Ecology of Pacific Northwest Forests. Island Press, Washington, DC.

Agee, J.K., 2002. The fallacy of passive management: managing for fire safe forest reserves. Conserv. Pract. 3, 18–26.

Agee, J.K., Skinner, C.N., 2005. Basic principles of forest fuel reduction treatments. For. Ecol. Manage. 211, 83–96.

Amoroso, M.M., Daniels, L.D., Bataineh, M., Andison, D.W., 2011. Evidence of mixed- severity fires in the foothills of the Rocky Mountains of west-central Alberta, Canada. For. Ecol. Manage. 262, 2240–2249.

Azuma, D.L., Donnegan, J., Gedney, D., 2004. Southwest Oregon Biscuit Fire: An Analysis of Forest Resources and Fire Severity. U.S. Forest Service Research Paper PNW-RP-560, Pacific Northwest Research Station, Portland, OR, USA.

Baker, W.L., 2009. Fire Ecology in Rocky Mountain Landscapes. Island Press, Washington DC.

Baker, W.L., 2012. Implications of spatially extensive historical data from surveys for restoring dry forests of Oregon's eastern Cascades. Ecosphere 3, 23.

Baker, W.L., 2014. Historical forest structure and fire in Sierran mixed-conifer forests reconstructed from General Land Office survey data. Ecosphere 5, 79.

Baker, W.L., 2015a. Historical Northern spotted owl habitat and old-growth dry forests maintained by mixed-severity wildfires. Landsc. Ecol. 30, 655–666.

Baker, W.L., 2015b. Are high-severity fires burning at much higher rates recently than historically in dry-forest landscapes of the Western USA? PLoS One 10, Article e0136147.

Baker, W.L., 2017. Restoring and managing low-severity fire in dry-forest landscapes of the western USA. PLoS One 12 (2), e0172288. https://doi.org/10.1371/journal.pone.0172288.

Baker, W.L., 2018. Historical fire regimes in ponderosa pine and mixed-conifer landscapes of the San Juan Mountains, Colorado, USA, from multiple sources. Fire 1, 23.

Baker, W.L., Ehle, D., 2001. Uncertainty in surface-fire history: the case of ponderosa pine forests in the western United States. Can. J. For. Res. 31, 1205–1226.

Baker, W.L., Hanson, C.T., 2017. Improving the use of early timber inventories in reconstructing historical dry forests and fire in the western United States. Ecosphere 8, e01935. Article.

Baker, W.L., Williams, M.A., 2015. Bet-hedging dry-forest resilience to climate-change threats in the western USA based on historical forest structure. Front. Ecol. Evol. https://doi.org/10.3389/fevo.2014.00088.

Baker, W.L., Williams, M.A., 2018. Land surveys show regional variability of historical fire regimes and dry forest structure of the western United States. Ecol. Appl. 28, 284–290.

Baker, W.L., Veblen, T.T., Sherriff, R.L., 2007. Fire, fuels and restoration of ponderosa pine-Douglas fir forests in the Rocky Mountains, USA. J. Biogeogr. 34, 251–269.

Baker, W.L., Hanson, C.T., Williams, M.A., 2018. Improving the use of early timber inventories in reconstructing historical dry forests and fire in the western United States: reply. Ecosphere 9, e02325. Article.

Baker, W.L., Hanson, C.T., Williams, M.A., DellaSala, D.A., 2023. Countering omitted evidence of variable historical forests and fire regime in western USA dry forests: the low-severity fire model rejected. Fire 6 (4), 146. https://doi.org/10.3390/fire6040146.

Balch, J.K., Bradley, B.A., Abatzoglou, J.T., Nagy, R.C., Fusco, E.J., Mahood, A.L., 2017. Human-started wildfires expand the fire niche across the United States. Proc. Natl. Acad. Sci. USA. www.pnas.org/cgi/doi/10.1073/pnas.1617394114.

Beaty, R.M., Taylor, A.H., 2001. Spatial and temporal variation of fire regimes in a mixed conifer forest landscape, Southern Cascades, USA. J. Biogeogr. 28, 955–966.

Beaty, R.M., Taylor, A.H., 2008. Fire history and the structure and dynamics of a mixed- conifer forest landscape in the northern Sierra Nevada, Lake Tahoe Basin, California, USA. For. Ecol. Manage. 255, 707–719.

Bekker, M.F., Taylor, A.H., 2001. Gradient analysis of fire regimes in montane forests of the southern Cascade Range, Thousand Lakes Wilderness, California, USA. Plant Ecol. 155, 15–28.

Bekker, M.F., Taylor, A.H., 2010. Fire disturbance, forest structure, and stand dynamics in montane forest of the southern Cascades, Thousand Lakes Wilderness, California, USA. Ecoscience 17, 59–72.

Bond, M.L., Lee, D.E., Siegel, R.B., Ward Jr., J.P., 2009. Habitat use and selection by California Spotted Owls in a postfire landscape. J. Wildl. Manag. 73, 1116–1124.

Bond, M.L., Siegel, R.B., Hutto, R.L., Saab, V.A., Shunk, S.A., 2012. A new forest fire paradigm: the need for high-severity fires. Wildlife Prof. 6, 46–49.

Brown, P., Kaufmann, M., Shepperd, W., 1999. Long-term, landscape patterns of past fire events in a montane ponderosa pine forest of central Colorado. Landsc. Ecol. 14, 513–532.

Buchalski, M.R., Fontaine, J.B., Heady III, P.A., Hayes, J.P., Frick, W.F., 2013. Bat response to differing fire severity in mixed-conifer forest, California, USA. PLoS One 8, e57884.

Burnett, R.D., Taillie, P., Seavy, N., 2010. Plumas Lassen Study 2009 Annual Report. U.S. Forest Service, Pacific Southwest Region, Vallejo, CA.

Calkin, D.E., Barrett, K., Cohen, J.D., Quarles, S.L., 2023. Wildland-urban fire disasters aren't actually a wildfire problem. Proc. Natl. Acad. Sci. U. S. A. 120 (51), e2315797120. https://doi.org/10.1073/pnas.2315797120.

Calkin, D.E., Cohen, J.D., Finney, M.A., Thompson, M.P., 2014. How risk management can prevent future wildfire disasters in the wildland-urban interface. Proc. Natl. Acad. Sci. U. S. A. 111, 746–751.

Clements, F.E., 1936. Nature and structure of the climax. J. Ecol. 24, 252–284.

Cocking, M.I., Varner, J.M., Knapp, E.E., 2014. Long-term effects of fire severity on oak- conifer dynamics in the southern Cascades. Ecol. Appl. 24, 94–107.

Colombaroli, D., Gavin, D.G., 2010. Highly episodic fire and erosion regime over the past 2,000 y in the Siskiyou Mountains, Oregon. Proc. Natl. Acad. Sci. USA 107, 18909–18915.

Conway, C.J., Kirkpatrick, C., 2007. Effect of forest fire suppression on buff-breasted flycatchers. J. Wildl. Manag. 71, 445–457.

Cooper, C.F., 1962. Pattern in ponderosa pine forests. Ecology 42, 493–499.

Covington, W.W., 2000. Helping western forests heal: the prognosis is poor for U.S. forest ecosystems. Nature 408, 135–136.

Covington, W.W., Moore, M.M., 1994. Southwestern ponderosa forest structure: changes since Euro-American settlement. J. For. 92, 39–47.

Crotteau, J.S., Varner III, J.M., Ritchie, M.W., 2013. Post-fire regeneration across a fire severity gradient in the southern Cascades. For. Ecol. Manage. 287, 103–112.

Dahl, K.A., Abatzoglou, J.T., Phillips, C.A., Ortiz-Partida, J.P., Licker, R., Merner, L.D., Ekwurzel, B., 2023. Quantifying the contribution of major carbon producers to increases in vapor pressure deficit and burned area in western US and southwestern Canadian forests. Environ. Res. Lett. 18 (6).

DellaSala, D.A., Hanson, C.T., 2019. Are wildland fires increasing large patches of complex early seral forest habitat? Diversity 11, 157. https://doi.org/10.3390/d11090157.

DellaSala, D.A., Bond, M.L., Hanson, C.T., Hutto, R.L., Odion, D.C., 2014. Complex early seral forests of the Sierra Nevada: what are they and how can they be managed for ecological integrity? Nat. reas J. 34, 310−324.

DellaSala, D.A., et al., 2017. Accommodating mixed-severity fire to restore and maintain ecosystem integrity with a focus on the Sierra Nevada of California, USA. Fire Ecol. 13, 148−171.

DellaSala, D.A., Baker, B., Hanson, C.T., Ruediger, L., Baker, W., 2022. Have western USA fire suppression and active management approaches become a contemporary Sisyphus? Biol. Conserv. 268. https://doi.org/10.1016/j.biocon.2022.109499.

Dennison, P.E., Brewer, S.C., Arnold, J.D., Moritz, M.A., 2014. Large wildfire trends in the western United States, 1984−2011. Geophys. Res. Lett. 41, 2928−2933.

Depro, B.M., et al., 2008. Public land, timber harvests, and climate mitigation: quantifying carbon sequestration potential on U.S. public timberlands. For. Ecol. Manage. 255, 1122−1134.

Dillon, G.K., Holden, Z.A., Morgan, P., Crimmins, M.A., Heyerdahl, E.K., Luce, C.H., 2011. Both topography and climate affected forest and woodland burn severity in two regions of the western US. 1984 to 2006. Ecosphere 2, 130.

Donato, D.C., Fontaine, J.B., Robinson, W.D., Kauffman, J.B., Law, B.E., 2009. Vegetation response to a short interval between high-severity wildfires in a mixed-evergreen forest. J. Ecol. 97, 142−154.

Egan, T., 2010. The Big Burn: Teddy Roosevelt and the Fire That Saved America. Houghton Mifflin Harcourt, New York.

Finney, M.A., Bartlette, R., Bradshaw, L., Close, K., Collins, B.M., Gleason, P., Min Hao, W., Langowski, P., McGinely, J., McHugh, C.W., Martinson, E., Omi, P.N., Shepperd, W., Zeller, K., 2003. Fire Behavior, Fuels Treatments, and Fire Suppression on the Hayman Fire. U.S. Forest Service Gen. Tech. Rpt. RMRS-GTR-114, Rocky Mountain Research Station, Missoula, Montana, USA.

Fulé, P.Z., Korb, J.E., Wu, R., 2009. Changes in forest structure of a mixed conifer forest, southwestern Colorado, USA. For. Ecol. Manage. 258, 1200−1210.

Fule, P.Z., Swetnam, T.W., Brown, P.M., Falk, D.A., Peterson, D.L., Allen, C.D., Aplet, G.H., Battaglia, M.A., Binkley, D., Farris, C., Keane, R.E., Margolis, E.Q., Grissino- Mayer, H., Miller, C., Hull Sieg, C., Skinner, C., Stephens, S.L., Taylor, A., 2014. Unsupported inferences of high severity fire in historical western United States dry forests: response to Williams and Baker. Global Ecol. Biogeogr. 23, 825−830.

Gibbons, P., van Bommel, L., Gill, A.M., Cary, G.J., Driscoll, D.A., Bradstock, R.A., Knight, E., Moritz, M.A., Stephens, S.L., Lindenmayer, D.B., 2012. Land management practices associated with house loss in wildfires. PLoS One 7, e29212.

Hagmann, R.K., Franklin, J.F., Johnson, K.N., 2013. Historical structure and composition of ponderosa pine and mixed-conifer forests in south-central Oregon. For. Ecol. Manage. 304, 492−504.

Haney, A., Apfelbaum, S., Burris, J.M., 2008. Thirty years of post-fire succession in a southern boreal forest bird community. Am. Midl. Nat. 159, 421−433.

Hanson, C.T., 2007. Post-fire management of snag forest habitat in the Sierra Nevada (Ph.D. dissertation). University of California at Davis, Davis, CA.

Hanson, C.T., 2013. Pacific Fisher habitat use of a heterogeneous post-fire and unburned landscape in the southern Sierra Nevada, California, USA. Open For. Sci. J. 6, 24−30.

Hanson, C.T., 2014. Conservation concerns for Sierra Nevada birds associated with high- severity fire. West. Birds 45, 204−212.

Hanson, C.T., Chi, T.Y., 2021. Impacts of postfire management are unjustified in spotted owl habitat. Front. Ecol. Evol. 9, 596282. Article.
Hanson, C.T., North, M.P., 2008. Postfire woodpecker foraging in salvage-logged and unlogged forests of the Sierra Nevada. Condor 110, 777−782.
Hanson, C.T., Odion, D.C., 2014. Is fire severity increasing in the Sierra Nevada mountains, California, USA? Int. J. Wildland Fire 23, 1−8.
Hanson, C.T., Odion, D.C., 2015a. Historical forest conditions within the range of the Pacific Fisher and Spotted Owl in the central and southern Sierra Nevada, California, USA. Nat. Area J.
Hanson, C.T., Odion, D.C., 2015b. Sierra Nevada fire severity conclusions are robust to further analysis: a reply to Safford et al. Int. J. Wildland Fire 24, 294−295.
Hanson, C.T., Odion, D.C., DellaSala, D.A., Baker, W.L., 2009. Overestimation of fire risk in the northern spotted owl recovery plan. Conserv. Biol. 23, 1314−1319.
Hawkins, L.R., Abatzoglou, J.T., Li, S., Rupp, D.E., 2022. Anthropogenic influence on recent severe autumn fire weather in the west coast of the United States. Geophys. Res. Lett. 49, e2021GL095496. https://doi.org/10.1029/2021GL095496.
Hessburg, P.F., Salter, R.B., James, K.M., 2007. Re-examining fire severity relations in pre-management era mixed conifer forests: inferences from landscape patterns of forest structure. Landsc. Ecol. 22, 5−24.
Hessburg, P.F., Prichard, S.J., Hagmann, R.K., Povak, N.A., Lake, F.K., 2021. Wildfire and climate change adaptation of western North American forests: a case for intentional management. Ecol. Appl. 31, e02432.
Heyerdahl, E.K., Brubaker, L.B., Agee, J.K., 2001. Spatial controls of historical fire regimes: a multiscale example from the Interior West, USA. Ecology 82, 660−678.
Heyerdahl, E.K., Lertzman, K., Wong, C.M., 2012. Mixed-severity fire regimes in dry forests of southern interior British Columbia, Canada. Can. J. For. Res. 42, 88−98.
Holden, Z.A., Swanson, A., Lucke, C.H., Affleck, D., 2018. Decreasing fire season precipitation increased recent western US forest wildfire activity. Proc. Natl. Acad. Sci. USA 115 (36). https://doi.org/10.1073/pnas.1802316115.
Huckaby, L.S., Kaufmann, M.R., Stoker, J.M., Fornwalt, P.J., 2001. Landscape Patterns of Montane Forest Age Structure Relative to Fire History at Cheesman Lake in the Colorado Front Range. U.S.D.A. Forest Service Proceedings RMRS-P-22, Rocky Mountain Research Station, Missoula, Montana, USA.
Hutto, R.L., 1995. Composition of bird communities following stand-replacement fires in Northern Rocky Mountain (U.S.A.) conifer forests. Conserv. Biol. 9, 1041−1058.
Hutto, R.L., 2006. Toward meaningful snag-management guidelines for postfire salvage logging in North American conifer forests. Conserv. Biol. 20, 984−993.
Hutto, R.L., 2008. The ecological importance of severe wildfires: some like it hot. Ecol. Appl. 18, 1827−1834.
Hutto, R.L., Conway, C.J., Saab, V.A., Walters, J.R., 2008. What constitutes a natural fire regime? Insight from the ecology and distribution of coniferous forest birds in North America. Fire Ecol. 4, 115−132.
Hutto, R.L., Bond, M.L., DellaSala, D.A., 2015. Using bird ecology to learn about the benefits of severe fire. In: DellaSala, D.A., Hanson, C.T. (Eds.), The Ecological Importance of Mixed-Severity Fires: Nature's Phoenix. Elsevier Inc., Amsterdam, Netherlands, pp. 55−88.
Hutto, R.L., Keane, R.E., Sherriff, R.L., Rota, C.T., Eby, L.A., Saab, V.A., 2016. Toward a more ecologically informed view of severe forest fires. Ecosphere 7, e01255. Article.

Hutto, R.L., Hutto, R.R., Hutto, P.L., 2020. Patterns of bird species occurrence in relation to anthropogenic and wildfire disturbance: Management implications. For. Ecol. Manag. 461, 117942.

Iniguez, J.M., Swetnam, T.W., Baisan, C.H., 2009. Spatially and temporally variable fire regime on Rincon Mountain, Arizona, USA. Fire Ecol. 5, 3–21.

Jenkins, S.E., Hull Sieg, C., Anderson, D.E., Kaufman, D.S., Pearthree, P.A., 2011. Late Holocene geomorphic record of fire in ponderosa pine and mixed-conifer forests, Kendrick Mountain, northern Arizona, USA. Int. J. Wildland Fire 20, 125–141.

Johnson, E.A., Gutsell, S.L., 1994. Fire frequency models, methods and interpretations. Adv. Ecol. Res. 25, 239–283.

Kasischke, E.S., Turetsky, M.R., 2006. Recent changes in the fire regimes across the North American boreal region—Spatial and temporal patterns of burning across Canada and Alaska. Geophys. Res. Lett. 33, L09703.

Kaufmann, M.R., Regan, C.M., Brown, P.M., 2000. Heterogeneity in ponderosa pine/Douglas-fir forests: age and size structure in unlogged and logged landscapes of central Colorado. Can. J. For. Res. 30, 698–711.

Keeley, J.E., Zedler, P.H., 2009. Large, high-intensity fire events in southern California shrublands: debunking the fine-grain age patch model. Ecol. Appl. 19, 69–94.

Keeley, J.E., Pausas, J.G., Rundel, P.W., Bond, W.J., Bradstock, R.A., 2011. Fire as an evolutionary pressure shaping plant traits. Trends Plant Sci. 16, 406–411.

Kinney, A., 1900. Forest and Water. Post Publishing Company, Los Angeles, California, USA.

Kitzberger, T., Araoz, E., Gowda, J., Mermoz, M., Morales, J., 2012. Decreases in fire spread probability with forest age promotes alternative community states, reduced resilience to climate variability and large fire regime shifts. Ecosystems 15, 97–112. https://doi.org/10.1007/s10021-011-9494-y.

Klenner, W., Walton, R., Arsenault, A., Kremsater, L., 2008. Dry forests in the Southern Interior of British Columbia: Historical disturbances and implications for restoration and management. For. Ecol. Manage. 256, 1711–1722.

Kotliar, N.B., Hejl, S.J., Hutto, R.L., Saab, V.A., Melcher, C.P., McFadzen, M.E., 2002. Effects of fire and post-fire salvage logging on avian communities in conifer-dominated forests of the western United States. Stud. Avian Biol. 25, 49–64.

Law, B., Bloemers, R., Colleton, N., Allen, M., 2023. Redefining the wildfire problem and scaling solutions to meet the challenge. Bull. At. Sci. 79 (6), 377–384. https://doi.org/10.1080/00963402.2023.2266941.

Law, B.E., Waring, R.H., 2015. Carbon implications of current and future effects of drought, fire and management on Pacific Northwest forests. For. Ecol. Manag. 355, 4–14. https://doi.org/10.1016/j.foreco.2014.11.023.

Law, B.E., et al., 2018. Land use strategies to mitigate climate change in carbon dense temperate forests. Proc. Natl. Acad. Sci. U. S. A. 115, 3663–3668.

Lesmeister, D.B., Sovern, S.G., Davis, R.J., Bell, D.M., Gregory, M.J., Vogeler, 2019. Mixed-severity wildfire and habitat of an old-forest obligate. Ecosphere 10 (4), e02696. Article.

Lesmeister, B.B., Davis, R.J., Sovern, S.G., Yang, Z., 2021. Northern spotted owl nesting forests as fire refugia: a 30-year synthesis of large wildfires. Fire Ecology 17, 32. https://doi.org/10.1186/s42408-021-00118-z.

Lindenmayer, D.B., Foster, D.R., Franklin, J.F., Hunter, M.L., Noss, R.F., Schmiegelow, F.A., Perry, D., 2004. Salvage harvesting policies after natural disturbance. Science 303, 1303.

Malison, R.L., Baxter, C.V., 2010. The fire pulse: wildfire stimulates flux of aquatic prey to terrestrial habitats driving increases in riparian consumers. Can. J. Fish. Aquat. Sci. 67, 570–579.

Marcoux, H., Gergel, S.E., Daniels, L.D., 2013. Mixed-severity fire regimes: How well are they represented by existing fire-regime classification systems? Can. J. For. Res. 43, 658–668.

Marlon, J.R., Bartlein, P.J., Gavin, D.G., Long, C.J., Anderson, R.S., Briles, C.E., Brown, K.J., Colombaroli, D., Hallett, D.J., Power, M.J., Scharf, E.A., Walsh, M.K., 2012. Long-term perspective on wildfires in the western USA. Proc. Natl. Acad. Sci. USA 109, E535–E543.

Mast, J.N., Veblen, T.T., Hodgson, M.E., 1997. Tree invasion within a pine/grassland ecotone: An approach with historic aerial photography and GIS modeling. For. Ecol. Manage. 93, 187–194.

Mast, J.N., Veblen, T.T., Linhart, Y.B., 1998. Disturbance and climatic influences on age structure of ponderosa pine at the pine/grassland ecotone, Colorado Front Range. J. Biogeogr. 25, 743–755.

McLauchlan, K.K., et al., 2020. Fire as a fundamental ecological process: research advances and frontiers. J. Ecol. https://doi.org/10.1111/1365-2745.13403.

Miller, J.D., Skinner, C.N., Safford, H.D., Knapp, E.E., Ramirez, C.M., 2012. Trends and causes of severity, size, and number of fires in northwestern California, USA. Ecol. Appl. 22, 184–203.

Minnich, R.A., Barbour, M.G., Burk, J.H., Sosa-Ramirez, J., 2000. Californian mixed-conifer forests under unmanaged fire regimes in the Sierra San Pedro Martir, Baja California, Mexico. J. Biogeogr. 27, 105–129.

Moritz, M.A., Batllori, E., Bradstock, R.A., Gill, A.M., Handmer, J., Hessburg, P.F., Leonard, J., McCaffrey, S., Odion, D.C., Schoennagel, T., Syphard, A.D., 2014. Learning to coexist with wildfire. Nature 515, 58–66.

Odion, D.C., Hanson, C.T., 2006. Fire severity in conifer forests of the Sierra Nevada, California. Ecosystems 9, 1177–1189.

Odion, D.C., Hanson, C.T., 2008. Fire severity in the Sierra Nevada revisited: conclusions robust to further analysis. Ecosystems 11, 12–15.

Odion, D.C., Hanson, C.T., 2013. Projecting impacts of fire management on a biodiversity indicator in the Sierra Nevada and Cascades, USA: the Black-backed Woodpecker. Open For. Sci. J. 6, 14–23.

Odion, D.C., Frost, E.J., Strittholt, J.R., Jiang, H., DellaSala, D.A., Moritz, M.A., 2004. Patterns of fire severity and forest conditions in the Klamath Mountains, northwestern California. Conserv. Biol. 18, 927–936.

Odion, D.C., Moritz, M.A., DellaSala, D.A., 2010. Alternative community states maintained by fire in the Klamath Mountains, USA. J. Ecol. 98, 96–105.

Odion, D.C., Hanson, C.T., Arsenault, A., Baker, W.L., DellaSala, D.A., Hutto, R.L., Klenner, W., Moritz, M.A., Sherriff, R.L., Veblen, T.T., Williams, M.A., 2014. Examining historical and current mixed-severity fire regimes in ponderosa pine and mixed- conifer forests of western North America. PLoS One 9, e87852.

Odion, D.C., et al., 2016. Areas of agreement and disagreement regarding ponderosa pine and mixed conifer forest fire regimes: a dialogue with Stevens et al. PLoS One. https://doi.org/10.1371/journal.pone.0154579.

Owen, S.M., Sieg, C.H., Sánchez Meador, A.J., Fulé, P.Z., Inigueza, J.M., Baggett, L.S., Fornwalt, P.J., Battaglia, M.A., 2017. Ponderosa pine regeneration in high-severity burn patches. For. Ecol. Manag. 405, 134–149.

Paritsis, J., Veblen, T.T., Holz, A., 2014. Positive fire feedbacks contribute to shifts from *Nothofagus pumilio* forests to fire-prone shrublands in Patagonia. J. Veg. Sci. 26, 89−101. https://doi.org/10.1111/jvs.12225.

Parks, S.A., Miller, C., Parisien, M.A., Holsinger, L.M., Dobrowski, S.Z., Abatzoglou, J., 2015. Wildland fire deficit and surplus in the western United States, 1984-2012. Ecosphere 6 (12). Article 275.

Paula, S., Arianoutsou, M., Kazanis, D., Tavsanoglu, A., Lloret, F., Buhk, C., Ojeda, F., Luna, B., Moreno, J.M., Rodrigo, A., Espelta, J.M., Palacio, S., Fernandez-Santos, B., Fernandes, P.M., Pausas, J.G., Michener, W.K., 2009. Fire-related traits for plant species of the Mediterranean Basin. Ecology 90, 1420-1420.

Pausas, J.G., 2015. Bark thickness and fire regime. Funct. Ecol. 29, 315−327.

Pausas, J.G., 2017. Bark thickness and fire regime: another twist. New Phytol. 213, 13−15.

Pausas, J.G., Keeley, J.E., 2009. A burning story: the role of fire in the history of life. Bioscience 59, 593−601.

Pausas, J.G., Keeley, J.E., Schwilk, D.W., 2017. Flammability as an ecological and evolutionary driver. J. Ecol. 105, 289−297.

Perry, D.A., Hessburg, P.F., Skinner, C.N., Spies, T.A., Stephens, S.L., Taylor, A.H., Franklin, J.F., McComb, B., Riegel, G., 2011. The ecology of mixed severity fire regimes in Washington, Oregon, and northern California. For. Ecol. Manage. 262, 703−717.

Perry, G.L.W., Wilmshurst, J.M., McGlone, M.S., McWethy, D.B., Whitlock, C., 2012. Explaining fire-driven landscape transformation during the Initial Burning Period of New Zealand's prehistory. Global Change Biol. 18, 1609−1621.

Pierce, J.L., Meyer, G.A., Jull, A.J.T., 2004. Fire-induced erosion and millennial-scale climate change in northern ponderosa pine forests. Nature 432, 87−90.

Platt, R.V., Schoennagel, T., 2009. An object-oriented approach to assessing changes in tree cover in the Colorado Front Range 1938−1999. For. Ecol. Manage. 258, 1342−1349.

Prichard, S.J., et al., 2021. Adapting western North American forests to climate change and wildfires: 10 common questions. Ecol. Appl. 31, e02433.

Pyne, S.J., 1982. Fire in America: A Cultural History of Wildland and Rural Fire (Cycle of Fire). University of Washington Press, Seattle.

Raphael, M.G., Morrison, M.L., Yoder-Williams, M.P., 1987. Breeding bird populations during twenty-five years of postfire succession in the Sierra Nevada. Condor 89, 614−626.

Romme, W.H., Despain, D.G., 1989. Historical perspective on the Yellowstone Fires of 1988. Bioscience 39, 695−699.

Roos, C.I., Swetnam, T.W., 2012. A 1416-year reconstruction of annual, multidecadal, and centennial variability in area burned for ponderosa pine forests of the southern Colorado Plateau region, Southwest USA. Holocene 22, 281−290.

Rota, C.T., 2013. Not All Forests Are Disturbed Equally: Population Dynamics and Resource Selection of Black-Backed Woodpeckers in the Black Hills, South Dakota. Ph.D. Dissertation. University of Missouri-Columbia, MO.

Schieck, J., Song, S.J., 2006. Changes in bird communities throughout succession following fire and harvest in boreal forests of western North America: literature review and meta-analyses. Can. J. For. Res. 36, 1299−1318.

Schoennagel, T., Sherriff, R.L., Veblen, T.T., 2011. Fire history and tree recruitment in the Colorado Front Range upper montane zone: implications for forest restoration. Ecol. Appl. 21, 2210−2222.

Schoennagel, T., et al., 2017. Adapt to more wildfire in western North American forests as climate changes. Proc. Natl. Acad. Sci. USA. www.pnas.org/cgi/doi/10.1073/pnas.1617464114.

Sestrich, C.M., McMahon, T.E., Young, M.K., 2011. Influence of fire on native and nonnative salmonid populations and habitat in a western Montana basin. Trans. Am. Fish. Soc. 140, 136—146.

Shatford, J.P.A., Hibbs, D.E., Puettmann, K.J., 2007. Conifer regeneration after forest fire in the Klamath-Siskiyous: how much, how soon? J. For. 105, 139—146.

Sherriff, R.L., Veblen, T.T., 2006. Ecological effects of changes in fire regimes in *Pinus ponderosa* ecosystems in the Colorado Front Range. J. Veg. Sci. 17, 705—718.

Sherriff, R.L., Veblen, T.T., 2007. A spatially explicit reconstruction of historical fire occurrence in the Ponderosa pine zone of the Colorado Front Range. Ecosystems 9, 1342—1347.

Sherriff, R.L., Platt, R.V., Veblen, T.T., Schoennagel, T.L., Gartner, M.H., 2014. Historical, observed, and modeled wildfire severity in montane forests of the Colorado Front Range. PLoS One 9, e106971.

Shinneman, D.J., Baker, W.L., 1997. Nonequilibrium dynamics between catastrophic disturbances and old-growth forests in ponderosa pine landscapes of the Black Hills. Conserv. Biol. 11, 1276—1288.

Siegel, R.B., Tingley, M.W., Wilkerson, R.L., Bond, M.L., Howell, C.A., 2013. Assessing Home Range Size and Habitat Needs of Black-Backed Woodpeckers in California: Report for the 2011 and 2012 Field Seasons. Institute for Bird Populations.

Steel, Z.L., Safford, H.D., Viers, J.H., 2015. The fire frequency-severity relationship and the legacy of fire suppression in California's forests. Ecosphere 6, 8.

Steel, Z.L., Fogg, A.M., Burnett, R., Roberts, J.L., 2022. When bigger isn't better—Implications of large high-severity wildfire patches for avian diversity and community composition. Divers. Distrib. 28, 439—453.

Stephens, S.L., Ruth, L.W., 2005. Federal forest fire policy in the United States. Ecol. Appl. 15, 532—542.

Stephens, S.L., Agee, J.K., Fule, P.Z., North, M.P., Romme, W.H., Swetnam, T.W., Turner, M.G., 2013. Managing forests and fire in changing climates. Science 342, 41—42.

Stephens, S.L., et al., 2021. Fire, water, and biodiversity in the Sierra Nevada: a possible triple win. Environ. Res. Comm. 3, 081004. Article.

Swanson, M.E., Franklin, J.F., Beschta, R.L., Crisafulli, C.M., DellaSala, D.A., Hutto, R.L., Lindenmayer, D.B., Swanson, F.J., 2011. The forgotten stage of forest succession: early-successional ecosystems on forest sites. Front. Ecol. Environ. 9, 117—125.

Taylor, A.H., Skinner, C.N., 1998. Fire history and landscape dynamics in a late-successional reserve, Klamath Mountains, California, USA. For. Ecol. Manage. 111, 285—301.

Turco, M., Abatzoglou, J.T., Herrera, S., Zhuang, Y., Jerez, S., Lucas, D.D., AghaKouchak, A., Cvijanovic, I., 2023. Anthropogenic climate change impacts exacerbate summer forest fires in California. Proc. Natl. Acad. Sci. U. S. A. 120 (25), e2213815120. https://doi.org/10.1073/pnas.2213815120.

van Wagtendonk, J.W., van Wagtendonk, K.A., Thode, A.E., 2012. Factors associated with the severity of intersecting fires in Yosemite National Park, California, USA. Fire Ecol 8, 11—32.

Veblen, T.T., 2003. Historic range of variability of mountain forest ecosystems: concepts and applications. For. Chron. 79, 223—226.

Veblen, T.T., Lorenz, D.C., 1986. Anthropogenic disturbance and recovery patterns in montane forests, Colorado Front Range. Phys. Geogr. 7, 1—24.

Veblen, T.T., Lorenz, D.C., 1991. The Colorado Front Range: A Century of Ecological Change. University of Utah Press, Salt Lake City, 186pp.

Veblen, T.T., Kitzberger, T., Donnegan, J., 2000. Climatic and human influences on fire regimes in ponderosa pine forests in the Colorado Front Range. Ecol. Appl. 10, 1178—1195.

Weaver, H.A., 1943. Fire as an ecological and silvicultural factor in the ponderosa pine region of the Pacific slope. J. For. 41, 7–15.
Westerling, A.L., 2016. Increasing western US forest wildfire activity: sensitivity to changes in the timing of spring. Phil. Trans. R. Soc. B 371, 20150178. https://doi.org/10.1098/rstb.2015.0178.
Westerling, A.L., Hidalgo, H.G., Cayan, D.R., Swetnam, T.W., 2006. Warming and earlier spring increases western US forest wildfire activity. Science 313, 940–943.
Whitlock, C., Marlon, J., Briles, C., Brunelle, A., Long, C., Bartlein, P., 2008. Long-term relations among fire, fuel, and climate in the north-western US based on lake-sediment studies. Int. J. Wildland Fire 17, 72–83.
Williams, J., 2012. Exploring the onset of high-impact mega-fires through a forest land management prism. For. Ecol. Manage. 294, 4–10.
Williams, M.A., Baker, W.L., 2010. Bias and error in using survey records for ponderosa pine landscape restoration. J. Biogeogr. 37, 707–721.
Williams, M.A., Baker, W.L., 2011. Testing the accuracy of new methods for reconstructing historical structure of forest landscapes using GLO survey data. Ecol. Monogr. 81, 63–88.
Williams, M.A., Baker, W.L., 2012a. Spatially extensive reconstructions show variable- severity fire and heterogeneous structure in historical western United States dry forests. Global Ecol. Biogeogr. 21, 1042–1052.
Williams, M.A., Baker, W.L., 2012b. Comparison of the higher-severity fire regime in historical (A.D. 1800s) and modern (A.D. 1984-2009) montane forests across 624,156ha of the Colorado Front Range. Ecosystems 15, 832–847.
Williams, M.A., Baker, W.L., 2013. Variability of historical forest structure and fire across ponderosa pine landscapes of the Coconino Plateau and south rim of Grand Canyon National Park, Arizona, USA. Landsc. Ecol. 28, 297–310.
Williams, M.A., Baker, W.L., 2014. High-severity fire corroborated in historical dry forests of the western United States: Response to Fulé et al. Global Ecol. Biogeogr. 23, 831–835.
Wright, C.S., Agee, J.K., 2004. Fire and vegetation history in the eastern Cascade mountains. Washington. Ecol. Appl. 14, 443–459.

Chapter 2

Ecosystem Benefits of Megafires

Dominick A. DellaSala[1] and Chad T. Hanson[2]
[1]Wild Heritage, A Project of Earth Island Institute, Berkeley, CA, United States; [2]John Muir Project of Earth Island Institute, Berkeley, CA, United States

2.1 JUST WHAT ARE MEGAFIRES?

Under extreme weather conditions (e.g., mega-droughts, heat domes, and strong winds), any wildfire start has the potential to blow up quickly, causing dramatic changes in plant and wildlife communities across landscapes that summarily provide renewed opportunities for colonization by fire-loving species uniquely adapted to the newly created postfire environments. Despite unprecedented increases in fire-suppression tactics, megafires have proven impossible to control and when in full force are responsible for the vast majority of area burning in active fire seasons. Negative perceptions about megafires and high costs of property damage have resulted in extreme controversy over fire management among the public, land managers, government officials, and even some conservation groups that typically view these fires as "catastrophes," "bad fires," to be somehow contained or reduce via widespread "active management."

By contrast, megafires are self-organizing natural forces that possess the sheer capacity to drive themselves into disturbance dynamics independent of suppression efforts (Clar et al., 1996). For instance, very large fires are affected by—and, in turn, generate—their own weather patterns with fire plumes ascending to over 9000 m (Figure 2.1). The plumes then create downward pressure gradients with high winds that direct fire spread, fire line propagation, and fire intensity over very large areas.

Megafires qualitatively differ from smaller fires such that, in very small fires, overall severity tends to be low and plant communities seldom change appreciably. In large fires, however, they experience varying natural succession trajectories at multiple spatial scales because the severity associated with larger fires tends to be quite variable in mixed-fire systems (Abrams et al., 1985; Kotliar and Wiens, 1990; Wu and Loucks, 1995; Odion et al., 2004, 2010; Kasischke and Turetsky, 2006; DellaSala et al., 2017). Most megafires

FIGURE 2.1 Smoke plume from a large 19th century fire in the San Gabriel Mountains of southern California (Kinney, 1900).

are ecologically beneficial because they result in high levels of beta (changes in species composition across fire severities) and alpha (within a particular fire severity patch) diversity (Whittaker, 1960), including a prolific pulse of critically important forest legacies (e.g., snags, downed logs, montane chaparral patches) over large landscapes.

In the western United States, megafires represented up to one-third of area burned from 1984 to 2010 (Stavros et al., 2014) and have been increasing more recently. In other (e.g., wetter) regions they are less common, depending on the frequency of regional droughts; but, when they do occur, they contribute to disproportionately high burn area (especially severe fire patches) compared with those of more frequent and less severe fires. Just how large a fire has to be to qualify as a megafire has been mainly related to socioeconomic factors (e.g., costs of fire suppression, property damage, loss of life), although there have been attempts to quantify these fires using statistical attributes related to fire size and frequency (Lin and Rinaldi, 2009). The popular press, decision makers, and some researchers (e.g., Forest Service Researcher Paul Hessburg; https://digitalcommons.usu.edu/rtw/2017/Oct18/7/; active February 12, 2023; Hessburg et al., 2021) have used the term "megafire" most often in a negative, ecologically misinformed sense and with a strong bias against researchers who disagree with their view and calls for massive intervention that will not work in a changing climate and would cause substantial collateral ecosystem damages (e.g., DellaSala et al., 2022). The term also increasingly has been used in science policy articles (Attiwill and Binkley, 2013; Williams, 2013), most often in a command-and-control sense aimed at preventing fires from reaching megafire status (Hessburg et al., 2021) even though this is impossible to achieve at scale despite such claims (DellaSala et al., 2022).

With the recognition that a certain amount of subjectivity is inherent in describing such a broad concept, we recommend that the determination of megafires be based on the spatiotemporal and ecologically based characteristics of large fires by comparing them over time at the regional scale. For instance, one spatial criterion for megafire determination might be when an individual fire is > 2 standard deviations above the average historical size of fires in a given region during a specified period (e.g., 25 years) and the high-severity component of that fire complex includes large patches (e.g., >400 ha, and perhaps some >1000 ha) consistent with the historical range of variation (see Odion et al., 2014; Odion et al., 2016; DellaSala and Hanson, 2019).

Statistical definitions can be cross-checked for ecological relevance by comparing fire characteristics to the regenerative properties of fire-dependent (pyrogenic) species (as in Turner and Dale, 1998). Megafire determinations also need to be based on evolutionary timelines, given the slow (relative to human time lines) postfire changes of abiotic factors (e.g., soils) and the resilience of surviving or regenerating seed sources as part of the typical postfire successional stages (see Swanson et al., 2011; DellaSala et al., 2014; DellaSala et al., 2017; DellaSala and Hanson, 2019). Moreover, megafire determinations should take into account the importance of large fires in closing the gap on historical regional fire deficits (Odion et al., 2014; Parks et al., 2015) and maintaining large patches of complex early seral forest—at more significant spatial and temporal scales—to provide habitat for rare and declining fire-dependent species (DellaSala et al., 2014, 2017; Hanson, 2014, 2021; DellaSala and Hanson, 2019).

In this chapter, we generally refer to a threshold value for megafires as landscape-scale fires of $\sim 50,000$ ha (500 km^2) (see, e.g., Keeley and Zedler, 2009) because this is most likely to be a biologically meaningful scale. Megafires may, however, occasionally occur over much larger areas and even regions. Some examples include the 2002 Biscuit fire ($\sim 220,000$ ha) in the southern Oregon portion of the Klamath-Siskiyou ecoregion (see Chapter 9); the 2002 Rodeo-Chedeski fire in Arizona (186,866 ha); the 2011 Wallow fire (217,741 ha) in the southwestern United States (Williams and Baker, 2012a); the 2013 Rim fire (104,176 ha) in the western Sierra Nevada (USDA, 2014); the simultaneous Labor Day (September 7, 2020) fires in Oregon (336,414 ha) and Washington (130,000 ha) (portions of Idaho and California also burned); the Creek fire of 2020 in the southern Sierra Nevada (153,738 ha); the August Complex fire of 2020 in northwestern California (417,898 ha); the Dixie fire of 2021 in the northern Sierra Nevada (389,837 ha), the >1 million ha fires (two) in Victoria, southeastern Australia (Lindenmayer and Taylor, 2020); and three million hectares in southeastern Australia over a 7-year period (Attiwill and Adams, 2013). The largest megafires (>50,000 ha), of interest herein, occur in forests/woodlands, shrubland, and grassland (herb) vegetation types.

Megafires with a large high severity component have been reoccurring in the western United States for centuries, but, for the most part, there has been no recent uptick (1990s forward) in large high-severity patches or conifer regeneration failures in large high-severity patches (DellaSala and Hanson, 2019; Hanson and Chi, 2021), despite claims of widespread ecosystem type shifts (Hessburg et al., 2021). This issue needs to be monitored closely. With climate change triggering more extreme fire-weather problems (Westerling, 2016; Abatzoglou and Williams, 2016), the occurrence of megafires is likely to go up with more fires becoming uncontrollable regardless of unprecedented fire suppression and massive amounts of spending on "active management" approaches (DellaSala et al., 2022).

2.2 MEGAFIRES AS GLOBAL CHANGE AGENTS

We describe some of the more general properties of megafires and why we believe they play a keystone role in maintaining biodiversity and ecosystem processes. We start with a global perspective and then zoom in on the western United States, where the term megafire ostensibly has its origins (Pyne, 2007). We also discuss whether megafires are currently increasing or have the capacity to do so as the planet further overheats and potential to remedy the ongoing fire deficit in many regions of western North America and the world (Niklasson and Granström, 2004; Odion et al., 2014).

Naturally occurring megafires (or very large fires, as described by others) have been reported in many dry regions of the world where such large fires are anticipated based on historical climate and the contemporary changing climatic conditions. Notable examples include Portugal (Tedim et al., 2013); northern China and southeastern Siberia (Cahoon et al., 1994); southern France (Ganteaume and Jappiot, 2013); Greece (San-Miguel-Ayanz et al., 2013); western United States (many examples (Stavros et al., 2014), most notably the greater Yellowstone ecosystem (Turner et al., 1998)); boreal forests of Canada, Alaska, and Russia (Kasischke and Bruhwiler, 2002); portions of Australia (Lindenmayer et al., 2010; Lindenmayer and Taylor, 2020); and sub-Saharan Africa (Bird and Cali, 1998). Anthropogenic megafires, such as those that have occurred in the Brazilian Amazon as part of deforestation policies for lumber and livestock, are not considered ecologically driven events and therefore are not included here.

Based on global patterns of fire behavior, the occurrence of megafires is governed primarily by top-down drivers such as increases in the sea surface temperature (Skinner et al., 2006), the Pacific Decadal Oscillation (when positive; Morgan et al., 2008), El Niño-Southern Oscillation (Skinner et al., 2006), midtropospheric surface-blocking events (high-pressure systems) during summer months (Johnson and Wowchuk, 1993), and the vapor pressure deficit (measure of atmospheric drying; see Westerling, 2016; Abatzoglou and Williams, 2016; Holden et al., 2018; Hawkins et al., 2022; Turco et al., 2023; Dahl et al., 2023). Local factors (terrain, vegetation) may contribute to the extent of a fire (Thompson and Spies, 2009; Ganteaume and Jappiot, 2013), especially areas with an industrial logging footprint that tend to burn severely because of homogeneous young tree plantations and logging slash (Zald and Dunn, 2018).

2.3 MEGAFIRES, LARGE SEVERE FIRE PATCHES, AND COMPLEX EARLY SERAL FORESTS

The relationship of megafires to native biodiversity warrants special attention by ecologists and land managers. Numerous studies have now addressed this issue in various ways and for very large, intense fires covering ~20,000–50,000 ha or larger. For example, after the fires of 1988, 2000, and 2003, some burned areas in the Intermountain West were transformed into complex early seral habitat for unique early-successional bird communities (Hutto 1995; Hutto, 2008, Smucker et al., 2005; see Chapter 3). In southwest Oregon's 2002 Biscuit fire, large high-severity fire patches and high-severity reburned areas were found to support high levels of native plant and bird species richness, with richness and abundance at least on par with unburned mature/old forest (Donato et al., 2009; Fontaine et al., 2009). In the southeastern Sierra Nevada mountains of California, greater richness of bird species was reported in the ~61,000 ha McNally fire of 2002 than in adjacent, unburned mature/old conifer forest (Siegel and Wilkerson, 2005), similar to bird richness levels in the ~18,000 ha Donner fire of 1960 in the northern Sierra Nevada (Raphael et al., 1987). Likewise, in the northern Sierra Nevada, the highest total bird abundance was found in unlogged high-severity fire areas within the ~23,000 ha Storrie fire of 2000 (Burnett et al., 2010). An investigation of moderate-/high-severity fire areas in the 104,176 ha Rim fire of 2013 in the central Sierra Nevada found high avian abundance and diversity just 1 year after the fire, including numerous species that were absent or nearly absent in adjacent unburned forest, concluding that these higher-severity areas in the Rim fire created "a rich habitat for early-successional birds that will sustain these rarer species on the Sierra landscape for years to come" (Fogg et al., 2014). As fire activity and high-severity fire areas increase, that can shift community composition toward more complex early seral forest, with more interior areas of large high-severity fire patches (Steel et al., 2022). This can provide vital ecological benefits, for example to the 13% of bird species that depend on the interiors of large high-severity fire patches in the Sierra Nevada (Steel et al., 2022), while interiors of large high-severity fire patches currently comprise only a small fraction of 1% of Sierra Nevada forests (DellaSala and Hanson, 2019). Importantly, thus far conifer establishment in the largest high-severity patches and high-severity fire rotations are operating within historical bounds (DellaSala and Hanson, 2019; Hanson and Chi, 2021) but the warming climate warrants close, ongoing monitoring.

More recent scientific research has further confirmed the ecological value and importance of very large fires that include large high-severity fire patches. The Rim fire in the forests of the central Sierra Nevada has been a particularly rich natural laboratory on response of birds to high severity fire patches. In the Rim fire, researchers found rare black-backed woodpeckers (*Picoides arcticus*) selecting the deeper interior locations of this megafire, and the interiors of larger high-severity fire areas (White et al., 2019), especially high-severity fire patches with very high densities of medium and large snags (Hanson and Chi, 2020). Endangered great gray owls (*Strix nebulosa*) increased following the Rim fire (Siegel et al., 2019). The Rim fire stimulated the proliferation of a super-abundance of pyrogenic morel mushrooms (Larson et al., 2016), providing a rich food source for mammalian wildlife, and ectomycorrhizal fungi networks remained intact (in places that were not logged) in the large high-severity fire patches (Glassman et al., 2016). In the even larger Wallow fire of 2011 in Arizona, scientists found that wolves (*Canis lupus*), black bears (*Ursus americanus*), mountain lions (*Puma concolor*), mule deer (*Odocoileus hemionus*), and elk (*Cervus canadensis*) all increased with higher levels of fire severity, due to enhanced food resources in complex early seral forest habitat created by higher-severity fire patches (Lewis et al., 2022). In the forests of the southern Sierra Nevada, gray wolves recently (2023) naturally colonized the area for the first time in a century, establishing a new pack territory in two adjacent megafires, the Windy fire of 2021 and the Castle fire of 2020. In a large fire in southern Oregon, Galbraith et al. (2019) found that, in high-severity fire areas, wild bee abundance was 20 times higher, and species richness was 11 times higher, than in lower-severity fire areas.

Megafires also influence habitat for rare and management-sensitive species such as the California spotted owl (*Strix occidentalis occidentalis*) and related spotted owl subspecies (Bond et al., 2022). This forest species mainly selected unlogged high-severity areas in the McNally fire area for foraging, more than lower fire severity areas and more than unburned mature/old forest (Bond et al., 2009). Additionally, after the 90,265 ha Horseshoe 2 fire of 2011 in ponderosa pine (*Pinus ponderosa*) and mixed-conifer forests of Arizona, Mexican spotted owl (*Strix occidentalis lucida*) populations and owl reproduction increased, particularly in territories with higher levels of high-severity fire (Moors, 2012, 2013), likely due to an enhanced small mammal prey base (Ganey et al., 2014). Similarly, at 1-year post-fire, but before post-fire logging, Lee and Bond (2015) found 92% occupancy of historical California spotted owl territories in the Rim fire, which is higher than average annual occupancy in unburned mature/old forest. Pair occupancy was not reduced in territories with mostly high-severity fire effects. This is similar to findings of Schofield et al. (2020), who reported that California spotted owl occupancy increased slightly after the Rim fire in Yosemite National Park, in the absence of post-fire logging. More recently, Bond et al. (2022) determined that the vast majority of northern spotted owl (*Strix occidentalis caurina*) territories from northern California to Washington experienced multiple logging entries before and after severe wildfires which, along with invasions from the superior competitive barred owl (*Strix varia*), likely contributed to site abandonment rather than the fires themselves. Hanson et al. (2021) similarly found no adverse impact of very large fires on spotted owl occupancy in territories where post-fire logging did not occur. And, in an extraordinary metaanalysis, using data mostly from very large fires, Lee (2020) found overall positive effects of wildfires on the three subspecies of spotted owls, after removing the confounding effect of post-fire logging that was found to consistently harm spotted owls.

In megafires, some rare, sensitive bat species also are most strongly associated with high-severity fire areas (Buchalski et al., 2013). Pacific fishers (*Pekania pennanti*), typically considered an old-growth associated species, also have been observed using unlogged moderate-high-severity fire areas within pre-fire dense, mature forest (McNally 2002 fire, Sequoia National Forest, CA) at levels comparable to their use of adjacent dense, old, unburned forest (Hanson, 2013). This is especially true for females (Hanson, 2015).

Megafires also trigger a "pulse" disturbance (Box 2.1) that replenishes complex early seral forests for abundant wildlife communities. Complex early

---

BOX 2.1 Pulse Versus Chronic Disturbance

Any large natural disturbance (e.g., landscape-level insect outbreaks, hurricanes, volcanic eruptions, big floods, coastal storms) triggers a pulse of biological activity (Turner and Dale, 1998). In the case of fires, a pulse disturbance occurs when a large fire of high intensity kills the majority of vegetation, leaving charred, fire-killed trees and other vegetation that may persist for decades to centuries as key structural elements in the new forest. Many terrestrial and aquatic species (see Chapters 3–6) depend on these postfire structural elements (Donato et al., 2012). By contrast, a chronic disturbance is a reoccurring one that can also repeatedly affect large landscapes. Most anthropogenic disturbances are chronic, accumulating in space and over time, and often exceed the capacity of disturbance-adapted species to regenerate. Anthropogenic chronic disturbances take place over short time lines that are outside historical bounds in terms of size and intensity, and they remove or inhibit, rather than create, biological legacies (DellaSala 2020). As such, they may "flip" ecosystem dynamics to altered states that carry irreparable consequences to biodiversity (Paine et al., 1998; Lindenmayer et al., 2011). Examples include logging, the application of herbicides, and planting of nursery-stock conifers that typically follow large fires; reseeding of postfire areas with nonnative plants for erosion abatement, where nonnative plants then outcompete the native plant establishment (Beyers, 2004); roads and chronic sediment input into streams that impact aquatic species (Trombulak and Frissell, 2000; Colombaroli and Gavin, 2010); and cattle grazing following reseeding in postfire areas, which compacts fragile soils and favors exotic over native species (Beschta et al., 2013; also see Chapter 9).

FIGURE 2.2 An example of complex early seral forest created by a large, high-severity fire patch in the Sierra Nevada. Note the preponderance of biological legacies (snags, shrubs).

seral habitat is the recently established vegetation stage in which intense fires kill most of the overstory trees (Swanson et al., 2011; DellaSala et al., 2014). This stage of forest succession produces biologically rich patches populated by abundant large snags, fallen logs, montane chaparral (patches of native flowering and berry-producing shrubs), and natural conifer regeneration of variable density that link successional stages across a temporal gradient (Swanson et al., 2011). Notably, the legacy pulse provided by large fires "lifeboats" essential structural elements from the unburned forests to the recent fire-created one, which then continues over decades to centuries as a forest matures (DellaSala, 2020). In this fashion, the precocity of complex early seral vegetation is evident in structural elements originating long before and immediately after the disturbance and persisting over time (Donato et al., 2012; see Figure 2.2). Dead trees, for instance, especially large ones, play a keystone role in complex early seral communities. When they are removed via logging (which is most often the case), most fire-dependent bird species decline, and many disappear altogether (Hutto, 2008; Hanson, 2014; also see Chapter 9).

This deep body of scientific evidence, indicating high ecological importance of very large fires, and large high-severity fire patches within such fires, soundly refutes misinformation widely used by land managers to push through intensive fuel reduction treatments involving logging. Omissions of this more comprehensive body of science as in Jones et al. (2022) has led to substantial impacts in spotted owl habitat from inappropriate logging and thinning.

2.4 HISTORICAL EVIDENCE OF MEGAFIRES

In fire-adapted ecosystems around the world, there is historical evidence that megafires have occurred for millennia. For instance, in Victoria, Australia, ~2 million ha burned in the "Black Friday Bushfires" in 1939, much of which occurred in montane-*Eucalyptus* forests (Attiwill and Adams, 2013). In 1825, in the forests of New Brunswick, Canada, the Miramichi fire spanned ~1.2 million hectares (Wein and Moore, 1977), the Saguenay fire in Quebec grew to ~ 390,000 ha in 1870 (NYT, 1870), and the "Great Fire of 1910" burned over 1.2 million hectares in northeastern Washington, northern Idaho, and western Montana.[1] Remarkably, the Great Fire burned most of the fire area in just 2 days, killing 87 people, mostly firefighters (Pyne, 2008; Egan, 2010). Large fires were also quite common in the Pacific Northwest during warm phases of the Holocene and pre-Holocene (up to 20,000 years ago; Whitlock, 1992).

In the western United States, the historical significance of large fires has been recorded by the National Interagency Fire Center (Appendix 2.1). Examination of historical records indicates effects of fire over large areas on property and loss of life, thereby understandably contributing to command-and-control responses (e.g., as in Hessburg et al., 2021 vs. DellaSala et al., 2022). Determining whether today's megafires are uncharacteristic requires a comprehensive understanding of a region's biological and cultural history of fire over timelines that span centuries, not decades. Thus, the historical context of large, high-severity fire patches is especially pertinent to the question: Are megafires categorically a modern anomaly or are such fires within the historical envelope, before modern fire suppression, land management practices, and anthropogenic climate change?

As a regional example we provide some relevant historical accounts and fire history reconstruction studies to address the occurrence of very large, high-severity fire patches in ponderosa pine and mixed-conifer forests of the western United States before the effects of fire suppression and logging. Readers should note that we are now focusing on large patches of high-severity fire within the megafire "matrix" that is a mosaic of varying patch sizes and severities. As such, we grouped historical megafires geographically to show broad-level influences from past fires and for later discussion regarding what the future may hold for megafires.

Rocky Mountain Region

Perhaps the most well-known example of a historical megafire is the "Big Burn" of 1910 in the northern Rocky Mountains, a high-severity fire of ~1.2 million hectares that occurred under extreme fire-weather conditions (e.g., drought and hot summer) in mostly remote, unlogged forest. The preburn forests included mostly lodgepole pine (*Pinus contorta*) and

---

1. For historical photos go to: https://www.google.com/search?q=photos+of+Big+Blow+out+1910+fires&espv=2&biw=1280&bih=629&tbm=isch&tbo=u&source=univ&sa=X&ei=JmNQVOPzGKaf8QG544GgCg&ved=0CDIQ7Ak; accessed February 20, 2024.

ponderosa pine/Douglas-fir (*Pseudotsuga menziesii*) (Leiberg, 1900a). To reiterate, the 1910 fire burned most of its landscape in just 2 days. But large, intense fires were not unusual for the region prior to mechanized fire suppression. For instance, high-severity fire patches hundreds to thousands of hectares were mapped by early US Geological Survey researchers in the late 1800s and early 1900s in ponderosa pine in similarly remote, unmanaged areas (Leiberg, 1900a).

In the central Rockies, within the ponderosa pine forests of Colorado's Front Range, numerous large, higher-severity fire patches (mixed-severity and higher-severity combined) were documented in the mid- to late 1800s (Williams and Baker, 2012b). Many of the higher-severity patches within the large fire perimeter were 1000–3000 ha; the maximum higher-severity patch size was 8331 ha. The mean and maximum historical high severity patch sizes were even larger than current patch sizes in this same area (Williams and Baker, 2012b). A recent analysis indicates that only a relatively minor proportion (16%) of the montane conifer forests of the Colorado Front Range have experienced an increase in fire severity such that they may exceed historical norms—and such areas are generally at the lowest elevations (Sherriff et al., 2014).

Eastern Cascades and Southern Cascades

High-severity fire patches of 1000–5000 ha and 5000–10,000 ha within very large fire complexes in the mixed-conifer forests of the Eastern Cascades of Washington State have been described using historical accounts (Perry et al., 2011) and reconstructions based on field surveys from the mid- to late 1800s (Baker, 2012). Similarly, pioneering historical ecologists (Leiberg, 1903, pp. 273–275, plate XL) mapped and reported a single high-severity fire patch ($\sim$14,000 ha) in predominantly ponderosa pine near Mount Pitt, south of Crater Lake, in unlogged forests of the eastern Oregon Cascades. In mixed-conifer and fir (*Abies* spp.) forests of the Southern Cascades of California, "widespread and high-severity fires" burned across vast areas, indicated by fire scars at numerous locations separated by > 30 km and occurring in the same years (1829, 1864, 1889) during dry conditions (Bekker and Taylor, 2010).

Oregon Coast Range and Klamath Region

In the western Oregon Coast Range, the 1849 Yaquina and 1853 Nestucca fires each spanned $\sim$200,000 ha (Gannett, 1902; Morris, 1934; see also Figure 2.3 below). In the Klamath region of southern Oregon, Leiberg (1900b) documented a high-severity fire area of $\sim$24,000 ha before the arrival of European colonizers and occurring in roughly equal areas of lodgepole and ponderosa pine.

(a) THE GREAT NESTUCCA BURN.

(b) THE GREAT YAQUINA BURN.

FIGURE 2.3 The Nestucca (a) and Yaquina (b) fires in the mid-1800s (Gannett, 1902).

Sierra Nevada

Leiberg (1902, plate VII) mapped numerous large, high-severity fire patches (defined as 75%—100% mortality of timber volume) occurring in the 1800s, including many patches 1000—5000 ha and numerous 5000—10,000 ha patches, within unlogged mixed-conifer forests (Hanson, 2007). Leiberg noted that "a large proportion" of these patches occurred in the early part of the 1800s, before European colonists. High-severity fire patches were mapped only if Leiberg and his team were able to locate evidence of the previous stands (e.g., the remains of fire-killed overstory trees and large downed logs). He also estimated the age of the naturally regenerating stands following high-severity fire (Leiberg, 1902). Because Leiberg mapped high-severity fire patches that occurred over the course of the nineteenth century, some of these large patches could have resulted from fires occurring in different years or decades, but in most cases, it would be exceedingly improbable that large, high-severity fire patches happened to occur immediately adjacent to previous large patches. Baker (2014), using General Land Office survey data from the mid- to late 1800s, found high-severity fire patches up to ~ 8000 ha and ~ 9000 ha in mixed-conifer and ponderosa pine forests of western Sierra Nevada.

Southwestern United States and Pacific Southwest

Fire history reconstructions have revealed contiguous areas with only high-severity fire effects over 500—1000 ha, and one >10,000 ha, within ponderosa pine forests in the Black Mesa and southeastern Mogollon Plateau areas of eastern-central and northern Arizona, respectively (Williams and Baker, 2012a). This was based on spatially extensive US General Land Office field plots from the mid- to late 1800s (Williams and Baker, 2012a) that were submitted to rigorous accuracy testing and have been found to be robust upon further examination and analysis (Williams and Baker, 2012a, 2014). Lang and Stewart (1910), in ponderosa pine forests of northern Arizona, observed "[v]ast denuded areas, charred stubs and fallen trunks and the general prevalence of blackened poles seem to indicate [fire] frequency and severity ...."

In the mountains of southern California, in what is now the northern portion of the Cleveland National Forest, the Santiago Canyon fire occurred in 1889, covering 125,000—200,000 ha of chaparral and dry conifer forest; the size varies among accounts (Keeley and Zedler, 2009).

Black Hills

In ponderosa pine-dominated forests of the Black Hills of South Dakota, a single high-severity fire patch of ~ 19,000 ha was documented (Dodge, 1876). Just over 2 decades later, US Geological Survey researchers found "a young pine forest springing up" in this area (Graves, 1899, p. 146) and photographed and described additional large, high-severity fire areas.

2.5 MEGAFIRES AND LANDSCAPE HETEROGENEITY

There is much hyperbole, misunderstanding, and fear conveyed in the media's reports of megafires. Many are based on unverified anecdotal representations from land managers, firefighters, timber industry officials, local community leaders, and politicians that claim overwhelmingly high-severity fire effects on forests or high-severity fire patches tens of thousands of hectares with no surviving trees and little or no potential for natural conifer regeneration, as in the misinformed case of the Rim fire of 2013 in the forests of the western Sierra Nevada of California (Cone, 2013; Jarvis, 2014). Once the smoke clears, however, the evidence generally indicates predominantly low- to moderate-severity fire effects over large areas, substantial heterogeneity even within large, high-severity fire patches, and forests naturally regenerating in ways that promote a diversity of successional stages and plant communities (Table 2.1) (DellaSala and Hanson, 2019).

Moreover, based on recent observations of high-severity fire patches within megafires, substantial intrapatch heterogeneity occurs at multiple scales. Where high-severity fire effects occur, large overstory trees often survive in varying densities, and the overall basal area mortality can be 100% in some portions of patches but often can be ~60%–80% as well (Table 2.2).

TABLE 2.1 High-Severity Fire Proportions in contemporary Megafires in the Western United States

Fire/location	Size (hectares)	Year	High severity area (%)	Source
Rim (California)	104,178	2013	20% (52% low, 28% moderate)	www.mtbs.gov
Biscuit (Oregon)	~200,000	2002	29% (41% low, 30% moderate)	www.mtbs.gov
Rodeo-Chedeski (Arizona)	186,866	2002	37%	Thompson and Spies (2009) Williams and Baker (2012a)
Hayman (Colorado)	53,212	2002	22%	Williams and Baker (2012a)
Wallow (Arizona)	217,741	2011	16%	Williams and Baker (2012a)

TABLE 2.2 Basal Area Mortality for Relative Delta Normalized Burn Ratio (RdNBR) Higher-Severity Fire Thresholds of 574 and 800 in Klamath and Sierra Nevada Plots

Region	RdNBR	Basal area mortality					
		Small trees included		Trees ≥30 cm DBH		Trees ≥50 cm DBH	
		Mean % (SD)	n	Mean % (SD)	n	Mean % (SD)	n
Klamath	574 ± 100	60.9 (35.6)	18	51.8 (39.0)	16	48.1 (40.1)	16
	800 ± 100	75.8 (24.8)	18	67.9 (33.4)	16	58.9 (37.1)	13
Sierra Nevada	574 ± 50	60.9 (35.1)	67	51.0 (39.5)	58	41.1 (44.4)	43
	800 ± 50	83.4 (27.2)	69	76.0 (35.5)	65	60.2 (46.2)	41

n represents the number of field validation plots.
Values are derived from US Forest Service field validation plot data and fire severity values from satellite imagery. Reproduced from Hanson et al. (2010).

In addition, patches of low- to moderate-severity fire, ranging from 0.1 ha to hundreds of hectares, occur throughout large, high-severity fire patches, such as in the largest (>4000 ha) high-severity fire patches within mixed-conifer forests during the McNally fire (Buchalski et al., 2013; Hanson, 2013: Figure 2.1) the Biscuit fire (Halofsky et al., 2011), and the ~104,000 ha Rim fire that occurred on the western slopes of the central Sierra Nevada mountains in 2013 (also see DellaSala and Hanson, 2019).

Photos taken in September 2014 (Figure 2.4a and c) show "flushing," the production of new, green needles from surviving terminal buds 1 year after a fire where there was 100% initial mortality of foliage (Hanson and North, 2009) of ponderosa pines hundreds of meters into the interior of one of the largest high-severity fire patches (~1000 ha) of the Rim fire. Surveys by one of us (C.T.H.) indicate, on average, >20 surviving trees per hectare in this large, high-severity fire patch, with live trees variably distributed in clumps, generally across 0.1–10 ha. We also observed natural postfire conifer regeneration 1 year after the fire within the same large, high-severity fire patch in the Rim fire (Figure 2.4b and d). Our postfire surveys indicate, on average, ~250 naturally regenerating conifers per hectare >200 m into the interior of this patch, with ponderosa and sugar pine (*Pinus lambertiana*) dominating interior regeneration (Figure 2.5), findings later verified by a much larger field study conducted at 5 years after the Rim fire (Hanson and Chi, 2021).

FIGURE 2.4 High-severity patches within the Rim Fire (2013) in the Sierra region, showing complex early seral forests and conifer establishment a year later. *Photos by Doug Bevington.*

Crotteau et al. (2013) found 715 naturally regenerated conifer seedlings per hectare in large, high-severity fire patches after the 2000 Storrie fire in the northern Sierra Nevada. In the same fire area, Cocking et al. (2014) found such high-severity fire patches to play a key role in the regeneration and maintenance of California black oak (*Quercus kelloggii*). This is also comparable to Haire and McGarigal (2008) who found regeneration of aspen (*Populus* sp.) and oak (*Quercus* spp.) in large, high-severity fire patches in the southwestern United States. In mixed evergreen forests of southwestern Oregon and northwestern California, Shatford et al. (2007) found several hundred conifer seedlings per hectare, even when plots were ≥300 m into high-severity fire patches and native shrub cover was very high to complete (i.e., the conifers grew up through the shrub cover). Additional conifer regeneration occurred in successive postfire years. Donato et al. (2009) made similar findings in large, high-severity fire patches of the Biscuit fire in the same region. One study by the US Forest Service found relatively little natural conifer regeneration in high-severity fire patches in the northern Sierra Nevada and advocated for postfire logging and plantation establishment (Collins and Roller, 2013). However, a visit to this area by one of us (C.T.H.) found that the study sites had generally been clearcut before the fires and thus there was little or no conifer seed source even before the fires occurred or were nearly pure black

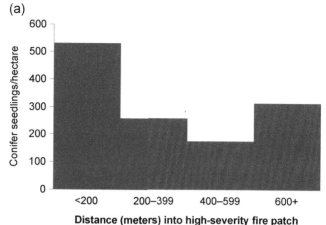

FIGURE 2.5 (a) Natural postfire conifer regeneration within a large, high-severity fire patch 1 year after the Rim Fire of 2013, in terms of seedling density and (b) the percentage of regeneration comprised of pine species (ponderosa and sugar pine) with increasing distance into the high-severity fire patch.

oak stands before the fires. More recent research found that natural postfire conifer regeneration in large, high-severity fire patches in the Rim fire on the Stanislaus National Forest, CA that was both vigorous and heterogeneous even >100 m, and >300 m, into the interior of the largest (>400 ha) high severity patches (DellaSala and Hanson, 2019; Hanson and Chi, 2021). Fire rotation intervals for high-severity patches also ranged from ~12 centuries to 4000 years and appear to be operating at low levels relative to historic bounds.

While natural postfire conifer regeneration can be relatively lower with increasing distance into large, high-severity fire patches (Haire and McGarigal, 2010; DellaSala and Hanson, 2019), the irregular nature of high-severity fire patch shapes leads to a surprisingly small proportion of the area of large, high-severity fire patches within megafires that is more than 100 or 200 m from the nearest live-tree edge (see Figure 2.4 in Halofsky et al., 2011; DellaSala and Hanson, 2019). The spatially scarce interior of most large, high-severity fire patches in megafires can play a vitally important role in maintaining complex early seral forest, dominated by native shrubs, for a relatively longer period of time before such habitat is replaced by young closed canopy conifer stands lacking the shrub and forb layers (Swanson et al., 2011; DellaSala et al., 2017; Hanson and Chi, 2021). This extended early natural-seral period can help to maintain populations of at-risk bird species associated with montane chaparral—many of which are declining as a result of fire suppression, postfire logging, and subsequent shrub removal and the establishment of artificial conifer plantations (Hanson, 2014; Phalan et al., 2019).

2.6 ARE MEGAFIRES INCREASING?

Because megafires have received considerable attention, in part due to exurban sprawl into "firesheds" (areas where homes and structures abut fire-prone forests; Chapter 10), the public is naturally concerned about whether they are increasing in size and frequency. But determining whether megafires are increasing depends on the historical baseline chosen for comparisons with current conditions. The historical baseline is a prime factor in determinations of what may be considered characteristic, or "natural," versus uncharacteristic, or anthropogenic, with respect to any fire. In Chapter 8, we illustrate how more recent historical versus longer time lines can result in significant bias about fire increases, leading to shifting baseline perspectives (i.e., the baseline for contemporary vs. historical comparisons is shifted to a more recent time line that may not reflect longer timelines and conditions under which fire-adapted communities evolved and consequentially may overstate "fuels reduction" needs). The selection of the baseline affects determinations about whether large fires are increasing or not as well.

For instance, megafires in the Algarve region of Portugal, where there has been no historical record of such fires, have within decades been considered the "new reality" ("new normal") that is overwhelming local firefighting resources (Tedim et al., 2013). Recent megafires in Amazonia have been attributed to unprecedented deforestation (slash and burn agriculture especially for beef production) and associated changes in regional climates from deforestation with far reaching global consequences. The total annual area of the boreal forests of Canada and Alaska affected by fire has doubled since the 1960s (Kasischke and Turetsky, 2006), and this increase has been correlated with global overheating (Gillett et al., 2004) and is not limited to those regions as similar recent trends are occurring in the Russian boreal as noted by Greenpeace Russia (https://greenpeace.ru/blogs/2021/09/17/2021-god-polnyj-rekordsmen-po-pozharam/; accessed March 3, 2023). Moreover, over the past several decades, the frequency of fire years with >1% of the region burned in boreal ecozones has increased from an average of 5 times per decade during the 1960s/1970s to that of 13 times per decade during the 1980s/1990s. But are these timelines sufficient to declare today's megafires as truly "uncharacteristic?"

In contrast to studies of short time lines, those covering longer intervals have concluded that, despite increases in fire in some areas of Canada's forests in recent decades, there is nevertheless currently far less fire than historically, such that there is now only about one-fourth as much annual fire as there was c.1850 (Bergeron et al., 2004, Chavardès et al., 2022; see also: https://www.preventionweb.net/news/forest-fires-north-americas-boreal-forests-are-burning-lot-less-150-years-ago). Other research has documented an ongoing deficit of wildfire in forests of the western United States (Law and Waring, 2015; Parks et al., 2015; Baker and Williams, 2018); however, this may be changing as in the Russian boreal as noted. Similarly, a meta-analysis of landscape-scale fires in forested regions of the western United States indicated much more variability in fire severity and extent in the historical record than previously reported (Odion et al., 2014, 2016). This was based on multiple lines of evidence, including early surveys (1880s General Land Office surveys), stand reconstructions (based on plot-level data), and charcoal evidence of large fires from thousands of years ago (Odion et al., 2014). To the surprise of many, based on historical comparisons, these researchers documented a current high-severity fire deficit rather than a surplus, as often claimed (Hessburg et al., 2021).

Using General Land Office records and other evidence, high-severity fire deficits have been documented for the Eastern Cascades, Northern and Central Rocky Mountains, Klamath, Sierra Nevada, and the southwestern United States (Odion and Hanson, 2013; Baker, 2014; Hanson and Odion, 2014; Odion et al., 2014). A stand-age analysis indicated that there is currently one-half to one-fourth as much high-severity fire, depending on the region, as there was before the early twentieth century in mixed-conifer and ponderosa pine forests of the western United States (Odion et al., 2014). While there are some equivocal indications of increases in fire severity (e.g., increases in total high-severity area, along with increases in fire as a whole, but not the proportion of high-severity fire effects in fire complexes) in some regions, such as portions of the southwestern United States and the southern Rockies (Dillon et al., 2011), most studies have found no increase in fire severity in most forested regions of the western United States (Hanson et al., 2009; Collins et al., 2009; Dillon et al., 2011; Miller et al., 2012; Hanson and Odion, 2014; Law and Waring, 2015; Baker and Williams, 2018; DellaSala and Hanson, 2019). One recent study reported an increase in fire severity in the Sierra Nevada (Miller and Safford, 2012), whereas a larger analysis found no such trend (Hanson and Odion, 2014, 2015). Thus, whether the gaps between current and historical fire occurrence are closing in these regions is unclear, suggesting that actions to allow or facilitate more managed wildland fire (fires caused by lightning that are allowed to burn for ecological benefit) should be encouraged where ongoing fire deficits exist and fires can burn under safe conditions. Given the tight association between megadroughts, heat domes, and megafires, this situation needs constant monitoring and updating as the climate-signal takes on an increasingly top-down behavioral influence on fires (DellaSala et al., 2022).

In Chapter 8, we discuss what the future of megafires might be in a changing climate which, based on regional, scaled-down climate change projections, may very well begin to close fire-area deficits across all severities over the coming decades. In most cases, however, postfire landscapes are not allowed to go through successional stages that result in high levels of biodiversity in complex early seral forests initially (Swanson et al., 2011; DellaSala et al., 2017) because they are so often logged after fires, followed by removal of native shrubs and the establishment of artificial conifer plantations (Chapter 9), putting numerous bird species at risk (Hanson, 2014; Thorn et al., 2018; Phalan et al., 2019). So, even though climate change may already be increasing megafire frequency and timing (Westerling, 2016), the ensuing postfire landscapes often are degraded by postfire management, rather than being enhanced by fire otherwise. In fact, in general, the larger the megafire, the more logging, resulting in extensive and intensive landscape-scale degradation that is at least on par with the degradation of green forests (see Chapter 9).

2.7 LANGUAGE MATTERS

The cultural fear of and misunderstanding about forest fires, as well as the climate of political and economic opportunism facilitated by postfire logging policies, has been used to send large sums of research dollars mostly appropriated by Congress and aimed at continuing destructive "active management" programs that result in casting doubt about those that question collateral ecosystem and climate damages from those practices (e.g., Hessburg et al., 2021). This has encouraged more logging at the expense of pyrodiverse, large-fire complexes and the climate (DellaSala et al., 2022). Thus, having an objective dialog about forest fires in general, and megafires in particular, remains even more challenging than during the first edition of this book. Highly charged and hyperbolic fire language so often used to describe fires (e.g., highly subjective "good" vs. "bad" fire), especially large, intense ones, has been escalating conflict and promotes attitudes to do something—anything—no matter the costs and damages. For a more ecologically literate public dialog to occur, and for forest conservation to be most effectively informed by current science not funded by the same agencies that are promoting destructive active management, we submit that the vernacular of fire must become less charged and much more ecologically based. Currently, major newspapers and television stations, policymakers, and land managers commonly describe a particular forest fire as having "destroyed," "damaged," "devastated," "nuked," "razed," "moonscaped," or "consumed" the area within the fire's perimeter, or they describe a certain area of forest as having been "lost" to fire, or "deforested" by fire. As an alternative, we suggest that, at a minimum, neutral language, such as "affected," "spanned," or "covered" (e.g., "The fire affected or spanned over 50,000 ha" or "The fire perimeter now encompasses some 50,000 ha"), or language that reflects the now well-documented ecological benefits of large mixed-severity fires, such as "restored" or "rejuvenated" by fire. Fire "risk," "hazard," or "threat" would become fire "chance" or "probability," "hazardous fuels" would become "post-fire habitat," and "stand-destroying" or "stand-replacing" fire might instead become "stand-initiating" or "stand-renewal" fire. In this vein, especially in regions with ongoing fire deficits relative to the historical spatiotemporal extent of fire, a larger-than-typical fire year would no longer be a "bad" year or "the worst" fire year on record but, rather, would be an "above-average" fire year closing the fire deficit and naturally reducing flammable vegetation at relevant scales. Fundamentally unscientific terms such as "catastrophic" would no longer be considered as legitimate public discourse, except when describing impacts to human communities and people.

Until we change the way we communicate about fire in forests, efforts to provide greater protections for postfire habitat and the many rare and imperiled wildlife species they harbor will be hampered, as will efforts to encourage greater use of managed wildland fire in more remote forested areas. We believe that such a shift in vernacular is more than warranted given the current state of scientific knowledge about the relative scarcity of postfire habitat (because it is often logged) and its great ecological importance to native biodiversity.

2.8 CONCLUSIONS

Postfire landscapes resulting from megafires that produce large, high-severity fire patches have become the forgotten seral stage (Swanson et al., 2011; DellaSala et al., 2017), devalued by most land managers, government officials, and even some scientists and conservation groups (e.g., The Nature Conservancy representatives, following the Rim fire, advocated for expansive postfire logging and tree planting based on Forest Service field trips to discuss postfire management; see Chapter 9). We encourage conservationists and ecologists to refrain from calling the ensuing postfire landscape a "catastrophe" in need of logging and tree planting, particularly given the fire deficit in some places, the rarity of unlogged postfire landscapes, and the prolific conifer establishment and native biodiversity that follows within year(s) of the burn. Megafires have been a top-down driver of ecological and evolutionary change for millennia and will continue to be a dominant natural force, despite command-and-control actions, because local weather effects directed by global overheating mostly govern these fires (DellaSala et al., 2022). Whether they are increasing requires reconstruction of appropriate historical baselines, and more comprehensive analysis of the periodicity, scale, and severity of large fires over long timelines and spatial gradients, relative to those previously used to describe these fires. Megafires may become more common as a result of overheating that has the potential to drive more of these fires over shorter timelines.

We recommend that ecologists and land managers set up permanent plots in megafire areas to better understand long-term effects on fire-dependent communities and resilience to fire, particularly periodic reburns (Donato et al., 2009). Land managers wishing to determine whether these fires are characteristic should use more comprehensive historical accounts (multiple lines of evidence and not just fire scar dating) to avoid shifting baseline perspectives that result in management actions that exacerbate widespread declines in species that are dependent on the pulse of complex early seral vegetation (Hanson, 2014; DellaSala et al., 2017; Phalan et al., 2019). Megafires may be the only large pulse of biological legacies that a particular area receives for many decades, or even centuries, and the structural elements and landscape heterogeneity produced by these events is not re-created by active management that instead degrades ecosystem dynamics (DellaSala, 2020). Megafire pulses create levels of native biodiversity and wildlife abundance comparable to those found in unburned old forest as noted.

By comparison, postfire management often is associated with chronic disturbances that operate outside the bounds of historical and evolutionary processes in terms of patch sizes, disturbance periodicity, and intensity (see Chapter 9; Lindenmayer et al., 2008; Thorn et al., 2018). Such events are not replacements for landscapes generated after fire because they typically remove most biological legacies created by large fires. Thus, from the standpoint of pyrophilous communities, large postfire landscapes of high ecological integrity (unlogged) are newly restored habitat areas rather than habitats "recovering" from an undisturbed state. Managers, scientists, and conservation groups wishing to maintain pyrodiversity should plan for these areas in reserve design and treat them on par with the more celebrated old-growth forests, given the comparability of biodiversity importance and relative rarity (DellaSala et al., 2014, 2017).

Managers also wishing to maintain biodiversity over large landscapes, such as in national parks, lands with wilderness character, and intact areas, should allow the occurrence of megafires to operate as a top-down ecosystem process via appropriate wildfire responses (e.g., work with fire for ecosystem benefits under safe conditions, as in the case of US national parks). Regardless of their immediate scenic change (green to charred), within 1–3 years many high-severity fire patches in conifer forests become rich, colorful landscapes with an abundance of flowers, shrubs, snags, downed logs, natural conifer regeneration, and the sights and sounds of prolific and diverse wildlife. Moreover, when it comes to these burned landscapes, beauty is in the eye of the beholder. Postfire landscapes shaped by megafires are dynamic places that are constantly shifting about as a result of natural successional processes; they are not biological wastelands, as often claimed or assumed (i.e., "bad fire").

Government officials and local communities living in firesheds need to prepare for megafires because these events will always be with us and may increase in frequency and extent in some regions as long as we keep treating the atmosphere as a dumping ground for greenhouse gas emissions, including deforesting/degrading large regions that otherwise sequester and store carbon for long periods even when severely burned (see Chapter 8; Harmon et al., 2022). Effective land-use planning (e.g., zoning) is needed to restrict exurban sprawl into firesheds and to limit increasing fire damage to human structures. That is, with more people and structures occupying firesheds, the prospects for even greater socioeconomic losses from megafires will escalate, triggering increased attempts at ecosystem-degrading command-and-control actions, unless proactive steps are taken to reduce fire risk in the home-ignition zone (see Chapter 10). Attempting to squelch small fires before they become megafires may succeed in non-drought years; however, this only perpetuates command-and-control ecosystem degradation given the loss of biological pulses and massive damages inflicted on fire-adapted communities from such actions (DellaSala, 2020; DellaSala et al., 2022). Once megafires do occur—generally under extreme weather conditions—they are self-reinforcing events that are extinguished when the weather changes (summer drought to fall rainfall, monsoonal rains in places). Regardless of what we attempt to do to them in the form of "active management" to suppress or slow them down (Hessburg et al., 2021), we cannot truly prevent megafires as extreme fire-weather is largely what drives them (DellaSala et al., 2022). To deny this is reality is a form of command-and-control hubris and climate change denialism (see Hessburg et al., 2021 vs. DellaSala et al., 2022 especially disagreements over the importance of the precautionary principle in conservation). A fundamental shift in thinking is needed to change public attitudes toward megafires in recognition of the substantial ecological benefits to forest ecosystems and biodiversity, particularly where ongoing fire deficits are occurring and megafires are burning safely in the backcountry, with the emphasis being strategic fire-risk reduction aimed at protecting homes and human communities through proven fire-safe measures (see Chapter 10).

APPENDIX 2.1 SOME FIRES OF HISTORICAL SIGNIFICANCE FROM RECORDS COMPILED BY THE NATIONAL INTERAGENCY FIRE CENTER (HTTP://WWW.NIFC.GOV/FIREINFO/FIREINFO_STATS_HISTSIGFIRES.HTML) FROM 1825 TO 2013

Date	Name	Location	Hectares	Significance
October 1825	Miramichi and Maine fires	New Brunswick and Maine	1.2 million	160 lives lost
1845	Great Fire	Oregon	600,000	Large area burned
1849	Yaquina	Oregon	180,000	Large area burned
1868	Coos	Oregon	120,000	Large area burned
October 1871	Peshtigo	Wisconsin and Michigan	1,512,000	1500 lives lost in Wisconsin
	Great Chicago	Illinois	Undetermined	250 lives lost
				17,400 structures destroyed
September 1881	Lower Michigan	Michigan	1,000,000	169 lives lost
				3000 structures destroyed
September 1894	Hinckley	Minnesota	64,000	418 lives lost
	Wisconsin	Wisconsin	Several million	Undetermined; some lives lost
February 1898	Series of South Carolina fires	South Carolina	1.2 million	Unconfirmed report of 14 lives lost and numerous structures and sawmills destroyed
September 1902	Yacoult	Washington and Oregon	≥400,000	38 lives lost
April 1903	Adirondack	New York	254,800	Large area burned
August 1910	Great Idaho	Idaho and Montana	1.2 million	85 lives lost

Continued

Date	Name	Location	Hectares	Significance
October 1918	Cloquet –Moose Lake	Minnesota	480,000	450 lives lost
				38 communities destroyed
September 1923	Giant Berkley	California	Undetermined	624 structures destroyed and 50 city blocks leveled
August 1933	Tillamook	Oregon	124,400	1 life lost
				Same area burned again in 1939
October 1933	Griffith Park	California	Undetermined	29 lives lost and 150 people injured
August 1937	Blackwater	Wyoming	Undetermined	15 lives lost and 38 people injured
July 1939	Northern Nevada	Nevada	Undetermined	5 lives lost
				First recorded firefighting fatality in a sage brush fuel type
October 1943	Hauser Creek	California	4000	11 US Marines killed and 72 injuries
				Fire was started by gunnery practice
October 1947	Maine	Maine	82,271	16 lives lost
1949	Mann Gulch	Montana	1736	13 smokejumpers killed
July 1953	Rattlesnake	California	Undetermined	15 lives lost
1956	Inaja	California	17,200	11 lives lost
November 1966	Loop	California	Undetermined	13 EL Cariso Hotshots lost their lives
1967	Sundance	Idaho	22,400	Burned 20,000 ha in just 9 h
September 1970	Laguna	California	70,170	382 structures destroyed

Date	Name	Location	Hectares	Significance
July 1972	Moccasin Mesa	New Mexico	1072	Fire suppression activities destroyed many archeological sites, which resulted in a national policy to include cultural resource oversight in wildland fires on federal lands
July 1976	Battlement Creek	Colorado	Undetermined	5 lives lost
July 1977	Sycamore	California	322	234 structures destroyed
November 1980	Panorama	California	9440	325 structures destroyed
1985	Butte	Idaho	Undetermined	72 firefighters deployed fire shelters for 1–2 h
1987	Siege of '87	California	256,000	Valuable timber lost in the Klamath and Stanislaus National Forests
1988	Yellowstone	Montana and Idaho	634,000	Large area burned
September 1988	Canyon Creek	Montana	100,000	Large area burned
June 1990	Painted Cave	California	1960	641 structures destroyed
	Dude Fire	Arizona	9667	6 lives lost
				63 homes destroyed
October 1991	Oakland Hills	California	600	25 lives lost and 2900 structures destroyed
August 1992	Foothills Fire	Idaho	102,800	1 life lost
1993	Laguna Hills	California	6800	366 structures destroyed in 6 h
July 1994	South Canyon Fire	Colorado	742	14 lives lost

Continued

Date	Name	Location	Hectares	Significance
	Idaho City Complex	Idaho	61,600	1 life lost
August 1995	Sunrise	Long Island	2000	Realization that the East can have fires similar to the West
August 1996	Cox Wells	Idaho	87,600	Largest fire of the year
June 1996	Millers Reach	Alaska	14,934	344 structures destroyed
July 1997	Inowak	Alaska	244,000	Threatened 3 villages
1998	Volusia Complex	Florida	44,452	Thousands of people evacuated from several counties
1998	Flagler/St. John	Florida	37,862	Forced the evacuation of thousands of residents
August 1999	Dunn Glen Complex	Nevada	115,288	Largest fire of the year
August–November 1999	Big Bar Complex	California	56,379	Series of fires caused several evacuations during a 3.5-month period
September–November 1999	Kirk Complex	California	34,680	Hundreds of people were evacuated by this complex of fires that burned for almost 3 months
May 2000	Cerro Grande	New Mexico	19,060	Originally a prescribed fire, 235 structures destroyed and the Los Alamos National Laboratory damaged
July 2001	Thirtymile	Washington	3720	14 fire shelters were deployed
				4 lives lost

Ecosystem Benefits of Megafires Chapter | 2   53

Date	Name	Location	Hectares	Significance
June 2002	Hayman	Colorado	54,400	600 structures destroyed
	Rodeo-Chediski	Arizona	184,800	426 structures destroyed
July 2003	Cramer	Idaho	5538	2 lives lost
October 2003	Cedar	California	110,000	2400 structures destroyed
				15 lives lost
2004	Taylor Complex	Alaska	522,237	Alaska fires during 2004 burned over 2.54 million hectares
June 2005	Cave Creek Complex	Arizona	99,324	11 structures destroyed
				Largest fire in the Sonoran Desert ever recorded
March 2006	East Amarillo Complex	Texas	362,898	80 structures destroyed
				12 lives lost
				Largest fire during the 2006 fire season
April 2007	Big Turnaround Complex	Georgia	155,207	Largest fire for the US Fish and Wildlife Service outside of Alaska
July 2007	Murphy Complex	Idaho	260,806	One of the largest fires in Idaho
2010	Long Butte	Idaho	120,000	
	Jefferson	Idaho	43,600	
2010	Four Mile Canyon	Colorado	2500	A wind-driven fire northwest of Boulder, Colorado, burned more than 170 structures and one fire engine
	Bastrop County Complex	Texas		1400 Residences burned in 3 days and two civilians were killed

Continued

Date	Name	Location	Hectares	Significance
June 2010	Schultz	Arizona	6000	Threatened hundreds of homes; a 12-year-old girl was tragically killed by flash floods that came out of the area burned by this fire
Jun 2011	Las Conchas	New Mexico	62,400	Threatened the Los Alamos National Laboratory
	Wallow	Arizona and New Mexico	215,200	Largest single fire ever recorded in the lower 48 states
August 2011	Pagami Creek	Minnesota	37,600	A significant 4-day wind event caused 32,800 ha to burn in late August and early September
May 2012	Whitewater-Baldy	New Mexico	119,138	Largest fire in New Mexico
June 2012	Waldo Canyon	Colorado	7579	346 homes burned
	White Draw	South Dakota	3600	C-130 Airtanker crash kills four crewmembers
	Long Draw	Oregon	223,051	One of the largest fires in Oregon
June 2013	Yarnell Hill	Arizona	3360	19 lives lost
August 2013	Rim	California	102,926	Largest fire in Sierra Nevada since formal records began in the 1930s

We Refer the Reader to This Weblink for More Recent Large Fires as the Dataset is Constantly Updated.

REFERENCES

Abatzoglou, J.T., Williams, A.P., 2016. Impact of anthropogenic climate change on wildfire across western US forests. Proc. Natl. Acad. Sci. USA. www.pnas.org/cgi/doi/10.1073/pnas.1607171113.

Abrams, M.D., Sprugel, D.G., Dickman, D.I., 1985. Multiple successional pathways on recently disturbed jack pine sites in Michigan. For. Ecol. Manag. 10, 31–48.

Attiwill, P.M., Adams, M.A., 2013. Megafires, inquiries and politics in the eucalypt forests of victoria, south-eastern Australia. For. Ecol. Manag. 294, 45–53.

Attiwill, P.M., Binkley, D., 2013. Exploring the megafire reality: a "forest ecology and management" conference. For. Ecol. Manag. 294, 1–3.

Baker, W.L., 2012. Implications of spatially extensive historical data from surveys for restoring dry forests of Oregon's eastern Cascades. Ecosphere 3. Article 23.

Baker, W.L., 2014. Historical forest structure and fire in Sierran mixed-conifer forests reconstructed from General Land Office survey data. Ecosphere 5. Article 79.

Baker, W.L., Williams, M.A., 2018. Land surveys show regional variability of historical fire regimes and dry forest structure of the western United States. Ecol. Appl. 28, 284–290.

Bekker, M.F., Taylor, A.H., 2010. Fire disturbance, forest structure, and stand dynamics in montane forest of the southern Cascades, Thousand Lakes Wilderness, California, USA. Ecoscience 17, 59–72.

Bergeron, Y., Gauthier, S., Flannigan, M., Kafka, V., 2004. Fire regimes at the transition between mixedwood and coniferous boreal forest in northwestern Quebec. Ecology 85, 1916–1932.

Beschta, R.L., Donahue, D.L., DellaSala, D.A., Rhodes, J.J., Karr, J.R., O'Brien, M.H., Fleischner, T.L., Williams, C.D., 2013. Adapting to climate change on western public lands: addressing the ecological effects of domestic, wild, and feral ungulates. Environ. Manag. 51, 474–491.

Beyers, J.L., 2004. Postfire seeding for erosion control: effectiveness and impacts on native plant communities. Conserv. Biol. 18, 947–956.

Bird, M.I., Cali, J.A., 1998. A million-year record of fire in sub-Saharan Africa. Nature 394, 767–769.

Bond, M.L., Lee, D.E., Siegel, R.B., Ward Jr., J.P., 2009. Habitat use and selection by California Spotted Owls in a postfire landscape. J. Wildl. Manag. 73, 1116–1124.

Bond, M.L., Y Chi, T., Bradley, C.M., DellaSala, D.A., 2022. Forest management, barred owls, and wildfire in Northern Spotted Owl territories. Forests 13, 1730. https://doi.org/10.3390/f13101730.

Buchalski, M.R., Fontaine, J.B., Heady III, P.A., Hayes, J.P., Frick, W.F., 2013. Bat response to differing fire severity in mixed-conifer forest, California, USA. PLoS One 8, e57884.

Burnett, R.D., Taillie, P., Seavy, N., 2010. Plumas Lassen Study 2009 Annual Report. U.S. Forest Service, Pacific Southwest Region, Vallejo, CA.

Cahoon, D.R., Stocks, B.J., Levine, J.S., Cofer, W.R., Pierson, J.M., 1994. Satellite analysis of the severe 1987 forest fires in northern China and southeastern Siberia. J. Geophys. Res. 99, 18627–18638.

Chavardès, R.D., et al., 2022. Converging and diverging burn rates in North American boreal forests from the Little Ice Age to the present. Int. J. Wildland Fire 31, 1184–1193.

Clar, S., Drossel, B., Schwabl, F., 1996. Forest fires and other examples of self-organized criticality. J. Phys. Condens. Matter 8, 6803.

Cocking, M.I., Varner, J.M., Knapp, E.E., 2014. Long-term effects of fire severity on oak- conifer dynamics in the southern Cascades. Ecol. Appl. 24, 94–107.

Collins, B.M., Roller, G.B., 2013. Early forest dynamics in stand-replacing fire patches in the northern Sierra Nevada, California, USA. Landsc. Ecol. 28, 1801−1813.
Collins, B.M., Miller, J.D., Thode, A.E., Kelly, M., van Wagtendonk, J.W., Stephens, S.L., 2009. Interactions among wildland fires in a long-established Sierra Nevada natural fire area. Ecosystems 12, 114−128.
Colombaroli, D., Gavin, D.G., 2010. Highly episodic fire and erosion regime over the past 2000 y in the Siskiyou Mountains, Oregon. Proc. Natl. Acad. Sci. U. S. A. 107, 18909−18914.
Cone, T., 2013. Nearly 40 Percent of Rim Fire Land a Moonscape. Associated Press news story, Fresno, California, USA.
Crotteau, J.S., Varner III, J.M., Ritchie, M.W., 2013. Post-fire regeneration across a fire severity gradient in the southern Cascades. For. Ecol. Manag. 287, 103−112.
Dahl, K.A., Abatzoglou, J.T., Phillips, C.A., Ortiz-Partida, J.P., Licker, R., Merner, L.D., Ekwurzel, B., 2023. Quantifying the contribution of major carbon producers to increases in vapor pressure deficit and burned area in western US and southwestern Canadian forests. Environ. Res. Lett. 18 (6).
DellaSala, D.A., Bond, M.L., Hanson, C.T., Hutto, R.L., Odion, D.C., 2014. Complex early seral forests of the Sierra Nevada: what are they and how can they be managed for ecological integrity? Nat. Area J. 34, 310−324.
DellaSala, D.A., Hutto, R.L., Hanson, C.T., Bond, M.L., Ingalsbee, T., Odion, D., Baker, W.L., 2017. Accommodating mixed-severity fire to restore and maintain ecosystem integrity with a focus on the Sierra Nevada of California, USA. Fire Ecology 13, 148−171.
DellaSala, D.A., Hanson, C.T., 2019. Are wildland fires increasing large patches of complex early seral forest habitat? Diversity 11, 157. https://doi.org/10.3390/d11090157.
DellaSala, D.A., 2020. Fire-mediated biological legacies in dry forested ecosystems of the Pacific Northwest, USA. In: Beaver, E.A., Prange, S., DellaSala, D.A. (Eds.), Disturbance Ecology and Biological Diversity. CRC Press Taylor and Francis Group, LLC, Boca Raton, FL, pp. 38−85.
DellaSala, D.A., Baker, B., Hanson, C.T., Ruediger, L., Baker, W., 2022. Have western USA fire suppression and active management approaches become a contemporary Sisyphus? Biol. Conserv. 268, 109499. Article.
Dillon, G.K., Holden, Z.A., Morgan, P., Crimmins, M.A., Heyerdahl, E.K., Luce, C.H., 2011. Both topography and climate affected forest and woodland burn severity in two regions of the western US, 1984 to 2006. Ecosphere 2, 130. Article.
Dodge, I.R., 1876. The Black Hills: A Minute Description of the Routes, Scenery, Soil, Climate, Timber, Gold, Geology, Zoology. Etc. James Miller Publisher, New York.
Donato, D.C., Fontaine, J.B., Robinson, W.D., Kauffman, J.B., Law, B.E., 2009. Vegetation response to a short interval between high-severity wildfires in a mixed-evergreen forest. J. Ecol. 97, 142−154.
Donato, D.C., Campbell, J.L., Franklin, J.F., 2012. Multiple successional pathways and precocity in forest development: can some forests be born complex? J. Veg. Sci. 23, 576−584.
Egan, T., 2010. The Big Burn: Teddy Roosevelt and the Fire That Saved America. Houghton Mifflin Harcourt, Boston.
Fogg, A., Burnett, R.D., Steel, Z.L., 2014. Short Term Changes in Avian Community Composition Within the Sierra Nevada's Massive Rim Fire. Point Blue Conservation Science, Petaluma, California, USA.
Fontaine, J.B., Donato, D.C., Robinson, W.D., Law, B.E., Kauffman, J.B., 2009. Bird communities following high-severity fire: response to single and repeat fires in a mixed evergreen forest, Oregon, USA. For. Ecol. Manag. 257, 1496−1504.

Galbraith, S.M., Cane, J.H., Moldenke, A.R., Rivers, J.W., 2019. Wild bee diversity increases with local fire severity in a fire-prone landscape. Ecosphere 10, e02668. Article.

Ganey, J.L., Kyle, S.C., Rawlinson, T.A., Apprill, D.L., Ward Jr., J.P., 2014. Relative abundance of small mammals in nest core areas and burned wintering areas of Mexican spotted owls in the Sacramento Mountains, New Mexico. Wilson J. Ornithol. 126, 47–52.

Gannett, H., 1902. The Forests of Oregon. U.S. Geological Survey, Government Printing Office, Washington, D.C.

Ganteaume, A., Jappiot, M., 2013. What causes large fires in southern France? For. Ecol. Manag. 294, 76–85.

Gillett, N.P., Weaver, A.J., Zwiers, F.W., Flannigan, M.D., 2004. Detecting the effect of climate change on Canadian forest fires. Geophys. Res. Lett. 31, L18211.

Glassman, S.I., Levine, C.R., DiRocco, A.M., Battles, J.J., Bruns, T.D., 2016. Ectomycorrhizal fungal spore bank recovery after a severe forest fire: some like it hot. ISME J. 10, 1228–1239.

Graves, H.S., 1899. The Black Hills forest reserve. In: The Nineteenth Annual Report of the Survey, 1897-1898. Part V. Forest Reserves. U.S. Geological Survey, Washington, D.C, pp. 67–164.

Haire, S.L., McGarigal, K., 2008. Inhabitants of landscape scars: succession of woody plants after large, severe forest fires in Arizona and New Mexico. Southwest. Nature 53, 146–161.

Haire, S.L., McGarigal, K., 2010. Effects of landscape patterns of fire severity on regenerating ponderosa pine forests (*Pinus ponderosa*) in New Mexico and Arizona, USA. Landsc. Ecol. 25, 1055–1069.

Halofsky, J.E., Donato, D.C., Hibbs, D.E., Campbell, J.L., Donaghy Cannon, M., Fontaine, J.B., Thompson, J.R., Anthony, R.G., Bormann, B.T., Kayes, L.J., Law, B.E., Peterson, D.L., Spies, T.A., 2011. Mixed-severity fire regimes: lessons and hypotheses from the Klamath-Siskiyou ecoregion. Ecosphere 2. Article 40.

Hanson, C.T., 2007. Post-Fire Management of Snag Forest Habitat in the Sierra Nevada. Ph. D. Dissertation. University of California at Davis, Davis, California, USA.

Hanson, C.T., 2013. Habitat use of Pacific Fishers in a heterogeneous post-fire and unburned forest landscape on the Kern Plateau, Sierra Nevada, California. Open For. Sci. J. 6, 24–30.

Hanson, C.T., 2014. Conservation concerns for Sierra Nevada birds associated with high- severity fire. West. Birds 45, 204–212.

Hanson, C.T., 2015. Use of higher-severity fire areas by female pacific Fishers on the kern plateau, Sierra Nevada, California, USA. Wildl. Soc. Bull. 39, 497–502.

Hanson, C.T., 2021. Smokescreen: Debunking Wildfire Myths to Save Our Forests and Our Climate. University Press Kentucky, Lexington, KY.

Hanson, C.T., Chi, T.Y., 2020. Black-backed woodpecker nest density in the Sierra Nevada, California. Diversity 12, 364. Article.

Hanson, C.T., Chi, T.Y., 2021. Impacts of postfire management are unjustified in spotted owl habitat. Front. Ecol. Evol. 9, 596282. Article.

Hanson, C.T., Lee, D.E., Bond, M.L., 2021. Disentangling post-fire logging and high-severity fire effects for spotted owls. Birds 2, 147–157.

Hanson, C.T., North, M.P., 2009. Post-fire survival and flushing in three Sierra Nevada conifers with high initial crown scorch. Int. J. Wildland Fire 18, 857–864.

Hanson, C.T., Odion, D.C., 2014. Is fire severity increasing in the Sierra Nevada mountains, California, USA? Int. J. Wildland Fire 23, 1–8.

Hanson, C.T., Odion, D.C., 2015. Sierra Nevada fire severity conclusions are robust to further analysis: a reply to Safford et al. Int. J. Wildland Fire 24, 294–295.

Hanson, C.T., Odion, D.C., DellaSala, D.A., Baker, W.L., 2009. Overestimation of fire risk in the northern spotted owl recovery plan. Conserv. Biol. 23, 1314–1319.

Hanson, C.T., Odion, D.C., DellaSala, D.A., Baker, W.L., 2010. More-comprehensive recovery actions for Northern Spotted Owls in dry forests: reply to Spies et al. Conserv. Biol. 24, 334–337.

Harmon, M.E., Hanson, C.T., DellaSala, D.A., 2022. Combustion of aboveground wood from live trees in megafires, CA, USA. Forests. Forests 13 (3), 391. https://doi.org/10.3390/f13030391.

Hawkins, L.R., Abatzoglou, J.T., Li, S., Rupp, D.E., 2022. Anthropogenic influence on recent severe autumn fire weather in the west coast of the United States. Geophys. Res. Lett. 49, e2021GL095496. https://doi.org/10.1029/2021GL095496.

Hessburg, P.F., Prichard, S.J., Hagmann, R.K., Povak, N.A., Lake, F.K., 2021. Wildfire and climate change adaptation of western North American forests: a case for intentional management. Ecol. Appl. 31, e0243.

Holden, Z.A., Swanson, A., Lucke, C.H., Affleck, D., 2018. Decreasing fire season precipitation increased recent western US forest wildfire activity. Proc. Natl. Acad. Sci. USA 115 (36). https://doi.org/10.1073/pnas.1802316115.

Hutto, R.L., 1995. Composition of bird communities following stand-replacement fires in Northern Rocky Mountain (U.S.A.) conifer forests. Conserv. Biol. 9, 1041–1058.

Hutto, R.L., 2008. The ecological importance of severe wildfires: some like it hot. Ecol. Appl. 18, 1827–1834.

Jarvis, B., 2014. Scorched Earth: Extreme Wildfires May Mean Forests without Trees. Sierra Magazine, San Francisco, California, USA.

Johnson, E.A., Wowchuk, D.R., 1993. Wildfires in the southern Canadian Rocky Mountains and their relationship to mid-trophospheric anomalies. Can. J. For. Res. 23, 1213–1222.

Jones, G.M., et al., 2022. Counteracting wildfire misinformation. Front. Ecol. Environ. 20, 392–393.

Kasischke, E.S., Bruhwiler, L.P., 2002. Emissions of carbon dioxide, carbon monoxide, and methane from boreal forest fires in 1998. J. Geophys. Res. 107, FFR2-1–FFR2-14.

Kasischke, E.S., Turetsky, M.R., 2006. Recent changes in the fire regimes across the North American boreal region—spatial and temporal patterns of burning across Canada and Alaska. Geophys. Res. Lett. 33, L09703.

Keeley, J.E., Zedler, P.H., 2009. Large, high-intensity fire events in southern California shrublands: debunking the fine-grain age patch model. Ecol. Appl. 19, 69–94.

Kinney, A., 1900. Forest and Water. Post Publishing Company, Los Angeles, CA, USA.

Kotliar, N.B., Wiens, J.A., 1990. Multiple scales of patchiness and patch structure: a hierarchical framework for the study of heterogeneity. Oikos 59, 253–260.

Lang, D.M., Stewart, S.S., 1910. Reconnaissance of the Kaibab National Forest. Northern Arizona University, Flagstaff. Unpublished report.

Larson, A.J., Cansler, C.A., Cowdery, S.G., Hiebert, S., Furniss, T.J., Swanson, M.E., Lutz, J.A., 2016. Post-fire morel (Morchella) mushroom abundance, spatial structure, and harvest sustainability. For. Ecol. Manag. 377, 16–25.

Law, B.E., Waring, R.H., 2015. Carbon implications of current and future effects of drought, fire and management on Pacific Northwest forests. For. Ecol. Manag. 355, 4–14. https://doi.org/10.1016/j.foreco.2014.11.023.

Lee, D.E., 2020. Spotted Owls and forest fire: Reply. Ecosphere 11, e03310. Article.

Lee, D.E., Bond, M.L., 2015. Occupancy of California spotted owl sites following a large fire in the Sierra Nevada, California. Condor 117, 228–236.

Leiberg, J.B., 1900a. Bitterroot forest reserve. In: USDI Geological Survey, Twentieth Annual Report to the Secretary of the Interior, 1898-99, Part V. Forest Reserves. US Government Printing Office, Washington, D.C., pp. 317−410

Leiberg, J.B., 1900b. Cascade range forest reserve, Oregon, from township 28 south to township 37 south, inclusive; together with the Ashland Forest Reserve and adjacent forest regions from township 28 south to township 41 south, inclusive, and from range 2 west to range 14 east, Willamette Meridian, inclusive. U.S. Geol. Surv Ann. Rep. 21 (Part V), 209−498.

Leiberg, J.B., 1902. Forest Conditions in the Northern Sierra Nevada, California. U.S. Government Printing Office, Washington, D.C. USDI Geological Survey, Professional Paper No. 8.

Leiberg, J.B., 1903. Southern part of Cascade range forest reserve. In: Langille, H.D., Plummer, F.G., Dodwell, A., Rixon, T.F., Leiberg, J.B., Gannett, H. (Eds.), Forest Conditions in the Cascade Range Forest Reserve, Oregon. Professional Paper No. 9, Series H, Forestry, 6. Department of the Interior, US Geological Survey, Government Printing Office, Washington, D.C., pp. 229−289

Lewis, J.S., LeSueur, L., Oakleaf, J., Rubin, E.S., 2022. Mixed-severity wildfire shapes habitat use of large herbivores and carnivores. For. Ecol. Manag. 506, 119933. Article.

Lin, J., Rinaldi, S., 2009. A derivation of the statistical characteristics of forest fires. Ecol. Model. 220, 898−903.

Lindenmayer, D.B., Burton, P.J., Franklin, J.F., 2008. Salvage Logging and its Ecological Consequences. Island Press, Washington D.C.

Lindenmayer, D., Blair, D., McBurney, L., Banks, S., 2010. Forest Phoenix: How a Great Forest Recovers After Wildfire. CSIRO Publishing, Victoria, Australia.

Lindenmayer, D.B., Hobbs, R.J., Likens, G.E., Krebs, C.J., Banks, S.C., 2011. Newly discovered landscape traps produce regimes shifts in wet forests. Proc. Natl. Acad. Sci. U. S. A. 108, 15887−15891.

Lindenmayer, D.B., Taylor, C., 2020. New spatial analyses of Australian wildfires highlight the need for new fire, resource, and conservation policies. Proc. Natl. Acad. Sci. USA 117 (22), 12481−12485. https://doi.org/10.1073/pnas.2002269117.

Miller, J.D., Safford, H., 2012. Trends in wildfire severity: 1984 to 2010 in the Sierra Nevada, Modoc Plateau, and southern Cascades, California, USA. Fire Ecol. 8, 41−57.

Miller, J.D., Skinner, C.N., Safford, H.D., Knapp, E.E., Ramirez, C.M., 2012. Trends and causes of severity, size, and number of fires in northwestern California, USA. Ecol. Appl. 22, 184−203.

Moors, A., 2012. Occupancy and Reproductive Success of Mexican Spotted Owls in the Chiricahua Mountains 2012. Report to the Coronado National Forest. Supervisor's office, Moors Wildlife Management Services, Globe, Arizona.

Moors, A., 2013. Occupancy and Reproductive Success of Mexican Spotted Owls in the Chiricahua Mountains 2013. Report to the Coronado National Forest. Supervisor's office, Moors Wildlife Management Services, Globe, Arizona.

Morgan, P., Heyerdahl, E.K., Gibson, C.E., 2008. Multi-season climate synchronized forest fires throughout the 20th-century, northern Rockies, USA. Ecology 89, 717−728.

Morris, W.G., 1934. Forest fires in western Oregon and western Washington. Oregon Hist. Q. 35, 313−339.

Niklasson, M., Granstrom, A., 2004. Short facts on Swedish fires with emphasis on fire history. In: Sigurgeirsson, A., Jõgiste, K. (Eds.), Natural Disturbances Dynamics as Components of Ecosystem Management Planning: Abstracts and Short Papers From the Workshop of the SNS Network, pp. 22−27. Geysir, Iceland, October 11-15, 2003.

NYT, 1870. The Great Saguenay Fire. New York Times, New York, USA. July 18, 1870.

Odion, D.C., Hanson, C.T., 2013. Projecting impacts of fire management on a biodiversity indicator in the Sierra Nevada and Cascades, USA: the Black-backed Woodpecker. Open For. Sci. J. 6, 14−23.

Odion, D.C., Frost, E.J., Strittholt, J.R., Jiang, H., DellaSala, D.A., Moritz, M.A., 2004. Patterns of fire severity and forest conditions in the western Klamath Mountains, California. Conserv. Biol. 18, 927−936.

Odion, D.C., Moritz, M.A., DellaSala, D.A., 2010. Alternative community states maintained by fire in the Klamath Mountains, USA. J. Ecol. 98, 96−105.

Odion, D.C., Hanson, C.T., Arsenault, A., Baker, W.L., DellaSala, D.A., Hutto, R.L., Klenner, W., Moritz, M.A., Sherriff, R.L., Veblen, T.T., Williams, M.A., 2014. Examining historical and current mixed-severity fire regimes in ponderosa pine and mixed- conifer forests of western North America. PLoS One 9, e87852.

Odion, D.C., et al., 2016. Areas of agreement and disagreement regarding ponderosa pine and mixed conifer forest fire regimes: a dialogue with Stevens et al. PLoS One. https://doi.org/10.1371/journal.pone.0154579.

Parks, S.A., Miller, C., Parisien, M.A., Holsinger, L.M., Dobrowski, S.Z., Abatzoglou, J., 2015. Wildland fire deficit and surplus in the western United States, 1984-2012. Ecosphere 6 (12). Article 275.

Paine, R.T., Tegner, M.J., Johnson, E.A., 1998. Compounded perturbations yield ecological surprises. Ecosystems 1, 535−545.

Perry, D.A., Hessburg, P.F., Skinner, C.N., Spies, T.A., Stephens, S.L., Taylor, A.H., Franklin, J.F., McComb, B., Riegel, G., 2011. The ecology of mixed severity fire regimes in Washington, Oregon, and northern California. For. Ecol. Manag. 262, 703−717.

Phalan, B.T., Northrup, J.M., Yang, Z., Deal, R.L., Rousseau, J.S., Spies, T.A., Betts, M.G., 2019. Impacts of the Northwest Forest Plan on forest composition and bird populations. Proc. Natl. Acad. Sci. USA. https://www.pnas.org/cgi/doi/10.1073/pnas.1813072116.

Pyne, S.J., 2007. Megaburning: the meaning of megafires and the means of management. In: Proceedings of the 4th International Wildland Fire Conference. Seville, Spain, May 13-17.

Pyne, S.J., 2008. Year of the Fires: The Story of the Great Fires of 1910. Mountain Press Publishing Company, Missoula, Montana, pp. 155−157. ISBN 978 0 87842 544 0.

Raphael, M.G., Morrison, M.L., Yoder-Williams, M.P., 1987. Breeding bird populations during twenty-five years of postfire succession in the Sierra Nevada. Condor 89, 614−626.

San-Miguel-Ayanz, J., Manuel Moreno, J., Camia, A., 2013. Analysis of large fires in European Mediterranean landscapes: lessons learned and perspectives. For. Ecol. Manag. 294, 11−22.

Shatford, J.P.A., Hibbs, D.E., Puettmann, K.J., 2007. Conifer regeneration after forest fire in the Klamath-Siskiyous: how much, how soon? J. For. 139−146.

Sherriff, R.L., Platt, R.V., Veblen, T.T., Schoennagel, T.L., Gartner, M.H., 2014. Historical, observed, and modeled wildfire severity in montane forests of the Colorado Front Range. PLoS One 9, e106971.

Schofield, L.N., Eyes, S.A., Siegel, R.B., Stock, S.L., 2020. Habitat selection by spotted owls after a megafire in Yosemite National Park. For. Ecol. Manag. 478, 118511. Article.

Siegel, R.B., Eyes, S.A., Tingley, M.W., Wu, J.X., Stock, S.L., Medley, J.R., Kalinowski, R.S., Casas, A., Lima-Baumbach, M., Rich, A.C., 2019. Short-term resilience of great gray owls to a megafire in California, USA. Condor 121, 1−13.

Siegel, R.B., Wilkerson, R.L., 2005. Short- and Long-Term Effects of Stand Replacing Fire on a Sierra Nevada Bird Community. The Institute for Bird Populations, Point Reyes Station, CA.

Skinner, W.R., Shabbar, A., Flannigan, M.D., Logan, K., 2006. Large forest fires in Canada and the relationship to global sea surface temperatures. J. Geophys. Res. Atmos. 111, 27.

Smucker, K.M., Hutto, R.L., Steele, B.M., 2005. Changes in bird abundance after wildfire: importance of fire severity and time since fire. Ecol. Appl. 15, 1535–1549.

Stavros, E.N., Abatzoglou, J., Larkin, N.K., McKenzie, D., Steel, E.A., 2014. Climate and very large wildland fires in the contiguous western USA. Int. J. Wildland Fire 23 (7), 899–914. https://doi.org/10.1071/WF13169.

Steel, Z.L., Fogg, A.M., Burnet, R., Roberts, L.J., Safford, H.D., 2022. When bigger isn't better- implications of large high-severity wildfire patches for avian diversity and community composition. Divers. Distrib. 28, 439–453.

Swanson, M.E., Franklin, J.F., Beschta, R.L., Crisafulli, C.M., DellaSala, D.A., Hutto, R.L., Lindenmayer, D., Swanson, F.J., 2011. The forgotten stage of forest succession: early successional ecosystems on forest sites. Front. Ecol. Environ. 9, 117–125.

Tedim, F., Remelgado, R., Borges, C., Carvalho, S., Martins, J., 2013. Exploring the occurrence of megafires in Portugal. For. Ecol. Manag. 294, 86–96.

Thompson, J.R., Spies, T.A., 2009. Vegetation and weather explain variation in crown damage within a large mixed-severity wildfire. For. Ecol. Manag. 258, 1684–1694.

Thorn, S., et al., 2018. Impacts of salvage logging on biodiversity: a meta-analysis. J. Appl. Ecol. https://doi.org/10.1111/1365-2664.12945.

Trombulak, S.C., Frissell, C.A., 2000. Review of ecological effects of roads on terrestrial and aquatic communities. Conserv. Biol. 14, 18–30.

Turco, M., Abatzoglou, J.T., Herrera, S., Cvijanovic, I., 2023. Anthropgenic climate change impacts exacerbate summer forest fires in California. Proc. Natl. Acad. Sci. USA 120 (25), e2213815120. https://doi.org/10.1073/pnas.221381512.

Turner, M.G., Dale, V.H., 1998. Comparing large, infrequent disturbances: what have we learned? Ecosystems 1, 493–496.

Turner, M.G., Baker, W.L., Peterson, C.J., Peet, R.K., 1998. Factors influencing succession: lessons from large, infrequent natural disturbances. Ecosystems 1, 511–523.

USDA, 2014. Rim Fire Recovery Project, Final Vegetation Report. U.S. Department of Agriculture, Forest Service, Stanislaus National Forest, Sonora, California, USA.

Wein, R.W., Moore, J.M., 1977. Fire history and rotations in the New Brunswick Acadian Forest. Can. J. For. Res. 7, 285–294.

Westerling, A.L., 2016. Increasing western US forest wildfire activity: sensitivity to changes in the timing of spring. Phil. Trans. R. Soc. B 371, 20150178. https://doi.org/10.1098/rstb.2015.0178.

White, A.M., Tarbill, G.L., Wilkerson, B., Siegel, R., 2019. Few detections of Black-backed Woodpeckers (*Picoides arcticus*) in extreme wildfires in the Sierra Nevada. Avian Conserv. Ecol. 14. Article 17.

Whitlock, C., 1992. Vegetational and climatic history of the Pacific Northwest during the Last 20,000 years: implications for understanding present-day biodiversity. Northwest Environ. J. 8, 5–28.

Whittaker, R.H., 1960. Vegetation of the Siskiyou Mountains, Oregon and California. Ecol. Monogr. 30, 279–338.

Williams, J., 2013. Exploring the onset of high-impact megafires through a forest land management prism. For. Ecol. Manag. 294, 4–10.

Williams, M.A., Baker, W.L., 2012a. Spatially extensive reconstructions show variable- severity fire and heterogeneous structure in historical western United States dry forests. Global Ecol. Biogeogr. 21, 1042–1052.

Williams, M.A., Baker, W.L., 2012b. Comparison of the higher-severity fire regime in historical (A.D. 1800s) and modern (A.D. 1984-2009) montane forests across 624,156 ha of the Colorado Front Range. Ecosystems 15, 832–847.

Williams, M.A., Baker, W.L., 2014. High-severity fire corroborated in historical dry forests of the western United States: Response to Fulé et al. Global Ecol. Biogeogr. 23, 831–835.

Wu, J., Loucks, O.L., 1995. From balance of nature to hierarchical patch dynamics: a paradigm shift in ecology. Q. Rev. Biol. 70, 439–466.

Zald, H.S.J., Dunn, C.J., 2018. Severe fire weather and intensive forest management increase fire severity in a multi-ownership landscape. Ecol. https://doi.org/10.1002/eap.1710. Applications.

FURTHER READING

Hanson, C.T., Bond, M.L., Lee, D.E., 2018. Effects of post-fire logging on California spotted owl occupancy. Nat. Conserv. 24, 93–105.

Chapter 3

Using Bird Ecology to Learn About the Benefits of Severe Fire

Richard L. Hutto[1], Monica L. Bond[2] and Dominick A. DellaSala[3]
[1]Division of Biological Sciences, University of Montana, Missoula, MT, United States; [2]Wild Nature Institute, Concord, NH, United States; [3]Wild Heritage, A Project of Earth Island Institute, Berkeley, CA, United States

3.1 INTRODUCTION

In this chapter, we do not provide an encyclopedic review of the more than 450 published papers that describe some kind of effect of fire on birds. In other words, we are not systematically proceeding through a litany of fire effects on birds of southeast pine forests, California chaparral, Australian eucalypt forests, South African fynbos, and so forth. Instead, we highlight underappreciated principles or lessons that emerge from selected studies of birds in conifer forests born of, and maintained by, mixed- to high-severity fire. Those lessons show how important and misunderstood basic fire ecology is when it comes to managing fire-dependent forest lands and shrublands, and the lessons may apply to other fire-dependent ecosystems that have historically experienced fires that are severe enough to stimulate an ecological succession of plant communities (as described in Chapter 1 and throughout much of this book). We also focus primarily on conifer forests of the western United States because they undergo an amazing transformation following severe fire and studies of these systems clearly reveal how birds evolved with severe fire. Insight that emerges from the study of bird populations is overlooked in management circles worldwide. This is unfortunate because the knowledge one can gain by studying the ecology of individual bird species argues strongly that severe fire needs to be maintained in the landscape if we hope to also maintain the integrity of fire-dependent systems.

Most studies of fire effects on birds are disappointingly "empty" because they are merely lists of birds that benefit from or are hurt by fire; they are not placed in the broader context of what a self-sustaining fire-dependent system looks like. To understand whether a particular change in abundance is "good" or "bad" requires insight into the patterns that occur under conditions that are as natural as possible for any given vegetation system. That, in turn, requires replicated study of what we can expect to find after "natural" fire in any given system. Thus, a study of the effects of, say, prescribed understory fire on birds is meaningless without knowing what a "natural" fire in that system would ordinarily produce. Many studies might show that bird species A increases after a prescribed fire, but is that a good thing? If bird species B increases after postfire salvage logging, is that a good thing? If bird diversity is higher in one fire treatment versus another, is that a good thing? For studies of fire effects to be useful, we need to address questions that inform management by tapping into a solid understanding of what constitutes a "natural" response to fire, and that requires knowing something about the fire regime under which a given system evolved. Only through distribution patterns and adaptations of individual species (not through effects on bird guilds or on diversity and similar composite metrics) can we begin to understand which kind of fire regime necessarily gave rise to specific patterns of habitat use and to adaptations that have evolved over millennia. Birds are excellent messengers; they carry all the information we need to reconstruct the historical conditions under which they evolved. All we have to do is listen.

3.2 INSIGHTS FROM BIRD STUDIES

Lesson 1: The Effects of Fire Are Context Dependent; Species Respond Differently to Fire Severities and Other Postfire Vegetation Conditions

One extremely important lesson that has emerged from studies of the fire effects on birds is that a given effect depends entirely on the vegetation type, the kind of fire, and the time since the fire (Recher and Christensen, 1981; Woinarski and Recher, 1997). For years, individual bird species have been labeled as "positive responders" or "negative responders" or "mixed responders" when, in fact, any species can be all of the above. The actual response of a bird species (or of any species) to fire, then, is dependent on context. The earliest papers on fire effects rarely provided details about the nature of the fire being studied, so the first attempt to conduct a meta-analysis based on a compilation of published results of fire effects (Kotliar et al., 2002) necessarily generated a lot of "mixed" responses by birds because some papers said a species was positively affected and others said the same species was negatively affected by fire. The seeming disagreement among studies was, in most cases, a simple result of researchers looking at different postfire vegetation conditions and times since fire. It was not until Smucker et al. (2005) separated their data into categories of fire severity and time since the fire that responses began to look much more consistent among studies that share a particular vegetation type, fire type, and time since the fire. As soon as one accounts for these factors, it becomes clear that the responses of most bird species are quite consistent and that most bird species benefit from severe fire (as we will more fully discuss below).

Time Since Fire

Species that benefit from severe fire are not only those that flourish during the first year or two following the disturbance event. The same can be said for species that are largely restricted to years 2—4, years 5—10, or even years 50—100 following severe fire. In fact, *most* plant and animal species are present only during a limited time following a disturbance. Therefore, *most* plant and animal species in disturbance-based systems depend on disturbance to periodically create the conditions they need. Many bird species that thrive after fire have been mislabeled as species hurt by fire because studies of bird response to fire typically involve only a brief period soon after the fire. For example, although Williamson's sapsucker (*Sphyrapicus thyroideus*) was labeled a "mixed responder" and brown creeper (*Certhia americana*), a "negative responder" in the meta-analysis by Kotliar et al. (2002), and the change in house wren (*Troglodytes aedon*) abundance was labeled "insignificant" in a study by Seavy and Alexander (2014), each of these species typically reaches its peak abundance several years after a fire, as revealed in an 11-year postfire study conducted after the Black Mountain fire, which burned near Missoula, Montana, in 2003 (Figure 3.1). Thus, each species clearly benefits from severe fire when viewed in the proper (and perhaps very restricted) time frame after fire.

By extending the duration of a postfire study beyond the first few years after a fire, most bird species reveal a unimodal response to time since fire, and most benefit from fire. That is, they reveal a greater probability of detection in the burned forest at some point during that postfire period than in the same forest before fire or in the surrounding unburned forest (Taylor and Barmore, 1980; Reilly, 1991a, 2000; Taylor et al., 1997; Hannon and Drapeau, 2005; Saab et al., 2007; Chalmandrier et al., 2013; Hutto and Patterson 2016; Lindenmayer et al., 2022). These results force one to appreciate that if conditions remain better for a brief period after fire than they are in very old plant communities near the end of the late seral stage of succession, then natural disturbance is periodically necessary to create the conditions needed by that species. Thus, a species being "hurt" in the short term by fire is not evidence that fire is somehow "bad" for that species and that it would have been better off without fire. In fact, once a system is beyond the ideal post-disturbance successional stage for a species, the only way to periodically "restore" conditions needed by that species is to disturb the system with another severe fire and then wait for the appropriate time period following disturbance again. The lesson is this: one cannot assess the effects of fire on any plant or animal species without examining whether the species is associated with or primarily restricted to a period of time preceding the oldest possible vegetation condition.

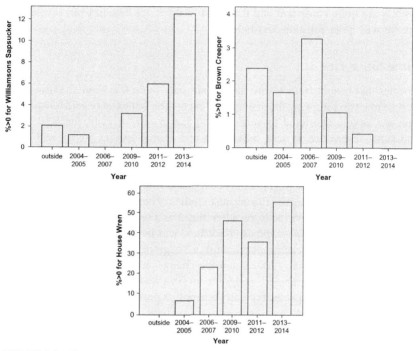

FIGURE 3.1 The probabilities of occurrence of Williamson's sapsucker, brown creeper, and house wren were significantly greater several years after the 2003 Black Mountain fire than they were either before the fire (as determined from survey data "outside" the burn perimeter in unburned, mixed-conifer forest of the same type) or during the first 2 years following the fire (R.L. Hutto, unpublished data; sample sizes exceed 150 point counts for each time period; $P < .05$, log linear analyses). Therefore, the benefit of severe fire for some species cannot be detected without restricting data collection to within a specific time period after the fire event.

A necessary consequence of different species occurring at different points in time following fire (in association with changes in vegetation type and structure) is that we must embrace natural severe disturbance processes because they create starting points for the development of the full range of vegetation-age categories, which, in turn, are needed for the maintenance of biological diversity (in particular beta diversity, the turnover in species number across gradients). Moreover, mixed-severity fires (which are always associated with high-severity fire events) help provide a variety of kinds of starting points, which, in turn, also help maintain biological diversity (Smucker et al., 2005; Haney et al., 2008; Rush et al., 2012; Sitters et al., 2014; also see Chapter 4 for mammals).

Old-Growth Forests

As already emphasized, most bird species clearly depend on severe fire to reset the successional clock, which stimulates development of the post-disturbance "age" to which they are best adapted. Still, many bird species are restricted in their habitat distribution to old growth. There are also ecosystems (e.g., eucalyptus forests, chaparral) where severe fire is natural but where there are fewer early fire-dependent bird species because many of the dominant plant species resprout, yielding a plant community structure and composition that "recovers" rapidly after fire (Figure 3.2). In these systems most bird species are associated with "mature" forms of those plant communities (e.g., Taylor et al., 2012; Legge et al., 2022).

In all vegetation types that undergo plant succession following mixed- to high-severity fire, there will always be some bird species that depend on long-unburned vegetation. Therefore, discovering that those species are absent in the short term or "hurt" by fire is not unexpected, nor is it a necessarily a problem that needs to be addressed (with the possible exception of climate-driven megafires as in Australia; Legge et al., 2022). The fact that fire temporarily removes large parts of a landscape from the pool of suitable conditions for those species is not a problem because the loss of suitable conditions is temporary, and there are usually nearby "refuges" of suitable conditions in places that have not burned for a long time (Bain et al., 2008; Leonard et al., 2014; Robinson et al., 2014; Winchell and Doherty, 2014). The caveat here is that a decline in suitable conditions for species associated with older vegetation will be temporary only in the absence of post-fire logging or any dramatic increase in fire frequency (i.e., as with climate change). Natural systems exist as an ever-changing mosaic of different postfire ages—all vegetation ages are present at some point in space all the time. Fire can become riskier, however, if humans

FIGURE 3.2 Resprouting eucalyptus trees following a severe fire that burned through the area only months earlier. *Photograph by Richard Hutto, taken in November 1999 near −34.284030°S, 150.725373°E in the tablelands above Wollongong, New South Wales, Australia.*

remove or degrade so much of the older vegetation through prefire logging or land conversion that older vegetation refuges become too few and far between. Understand clearly, however, that the absence of late-succession forest refuges is a problem that stems from excessive logging or development, not from the presence of fire per se.

Now that we are down to the last remaining old-growth forest remnants in California and Oregon, some believe that we should thin the forests around those remnants to protect them from fire. The effect of altering mature forest structure and processes surrounding the last remaining old-growth remnants on the remnants themselves is, however, most concerning as many so-called "fuel management" projects remove large fire-resistant trees (see Chapters 10 and 11), and the most comprehensive recent research finds that more open, low-density forests burn more severely in wildfires, while denser, older forests burn less severely, due to microclimate factors (Lesmeister et al., 2021). Moreover, as has been discussed in reference to eucalyptus forest systems, many old-growth forest patches are old precisely because they are situated in places that are relatively immune to severe fire (Bowman, 2000) (i.e., fire refugia); the same is undoubtedly true of many old-growth mixed-conifer forest patches. Unburned forest patches surrounding unburned, old-growth forest patches also have been suggested to be important as dispersal corridors across which old-growth species may recolonize recently burned areas as succession proceeds toward later stages (Pyke et al., 1995; Robinson et al., 2014; Seidl et al., 2014) (i.e., source-sink populations). Therefore, proposals to thin the forest around remaining old-growth stands may be well intentioned but reflect a lack of appreciation for the resilience associated with plant communities born of, and maintained by, natural disturbance processes (a case in point is the spotted owl [*Strix occidentalis*]; see Box 3.1 and Bond et al., 2022).

Postfire Vegetation Conditions

One must account not only for time since fire but also for fire severity and other forest conditions (e.g., vegetation composition and tree density) to adequately assess fire effects on animal species. Smucker et al. (2005) accounted for both time since fire and fire severity in an analysis of bird occurrence patterns following the Bitterroot fires of 2000 in Montana, and the results were profound. Once they accounted for fire severity alone, it became abundantly clear that many of the same bird species that had been labeled as "mixed responders" to fire by others (e.g., Kotliar et al., 2002) were not at all mixed in their response to fire. The importance of fire severity is strikingly apparent in even the simplest graphs of percentage occurrence across severity categories (Figure 3.3).

Lesson 2: Given the Appropriate Temporal and Vegetation Conditions, a Surprising Number of Bird Species Can be Seen to Benefit from Severe Fire

After we combine information on the time since fire, fire severity, and perhaps one or two additional vegetation variables, most bird species appear to

BOX 3.1 Old-Growth Species and Severe Disturbance Events

There are a number of old-growth–dependent species in North American conifer forests, but severe fire may not pose anywhere near the threat to those species that one might suppose. Consider the spotted owl, one of the most iconic old-growth-dependent bird species in the Pacific Northwest, California, and Southwest (extending into northern Mexico). This federally listed threatened raptor typically nests, roosts, and forages in dense conifer and mixed-conifer-oak forests dominated by large (>50-cm diameter at breast height), older trees and peppered with big decadent snags and fallen logs. High levels of canopy cover (generally >60%) from overhead foliage is an important component of nesting and roosting stands; thus, spotted owls were long presumed to be seriously harmed where severe fire burned the forest canopy. Indeed, over the past several decades, most forest management efforts in the range of the spotted owl (a Forest Service management indicator species) has been driven by logging to prevent or reduce fire to "save" the owl, including the latest US Fish and Wildlife Service recovery plans for the northern and Mexican spotted owls (Bond et al., 2022). Yet, the forests where the owl dwells have experienced mixed- and high-severity fire for millennia (Baker 2015a). So how do these birds actually respond when severe fire affects habitat within their home ranges?

Several studies have demonstrated that all three subspecies of spotted owl can survive and thrive (i.e., successfully reproduce) within territories that have experienced moderate- and high-severity fire (Bond et al., 2002, 2022; Jenness et al., 2004; Roberts et al., 2011; Lee et al., 2012, 2013; Lee, 2018). High levels of severe fire in a nest stand can cause spotted owls to abandon that territory (Lee et al., 2013), but only a small fraction of sites ever exceed that threshold in any given fire (Lee, 2018). Moreover, a higher probability of abandonment after fire was documented only in a small geographical region where prefire forest patches were limited or isolated (Lee et al., 2013) and in areas that were logged after fire (Lee et al., 2012; Clark et al., 2013); reduced occupancy did not occur in unlogged areas where prefire forest cover was more abundant (Lee et al., 2012, 2013; Lee, 2018; Bond et al., 2022). For example, the year after the 2013 Rim Fire—one of the largest fires to occur in California within the past century—at least six pairs of California spotted owls (*S. occidentalis occidentalis*) were detected in sites where >70% of the "suitable habitat" around their nest stands burned at high severity. (At one occupied site severe fire burned 96% of the habitat!)

Why do they stick around in burned territory? One study found California spotted owls selectively hunted (mostly for woodrats and gophers) in stands recently burned by severe fire when those burned forests were available to them and relatively near the nest or roost stand (Bond et al., 2009, 2013). Another study showed that during winter, Mexican spotted owls (*S. occidentalis lucida*) moved up to 14 km into burned forests where prey biomass was 2–6 times greater than in their breeding-season nesting areas (Ganey et al., 2014). Spotted owls are perch-and-pounce predators, so it is not surprising that they avoided foraging in areas that were logged after fire, as there were no longer any perch trees (Bond et al., 2009), nor is it surprising that postfire logging reduced site occupancy and survival rates (Clark et al., 2013; Lee et al., 2013; Bond et al., 2022). In these studies,

Continued

BOX 3.1 Old-Growth Species and Severe Disturbance Events—cont'd

spotted owls still nested and roosted in live trees, underscoring the importance of unburned/low-severity refuges within the larger landscape mosaic of mixed-severity fire. Still, the point is that where severe fire is natural, even old-growth species can partake of its bounty. The spotted owl, too, is sending a message here: A natural fire regime provides a bedroom, nursery, and kitchen for even old-growth-dependent species, as long as the burned forest is left standing.

Despite this evidence, the US Fish andand Wildlife Service has called for aggressive, large-scale thinning in northern spotted owl habitat in dry forests as a means of reducing fire intensity (U.S. Fish and Wildlife Service, 2011). This "recovery" objective for the owl was developed over objections raised by scientists (Hanson et al., 2009, 2010) and professional societies such as The Wildlife Society and Society for Conservation Biology. Notably, Odion et al. (2014b) simulated changes in owl habitat over a four-decade period following fire and the kind of thinning proposed by federal land managers. The simulation study showed that thinning over large landscapes would remove 3.4–6.0 times more of their dense, late-successional habitat in the Klamath and dry Cascades, respectively, than forest fires would, even given a future increase in the amount of high-severity fire. Further, Baker (2015a) documented that before extensive Euro-American settlement, mixed- and high-severity fires shaped dry forests in the Eastern Cascades of Oregon and provided important habitat for northern spotted owls there. These studies challenge the paradigm that severe fire is a serious threat to spotted owls, which evolved in landscapes shaped by such fire, and that extensive logging is needed to ameliorate this widely believed but overstated threat.

---

benefit from severe fire. For each species there is a particular combination of burned forest variables that creates ideal conditions for that species, as evidenced by an abundance that exceeds that in a long-unburned patch of the same vegetation type. Indeed, when Hutto and Patterson (2015) considered just two fire-context variables (time since fire and fire severity), they found 46 of 50 species to be more abundant in some combination of those two variables than in long-unburned stands (Figure 3.4). Thus, not only are most species relatively abundant in one burned forest condition or another but the average point in space and time occupied by each species is also species specific (Figure 3.5). A more recent analysis based on data drawn from more than 7000 point counts across western Montana showed that about half of 68 conifer-forest bird species were significantly more abundant in burned forest at some combination of time-since-fire and fire severity than in unburned conifer forest (Hutto et al., 2020).

As an introduction to some of the fascinating biology surrounding severely burned forests, consider the black-backed woodpecker (*Picoides arcticus*), American three-toed woodpecker (*Picoides dorsalis*), hairy woodpecker

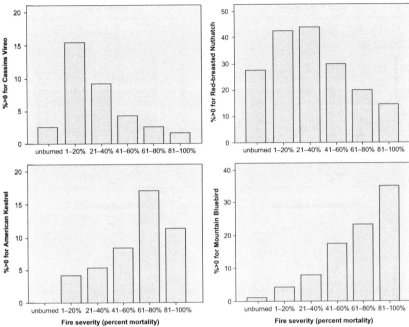

FIGURE 3.3 Example plots of the percentage occurrence of four mixed-conifer bird species in relation to fire severity in the first few years after fire. Data were drawn from 7043 survey points distributed across 110 different fires that burned since 1988 in western Montana. Sample sizes exceed 700 point counts per severity category. All patterns are significant ($P < 0.05$, log linear analyses). Note that each species is more abundant in burned than in unburned forest, and each is relatively more abundant at a level of burn severity (percentage of tree mortality) that differs from that occupied by the other species. Scientific names for birds from top left clockwise to bottom right are *Vireo cassinii, Sitta canadensis, Sialia currucoides, Falco sparverius*.

(*Picoides villosus*), northern flicker (*Colaptes auratus*), and Lewis's woodpecker (*Melanerpes lewis*). Each is more abundant in severely burned than in unburned mixed-conifer forest (see patterns of habitat occurrence for four of the five species in Figures 3.11 and 3.12) because of an abundance of food (beetle larvae and ants) and potential nest sites associated with standing dead trees. The Williamson's sapsucker and olive-sided flycatcher (*Contopus cooperi*) find the abrupt edges between severely burned and unburned forest to be ideal nest locations (Figure 3.6). A host of secondary cavity-nesting and snag-nesting species (e.g., northern hawk owl [*Surnia ulula*], great gray owl [*Strix nebulosa*], mountain bluebird [*Sialia currucoides*], western bluebird [*Sialia mexicana*], house wren, and tree swallow [*Tachycineta bicolor*]) benefit from new forest openings. Here they find a mature-forest legacy of already existing broken-top snags (Figure 3.7), where a disproportionately large number of nest sites are located (Hutto, 1995). These species depend on cavities drilled by woodpeckers that colonize recently burned forests

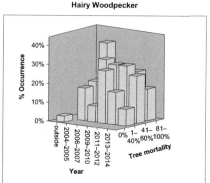
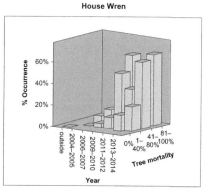
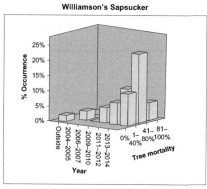
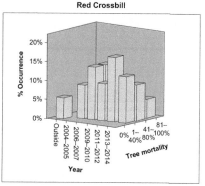

FIGURE 3.4 Example plots of percentage occurrence for various mixed-conifer bird species in relation to both time since fire and fire severity after the 2003 Black Mountain fire near Missoula, Montana (R.L. Hutto, unpublished; sample sizes exceed 35 point counts for each time-by-severity category; all patterns are significantly nonrandom as determined by log linear analyses [$P < 0.05$]). The examples were selected to illustrate that each species is more abundant in burned than in unburned forest (the occurrence rate in unburned forest shown in the first time period), and each is most abundant in a different combination of time since fire and burn severity (percentage of tree mortality).

(Tarbill et al., 2015) and by already existing cavities previously in snags that persist in mature- to old-growth forests that burn in a severe fire. A variety of species (e.g., flammulated owl [*Psiloscops flammeolus*], mountain bluebird, Townsend's solitaire [*Myadestes townsendi*], and dark-eyed junco [*Junco hyemalis*]) make use of the cavities created by burned-out root wads or uprooted trees that happen to blow down in the first few years after severe fire (Figure 3.8). Many species (e.g., Clark's nutcracker [*Nucifraga columbiana*], Cassin's finch [*Haemorhous cassinii*], red crossbill [*Loxia curvirostra*], and pine siskin [*Spinus pinus*]) take advantage of seeds that are released or made available in cones that open after severe fire (Figure 3.9). Still more bird species (e.g., calliope hummingbird [*Selasphorus calliope*], lazuli

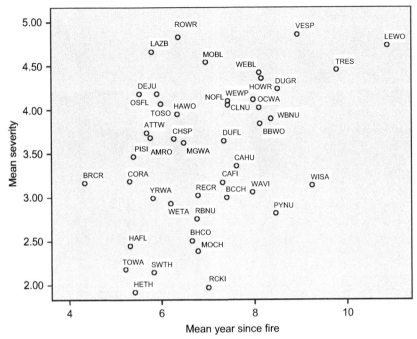

FIGURE 3.5 In combination, the mean time since fire and mean fire severity at points of occurrence for each of 46 (mnemonically coded) species differs from that of every other species. Mean values were calculated from the kind of data presented in Figure 3.4.

bunting [*Passerina amoena*], and MacGillivray's warbler [*Geothlypis tolmiei*]) use the shrub-dominated early seral stage for feeding and nesting and as display sites (Hutto, 2014).

Lesson 3: Not Only Do Most Bird Species Benefit from Severe Fire, but Some also Appear to *Require* Severe Fire to Persist

The black-backed woodpecker has become an iconic indicator of severely burned forests because its distribution is nearly restricted to such conditions. Bent (1939) provided the first description of the unusual association between this woodpecker species and burned forests when he noted that Manly Hardy wrote to Major Bendire in 1895 about finding the woodpecker to be "... so abundant in fire-killed timber areas that I once shot the heads off six in a few minutes when short of material for a stew." This anecdote, reflecting the importance of severe fire, went largely unnoticed until the 1970s, when Dale Taylor studied birds in relation to time since fire in the Yellowstone and Grand Teton National Parks. His more systematic study uncovered the same remarkable pattern. Taylor was the first person to evaluate data drawn from a series of burned conifer forest stands of differing ages, and he found the appearance of the black-backed woodpecker to be restricted to the first few

FIGURE 3.6 Williamson's sapsucker (*left*) and olive-sided flycatcher (*right*) are known to nest disproportionately often near the abrupt edges between severely burned and unburned forest. *Photographs by Richard Hutto.*

years after fire (Taylor and Barmore, 1980). A subsequent before-and-after fire study by Apfelbaum and Haney (1981) and studies of burned versus adjacent unburned forest by Niemi (1978), Pfister (1980), and Harris (1982) provided additional evidence that this bird species is strongly associated with burned forest conditions. Following the Rocky Mountain fires of 1988, Hutto (1995) conducted a more comprehensive study of the distribution of black-backed woodpeckers across a broad range of vegetation types. That study served to reinforce the notion that this species is an ideal indicator of severely burned mixed-conifer forest. More specifically, Hutto provided a meta-analysis of his own and already published bird survey data collected from burned forests and from eight unburned vegetation types that showed the black-backed woodpecker to be relatively restricted to burned forests. To address the potential problem of putting too much faith in distribution patterns derived from bird occurrence rates that were based on a variety of study durations and methods, Hutto subsequently coordinated the collection of standardized bird survey data from more than 18,000 points distributed across every major vegetation type in the US Forest Service Northern Region. The results (Hutto, 2008) were strikingly similar to what earlier studies showed: one is hard pressed to find a black-backed woodpecker anywhere but in a recently burned forest (Figure 3.10).

FIGURE 3.7 Compared with burned trees with intact tops, broken-top snags that were already snags before the fire burned are used disproportionately more often as nest sites by cavity-nesting bird species. The black-backed woodpecker also roosts almost entirely in burned-out hollows, forked trunks, or other relatively unusual structures that create crevices in "deformed" snags that existed before the forest burned (Siegel et al., 2014). Pictured (*left to right*) are a young hairy woodpecker in its nest cavity, an American robin (*Turdus migratorius*) nest, and a northern flicker nest. The implications are profound—old-growth elements (snags) are really important to birds that depend on burned forest conditions, so burned, old-growth forests are as valuable to wildlife as unburned old-growth forests. *Photographs by Richard Hutto.*

Numerous studies (most published just in the past decade) provide additional detail that can help us better understand this remarkable association between the black-backed woodpecker and severely burned forests. Here we list some of the insights we have gained:

1. The magical appearance of woodpeckers within weeks of a fire (Blackford, 1955; Uxley, 2014) suggests that either smoke, or perhaps the fire or burned landscape itself, provides a stimulus for birds to colonize newly burned forests. Dispersal by young birds from other burned-forest conditions accounts for most of the colonization of newly burned areas (Siegel et al., 2016; Stillman et al., 2021). Black-backed woodpeckers that hatch in severely burned areas subsequently select interior areas of the fire with significantly higher proportions of high-severity fire (White et al., 2019), and very high snag basal area (Hanson and Chi, 2020).

FIGURE 3.8 The architecture of a burned forest becomes modified after trees begin to blow down in the first few years after a fire, and a number of bird species make use of the root wads as nest sites. A Townsend's solitaire nest is highlighted here. *Photograph by Richard Hutto.*

FIGURE 3.9 Few people seem to realize how important Clark's nutcrackers are as seed dispersers after severe fire in ponderosa pine forests. Pictured here are examples of a nutcracker extracting seeds from a ponderosa pine (*Pinus ponderosa*) cone that opened after fire (*left*) and a nutcracker with a throat pouch full of seeds in the scorched ground beneath a ponderosa pine canopy. *Photographs by Richard Hutto.*

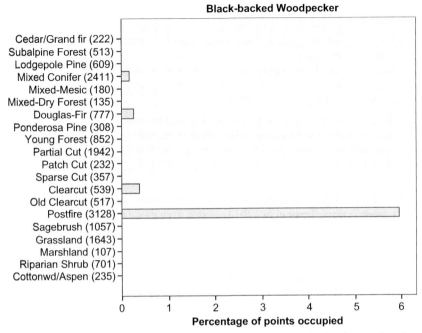

FIGURE 3.10 Histogram bars indicate the percentage of points (sample sizes in parentheses) at which the black-backed woodpecker was detected in each of 21 distinct vegetation types within northern Idaho and western Montana. The distribution is nonrandom ($X^2 = 559.43$; df = 19; $P < 0.0001$) and reveals that the black-backed woodpecker is highly specialized in its use of burned conifer forest. *Data from Hutto (2008).*

2. Breeding and nest densities increase more rapidly than expected based on recruitment alone (Yunick, 1985; Youngman and Gayk, 2011), which suggests that the process of immigration after fire is significant (i.e., source—sink relations).
3. Woodpecker diet, which is based mainly on wood-boring beetle larvae that feed almost exclusively on recently burned and killed trees (Murphy and Lehnhausen, 1998; Powell et al., 2002; Fayt et al., 2005), reflects the broad postfire change in animal community composition that accompanies severe fire.
4. The woodpecker's nonrandom use of forest patches containing dense, larger-diameter trees (Saab and Dudley, 1998; Saab et al., 2002, 2009; Nappi and Drapeau, 2011; Dudley et al., 2012; Seavy et al., 2012) that have burned at high rather than low severity (Schmiegelow et al., 2006; Koivula and Schmiegelow, 2007; Hanson and North, 2008; Hutto, 2008; Nappi and Drapeau, 2011; Youngman and Gayk, 2011; Siegel et al., 2013) is striking and consistent among studies.
5. The window of opportunity for occupancy by this species is not only soon after fire, but generally lasts only about a half-dozen years before the birds

(and the abundant native beetle populations) disappear (Taylor and Barmore, 1980; Apfelbaum and Haney, 1981; Murphy and Lehnhausen, 1998; Hoyt and Hannon, 2002; Saab et al., 2007; Nappi and Drapeau, 2009; Saracco et al., 2011).

6. The size of the home ranges of black-backed woodpeckers within burned forests is significantly smaller (indicating better quality habitat) than those outside burned forests (Rota et al., 2014b; Tingley et al., 2014). Even more telling is that nest success is significantly higher inside than outside burned forests (Nappi and Drapeau, 2009; Rota et al., 2014a).

7. Estimated population growth rates are insufficient to maintain a growing population outside burned forests (Rota et al., 2014a). Thus, although one could argue that low woodpecker densities in green-tree forests multiplied by a much larger unburned forest area might yield even more woodpeckers in green forests (Fogg et al., 2014), a sink area alone (no matter how large) can never yield a viable population of woodpeckers (Odion and Hanson, 2013).

8. The importance of severely burned forests as foraging locations for wintering black-backed woodpeckers is virtually unknown; the only detailed work so far (Kreisel and Stein, 1999) revealed densities that were an order of magnitude greater in burned than in unburned forests.

The biology surrounding this single bird species clearly reflects not only the ecological importance but also the necessity of severely burned forests. Unfortunately, major environmental organizations have yet to focus conservation efforts on burned forests (Schmiegelow et al., 2006), and management guidelines developed by state agencies to designate important wildlife habitats (e.g., https://wildlife.ca.gov/Data/CWHR/Wildlife-Habitats) do not even have burned conifer forests on their radar.

The distributional stronghold of the black-backed woodpecker might be considered to lie within the boreal forests of Canada, which nobody doubts are among the most severe fire dependent ecosystems in the world, but the bird's distribution south into the California Sierras and Rocky Mountains of the Intermountain West confirms that severe fires in those areas have been historically important as well. A North American forest bird species that is more narrowly restricted to a single forest condition does not exist; the black-backed woodpecker is the definition of a specialist. Everything about this bird species, including its distribution, territory size, breeding success, and even coloration pattern (which matches blackened trees), all indicate that this species needs expansive patches of high-severity fire, occurring within dense, mature forests (Hanson and Chi, 2020), to persist (Figure 3.11).

We have taken the liberty to provide extensive detail on this particular species because its ecological story carries significant management implications. Because public land managers have a legal responsibility to manage for the maintenance of biodiversity and ecosystem integrity, finding even a single

FIGURE 3.11 Black-backed woodpecker—a species that is relatively restricted in its distribution to severely burned forests. *Photograph by Richard Hutto.*

species that depends on severe fire should be enough to raise their awareness that severely burned mixed-conifer forests provide necessary habitat as well. Thus, the black-backed woodpecker is an ideal focal species for bringing attention to the fact that burned forest conditions are important to maintain in the landscape (DellaSala et al., 2014, 2017). The evolutionary history that has led to a strong association between burned forests and the woodpecker also raises questions about whether (as many assume) severe fires in mixed-conifer forests are really beyond the historical natural range of variation, whether we need to be thinning forests far removed from homes to reduce fire severity, whether we need to be suppressing fire outside the wildland—urban interface, and whether we should "salvage" log trees (including important legacy trees; see Chapter 11) after fire. Yes, the story surrounding this focal species is important.

Bird Species in Other Regions That Seem to Require Severe Fire (Lesson 3 Continued)

Do any other bird species seem not only to benefit from but also to require severe fire to persist? The presence of a species in a specific environment and its absence elsewhere would be a clear indication that it depends on that particular environment. For species that occur across a range of environmental conditions, the places where they are relatively abundant are also likely to represent places that are required for population persistence because they persist in source areas and they are generally less abundant in, and their abundance is more variable through time in, more marginal or sink areas (Pulliam, 1988; Sergio and Newton, 2003). Although the same level of biological detail that has been amassed for the black-backed woodpecker has not been collected for most other fire-associated bird species, the habitat distribution patterns of numerous bird species reveal that they are nowhere more abundant than in recently burned forests. For example, Hutto (1995) listed 15 species that were more abundant in recently burned forests than in any of 14 other vegetation types. Graphs generated from surveys conducted across an even broader range of vegetation types show just how striking these habitat distribution patterns can be: numerous species are nowhere more abundant than they are in severely burned forests (Hutto and Young, 1999; Hutto and Patterson 2016) (Figure 3.12).

Many mixed-conifer bird species (e.g., black-backed woodpecker, American three-toed woodpecker, hairy woodpecker, northern flicker, olive-sided flycatcher, western wood-pewee [*Contopus sordidulus*], dusky flycatcher [*Empidonax oberholseri*], mountain bluebird, Townsend's solitaire, house wren, tree swallow, lazuli bunting, Clark's nutcracker, red crossbill) fall consistently into a short-term "benefit" category, as revealed either by some measure of abundance or nest success in studies of burned versus unburned or before versus after fire (Bock and Lynch, 1970; Bock et al., 1978; Taylor and Barmore, 1980; Apfelbaum and Haney, 1981; Raphael et al., 1987; Hutto, 1995; Kotliar et al., 2002; Hannah and Hoyt, 2004; Smucker et al., 2005; Mendelsohn et al., 2008; Seavy and Alexander, 2014; Hutto and Patterson 2016; Hutto et al., 2020). Even severely burned patches within conifer forests that we have come to associate with low-severity fire can provide critically important habitat for species like the buff-breasted flycatcher [*Moucherolle beige*] (Kirkpatrick et al., 2006; Conway and Kirkpatrick, 2007; Hutto et al., 2008).

One of the most celebrated examples of a fire specialist involves the federally endangered Kirtland's warbler (*Setophaga kirtlandii*). It occurs almost exclusively in young (5- to 23-year-old) jack pine (*Pinus banksiana*) forest historically created by severe fire (Walkinshaw, 1983). In addition, pairing success is significantly higher in burned than in unburned forests (98% vs. 58% success; Probst and Hayes, 1987). The need for severe fire is obvious not only because, historically, it must have taken severe fires to stimulate forest succession but also because of how its critically endangered population increased dramatically after a fire accidently escaped within its breeding range (James and McCulloch, 1995). Managers have had difficulty trying to recreate conditions that mimic natural postfire conditions through the use of logging techniques (Probst and Donnerwright, 2003; Spaulding and Rothstein, 2009),

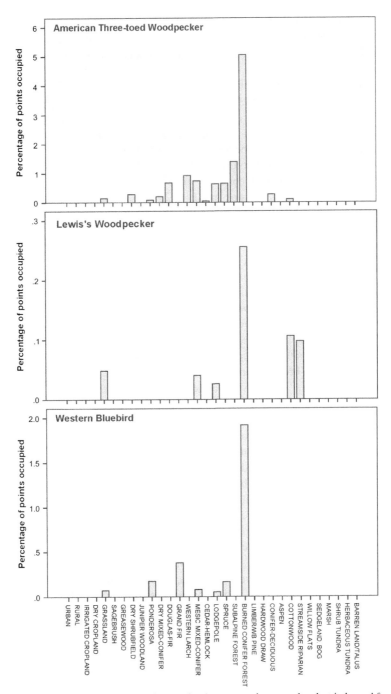

FIGURE 3.12 Several graphs depicting species that seem to be more abundant in burned forests than in any other vegetation type in the northern Rocky Mountains. Data were drawn from a subset of the Northern Region Landbird Monitoring Program database consisting of 20,000 survey points distributed across northern Idaho and western Montana.

and efforts to use these artificial means to maintain warbler populations miss the point. Conservation efforts should be directed toward maintaining severely burned forests, not toward finding a way around the natural fire disturbance process.

In the hardwood forests of the southern Appalachians, total bird abundance, diversity, and species richness were found to be highest in high-severity fire areas, where these measures were approximately twice as high in high-severity fire patches compared to unburned forests (Greenberg et al., 2023).

In Australia, where few species are thought to be restricted to recently burned shrubland or burned forest conditions, early colonists are viewed as generalists, and management concerns are focused on postfire decreases in late-succession specialists (Serong and Lill, 2012; Lindenmayer et al., 2022). Nevertheless, data from Lindenmayer et al. (2014) show that a number of bird species decline in abundance 1—2 years after moderate to severe fire but then return to levels comparable to, or *higher* than, levels in unburned forests within 3 years following fire (also see Lindenmayer et al., 2022 for more recent results). Indeed, upon further inspection, we found that the superb fairywren (*Malurus cyaneus*), gray fantail (*Rhipidura albiscapa*), yellow-faced honeyeater (*Lichenostomus chrysops*), white-fronted honeyeater (*Purnella albifrons*), dusky robin (*Melanodryas vittata*), flame robin (*Petroica phoenicea*), willie wagtail (*Rhipidura leucophrys*), gray shrike-thrush (*Colluricincla harmonica*), varied sittella (*Daphoenositta chrysoptera*), apostlebird (*Struthidea cinerea*), white-browed scrubwren (*Sericornis frontalis*), brown thornbill (*Acanthiza pusilla*), spotted pardalote (*Pardalotus punctatus*), welcome swallow (*Hirundo neoxena*), dusky woodswallow (*Artamus cyanopterus*), black-faced woodswallow (*Artamus cinereus*), and silver-eye (*Zosterops lateralis*) each have been shown by one or more authors to be more abundant in severely burned than in long unburned, dry sclerophyll forests (Christensen and Kimber, 1975; McFarland, 1988; Reilly, 1991a, 1991b, 2000; Turner, 1992; Taylor et al., 1997; Fisher, 2001; Leavesley et al., 2010; Recher and Davis, 2013; Lindenmayer et al., 2014). Thus, many eucalyptus forest species also seem to require severe fire to create the early successional forest conditions within which they are most abundant, but most of those species are not restricted to conditions that occur during the first year or two after fire. In comparison with the dramatic change in bird species composition following severe fire in mixed-conifer forests, there is, in fact, a notable lack of turnover in bird species composition following severe fire in eucalyptus forests (compare before-and-after fire data from Australia and the western United States in Table 3.1). This difference in response to fire is presumably because eucalyptus trees resprout rapidly from epicormic shoots (Figure 3.2). Lindenmayer et al. (2014) also note that in montane ash forests, "… very rapid vegetation regeneration and canopy closure on severely burned sites … may limit the influx of open-country birds and preclude the evolutionary

TABLE 3.1 Probabilities of the Occurrence of Bird Species in Burned and Unburned Australian Eucalypt Forests in the Tablelands Above Wollongong, New South Wales, and in Burned and Unburned Mixed-Conifer Forests in Western Montana (R.L. Hutto, Unpublished Data)

Australian eucalyptus forest			Western North American mixed-conifer forest		
Species	unburned (n = 39)	burned (n = 35)	Species	unburned (n = 1143)	burned (n = 638)
New Holland Honeyeater	0.161	0	Townsend's Warbler	0.4	0.03
Little Wattlebird	0.095	0	Solitary Vireo	0.238	0.021
Scarlet Robin	0.019	0	Golden-crowned Kinglet	0.235	0.021
Yellow-faced Honeyeater	0.040	0.231	Gray Jay	0.084	0.009
Painted Button-Quail	0	0.035	Pileated Woodpecker	0.052	0.006
Grey Shrike-thrush	0	0.058	Swainson's Thrush	0.43	0.062
Olive-backed Oriole	0	0.131	Varied Thrush	0.103	0.015
			White-breasted Nuthatch	0.017	0.003
			Black-capped Chickadee	0.053	0.012
			Red-breasted Nuthatch	0.591	0.145
			Ruby-crowned Kinglet	0.316	0.086
			Hammond's Flycatcher	0.091	0.027
			Hermit Thrush	0.048	0.015
			Orange-crowned Warbler	0.098	0.036
			Western Tanager	0.398	0.163
			Mountain Chickadee	0.219	0.092
			MacGillivray's Warbler	0.201	0.095
			Yellow-rumped Warbler	0.521	0.249
			Warbling Vireo	0.145	0.098
			Clark's Nutcracker	0.022	0.047
			Pine Siskin	0.111	0.257
			Rufous Hummingbird	0.014	0.038
			Northern Flicker	0.076	0.21
			Calliope Hummingbird	0.01	0.03
			Song Sparrow	0.004	0.015
			Olive-sided flycatcher	0.025	0.107
			Rufous-sided Towhee	0.01	0.044
			Cassin's Finch	0.029	0.13
			American Kestrel	0.003	0.015
			Mourning Dove	0.004	0.021
			Hairy Woodpecker	0.021	0.124
			Three-toed Woodpecker	0.007	0.056
			Northern Waterthrush	0.003	0.033
Australian eucalyptus forest			**Western North American mixed-conifer forest**		
Species	unburned (n = 39)	burned (n = 35)	Species	unburned (n = 1143)	burned (n = 638)
			Green-tailed Towhee	0.001	0.012
			White-crowned Sparrow	0.002	0.027
			Lazuli Bunting	0.01	0.148
			House Wren	0.004	0.086
			Western Wood-pewee	0.003	0.104
			Mountain Bluebird	0.004	0.281
			American Robin	0.185	0.441
			Lincoln's Sparrow	0	0.015
			Tree Swallow	0	0.089
			Rock Wren	0	0.044
			Black-backed Woodpecker	0	0.05

Numbers of survey points are given in parentheses. Birds are ordered by the unburned-to-burned ratio of abundance, and species that are completely absent from or are significantly (Mann-Whitney U tests) less abundant in the opposite condition are highlighted in yellow. In both locations are bird species restricted to either early or later successional stages, but the amount of species turnover (degree of replacement of late with early succession specialists) is less pronounced after severe fire in Australia than after severe fire in the western United States development of early successional species" (p. 474). Nevertheless, the bird species listed above suggest that many may still depend on intermediate stages of succession before the development of a fully mature forest and that a slightly different perspective might be needed to expose the ecological importance of severe fire to birds of Australian eucalypt forests.

Taken together, we hope we have provided enough ecological information derived from birds to solidify the notion that severe fire in most severe fire-dependent shrublands and forests is both natural and necessary for maintenance of the ecological integrity of such systems.

3.3 POSTFIRE MANAGEMENT IMPLICATIONS

Severe fire is natural and necessary in most—not relatively few—conifer forests and in many other vegetation types worldwide as well (see Chapters 1 and 2). Current management practices designed to suppress fire, mitigate fire severity, "restore" or "rehabilitate" burned forests after fire, and "mimic" the effects of severe fire are incompatible with the maintenance of ecosystem integrity (Chapters 9-12). Below we use results from bird research as evidence to support this statement, and we offer positive suggestions about what land managers could be doing differently.

3.4 FIRE RISK REDUCTION SHOULD BE FOCUSED ON HUMAN POPULATION CENTERS

The dependence of so many bird (and many other plant and animal) species on conditions created by severe fire is clear. It necessarily follows that we cannot prevent fire and still retain anything close to a natural world. The obvious alternative is to focus fire prevention efforts toward human population centers that are most at risk from severe fire so that fire can be left to periodically restore forest conditions elsewhere (Schoennagel et al., 2017). Smokey Bear needs to refine his message so that it reflects a desire to save human lives and property, *not* a desire to save trees from fire in our wildlands (see Chapters 10 and 12).

3.5 FIRE SUPPRESSION SHOULD BE FOCUSED ON HUMAN COMMUNITIES

Because many species depend on severe fire, it also necessarily follows that we should focus suppression efforts on areas immediately adjacent to human settlements (see Chapters 10 and 12). Wildland firefighters should serve primarily as support for firefighters who defend homes and human lives. Efforts to suppress fire beyond settled areas should be viewed as little more than efforts to save the forest from itself—dry forests need fire in the same way that they need sunlight and rain.

3.6 HIGH-SEVERITY FIRES BEGET MIXED-SEVERITY RESULTS

In contrast with high-severity fire, low-severity understory fires cannot create as broad a range of postfire conditions as severe fires can, nor can they stimulate the postfire process of ecological succession like a severe fire. Therefore, managing for the maintenance of biodiversity requires including severe fires and the mixed-severity landscape effects that result from such fires (Nappi et al., 2010; Taylor et al., 2012; DellaSala et al., 2017; Hutto et al., 2020).

3.7 MITIGATE FIRE SEVERITY THROUGH THINNING ONLY WHERE ECOLOGICALLY APPROPRIATE

Because many species depend on severe fire, it necessarily follows that we should focus forest-thinning efforts adjacent to human communities (e.g., closest to homes—home ignition zone—see Chapter 10), and prioritize fire in forest wildlands. The distributions of black-backed woodpeckers and many other fire-dependent plant and animal species make it abundantly clear that forest thinning to reduce fire severity—if even possible—is ecologically justified in only a very small proportion of the American West (Odion et al., 2014a; Sherriff et al., 2014). The presence of numerous fire-dependent species in most conifer forests throughout the American West (as illustrated by the abundance of bird research results considered in this chapter) is the strongest possible indication that the same forests have burned severely for millennia and are likely still within their historical range of natural variation (Odion et al., 2014a; Law and Waring, 2015; Parks et al., 2015; Baker 2015b; DellaSala and Hanson, 2019; Baker et al., 2023).

The distribution of birds like the black-backed woodpecker and other fire-dependent plant and animal species, which blanket most of the forested land in the American West, are clearly at odds with claims that as much as 40% of public forested lands in parts of the United States are in need of restoration to prevent or mitigate the effects of severe fire (e.g., Haugo et al., 2015; Kelsey 2019). Lower-severity fires do not produce the mixed- and high-severity conditions needed by the most fire-dependent bird species, so efforts to mitigate fire severity in most places is incompatible with maintenance of the ecological integrity of most conifer forest systems (Odion et al., 2014a). So, what should we be doing differently? We could realize that modeled estimates indicating that our forests are in conditions that lie beyond the historical natural range of variation are just that—modeled estimates that rest strongly on many untested assumptions. We should always compare modeled results with insight gained by ecologists who can also draw strong inferences about historical conditions and, more specifically, about the kind of environments that necessarily led to adaptations of plants and animals that reflect the distant past much more accurately than other methods commonly used to reconstruct natural fire regimes.

3.8 POSTFIRE "SALVAGE" LOGGING IN THE NAME OF RESTORATION OR REHABILITATION IS ALWAYS ECOLOGICALLY INAPPROPRIATE AND MISDIRECTED

Postfire "salvage" logging, seeding, planting, and shrub removal have overwhelmingly negative effects on natural systems (Lindenmayer et al., 2004; Lindenmayer and Noss, 2006; McIver and Starr, 2006; Swanson et al., 2011; DellaSala et al., 2014; Hanson, 2014; Thorn et al., 2018; also see Chapter 9), and birds have been instrumental in uncovering that fact. There is nothing as obvious to a birdwatcher as the negative effect of postfire salvage logging on the most fire-dependent birds (Uxley, 2014), and these anecdotal impressions are backed up by the strongest and most consistent scientific results ever published on any wildlife management issue (Hutto, 1995, 2006, 2008; Morissette et al., 2002; Nappi et al., 2004; Hutto and Gallo, 2006; Koivula and Schmiegelow, 2007; Hanson and North, 2008; Cahall and Hayes, 2009; Saab et al., 2009; Rost et al., 2013; Thorn et al., 2018; Hutto et al., 2020). One look at, or one walk through, a postfire logged forest (Figure 3.13, also see Chapter 9) after knowing something about the biological wonder associated with a severely burned forest should be enough to convince any thinking person that there is no justification for this kind of land management activity.

It is bad enough that forests logged after fire are made unsuitable for black-backed woodpeckers and other early postfire specialists, but much worse is that postfire logging and shrub removal through mechanical or chemical means may also act as an "ecological trap" (Robertson and Hutto, 2006). This can

FIGURE 3.13 A vivid view of what can only be described as an ecological disaster following this postfire salvage logging operation, which took place after the 1988 Combination fire in Montana. *Photograph by Richard Hutto.*

occur when birds are attracted to burned areas that seem to be suitable and then those areas are suddenly transformed by logging or shrub removal into unsuitable habitat in an unnaturally rapid period of time. This is the most reasonable explanation for why black-backed woodpeckers are more abundant in dense, burned forests that are logged after fire than they are in burned forests that are logged before fire.Birds are not attracted to the latter, where tree densities are too low and sizes are too small to provide suitable habitat, but they are attracted to the former before the trees are unexpectedly removed (Hutto, 2008).

The adverse impacts of postfire salvage logging have been documented in globally different ecosystems (Rost et al., 2012; Thorn et al., 2018; Georgiev et al., 2020), and there is absolutely no ecological justification for this kind of logging in the mixed-conifer forests of the western United States. In addition, there is no economic justification to salvage log after fire, because there are always better places to harvest timber without anywhere near the negative ecological consequences associated with postfire salvage logging. There is also a huge ecotourism potential associated with unlogged, burned forest because of the once-in-a-lifetime experiences people can have if they are directed to enjoy the birds, wildflowers and everything else that comes with a beautifully blackened forest. This is a matter of setting priorities for timber harvest—burned forests, along with mature/old-growth forests, should be at the bottom of the list. Burned forests not only provide unique ecological value but also set the stage for the development of future forest conditions (including future old growth)—conditions that are much more varied than those associated with development after artificial disturbance from logging. Forests have their own rules and timetables associated with the natural process of ecological succession, and we should embrace that variety and complexity. What could be done differently? Postfire rehabilitation should focus on roads, culverts, and other infrastructure issues, and nothing else. We need to recognize that new forest conditions get created after fire, and a disturbance-dependent forest does not need to be "fixed" after disturbance takes place.

3.9 WE CAN DO MORE HARM THAN GOOD TRYING TO "MIMIC" NATURE

Prescribed burning, forest thinning, and the use of other forms of artificial disturbance in an effort to mimic nature are often poor substitutes for natural disturbance processes (DellaSala et al., 2022). Prescribed burning is usually done out of season, too frequently, and in a manner that is far too low severity to have the necessary effects in most systems that evolved with mixed severity fire (England, 1995; Tucker and Robinson, 2003; Penman and Towerton, 2008; Peters and Sala, 2008; Arkle and Pilliod, 2010; Rota et al., 2014a). Thinning forests in a manner thought to mimic disturbance effects is also likely to be problematic because natural disturbance (the process of fire itself) produces effects that cannot be emulated through artificial means (Schieck and Song, 2006; Reidy et al., 2014; DellaSala et al., 2022). Moreover, a thinned forest that subsequently burns in a natural fire event will not be suitable as postfire habitat for early postfire specialists because of the reduction in tree densities and sizes (Hutto, 2008) that otherwise act as biological legacies in severely burned patches (DellaSala, 2020). Finally, the use of forest thinning in the name of forest restoration is inappropriately applied to relatively mesic mixed-conifer forests that are unlikely to need restoration, as indicated by a lack of posttreatment change in bird communities toward what one would expect if the forests were actually outside the historical range of natural variation (Hutto et al., 2014).

Except in the case of an endangered species, the worst management approach is one that focuses narrowly on creating *artificial* conditions needed by a single species. This is "single-species management," which is not the same thing as using a "management indicator approach." Management indicators are not meant to be tools that enable land managers to artificially modify land conditions to benefit a single species. Instead, a management indicator species should be used as an indication of a particular kind of "natural" condition that needs to be maintained on the landscape and as a check that the land condition is indeed acceptable to a species that requires such conditions. Even for an endangered species, we should always be thinking about maintaining the "natural" conditions that historically maintained its population. For example, although artificial tree plantations may provide conditions used by Kirtland's warbler (Spaulding and Rothstein, 2009), the bird historically nested beneath the canopy of young trees born of fire. Therefore, we should create conditions safe enough to allow natural severe fire events to unfold throughout most of its historical range. As clearly stated in the Endangered Species Act (ESA, Section 2), "the purposes of this act are to provide a means whereby the *ecosystems upon which endangered species and threatened species depend* may be conserved ..." (our italics). Conservation should be about the larger system (e.g., maintaining a fire disturbance-based jack pine forest system), more so than finding a way to maintain a species through artificial means. The black-backed woodpecker is one such "indicator" or "focal species" that should be used to inform us about a critically important "natural" disturbance processes and vegetation conditions that we need to maintain in severely burned forests and all the associated organisms that thrive within them.

What could we be doing differently? We need to trust that disturbance-dependent systems need severe disturbance (yes, that means a lot of tree death) to stimulate ecological succession in a manner that is indeed naturally biodiverse. We also need to appreciate that modeled *means and standard deviations* associated with measures of forest structure are not the same things as historical *ranges of variation* associated with the same measures. While some places have tree densities that exceed some estimated historical average value, it does not mean they fall outside the historical range of natural variation. Land managers need to relax in response to severe fire. As long as we can reduce the frequency of human-caused fires and remain safe during naturally ignited fire events, a management option that is based on working with fire for ecosystem benefits will be best for biodiversity (Gill, 2001; Bradstock, 2008, see Chapter 13). In this context, noting that safety is best achieved through home hardening and mechanical treatments in small areas immediately adjacent to structures (Cohen, 2000; Cohen and Stratton, 2008; Winter et al., 2009; Stockmann et al., 2010; Gibbons et al., 2012; Syphard et al., 2014), and not through mechanical treatments in more remote wildlands, is important (Schoennagel et al., 2017). Given this fact, it is puzzling why mechanical forest management in relatively remote, publicly owned wildlands have become the most commonly used tactic to reduce wildfire risk (Schoennagel et al., 2009). DellaSala et al. (2022) provide more recent discussions of the futility of unmitigated command-and-control approaches toward wildfire and nature.

3.10 CONCLUDING REMARKS

The most important ecological lessons we can take away from the bird research described in this chapter are that (1) many species have evolved to require severe fire to create the conditions that they need, and (2) even though some ecological systems may have departed significantly from what are believed to be historical conditions (e.g., tree plantations in the Pacific Northwest), birds are telling us through their behavior and distribution patterns that the majority of fire-dependent ecosystems are still within the historical range of natural variation, are plenty "resilient," and are fully capable of proceeding quite naturally through the process of succession following a severe fire event (Hanson, 2018; Hanson and Chi, 2021). Therefore, thinning forests in the name of restoration is largely unnecessary (DellaSala et al., 2022). If this were not true, the world would be full of places that experienced a severe fire disturbance and then underwent an unnatural transformation or "type conversion" following the disturbance event, never to return to what was there before disturbance. It is most telling that those kinds of places are rare indeed (Hanson and Chi, 2021).

For those who would like to read, view, or hear more about the relationship between birds and severe fire, there are excellent children's books (e.g., Peluso, 2007; Collard, 2015) and a fire-story web page (https://www.fireecologystory.com/) devoted to building an appreciation for the role of severe fire in our forests.

REFERENCES

Apfelbaum, S., Haney, A., 1981. Bird populations before and after wildfire in a Great Lakes pine forest. Condor 83, 347—354.

Arkle, R.S., Pilliod, D.S., 2010. Prescribed fires as ecological surrogates for wildfires: a stream and riparian perspective. For. Ecol. Manag. 259, 893—903.

Bain, D.W., Baker, J.R., French, K.O., Whelan, R.J., 2008. Post-fire recovery of eastern bristlebirds (*Dasyornis brachypterus*) is context-dependent. Wildl. Res. 35, 44–49.

Baker, W.L., 2015a. Historical Northern Spotted Owl habitat and old-growth dry forests maintained by mixed-severity wildfires. Landscape Ecol. 30, 665–666.

Baker, W.L., 2015b. Are high-severity fires burning at much higher rates recently than historically in dry-forest landscapes of the Western USA? PLoSOne 10 (9), e0136147. https://doi.org/10.1371/journal.pone.0136147.

Baker, W.L., Hanson, C.T., DellaSala, D.A., 2023. Nature-based solutions to extreme wildfires. Fire 6, 428. https://doi.org/10.3390/fire6110428.

Bent, A.C., 1939. Life histories of North American woodpeckers. U.S. Natl. Mus. Bull. 174, 334.

Blackford, J.L., 1955. Woodpecker concentration in burned forest. Condor 57, 28–30.

Bock, C.E., Lynch, J.F., 1970. Breeding bird populations of burned and unburned conifer forest in the Sierra Nevada. Condor 72, 182–189.

Bock, C.E., Raphael, M., Bock, J.H., 1978. Changing avian community structure during early post-fire succession in the Sierra Nevada. Wilson Bull. 90, 119–123.

Bond, M.L., Gutierrez, R.J., Franklin, A.B., LaHaye, W.S., May, C.A., Seamans, M.E., 2002. Short-term effects of wildfires on spotted owl survival, site fidelity, mate fidelity, and reproductive success. Wilson Bull. 30, 1022–1028.

Bond, M.L., Lee, D.E., Siegel, R.B., Ward, J.P., 2009. Habitat use and selection by California Spotted Owls in a postfire landscape. J. Wildl. Manag. 73, 1116–1124.

Bond, M.L., Lee, D.E., Siegel, R.B., Tingley, M.W., 2013. Diet and home range size of California Spotted Owls in a burned forest. Western Birds 44, 114–126.

Bond, M.L., Chi, T.Y., Bradley, C.M., DellaSala, D.A., 2022. Forest management, barred owls, and wildfire in Northern Spotted Owl territories. Forests 13, 1730. https://doi.org/10.3390/f13101730.

Bowman, D.M.J.S., 2000. Australian Rainforests: Islands of Green in a Land of Fire. Cambridge University Press, Cambridge, UK.

Bradstock, R.A., 2008. Effects of large fires on biodiversity in south-eastern Australia: disaster or template for diversity? Int. J. Wildland Fire 17, 809–822.

Cahall, R.E., Hayes, J.P., 2009. Influences of postfire salvage logging on forest birds in the Eastern Cascades, Oregon, USA. For. Ecol. Manag. 257, 1119–1128.

Chalmandrier, L., Midgley, G.F., Barnard, P., Sirami, C., 2013. Effects of time since fire on birds in a plant diversity hotspot. Acta Oecol. 49, 99–106.

Christensen, P., Kimber, P.C., 1975. Effects of prescribed burning on the flora and fauna of southwest Australian forests. Proc. Ecol. Soc. Aust. 9, 85–106.

Clark, D.A., Anthony, R.G., Andrews, L.S., 2013. Relationship between wildfire, salvage logging, and occupancy of nesting territories by northern spotted owls. J. Wildl. Manag. 77, 672–688.

Cohen, J.D., 2000. Preventing disaster: home ingnitability in the wildland-urban interface. J. For. 98, 15–21.

Cohen, J.D., Stratton, R.D., 2008. Home Destruction Examination: Grass Valley Fire. USDA Forest Service, pp. 1–26. R5-TP-026b.

Collard III, S.B., 2015. Fire Birds: Valuing Natural Wildfires and Burned Forests. Mountain Press for Bucking Horse Books, Missoula, MT.

Conway, C.J., Kirkpatrick, C., 2007. Effect of forest fire suppression on Buff-breasted Flycatchers. J. Wildl. Manag. 71, 445–457.

DellaSala, D.A., Bond, M.L., Hanson, C.T., Hutto, R.L., Odion, D.C., 2014. Complex early seral forests of the Sierra Nevada: what are they and how can they be managed for ecological integrity? Nat. Areas J. 34, 310–324.

DellaSala, D.A., Hutto, R.L., Hanson, C.T., Bond, M.L., Ingalsbee, T., Odion, D., Baker, W., 2017. Accommodating mixed-severity fire to restore and maintain ecosystem integrity with a focus on the Sierra Nevada of California, USA. Fire Ecol. 13, 148−171.

DellaSala, D.A., Hanson, C.T., 2019. Are wildland fires increasing large patches of complex early seral forest habitat? Diversity 11, 157. https://doi.org/10.3390/d11090157.

DellaSala, D.A., 2020. Fire-mediated biological legacies in dry forested ecosystems of the Pacific Northwest, USA. In: Beaver, E.A., Prange, S., DellaSala, D.A. (Eds.), Disturbance Ecology and Biological Diversity. CRC Press Taylor and Francis Group, LLC, Boca Raton, FL, pp. 38−85.

DellaSala, D.A., Baker, B., Hanson, C.T., Ruediger, L., Baker, W., 2022. Have western USA fire suppression and active management approaches become a contemporary Sisyphus? Biological Conservation 268. https://doi.org/10.1016/j.biocon.2022.109499.

Dudley, J.G., Saab, V.A., Hollenbeck, J.P., 2012. Foraging-habitat selection of Black-backed Woodpeckers in forest burns of southwestern Idaho. Condor 114, 348−357.

England, A.S., 1995. Avian Community Organization along a Post-fire Age Gradient in California Chaparral. Ph.D. Thesis. University of California-Davis.

Fayt, P., Machmer, M.M., Steeger, C., 2005. Regulation of spruce bark beetles by woodpeckers—a literature review. For. Ecol. Manag. 206, 1−14.

Fisher, A.M., 2001. Avifauna changes along a *Eucalyptus* regeneration gradient. EMU 101, 25−31.

Fogg, A.M., Roberts, L.J., Burnett, R.D., 2014. Occurrence patterns of Black-backed Woodpeckers in green forest of the Sierra Nevada Mountains, California, USA. Avian Conserv. Ecol. 9, 3.

Ganey, J.L., Kyle, S.C., Rawlinson, T.A., Apprill, D.L., Ward, J.P., 2014. Relative abundance of small mammals in nest core areas and burned wintering areas of Mexican Spotted Owls in the Sacramento Mountains. New Mexico. Wilson J. Ornith. 126, 47−52.

Georgiev, K.B., Chao, A., Castro, J., Chen, Y.-H., Choi, C., Fontaine, J.B., Hutto, R.L., Lee, E.-J., Müller, J., Rost, J., Żmihorski, M., Thorn, J., 2020. Salvage logging changes the taxonomic, phylogenetic and functional successional trajectories of forest bird communities. J.Appl.Ecol. 57, 1103−1112.

Gibbons, P., van Bommel, L., Gill, A.M., Cary, G.J., Driscoll, D.A., Bradstock, R.A., Knight, E., Moritz, M.A., Stephens, S.L., Lindenmayer, D.B., 2012. Land management practices associated with house loss in wildfires. PLoS One 7, e29212.

Gill, A.M., 2001. Economically destructive fires and biodiversity conservation: an Australian perspective. Conserv. Biol. 15, 1558−1560.

Greenberg, C.H., Moorman, C.E., Elliott, K.J., Martin, K., Hopey, M., Caldwell, P.V., 2023. Breeding bird abundance and species diversity greatest in high-severity wildfire patches in central hardwood forests. For. Ecol. Manage. 529, 120715.

Haney, A., Apfelbaum, S., Burris, J.M., 2008. Thirty years of post-fire succession in a southern boreal forest bird community. Am. Midl. Nat. 159, 421−433.

Hannah, K.C., Hoyt, J.S., 2004. Northern Hawk Owls and recent burns: does burn age matter? Condor 106, 420−423.

Hannon, S.J., Drapeau, P., 2005. Bird responses to burning and logging in the boreal forest of Canada. Stud. Avian Biol. 30, 97−115.

Hanson, C.T., 2014. Conservation concerns for Sierra Nevada birds associated with high-severity fire. Western Birds 45, 204−212.

Hanson, C.T., 2018. Landscape heterogeneity following high-severity fire in California's forests. Wildlife Soc. Bulletin 42, 264−271.

Hanson, C.T., Chi, T.Y., 2020. Black-Backed Woodpecker Nest Density in the Sierra Nevada, California. Diversity vol 12, 364.

Hanson, C.T., Chi, T.Y., 2021. Impacts of postfire management are unjustified in spotted owl habitat. Frontiers Ecol. Evol. 9, 596282.

Hanson, C.T., North, M.P., 2008. Postfire Woodpecker foraging in salvage-logged and unlogged forests of the Sierra Nevada. Condor 110, 777−782.

Hanson, C.T., Odion, D.C., DellaSala, D.A., Baker, W.L., 2009. Overestimation of fire risk in the Northern Spotted Owl recovery plan. Conserv. Biol. 23, 1314−1319.

Hanson, C.T., Odion, D.C., DellaSala, D.A., Baker, W.L., 2010. More-comprehensive recovery actions for Northern Spotted Owls in dry forests: reply to Spies et al. Conserv. Biol. 24, 334−337.

Harris, M.A., 1982. Habitat Use Among Woodpeckers in Forest Burns. M.S. Thesis. University of Montana, Missoula, MT.

Haugo, R., Zanger, C., DeMeo, T., Ringo, C., Shlisky, A., Blankenship, K., Simpson, M., Mellen-McLean, K., Kertis, J., Stern, M., 2015. A new approach to evaluate forest structure restoration needs across Oregon and Washington, USA. For. Ecol. Manag. 335, 37−50.

Hoyt, J.S., Hannon, S.J., 2002. Habitat associations of Black-backed and Three-toed woodpeckers in the boreal forest of Alberta. Can. J. For. Res. 32, 1881−1888.

Hutto, R.L., 1995. Composition of bird communities following stand-replacement fires in northern Rocky Mountain (U.S.A.) conifer forests. Conserv. Biol. 9, 1041−1058.

Hutto, R.L., 2006. Toward meaningful snag-management guidelines for postfire salvage logging in North American conifer forests. Conserv. Biol. 20, 984−993.

Hutto, R.L., 2008. The ecological importance of severe wildfires: some like it hot. Ecol. Appl. 18, 1827−1834.

Hutto, R.L., 2014. Time budgets of male Calliope Hummingbirds on a dispersed lek. Wilson J. Ornithol. 126, 121−128.

Hutto, R.L., Gallo, S.M., 2006. The effects of postfire salvage logging on cavity-nesting birds. Condor 108, 817−831.

Hutto, R.L., Young, J.S., 1999. Habitat Relationships of Landbirds in the Northern Region, USDA Forest Service. USDA Forest Service General Technical Report RMRS-GTR-32, pp. 1−72.

Hutto, R.L., Patterson, D.A., 2015. Hidden Positive Fire Effects on Birds Exposed Only After Controlling for Fire Severity and Time Since Fire (unpublished MS).

Hutto, R.L., Patterson, D.A., 2016. Positive effects of fire on birds may appear only under narrow combinations of fire severity and time-since-fire. Int. J. Wildland Fire 25, 1074−1085.

Hutto, R.L., Conway, C.J., Saab, V.A., Walters, J.R., 2008. What constitutes a natural fire regime? insight from the ecology and distribution of coniferous forest birds in North America. Fire Ecol. 4, 115−132.

Hutto, R.L., Flesch, A.D., Fylling, M.A., 2014. A bird's-eye view of forest restoration: do changes reflect success? For. Ecol. Manag. 327, 1−9.

Hutto, R.L., Hutto, R.R., Hutto, P.L., 2020. Patterns of bird species occurrence in relation to anthropogenic and wildfire disturbance: management implications. Forest Ecol. Manag. 461, 117942.

James, F.C., McCulloch, C.E., 1995. The strength of inferences about causes of trends in populations. In: Martin, T.E., Finch, D.M. (Eds.), Ecology and Management of Neotropical Migratory Birds: A Synthesis and Review of Critical Issues. Oxford University Press, New York, NY, pp. 40−51.

Jenness, J.S., Beier, P., Ganey, J.L., 2004. Associations between forest fire and Mexican Spotted Owls. For. Sci. 50, 765−772.

Kelsey, R., 2019. Wildfires and Forest Resilience: The Case for Ecological Forestry in the Sierra Nevada. Pages 1-12 unpublished report. The Nature Conservancy, Sacramento, CA.

Kirkpatrick, C., Conway, C., Jones, P.B., 2006. Distribution and relative abundance of forest birds in relation to burn severity in southeastern Arizona. J. Wildl. Manag. 70, 1005–1012.

Koivula, M.J., Schmiegelow, F.K.A., 2007. Boreal woodpecker assemblages in recently burned forested landscapes in Alberta, Canada: effects of post-fire harvesting and burn severity. For. Ecol. Manag. 242, 606–618.

Kotliar, N.B., Hejl, S.J., Hutto, R.L., Saab, V.A., Melcher, C.P., McFadzen, M.E., 2002. Effects of fire and post-fire salvage logging on avian communities in conifer-dominated forests of the western United States. Stud. Avian Biol. 25, 49–64.

Kreisel, K.J., Stein, S.J., 1999. Bird use of burned and unburned coniferous forests during winter. Wilson Bull. 111, 243–250.

Law, B.E., Waring, R.H., 2015. Carbon implications of current and future effects of drought, fire and management on Pacific Northwest forest. Forest Ecol. Manag. 355, 4–14.

Leavesley, A.J., Cary, G.J., Edwards, G.P., Gill, A.M., 2010. The effect of fire on birds of mulga woodland in arid central Australia. Int. J. Wildland Fire 19, 949–960.

Lee, D.E., 2018. Spotted owls and forest fire: a systematic review and meta-analysis of the evidence. Ecosphere 9 (7). https://doi.org/10.1002/ecs2.2354. Article e02354.

Lee, D.E., Bond, M.L., Siegel, R.B., 2012. Dynamics of breeding-season site occupancy of the California Spotted Owl in burned forests. Condor 114, 792–802.

Lee, D.E., Bond, M.L., Borchert, M.I., Tanner, R., 2013. Influence of fire and salvage logging on site occupancy of spotted owls in the San Bernardino and San Jacinto Mountains of Southern California. J. Wildl. Manag. 77, 1327–1341.

Legge, S., et al., 2022. The conservation impacts of ecological disturbance: time-bound estimates of population loss and recovery for fauna affected by the 2019-2020 Australian megafires. Global Ecol. Biogeogr. https://doi.org/10.1111/geb.13473.

Lesmeister, D.B., Davis, R.J., Sovern, S.G., Yang, Z., 2021. Northern spotted owl nesting forests as fire refugia: a 30-year synthesis of large wildfires. Fire Ecol 17. Article 32.

Leonard, S.W.J., Bennett, A.F., Clarke, M.F., 2014. Determinants of the occurrence of unburnt forest patches: potential biotic refuges within a large, intense wildfire in south-eastern Australia. For. Ecol. Manag. 314, 85–93.

Lindenmayer, D.B., Noss, R.F., 2006. Salvage logging, ecosystem processes, and biodiversity conservation. Conserv. Biol. 20, 949–958.

Lindenmayer, D.B., Foster, D.R., Franklin, J.F., Hunter, M.L., Noss, R.F., Schmiegelow, F.A., Perry, D., 2004. Salvage harvesting policies after natural disturbance. Science 303, 1303.

Lindenmayer, D.B., Blanchard, W., McBurney, L., Blair, D., Banks, S.C., Driscoll, D.A., Smith, A.L., Gill, A.M., 2014. Complex responses of birds to landscape-level fire extent, fire severity and environmental drivers. Divers. Distrib. 20, 467–477.

Lindenmayer, D.B., Blanchard, W., Bowd, E., Scheele, B.C., Foster, C., Lavery, T., McBurney, L., Blair, D., 2022. Rapid bird species recovery following high-severity wildfire but in the absence of early successional specialists. Divers. Distrib. https://doi.org/10.1111/ddi.13611.

McFarland, D., 1988. The composition, microhabitat use and response to fire of the avifauna of subtropical heathlands in Cooloola National Park, Queensland. EMU 88, 249–257.

McIver, J.D., Starr, L., 2006. A literature review on the environmental effects of postfire logging. West. J. Appl. For. 16, 159–168.

Mendelsohn, M.B., Brehme, C.S., Rochester, C.J., Stokes, D.C., Hathaway, S.A., Fisher, R.N., 2008. Responses in bird communities to wildland fires in southern California. Fire Ecol. 4, 63–82.

Morissette, J.L., Cobb, T.P., Brigham, R.M., James, P.C., 2002. The response of boreal forest songbird communities to fire and post-fire harvesting. Can. J. For. Res. 32, 2169–2183.

Murphy, E.G., Lehnhausen, W.H., 1998. Density and foraging ecology of woodpeckers following a stand-replacement fire. J. Wildl. Manag. 62, 1359–1372.

Nappi, A., Drapeau, P., 2009. Reproductive success of the Black-backed Woodpecker (*Picoides arcticus*) in burned boreal forests: are burns source habitats? Biol. Conserv. 142, 1381–1391.

Nappi, A., Drapeau, P., 2011. Pre-fire forest conditions and fire severity as determinants of the quality of burned forests for deadwood-dependent species: the case of the black- backed woodpecker. Can. J. For. Res. 41, 994–1003.

Nappi, A., Drapeau, P., Savard, J.-P.L., 2004. Salvage logging after wildfire in the boreal forest: is it becoming a hot issue for wildlife? For. Chron. 80, 67–74.

Nappi, A., Drapeau, P., Saint-Germain, M., Angers, V.A., 2010. Effect of fire severity on long-term occupancy of burned boreal conifer forests saproxylic insects and wood-foraging birds. Int. J. Wildland Fire 19, 500–511.

Niemi, G.J., 1978. Breeding birds of burned and unburned areas in northern Minnesota. Loon 50, 73–84.

Odion, D.C., Hanson, C.T., 2013. Projecting impacts of fire management on a biodiversity indicator in the Sierra Nevada and Cascades, USA: the black-backed woodpecker. Open Forest Sci. J. 6, 14–23.

Odion, D.C., Hanson, C.T., Arsenault, A., Baker, W.L., DellaSala, D.A., Hutto, R.L., Klenner, W., Moritz, M.A., Sherriff, R.L., Veblen, T.T., Williams, M.A., 2014a. Examining historical and current mixed-severity fire regimes in ponderosa pine and mixed- conifer forests of western North America. PLoS One 9, e87852, 87851–87814.

Odion, D.C., Hanson, C.T., DellaSala, D.A., Baker, W.L., Bond, M.L., 2014b. Effects of fire and commercial thinning on future habitat of the Northern Spotted Owl. Open Ecol. J. 7, 37–51.

Parks, S.A., Miller, C., Parisien, M.-A., Holsinger, L.M., Dobrowski, S.Z., Abatzoglou, J., 2015. Wildland fire deficit and surplus in the western United States, 1984-2012. Ecosphere 6. Article 275 1-27513.

Peluso, B.A., 2007. The Charcoal Forest: How Fire Helps Animals and Plants. Mountain Press, Missoula, MT.

Penman, T.D., Towerton, A.L., 2008. Soil temperatures during autumn prescribed burning: implications for the germination of fire responsive species? Int. J. Wildland Fire 17, 572–578.

Peters, G., Sala, A., 2008. Reproductive output of ponderosa pine in response to thinning and prescribed burning in western Montana. Can. J. For. Res. 38, 844–850.

Pfister, A.R., 1980. Postfire Avian Ecology in Yellowstone National Park. M.S. Thesis. Washinton State University, Pullman, WA.

Powell, H.D.W., Hejl, S.J., Six, D.L., 2002. Measuring woodpecker food: a simple method for comparing wood-boring beetle abundance among fire-killed trees. J. Field Ornithol. 73, 130–140.

Probst, J.R., Donnerwright, D., 2003. Fire and shade effects on ground cover structure in Kirtland's Warbler habitat. Am. Midl. Nat. 149, 320–334.

Probst, J.R., Hayes, J.P., 1987. Pairing success of Kirtland's Warblers in marginal vs. suitable habitat. Auk 104, 234–241.

Pulliam, H.R., 1988. Sources, sinks, and population regulation. Am. Nat. 132, 652–661.

Pyke, G.H., Saillard, R., Smith, J., 1995. Abundance of Eastern Bristlebirds in relation to habitat and fire history. EMU 95, 106–110.

Raphael, M.G., Morrison, M.L., Yoder-Williams, M.P., 1987. Breeding bird populations during twenty-five years of post-fire succession in the Sierra Nevada. Condor 89, 614–626.

Recher, H.F., Christensen, P.E., 1981. Fire and the evolution of the Australian biota. In: Keast, A., Junk, D.W. (Eds.), Ecological Biogeography of Australia. D.W. Junk, The Hague, pp. 135–162.

Recher, H.F., Davis Jr., W.E., 2013. Response of birds to a wildfire in the Great Western Woodlands, Western Australia. Pac. Conserv. Biol. 19, 188–203.

Reidy, J.L., Thompson III, F.R., Kendrick, S.W., 2014. Breeding bird response to habitat and landscape factors across a gradient of savanna, woodland, and forest in the Missouri Ozarks. For. Ecol. Manag. 313, 34–46.

Reilly, P., 1991a. The effect of wildfire on bird populations in a Victorian coastal habitat. EMU 91, 100–106.

Reilly, P., 1991b. The effect of wildfire on bush bird populations in six Victorian coastal habitats. Corella 15, 134–142.

Reilly, P., 2000. Bird populations in a Victorian coastal habitat twelve years after a wildfire in 1983. EMU 100, 240–245.

Roberts, S.L., van Wangtendonk, J.W., Miles, A.K., Kelt, D.A., 2011. Effects of fire on spotted owl site occupancy in a late-successional forest. Biol. Conserv. 144, 610–619.

Robertson, B.A., Hutto, R.L., 2006. A framework for understanding ecological traps and an evaluation of existing ecological evidence. Ecology 87, 1075–1085.

Robinson, N.M., Leonard, S.W.J., Bennett, A.F., Clarke, M.F., 2014. Refuges for birds in fire-prone landscapes: the influence of fire severity and fire history on the distribution of forest birds. For. Ecol. Manag. 318, 110–121.

Rost, J., Clavero, M., Brotons, L., Pons, P., 2012. The effect of postfire salvage logging on bird communities in Mediterranean pine forests: the benefits for declining species. J. Appl. Ecol. 49, 644–651.

Rost, J., Hutto, R.L., Brotons, L., Pons, P., 2013. Comparing the effect of salvage logging on birds in the Mediterranean Basin and the Rocky Mountains: common patterns, different conservation implications. Biol. Conserv. 158, 7–13.

Rota, C.T., Millspaugh, J.J., Rumble, M.A., Lehman, C.P., Kesler, D.C., 2014a. The role of wildfire, prescribed fire, and mountain pine beetle infestations on the population dynamics of Black-backed Woodpeckers in the Black Hills, South Dakota. PLoS One 9, e94700.

Rota, C.T., Rumble, M.A., Millspaugh, J.J., Lehman, C.P., Kesler, D.C., 2014b. Space-use and habitat associations of Black-backed Woodpeckers (Picoides arcticus) occupying recently disturbed forests in the Black Hills, South Dakota. For. Ecol. Manag. 313, 161–168.

Rush, S., Klaus, N., Keyes, T., Petrick, J., Cooper, R., 2012. Fire severity has mixed benefits to breeding bird species in the southern Appalachians. For. Ecol. Manag. 263, 94–100.

Saab, V.A., Dudley, J.G., 1998. Responses of Cavity-Nesting Birds to Stand-Replacement Fire and Salvage Logging in Ponderosa pine/Douglas-fir Forests of Southwestern Idaho. USDA Forest Service Research Paper RMRS-RP-11, pp. 1–17.

Saab, V., Brannon, R., Dudley, J., Donohoo, L., Vanderzanden, D., Johnson, V., Lachowski, H., 2002. Selection of Fire-Created Snags at Two Spatial Scales by Cavity-Nesting Birds. USDA Forest Service General Technical Report PSW-GTR-181, pp. 835–848.

Saab, V.A., Russell, R.E., Dudley, J., 2007. Nest densities of cavity-nesting birds in relation to post-fire salvage logging and time since wildfire. Condor 109, 97–108.

Saab, V.A., Russell, R.E., Dudley, J.G., 2009. Nest-site selection by cavity-nesting birds in relation to postfire salvage logging. For. Ecol. Manag. 257, 151–159.

Saracco, J.F., Siegel, R.B., Wilkerson, R.L., 2011. Occupancy modeling of Black-backed Woodpeckers on burned Sierra Nevada forests. Ecosphere 2, 1–17.

Schieck, J., Song, S.J., 2006. Changes in bird communities throughout succession following fire and harvest in boreal forests of western North America: literature review and meta-analyses. Can. J. For. Res. 36, 1299–1318.

Schmiegelow, F.K.A., Stepnisky, D.P., Stambaugh, C.A., Koivula, M., 2006. Reconciling salvage logging of boreal forests with a natural-disturbance management model. Conserv. Biol. 20, 971–983.

Schoennagel, T., Nelson, C.R., Theobald, D.M., Carnwath, G.C., Chapman, T.B., 2009. Implementation of National Fire Plan treatments near the wildland-urban interface in the western United States. Proc. Natl. Acad. Sci. 106, 10706–10711.

Schoennagel, T., et al., 2017. Adapt to more wildfire in western North American forests as climate changes. PNAS. www.pnas.org/cgi/doi/10.1073/pnas.1617464114.

Seavy, N.E., Alexander, J.D., 2014. Songbird response to wildfire in mixed-conifer forest in southwestern Oregon. Int. J. Wildland Fire 23, 246–258.

Seavy, N.E., Burnett, R.D., Taille, P.J., 2012. Black-backed woodpecker nest-tree preference in burned forests of the Sierra Nevada, California. Wildl. Soc. Bull. 36, 722–728.

Seidl, R., Rammer, W., Spies, T.A., 2014. Disturbance legacies increase the resilience of forest ecosystem structure, composition, and functioning. Ecol. Appl. 24, 2063–2077.

Sergio, F., Newton, I., 2003. Occupancy as a measure of territory quality. J. Anim. Ecol. 72, 857–865.

Serong, M., Lill, A., 2012. Changes in bird assemblages during succession following disturbance in secondary wet forests in south-eastern Australia. EMU 112, 117–128.

Sherriff, R.L., Platt, R.V., Veblen, T.T., Schoennagel, T.L., Gartner, M.H., 2014. Historical, observed, and modeled wildfire severity in montane forests of the Colorado Front Range. PLoS One 9, e106971.

Siegel, R.B., Tingley, M.W., Wilkerson, R.L., Bond, M.L., Howell, C.A., 2013. Assessing Home Range Size and Habitat Needs of Black-Backed Woodpeckers in California. The Institute for Bird Populations, Point Reyes Station, California.

Siegel, R.B., Wilkerson, R.L., Tingley, M.W., Howell, C.A., 2014. Roost sites of the Black-backed Woodpecker in burned forest. Western Birds 45, 296–303.

Siegel, R.B., Tingley, M.W., Wilkerson, R.L., Howell, C.A., Johnson, M., Pyle, P., 2016. Age structure of Black-backed Woodpecker populations in burned forests. Auk 133, 69–78.

Sitters, H., Christie, F.J., Di Stefano, J., Swan, M., Penman, T., Collins, P.C., York, A., 2014. Avian responses to the diversity and configuration of fire age classes and vegetation types across a rainfall gradient. For. Ecol. Manag. 318, 13–20.

Smucker, K.M., Hutto, R.L., Steele, B.M., 2005. Changes in bird abundance after wildfire: importance of fire severity and time since fire. Ecol. Appl. 15, 1535–1549.

Spaulding, S.E., Rothstein, D.E., 2009. How well does Kirtland's warbler management emulate the effects of natural disturbance on stand structure in Michigan jack pine forests? For. Ecol. Manag. 258, 2609–2618.

Stillman, A.N., Lorenz, T.J., Siegel, R.B., Wilkerson, R.L., Johnson, M., Tingley, M.W., 2021. Conditional natal dispersal provides a mechanism for populations tracking resource pulses after fire. Behav. Ecol.

Stockmann, K., Burchfield, J., Calkin, D., Venn, T., 2010. Guiding preventative wildland fire mitigation policy and decisions with an economic modeling system. For. Policy Econ. 12, 147–154.

Swanson, M.E., Franklin, J.F., Beschta, R.L., Crisafulli, C.M., DellaSala, D.A., Hutto, R.L., Lindenmayer, D.B., Swanson, F.J., 2011. The forgotten stage of forest succession: early-successional ecosystems on forest sites. Front. Ecol. Environ. 9, 117–125.

Syphard, A.D., Brennan, T.J., Keeley, J.E., 2014. The role of defensible space for residential structure protection during wildfires. Int. J. Wildland Fire 23, 1165−1175.
Tarbill, G.L., Manley, P.N., White, A.M., 2015. Drill, baby, drill: the influence of woodpeckers on post-fire vertebrate communities through cavity excavation. J. Zool. 296, 95−103.
Taylor, D.L., Barmore, W.J., 1980. Post-fire succession of avifauna in coniferous forests of Yellowstone and Grand Teton National Parks, Wyoming. In: DeGraaf, R.M. (Ed.), Workshop Proceedings: Management of Western Forests and Grasslands for Nongame Birds. USDA Forest Service General Technical Report INT-86, Ogden, UT, pp. 130−145.
Taylor, R., Duckworth, P., Johns, T., Warren, B., 1997. Succession in bird assemblages over a seven-year period in regrowth dry sclerophyll forest in south-east Tasmania. EMU 97, 220−230.
Taylor, R.S., Watson, S.J., Nimmo, D.G., Kelly, L.T., Bennett, A.F., Clarke, M.F., 2012. Landscape-scale effects of fire on bird assemblages: does pyrodiversity beget biodiversity? Divers. Distrib. 18, 519−529.
Thorn, S., et al., 2018. Impacts of salvage logging on biodiversity: a meta-analysis. Journal of Applied Ecology 55, 279−289.
Tingley, M.W., Wilkerson, R.L., Bond, M.L., Howell, C.A., Siegel, R.B., 2014. Variation in home-range size of Black-backed Woodpeckers. Condor 116, 325−340.
Tucker Jr., J.W., Robinson, W.D., 2003. Influence of season and frequency of fire on Henslow's Sparrows (*Ammodramus henslowii*) wintering on gulf coast pitcher plant bogs. Auk 120, 96−106.
Turner, R.J., 1992. Effect of wildfire on birds at Weddin Mountain, New South Wales. Corella 16, 65−74.
U.S. Fish and Wildlife Service, 2011. Revised Recovery Plan for the Northern Spotted Owl (*Strix Occidentalis Caurina*). USFWS, Portland, Oregon.
Uxley, W., 2014. Firebird. Bird Watching J. October, 26−31October.
Walkinshaw, L.H., 1983. Kirtland's Warbler, the Natural History of an Endangered Species. Cranbrook Institute of Science, Bloomfield Hills, MI.
White, A.M., Tarbill, G.L., Wilkerson, B., Siegel, R., 2019. Few detections of Black-backed Woodpeckers (*Picoides arcticus*) in extreme wildfires in the Sierra Nevada. Avian Cons. Ecol. 14. Article 17.
Winchell, C.S., Doherty, P.F., 2014. Effects of habitat quality and wildfire on occupancy dynamics of Coastal California Gnatcatcher (*Polioptila californica californica*). Condor 116, 538−545.
Winter, G., McCaffrey, S., Vogt, C.A., 2009. The role of community policies in defensible space compliance. For. Policy Econ. 11, 570−578.
Woinarski, J.C.Z., Recher, H.F., 1997. Impact and response: a review of the effects of fire on the Australian avifauna. Pac. Conserv. Biol. 3, 183−205.
Youngman, J.A., Gayk, Z.G., 2011. High density nesting of Black-backed Woodpeckers (*Picoides arcticus*) in a post-fire Great Lakes jack pine forest. Wilson J. Ornithol. 123, 381−386.
Yunick, R.P., 1985. A review of recent irruptions of the Black-backed Woodpecker and Three-toed Woodpecker in eastern North America. J. Field Ornithol. 56, 138−152.

Chapter 4

Mammals and Mixed- and High-Severity Fire

Monica L. Bond
Wild Nature Institute, Concord, NH, United States

4.1 INTRODUCTION

Mammals are ecologically and economically important members of the landscapes in which they live. Large herbivores like deer (*Odocoileus* spp.) and elk (*Cervus elaphus*), and predators like bears (*Ursus* spp.) and wolves (*Canis lupus*), are highly conspicuous and well-known "flagship" mammal species, whereas rodents, bats, and mustelids are cryptic but no less important in their ecosystems. Many species have developed broad ecological tolerance from exposure to environmental variation and natural disturbances over long time periods (Lawler, 2003). However, widespread hunting and excessive habitat fragmentation of landscapes by modern-day humans are qualitatively and quantitatively different from the natural disturbances to which many mammals were exposed in the past (Spies and Turner, 1999). Recent disturbances have resulted in contraction of historical ranges and population declines. In North America alone, notable population declines include elk, grizzly bears (*Ursus arctos*), gray wolves, Canada lynx (*Lynx canadensis*), bighorn sheep (*Ovis canadensis*), beaver (*Castor canadensis*), the larger species of forest mustelids, and several heteromyid rodents.

As also discussed in the chapters of this book, mixed- and high-severity wildfire is a natural disturbance in many vegetation systems of North America, the Mediterranean, Australia, and Africa. The effects of severe fire on organisms vary spatially and temporally, by habitat type, and by species, but how do these disturbances specifically impact mammals? As with any natural disturbance, some species are adversely affected ("fire-averse" species), others benefit ("fire-loving" or pyrophilous species), and still others have a neutral response to fires, at least initially.

The dynamics of populations and communities of mammals after severe fire depend on factors such as the degree of ecological change, time since fire (see Chapter 3 for birds as well), size and spatial configuration of burned and unburned areas, extent of edge, isolation of habitat patches by urbanization and roads, and invasion of nonnative species (Smith, 2000; Shaffer and Laudenslayer, 2006; Arthur et al., 2012; Diffendorfer et al., 2012; Fontaine and Kennedy, 2012). In theory, mammalian populations should be stable and resilient across the landscape wherever prefire populations and critical habitats are not greatly reduced and/or fragmented by human activities, and where severe fires occur in a spatial and temporal pattern in which a species has evolved (Shaffer and Laudenslayer, 2006). The capability of fire-loving individuals to utilize severely burned areas or for fire-averse populations to recover after fire, however, can be compromised when prefire habitat fragmentation has resulted in small and/or isolated populations. Postfire management actions, such as logging of burned trees and use of herbicides and pesticides, also adversely influence population dynamics and habitat use (see Chapter 9).

In this chapter, I provide an overview of published studies about mammalian responses to mixed- and high-severity fires in forests, woodlands, shrublands, deserts, and grasslands around the world. I describe research on the effects of severe fire on four major taxonomic groups of mammals: bats, small mammals, carnivores, and ungulates. I emphasize peer-reviewed publications, particularly those with robust methodologies and analyses, because these are the accepted standard in science. I also use nonpeer-reviewed data when necessary to supplement information from the peer-reviewed literature. I do not cite every published study but instead provide a balanced overview of severe-fire effects on these taxa. I encourage readers to investigate further the scientific literature on habitat use and population responses of mammals to severe fire because the state of the science is constantly evolving.

Few studies have documented direct effects of fire on wildlife (e.g., mortality from asphyxiation, heat stress, burning, or physiological stress; however, see Singer et al., 1989), but wildlife biologists, at least in North America, generally agree that direct mortality from fire is typically very low and does not significantly influence populations (Smith, 2000; Jolly et al., 2022). Thus, I focus here on the indirect responses to severe fire, such as postfire occupancy, abundance or density, survival, reproduction, and use of habitat (e.g., breeding, resting, foraging). I define "significant effects" according to the generally accepted scientific definition of statistical significance (i.e., at the 0.05 probability level). I exclude studies that simulated or modeled fires, choosing instead to focus on observations of real systems responding to severe wildfire. Further, many published studies on the topic do not account for fire severity or postfire salvage logging, and these were also excluded unless I otherwise specified.

Appendix 4.1 is a summary of published studies by mammalian taxa and directional response to severe wildfire (negative, neutral, positive) over three time periods after fire. I present results from studies comparing unburned habitats with high-severity burn from wildfire (rather than prescribed fire) and without the confounding effect of postfire logging. For small mammals, only species with enough detections to determine directional response were included in the appendix.

4.2 BATS

Bats perform unique and critical ecosystem services by consuming vast quantities of insects, thereby transferring nutrients, most notably nitrogen, from foraging to roosting areas via their feces (Gruver and Keinath, 2006). Bats are predators of adult mosquitoes and thus play an important role in controlling mosquito populations and reducing disease transmission (Reiskind and Wund, 2009). Further, nectar-feeding bats are primary pollinators of many plant species throughout the world (Molina-Freaner and Eguiarte, 2003).

The current literature on the effects of fire on bats strongly suggests that mixed- and high-severity fires are explicitly beneficial, particularly to edge- and open-adapted species (Blakey et al., 2019). In a study comparing the relative activity of six phonic groups of mostly rare and sensitive bat species across unburned and moderate- and high-severity burned mixed-conifer stands 1 year after fire in the southern Sierra Nevada, bat activity in burned areas was equivalent to or greater than activity in unburned areas for all groups based on echolocation frequencies (Buchalski et al., 2013). Indeed, two of the phonic groups showed a positive response to high-severity fire but a neutral response to moderate-severity fire, demonstrating the importance of severity-specific responses. In dry forests and woodlands of the Pilliga region in New South Wales, Australia, total bat species composition was unaffected by fire 1 and 4 years following high-severity fire compared with 2 years prior to fire, with activity of individual species either neutral or positively affected (Law et al., 2018). The positive response to mixed- and high-severity fire by bats mirrors findings for a range of bird species (see Chapter 3) and provides evidence of a long evolutionary relationship between bats and severe fire.

Several studies have documented how roosting bats use basal hollows of large trees (Gellman and Zielinski, 1996; Zielinski and Gellman, 1999; Fellers and Pierson, 2002; Mazurek, 2004) (Figure 4.1). Basal hollows are cavities formed by repeated fire scarring and healing (Zielinski and Gellman, 1999). For bats that roost in basal hollows of large trees, high-severity fire may not only destroy or reduce the longevity of existing roost trees but it also creates new roost trees. In addition, fire creates gaps in the canopy that increase the amount of solar radiation reaching the subcanopy where bats roost. These warmer temperatures may facilitate thermoregulation (Brigham et al., 1997; Boyles and Aubrey, 2006) and are particularly beneficial to reproductive females because increased temperatures are associated with greater fetal and neonate growth (Brigham et al., 1997; Johnson et al., 2009). Finally, high-severity fire creates a "pulse" of insect prey (e.g., aquatic insects (Malison and Baxter, 2010), moths, beetles, and flies (Schwab, 2006)), as well as new

FIGURE 4.1 Basal hollows in large trees are created by periodic fire scarring and healing, creating important roost sites for bats. A Townsend's big-eared bat (*Corynorhinus townsendii*) roost tree in a coast redwood (*Sequoia sempervirens*) in Grizzly Creek State Park, northern California. *Photo by M.J. Mazurek taken in 2015.*

natural edge habitat that provides novel foraging opportunities (Fellers and Pierson, 2002).

Comparisons of food web components between unburned watersheds and areas of low- and high-severity fires 5 years after fire in Douglas-fir (*Pseudotsuga menziesii*) and ponderosa pine (*Pinus ponderosa*) forests in central Idaho showed high insect biomass in heavily burned areas and correspondingly high bat detection rates (Malison and Baxter, 2010). Notably, high-severity sites had almost five times more biomass of zoobenthic insects and more than three times the number of emerging adult aquatic insects than low-severity sites (and twice as many as unburned areas). The frequency of bat echolocation calls also was significantly greater at high-severity sites than at unburned sites, because aquatic insects emerging from streams into the terrestrial environment are an important food source for bats. In a review of the responses of stream benthic macroinvertebrates to fire, Minshall (2003) concluded that "results for macroinvertebrates generally support the belief that fire and similar natural disturbance events are not detrimental to the sustained maintenance of diverse and productive aquatic ecosystems (i.e., those found in undisturbed forests)" (p. 159). While individual taxa respond differently to the physical changes in stream structure and short-term and long-term postfire changes in vegetation, Minshall noted that streams are inherently unstable and dynamic environments in which disturbance, including high-severity fire, is a regular occurrence, and many species are opportunistic and can shift food resources in response to fire.

In mid-elevation forests burned at mixed and high severity in western Montana, Schwab (2006) characterized roost sites and sampled potential prey sources for two forest-dwelling, insectivorous bat species, the little brown bat (*Myotis lucifugus*) and the long-eared myotis (*Myotis evotis*). These species roosted in larger-diameter snags (standing dead trees) in high-density stands of fire-killed trees. Proximity to perennial streams also was important in roost site selection for these two species in burned forests. Wildland fire apparently created an abundance of roosting sites and insect prey for bats. Although the abundance of Lepidoptera (moths) and Trichoptera (caddis flies) was similar in burned and unburned forests, the abundance of Diptera (flies) and Coleoptera (beetles) was significantly higher in burned forests. Overall, the median capture rate of all insects in the burn was 1.78 times higher than the median capture rate in unburned forests—but there was considerable variability in the composition and abundance of particular species. Eight of the 11 orders of insects were more abundant in burned sites. In addition, beetles, flies, and caddis flies were significantly more abundant in burned than unburned sites in the first year after fire, although they decreased significantly the second year after fire. Thus, retention of burned trees after fire is important for insectivorous bats. In fact, removing burned trees decreased mammalian (and avian) predation on the abundance of insects that occurred 1 year after fire. Unsurprisingly, tree size also matters for bats that select larger-diameter trees for roosting. A near-complete lack of snags $\geq$30 cm diameter-at-breast height (dbh) may have precluded the use of severely burned boreal forest patches by little brown bats 10 years after the Fox Lake Fire in Yukon, Canada (Jung, 2020). Snags in unburned forests can be recruited from existing green trees, but in severely burned forests postfire logging eliminates both existing and future snags for nearly a century because few trees are available for snag recruitment until large-diameter trees have regrown (Schwab, 2006).

As with many bird species, mixed- and high-severity fire in forest ecosystems likely enhances foraging opportunities for bats (Buchalski et al., 2013; Doty et al., 2016). Many insect species inhabiting coniferous forests are highly evolved to exploit severely burned forests and are aptly termed "pyrophilous." Certain beetle species in particular are strongly attracted to highly burned forests. Saint-Germain et al. (2004) noted that, "some insect groups have adapted to recurrent forest fires by evolving sensory organs and life strategies that allow them to exploit these high-quality habitats efficiently. Pyrophilous Buprestids of the genera *Oxypteris* and *Merimna* and the Cerambycid

Arhopalus tristis (F.) have been shown to respond physiologically to smoke and/or heat generated by fire, and use them as signals leading toward the newly created habitat ... Several other Coleoptera species uncommon in mature forests congregate in exceptionally high densities in burned stands" (p. 583).

In a study of fire-loving beetle communities in a large fire that burned boreal black spruce (*Picea mariana*) forest in Quebec, Canada, more than half of the 86 taxa captured were restricted to burned stands (Saint-Germain et al., 2004). Moreover, total captures and species richness were higher in burned stands, especially the oldest severely burned forests. Captures were significantly lower the second year after the fire for all burned stands, indicating that the utility of burned forests for these beetles is greatest in the first year following fire.

Insects utilizing dead trees occur at much lower abundances in low-severity sites, which by definition have far fewer fire-killed trees than high-severity sites. Malison and Baxter (2010) stated that, "our results suggest that high severity fires do not play the same ecological role as low severity fires and allowing high severity fires to burn (rather than suppressing them) in certain forest types could be important in maintaining ecosystem function" (p. 577). Similarly, in his severely burned study site, Schwab (2006) noted, "26% of all [insect] families captured were restricted to sites within the burn suggesting a unique environment created only after fire." Thus, ecological changes caused by mixed- and high-severity fires cannot be mimicked by low-severity prescribed burns (Box 4.1).

BOX 4.1

(1) Bats preferentially roost and forage in burned forests.
(2) High-severity fire creates a superabundance of native insect prey.
(3) Bats select denser stands of fire-killed trees for roosting in burned forests and forage significantly more in forests burned by high-severity fire than in unburned and low-severity fire-affected forests.
(4) Large burned trees for roosting have significant positive benefits for bats.
(5) Postfire logging removes roost trees, reduces the abundance of prey, and reduces habitat suitability for bats.

4.3 SMALL MAMMALS

Small mammals are critically important to ecosystems because they can influence vegetation structure and composition by dispersing seeds and ectomycorrhizal fungi and by aerating soils (Maser et al., 1978). They also provide an essential prey base for carnivores, and the distribution of small mammals can affect the use of space and the habitat selection of their predators (Carey et al., 1992; Ward et al., 1998). Small mammals have comparatively small home ranges and therefore are quite sensitive to habitat change, making them good biological indicators (Haim and Izhaki, 1994). Assemblages include rodents and insectivores of the families Soricidae (shrews), Talpidae (moles), Aplodontidae (mountain beavers), Sciuridae (squirrels, chipmunks, and marmots), Geomyidae (gophers), Heteromyidae (pocket mice and kangaroo rats), Muridae (voles, mice, rats, and woodrats), and Dipodidae (jumping mice). Larger-bodied small mammals include rodents in the Castoridae (beaver) and Erethizontidae (porcupine) families, as well as lagomorphs (pika, hares, and rabbits), and Australian and American marsupials (Marsupialia).

The occupation of severely burned areas by small mammals is related to regrowth of the vegetation structure with which various species are associated (Torre and Díaz, 2004; Lee and Tietje, 2005; Vamstad and Rotenberry, 2010; Diffendorfer et al., 2012; Kelly et al., 2012; Borchert and Borchert, 2013), as well as with seed and insect production and availability (Coppeto et al., 2006), and cavities created by woodpeckers in snags (Tarbill, 2010). Further, some small mammals cope with surviving immediately after severe fire by reducing energy expenditure (Matthews et al., 2017; Nowak et al., 2016). I discuss fire effects on small mammals according to habitat type but give special attention to the deer mouse (*Peromyscus maniculatus*)—an exceptionally "fire-loving" species—in its own section (Figure 4.2).

FIGURE 4.2 Deer mice increase after severe fire in a variety of habitats. A deer mouse captured 2 years after forest dominated by Douglas-fir with some lodgepole pine (*Pinus contorta*), western larch (*Larix occidentalis*), and ponderosa pine burned severely in the 2005 Tarkio Fire, Montana. *Photo by Rafal Zwolak taken in 2005.*

Chaparral and Coastal Sage Scrub

The chaparral and coastal sage scrub vegetation types in central and southern California support an exceptionally rich diversity of rodents that are well-adapted to a regime of periodic, very-high-intensity fire (see Chapter 6). Many studies have examined small-mammal communities after both prescribed and wildfire in these vegetative types. During intense fires, some individuals among small, less vagile animals may suffer mortality, but many others survive in rock crevices, riparian areas, large downed logs, and underground burrows where temperatures remain cool and the air clean (Chew et al., 1959; Quinn, 1979; Lawrence, 1966; Wirtz, 1995; Smith, 2000). Following fire, small-mammal communities change over time (Diffendorfer et al., 2012; Arthur et al., 2012; Borchert and Borchert, 2013) and space (Schwilk and Keeley, 1998), depending on the vegetation associations of the various species. Species preferring open habitat, including pocket mice (*Chaetodipus* spp.), California voles (*Microtus californicus*), harvest mice (*Reithrodontomys megalotis*), and, especially, kangaroo rats (*Dipodomys* spp.) and deer mice can increase quite dramatically and quickly after severe shrubland fire. Over a period of several years, as shrubs resprout and grow denser and as different food sources become available, small-mammal species preferring a shrubby overstory, including woodrats (*Neotoma* spp.), California mice (*Peromyscus californicus*), brush mice (*Peromyscus boylii*), and cactus mice (*Peromyscus eremicus*), increase in number (Cook, 1959; Wirtz, 1977; Price and Waser, 1984; Brehme et al., 2011; Borchert and Borchert, 2013). Compared with unburned chaparral and grassland, severely burned chaparral had the highest rodent diversity 4 years after a high-intensity wildfire near Mount Laguna in San Diego County (Lillywhite, 1977). Published data are not currently available for lagomorphs in chaparral wildfires, but prescribed burning of chamise (*Adenostoma fasciculatum*) chaparral in northern California increased black-tailed jackrabbit (*Lepus californicus*) densities by 500% −1000% the year following fire (Howard, 1995).

Forests and Woodlands

Forests offer important habitats for small mammals, especially shrews, mice, tree voles, and squirrels. Mixed- and high-severity fire in forested habitats can have pronounced effects on small-mammal populations by creating or transforming habitat structures such as live and dead trees, shrubs, and coarse woody debris. While some studies have shown that severely burned conifer forests in North America support fewer individuals of some rodents and insectivores immediately after fire compared with adjacent unburned sites (e.g., pinyon mice [*Peromyscus truei*; Borchert et al., 2014] and masked shrews (*Sorex cinereus*) and southern red-backed voles [*Myodes gapperi*; Zwolak and Forsman, 2007]), numbers begin to rebound several years after fire, often by individuals surviving in unburned refuges within the larger burn perimeter. Northern red-backed voles (*Myodes rutilus*), considered old-growth specialists, began repopulating an intense burn in boreal Alaska from surrounding unburned forest and started reproducing 3 years thereafter (West, 1982). In boreal forests of northern Sweden, severe fire resulted in reduced densities of bank voles (*Myodes glareolus*), but this effect persisted nearly a decade after fire (Ecke et al., 2019). Bank voles feed on forbs, fungi, berries, and tree lichens that were destroyed by the fire, suggesting postfire recovery of this species in northern boreal forests may be prolonged compared with other areas. In meadows within pine-oak forests of east-central Arizona, high intensity wildfire had no effect on small mammal communities (dominated by deer mice and Arizona montane vole, *Microtus montanus arizonensis*) up to 2 years postfire (Horncastle et al., 2019).

Unburned refuges and vegetation changes over time also mediate postfire mammal population dynamics in other forests types, notably *Eucalyptus* forests in Australia. Eucalypt forests are generally resilient to high-severity fire as the dominant tree species resprout epicormically (Campbell-Jones et al., 2022). The ground-dwelling small mammal community in the most intensely burned eucalypt forests in southeastern New South Wales, Australia recovered to half the prefire number of species within 3 months postfire, and number of species was comparable to low-burned sites within 6 months postfire (Mikac et al., 2023). Numbers of bush rats (*Rattus fuscipes*) and agile antechinus (*Antechinus agilis*) were reduced compared with populations in adjacent unburned areas 6 months after severe fire in a mountain ash (*Eucalyptus regnans*) forest, but the population in the burned area was composed of residual animals that had survived the fire rather than animals recolonizing from adjacent forests (Banks et al., 2011). Similarly, in situ survival rather than colonization from surrounding unburnt areas was the likely mechanism for postfire increases in small mammals after large intense wildfires in various vegetation types in Grampians National Park of southeastern Australia (Hale et al., 2021). In northwestern Australian tropical savanna woodlands dominated by tussock and hummock grasses, capture rates of pale rats (*Rattus tunneyi*) 6 weeks after fire was lower with increasing fire severity, but by 1 year there were no differences from prefire levels and unburnt areas (Shaw et al., 2021). Recovery of the pale rat population after lower severity fires was due to in situ survival and reproduction, compared to recolonization of the severely burned areas.

Long-term studies are especially useful because responses relative to time since fire can be quantified. One study examined marsupial population dynamics over a 28-year period following severe wildfire in a southeastern Australia *Eucalyptus* forest reserve (Arthur et al., 2012). Bandicoots (*Isoodon obesulus* and *Perameles nasuta*) increased immediately following the fire, peaked 15 years later, and then declined, associated with an increase and decline of shrub cover. The potoroo (*Potorous tridactylus*) population was similar before and immediately after the fire but began to increase a decade later as tree cover increased. Wombats (*Vombatus ursinus*) exhibited a stable population trend for the first decade after the fire, then slowly declined along with a decline in ground litter cover. Finally, larger macropods (eastern gray kangaroo [*Macropus giganteus*], red-necked wallaby [*Macropus rufogriseus*], and swamp wallaby [*Wallabia biocolor*]) remained at high densities after the fire then declined a decade later as vegetation cover increased. Another study quantified arboreal mammal species richness (possums and gliders) in a eucalypt forest in southeastern Australia 3 and 10 years after severe fire (Campbell-Jones et al., 2022). Greater glider (*Petauroides volans*) occurrence decreased as fire severity increased, but increased with time since fire, as this species is strongly associated with hollows and leaves of eucalypts. The richness of arboreal mammals increased over time and was greater in unburnt or less intensively burned forests. Thus, even in highly resilient forest types, arboreal native mammals can be adversely affected by severe fire, at least temporarily.

Rabbits and hares are associated with shrubs and small conifers that provide cover (Ream, 1981; Howard, 1995). Severe fire temporarily eliminates this habitat structure, but it quickly returns as the vegetation regrows, stimulated by intense fire. Snowshoe hares (*Lepus americanus*) in a boreal forest in Alberta, Canada, moved out of intensely burned sites to surrounding habitat immediately after fire but returned the second summer after the fire when shrubs resprouted, and the postfire population trajectory increased above prefire numbers (Keith and Surrendi, 1971). Snowshoe hares were more abundant in sampled stands within large burns 12 years postfire compared to smaller burns and unburned mature forests in southern British Columbia, because these stands in large fires supported higher densities of lodgepole pine (*Pinus contorta*) saplings (Hutchen and Hodges, 2019); however, fire severity was not analyzed in this study.

Tree squirrels, including Douglas squirrels (*Tamiasciurus douglasii*) and northern flying squirrels (*Glaucomys sabrinus*), typically are associated with late-successional coniferous forests in California and the Pacific Northwest in the United States (Carey, 2000). Thus, they may be adversely affected by intense fire (Zwolak and Forsman, 2007), but few data currently are available to refute or support this hypothesis. Federally endangered Mount Graham red squirrels (*Tamiasciurus fremonti grahamensis*) largely survived intense fires in spruce-fire forests in southeastern Arizona by sheltering in nests and underground burrows, or moving to new locations. Populations were presumed to have markedly decreased immediately following the largest and most severe fire in more than 25 years, which burned more than half the known middens, but actual postfire occupancy levels have yet to be verified by field data

(Merrick et al., 2021). Chipmunks and ground squirrels can occupy forests after severe fire where shrubs provide cover and food (Borchert et al., 2014). Townsend's chipmunks (*Neotamias townsendii*) were abundant in early seral forests with dense shrub cover (Campbell and Donato, 2014). Gray-collared chipmunks (*Tamias cinereicollis*) and least chipmunks (*Tamias minimus*) showed no significant response to wildfire in ponderosa pine forests of the southwestern United States (Converse et al., 2006), and the proportion and composition of two chipmunk species, *Tamias amoenus* and *Tamias ruficaudus*, did not differ between severely burned and unburned conifer forest in Montana (Zwolak and Forsman, 2007).

The increase in the availability, amount, and quality of forage for herbivorous small mammals is an important determinant of the post-severe-fire community. In plots recently burned by large, intense wildfires in a Mediterranean pine-oak woodland in Spain, the abundance of small mammals—mostly mice and shrews—was higher than expected based on vegetation characteristics alone (Torre and Díaz, 2004). The authors attributed small-mammal increases to large quantities of seeds and seedlings in burned sites.

Deserts

The role of severe fire and its effects on small mammals in desert grasslands is somewhat controversial (Killgore et al., 2009; Vamstad and Rotenberry, 2010). Most desert systems are not adapted to frequent fire because many species of long-lived perennial desert plants have low recruitment rates and long life spans and lack the ability to resprout. Fire size and frequency in some areas have increased recently because of the invasion of exotic grasses from livestock grazing (Brooks, 2000) and other causes (Burbidge and McKenzie, 1989). In general, most research shows a lack of significant long-term effects of intense fire on the abundance of desert small mammals, although fire can alter community composition. Similar to shrub types in southern California, rodents in the family Heteromyidae increased following a large, intense wildfire in a perennial grassland in southeastern Arizona, whereas species in the family Cricetidae declined immediately after fire, began increasing 4 years after fire, and returned to prefire levels by the sixth year (Bock et al., 2011). Rodent abundance and species richness were no different between burned and unburned plots after wildfires in Joshua tree (*Yucca bevifolia*) woodlands of the Mojave Desert in the American Southwest (Vamstad and Rotenberry, 2010). Merriam's kangaroo rat (*Dipodomys merriami*) dominated the burned sites. As postfire vegetation changed from annuals to sub-shrubs and then to long-lived perennials, however, the composition of rodent species changed, and the diversity of rodents increased over time.

Habitat type is important to fire effects in deserts. In Australia, wildfires in stony desert habitats with sparse grasses have less effect on habitat structure and small mammals than wildfires in sandy desert habitats with denser hummock grass spinifex (*Triodia* spp.) (Pastro et al., 2014). For example, an intense wildfire did not affect the total abundance and species richness of small mammals in the stony (gibber) desert in central Australia, although some species increased and others decreased immediately following fire (Letnic et al., 2013). By contrast, 9 months after intense wildfire in a spinifex grassland in the same region, small-mammal diversity declined compared with before the fire and with prescribed burned areas, although the abundance of animals captured was similar (Pastro et al., 2011). Data were unavailable from wildfires, but hare (*Lepus* spp.) abundance increased by 300% after prescribed burning in East African savanna grasslands (Ogen-Odoi and Dilworth, 1984).

Deer Mice

In North America, generalist deer mice are often the most abundant rodent after severe fire in a variety of vegetation types (Borchert et al., 2014). This species responds strongly and positively to high-intensity fire in both shrubland and conifer forests. Deer mice increased significantly over time in moderately and severely burned mixed-conifer forests in the San Bernardino Mountains of southern California over a 5-year period after fire (Borchert et al., 2014). During 2 years subsequent to intense fire, deer mice were invariably the most numerous species in burned study sites in a Douglas-fir-Western larch forest in Montana (Zwolak and Forsman, 2008). Converse et al. (2006) attributed increased abundance of deer mice after wildfire in southwestern United States ponderosa pine forests to greater seed production or detectability of seeds after fire.

Dramatic increases in deer mice in severely burned conifer forests were not simply a result of colonization of the burn by animals from surrounding unburned forests. When population densities were low, the vast majority of individually ear-tagged deer mice were found in forest areas after severe fire, and mice appeared regularly in unburned forests only when population densities were high (Zwolak and Forsman, 2008). This finding indicated that severely burned forest was *preferred* deer mouse habitat and that the postfire population increase was intrinsic to the burn; thus the burn itself was a source habitat.

Overall, these observations from small-mammal studies in mixed- and severely burned shrublands, forests, and grasslands underscore the important roles played by high-severity fire patches, unburned refuges within a fire area, and the time since fire in population dynamics after severe fire (Box 4.2).

BOX 4.2

(1) After intense wildfire, small-mammal communities are dynamic and associated with vegetation structure at different successional stages.

(2) Intense fire may increase the availability and abundance of seeds and seedlings for herbivorous small mammals, but those that feed on lichens in northern boreal forests may be adversely affected for longer periods.

(3) Unburned refuges and time since fire are important determinants of small-mammal communities following intense fire.

(4) The richness and abundance of small-mammal species is high following intense fire in chaparral and coastal sage scrub communities of southern California. Heteromyid rodents and deer mice often dominate severely burned shrublands, and heteromyids dominate postburn desert grasslands.

(5) Some small-mammal species decrease shortly after intense fire in North American conifer forests, but they can recover to prefire levels within 1 to several years after fire. Deer mice dramatically increase following intense fire.

4.4 CARNIVORES

Carnivores are critically important "top-down" regulators of ecosystem processes. Elimination of top carnivores unleashes a cascade of adverse effects, including relaxation of predation as a selective force on prey species, spread of disease, explosions of herbivore populations, and subsequent reproductive failure and local extinction of some plants, birds, herptiles, and rodents (Crooks and Soulé, 1999; Terborgh et al., 2001). Large carnivores include ursids (bears), canids (wolves), and larger felids (puma, lions, and jaguars). Medium-sized carnivores, or "mesocarnivores," include canids (coyotes and foxes), Procyonidae (ringtails and raccoons), mustelids (wolverine, marten, fisher, weasels, mink, and badgers), Mephitidae (skunks), and smaller felids (lynx and bobcats). Currently published research on carnivores in mixed and severe wildfires is limited primarily to forested habitats.

Mesocarnivores and Large Cats

Many mesocarnivores are associated with forested habitats. Some are habitat generalists, whereas others are forest specialists, riparian associates, or semi-aquatic (Buskirk and Zielinski, 2003). Martens (*Martes* spp.) occur in dense coniferous or deciduous forests across the northern hemisphere. They also regularly use severely burned habitats. Some evidence suggests martens use burns only when postfire trees are not logged. For instance, stone marten (*Martes foina*) were not detected in an intensely burned but extensively postfire-logged Aleppo pine (*Pinus halepensis*) forest in Greece the second and third years after wildfire and logging (Birtsas et al., 2012). These martens were found only in Turkish red pine (*Pinus brutia*) forests burned by wildfire 9 years earlier and not in nearby unburned forests (Soyumert et al., 2010). In coniferous forests of the Alaskan taiga, resident and transient American martens (*Martes americanus*) were captured in a 6-year-old unlogged burn more often than in an island of unburned mature forest surrounded by the burn (Paragi et al., 1996). The authors did not quantify burn severity in their study area but described fire-affected sites as having portions of "severe" burn, and most of the vegetation was in early to mid-seral stages, with dead, fire-scarred trees still standing, consistent with mixed- and high-severity fire. There was no age difference between martens trapped live in the mature forests and those trapped in the burn, and marten foraging intensity was greatest in the recently burned area (Paragi et al., 1996). Conversely, another study found martens avoided stands of boreal forests burned from 2 to 20 years prior (Gosse et al., 2005), but the study did not quantify or describe burn severity nor specify whether the burned forest was logged.

Larger cousins to the marten, fisher (*Martes pennanti* or *Pekania pennanti*) are rare mesocarnivores associated with dense, mature, boreal and mixed conifer-hardwood forests of North America (Powell and Zielinski, 1994). A recent study in the southern Sierra Nevada, however, used scat sampling to detect fisher habitat preferences and demonstrated that the species used denser, mature forests that had experienced moderate- and high-severity fire 10 and 12 years prior and that were not logged after fire (Hanson, 2013) (Figure 4.3). It is likely that both martens and fishers use severely burned forests for foraging rather than denning.

FIGURE 4.3 Representative foraging location based upon global positioning system coordinates for a confirmed female Pacific fisher scat detection site several hundred meters into the interior of the largest high-severity fire patch (>5000 ha) in the McNally Fire of 2002, Sequoia National Forest, California. *Photo by Chad Hanson taken in 2014.*

Another forest specialist, the Canada lynx, was radio-collared and tracked in and around a recent and older wildfire on the eastern slopes of the North Cascade Mountains in Washington. Lynx were detected in freshly burned areas regularly, but were likely moving between unburned mature forest refugia (Vanbianchi et al., 2017). However, avoidance of severely burned areas up to 6 years after the Tripod Fire might have been the result of logging that occurred 2 years postfire (quantified in Lewis et al., 2012) and which was not included in the analysis. Conversely, lynx selected older burned forest. The authors concluded that forest structure in the form or residual living trees or patches of denser forest regeneration allowed lynx to utilize new burns and thrive in older burns. These results provide intriguing evidence that even old-forest specialist species are adapted to and can exploit postfire conditions in regions where mixed- and high-severity fire is natural (see Chapter 3, Box 3.1: spotted owls).

Another large cat, the mountain lion (*Puma concolor*), was found to select areas with higher fire severity in the Wallow fire of 2011, as did Mexican gray wolves (*Canis lupus baileyi*) (Lewis et al., 2022).

Foxes apparently prefer severely burned forest areas over unburned areas, but they may be less tied to forest structure than martens, fishers, and lynx and thus less sensitive to postfire logging. Red fox (*Vulpes vulpes*) in Turkish red pine forests were detected more often in the 9-year-old unlogged wildfire area (Soyumert et al., 2010); in postfire-logged Aleppo pine forests in Greece, red foxes were detected most often in severely burned areas, rather than moderately and unburned areas (Birtsas et al., 2012). In 3 of 4 years after intense wildfire in mixed-conifer forests of the San Bernardino Mountains in southern California, gray foxes (*Urocyon cinereoargenteus*) were detected more often in mixed-severity burned over unburned areas, and in two of the years no foxes at all were captured in the unburned area, but coyote (*Canis latrans*) were detected more often in unburned forests (Borchert, 2012). Both gray fox and coyote scats were more numerous in areas burned by intense wildfire than in unburned areas 2 years after fire in interior chaparral, Madrean evergreen woodland, and ponderosa pine forest in Arizona (Cunningham et al., 2006).

Striped skunk (*Mephitis mephitis*), ringtail (*Bassariscus astutus*), and raccoon (*Procyon lotor*) were photocaptured only in mixed-conifer forests in southern California burned by high-intensity fire, but each were photographed only once (Borchert, 2012). Bobcat (*Lynx rufus*) were photocaptured in similar numbers in severely burned and unburned forest, but captures in the burned area decreased over time over the 4 years of the study. Finally, mountain lions were photocaptured more often in severely burned forest, but the overall sample was small (four lion in burned areas, one lion in unburned areas).

Bears

Although grizzly bears are flexible in the habitats they use, in British Columbia, Canada, radio-collared grizzly bears strongly selected open forest

BOX 4.3 Seed Dispersal by Carnivores

Fleshy fruits are an important component of the diet of many carnivores, especially during certain seasons when other resources are scarce. Indeed, the germination of many seeds is facilitated by passage through the carnivore gut because it removes the fruit pericarp and scarifies the seed coat (Herrera, 1989). Carnivores are important dispersers of seeds because they have relatively large home ranges and long gut retention times, thus spreading the seeds far from the parent plant. This may be an important mechanism whereby early seral habitats are seeded. For example, in experimental and field tests in severely burned Aleppo pine forest in Spain, Rost et al. (2012) demonstrated that carnivores, including red fox, stone marten, and European badger (*Meles meles*), were important dispersers of Mediterranean hackberry (*Celtis australis*) seeds into the burned areas. These carnivores traveled long distances into the fire area, dispersing seeds more than 1 km from the parent plant. Moreover, seeds collected from scat (i.e., that had passed through the gut) in the burned study area had a significantly greater germination rate than unscarified seeds, both in the greenhouse and in the field.

---

burned by wildfires 50—70 years earlier at high elevations because these sites supported prolific huckleberries (McLellan and Hovey, 2001) (see Box 4.3). Wildfire also promotes the regeneration of whitebark pine (*Pinus albicaulis*) seeds, another important food source for bears (Kunkel, 2003). Wildfire is not equivalent to logging, as regenerating timber harvests were rarely used by bears in any season (McLellan and Hovey, 2001).

One study compared the demographics and physiology of black bears (*Ursus americanus*) occupying burns of two ages, 13 and 35 years old, in spruce (*Picea* spp.) and aspen (*Populus tremuloides*) forests of the Kenai Peninsula of Alaska (Schwartz and Franzmann, 1991). The authors did not specify burn intensity, but they noted that 5% of the older burn was logged after fire for "improvement" of moose (*Alces alces*) habitat, and they pointed out that the more recent fire burned at a greater intensity than the older fire. The density of bears and the percentage of cubs born were similar between the two sites, but all age groups of bears were significantly larger in the recent burn area. Bears in the older burn area consumed more cranberries (Box 4.3), whereas the number of moose calves consumed per bear was much larger in the recent burn area, likely explaining the larger size of the bears. Females in the recent burn area also produced litters at a younger age and had a shorter interval between weaning of yearlings than females in the older burn area. Moreover, cub survival was significantly higher in the recent burn area. The vigor of black bear populations was associated with moose abundance, which was significantly enhanced in the 13-year-old fire area.

Another study compared the demography of a population of black bears in interior chaparral, Madrean evergreen woodland, and ponderosa pine forest burned by high-intensity wildfire for 3 years after fire using (1) the population in a nearby unburned site for 3 years and (2) results from earlier demographic research on the fire site from 20 years earlier, conducted over a 6-year period (Cunningham and Ballard, 2004). The sex ratio at the 3-year-old burned site was more skewed toward males than in either the unburned reference site or 20 years before the burn. The authors presumed that the fire had reduced the adult female population. However, it is also possible that the female population already had been reduced in the 20 years before the fire occurred, when the population was not monitored. Indeed, an alternative scenario could be that the population of both adult females and males had been declining at Four Peaks before fire, and the fire actually attracted males to the site, who have larger home ranges, thus skewing the sex ratio.

The above study reported complete reproductive failure in the 3 years after fire at the burned site compared with 36% of cubs surviving to 1 year of age on the unburned control site (Cunningham and Ballard, 2004). More cubs had survived to year 1 at the burned site 20 years before the fire. During the 1970s, however, complete reproductive failure also occurred in the absence of fire during 3 of the 6 years of study. Thus, years of complete reproductive failure in that study area were not unusual. Overall, reproductive success was lowest in the burned forest compared with the same site 20 years before fire and an unburned reference site, suggesting the possibility of negative short-term effects of high-intensity fire on black bear reproduction. The mortality of adult bears from hunting, however, was 2.5 times higher in the fire area than in the unburned area (Cunningham et al., 2001), which would be expected to influence cub survival, potentially confounding results. The overall density of black bears in the fire area was higher than prefire densities in the area (Cunningham et al., 2001).

Radio-collared black bears studied before and after the 2011 Wallow Fire in Arizona temporarily reduced their use of high-severity burn area 1 year after fire but then used these areas in proportion to their availability by the second year after the fire (Crabb et al., 2022). However, some individual bears used the severely burned areas extensively even immediately postfire, possibly to forage on aerially seeded barley. Further, both male and female home range sizes were smaller postfire than prefire. Postfire logging occurred in the Wallow Fire, but affected just 3% of the burn (see Wan et al., 2020) (Box 4.4). At 7 years postfire in the Wallow fire, black bears selected areas with higher fire severity (Lewis et al., 2022).

BOX 4.4

(1) Grizzly bears use areas burned by intense wildfire because of increases in berry production, and black bears continued to use severely burned areas in proportion to availability, although results from studies of the effects of intense fire on black bear demographics are equivocal.

(2) Martens and fisher are mesocarnivores that are dense, mature forest specialists for denning and resting but use severely burned forests that were not logged after fire, most likely for foraging.

(3) Foxes regularly use severely burned forests (regardless of postfire logging for one Mediterranean species), but results from research on coyotes are equivocal.

(4) Carnivores are important dispersers of seeds deep into severely burned forest areas.

4.5 UNGULATES

As major herbivorous components of ecosystems, ungulates can act as keystone species with profound effects on vegetation development and productivity in forests, woodlands, and grassland ecosystems throughout the world (Hobbs, 1996; Wisdom et al., 2006). Hobbs (1996) stated, "ungulates are not merely outputs of ecosystems, they may also serve as important regulators of ecosystem processes at several scales of time and space" (p. 695). Ungulates, Hobbs further noted, are "important agents of environmental change, acting to create spatial heterogeneity, accelerate successional processes, and control the switching of ecosystems between alternative states." Ungulates regulate nitrogen cycling and influence plant size and morphology (Singer et al., 2003). Because grazing and browsing by ungulates affects the biomass, structure, and type of vegetation available to burn, these animals can actually regulate the dynamics of fire (Hobbs, 1996; Wisdom et al., 2006).

Episodic disturbance agents such as fire strongly interact with ungulate herbivory over space and time. For example, removal of fine fuels by ungulate grazers may reduce the frequency of ground fires but can increase crown fires by enhancing the development of ladder trees, especially when combined with a relatively long absence of fire (Hobbs, 1996). Further, postfire plant regeneration provides forage species that are highly palatable to ungulates (Carlson et al., 1993), which attracts ungulates to burned areas, where they influence vegetation regrowth after fire (Canon et al., 1987; Wan et al., 2014). Moose rapidly immigrated to burned areas after a large wildfire in mixed coniferous-deciduous forests of northern Minnesota (Peek, 1974). In fact, fire size can moderate the adverse effects of ungulate herbivory on vegetation recovery. Compared with small fires, large fires "swamp" the effects of ungulate

BOX 4.5
(1) Ungulates interact strongly with episodic disturbances. Many are attracted to severely burned areas because of increased forage palatability and availability, where in turn they influence vegetation regrowth.
(2) Elk, bighorn sheep, and mule deer generally increase after intense fire in shrublands and forests.
(3) The larger the area of high-severity fire, the lower the adverse impact on regrowth of aspen forests from ungulate herbivory.
(4) Caribou may be adversely affected when intense fire reduces lichen used for winter forage.

herbivory, for example, by providing sufficient new grass production to offset browsing, and enabling woody species such as aspen (*Populus* sp.) to grow to tree height (Biggs et al., 2010). In intensively burned ponderosa pine, mixed-conifer, and spruce-fir forests of northern New Mexico, elk selectively foraged on grasses over shrubs (Biggs et al., 2010). In 25 wildfires throughout five national forests in Utah, larger areas of aspen forest that burned with greater severity had the highest growth potential for aspen regeneration, and these high burn-severity conditions stimulated defensive chemicals in plants that lowered the levels of damage done by ungulate browsing (Wan et al., 2014). Wan et al. noted that "this effect may be particularly strong if amplified over large post-fire landscapes by saturating the browse capacity of the ungulate community." (See Box 4.5). At 7 years postfire in the Wallow fire of 2011 in Arizona, mule deer and elk selected areas with higher fire severity (Lewis et al., 2022).

Positive effects of high-severity fire on ungulates likely are most pronounced in vegetation types that are most adapted to high-intensity fires, such as aspen forests and shrublands. Mountain or bighorn sheep selected intensely burned shrublands up to 15 years after fire in Montana (DeCesare and Pletscher, 2006) and in southern California mountains (Bleich et al., 2008). Wildfire increased the carrying capacity of southern California desert bighorn sheep (*Ovis canadensis nelsoni*) in the San Gabriel Mountains, dramatically increasing the number of animals in this endangered population (Holl et al., 2004). A large natural fire on the eastern slopes of the Sierra Nevada mountains in California improved the winter range of Sierra Nevada bighorn sheep (*Ovis canadensis sierrae*) by increasing green forage availability, shifting diet composition to include more forbs, and possibly decreasing predation risk from mountain lions by increasing visibility (Greene et al., 2012). Overall, large, high-severity fire in bighorn sheep shrubland/forest habitats increases forage quality and availability as well as visual openness, which is critical because several populations are listed as endangered.

Studies investigating the impact of fire on mule deer (*Odocoileus hemionus*), a common herbivore in the western United States, indicate that populations tend to increase after severe fire, especially in chaparral communities. In a review of the literature on ungulate responses to fire, Smith (2000) reported mule deer density in intensely burned chaparral was more than twice as high as that in mature chaparral in California, and it increased 400% the first year after high-intensity fire in chamise chaparral. Density then decreased each year afterward until preburn levels were reached 5–12 years later. Chamise chaparral burned by a large wildfire in California had more deer use per square mile than unburned chamise chaparral (Bendell, 1974). In northern coastal California, mule deer densities in chaparral burned by high-intensity wildfire the year before were four times greater than in unburned chaparral (Taber and Dasmann, 1957). Because the fire described in this study was relatively small, deer may have moved from one area to another rather than actually increasing the population via higher birth rates. Similarly, mule deer in central coastal California strongly preferred burned habitat, with a 400% increase in the density of deer in prescribe-burned chaparral near oak woodlands, relative to preburn density, by the second growing season (Klinger et al., 1989). Here the increase in the use of burned chaparral was attributed to movements of deer from adjacent oak woodlands rather than an intrinsic increase in population size. Heavy use of prescribe-burned chamise chaparral by mule deer was reported in the San Jacinto Mountains of southern California (Roberts and Tiller, 1985).

Other studies documented postfire increases in the number of mule deer in conifer forests. Visual observations of 543 mule deer indicated a preference for burned over unburned Douglas-fir/ninebark (*Physocarpus* spp.) and burned ponderosa pine/bluebunch wheatgrass (*Pseudoroegneria spicata*) habitat types during winter and spring in the Selway-Bitterroot Wilderness of Idaho, although the authors did not specifically define the burn severity of sites used by deer (Keay and Peek, 1980). Two other studies that documented increases in mule deer in burned forests hypothesized that postfire logging removes protective cover, a critical habitat element for mule deer. Significantly more deer droppings were located in pinyon-juniper woodlands of Arizona burned by high-intensity fire 13 years earlier than in adjacent unburned areas (McCulloch, 1969). The author surmised that the standing forest of dead trees and fallen trunks provided some cover for deer from predators. Both mule deer and elk used intensely burned lodgepole pine forests at two sites in Wyoming significantly more than paired clearcut sites of the same ages (9 and 5 years old), based on fecal pellet counts (Davis, 1977). Davis (1977, p. 787) stated: "Deer and elk use was greater in burned areas with standing dead timber than in clearcut areas without it. In the Sierra Madre study area, the burned and clearcut plots both had the same number of plant species present, and they both had standing dead timber. However, the burned plot with much more standing dead timber had more deer and elk use. Fire opened up the canopy allowing

FIGURE 4.4 Mule deer respond positively to high-severity fire in forests. In this photo, mule deer forage on fresh vegetation growing in the first postfire year following the Rim fire of 2013 on the Stanislaus National Forest, central Sierra Nevada. *Photo by Chad Hanson taken in 2014.*

light to enter, stimulating growth of forage plants, while the dead trees left standing provided good protective cover" (see Figure 4.4).

Available studies generally report increases in the reproductive rates and body condition of female mule deer in burned habitats. The reproductive rate was 1.32 fawns per doe in the first year after wildfire in northern coastal California, compared with 0.77 fawns per doe in unburned chaparral (Taber and Dasmann, 1957). After 3 years, the reproductive rate of deer at the burned site declined to that of deer in the unburned site. Chamise chaparral burned by a large wildfire produced heavier deer, and does had a higher frequency of ovulation, gave birth to more fawns, and wintered in better condition than does in dense, unburned chamise (Bendell, 1974). Another study, however, documented no difference in fawn-to-doe ratios between burned and unburned chaparral interspersed with oak woodlands in central California (Klinger et al., 1989).

Foraging studies indicate that mule deer populations in chaparral habitats burned by high-intensity fire often increase as a result of the increased availability of browse. *Ceanothus*—a high-quality food for ungulates (Hobbs, 1996)—is abundant after fire because it reproduces from seed that is scarified by burning (Smith, 2000). Thimbleberry (*Rubus parviflorus*) also generally increases after fire (Smith, 2000). Moreover, fire can increase the palatability of foliage for deer as well as the crude protein content (Smith, 2000). The improved quantity and quality of browse may be related to the fire-caused increase in available nutrients in the soil. As such, deer populations often benefit from the increased food production and nutritional value of their food in recently burned areas. Length and surface enlargement factor of papillae (the surface area within the intestine for absorbing nutrients) of necropsied mule deer were greater in those from high-intensity burned than unburned ponderosa pine habitat in the southern Black Hills of South Dakota (Zimmerman et al., 2006). These physiological factors indicate higher forage quality, such as greater concentration of volatile fatty acids. The authors concluded that fire was beneficial at the mucosal level for mule deer: the increase in forage quality from burning caused a rapid change in papillary morphology, allowing the deer to take up more nutrients.

Lichens in boreal habitats are preferred winter forage for caribou (*Rangifer tarandus*), yet large wildfires that depleted lichens had no effect on home-range size, range fidelity, or the survival and fecundity of woodland caribou (*Rangifer tarandus caribou*) in Alberta, Canada (Dalerum et al., 2007). Caribou avoided foraging in burned compared with unburned areas (Dalerum et al., 2007; Joly et al., 2010), and especially avoided severely burned areas during the winter, with the avoidance lasting almost 30 years after fire (Palm et al., 2022). Lichens are significantly reduced by wildfire and take decades to recover to prefire abundance (Joly et al., 2010; Palm et al., 2022) (Box 4.5).

4.6 MANAGEMENT AND CONSERVATION RELEVANCE

The abundance of certain mammal species after fire has direct benefits to land managers in the form of irreplaceable ecosystem and economic services. Bats are voracious predators of insects—many of them consume crop and forest pests—and as such are important regulators of insect populations, including disease-carrying mosquitoes (Reiskind and Wund, 2009). Bats are also critical pollinators of many plants (Molina-Freaner and Eguiarte, 2003). The loss of bats in North America could cost the economy $3.7 billion per year in agricultural losses alone (Boyles et al., 2011). Small mammals aerate the soil and, along with many carnivores, are important dispersers of seeds and fungi (Maser et al., 1978; Rost et al., 2012). Large carnivores are top-down regulators of smaller carnivores and ungulates, and are vital to the health and function of natural ecosystems. Ungulates help to cycle nitrogen and provide big-game hunting opportunities and food for humans. Indeed, in 2001 alone, hunting of ungulates and large carnivores in the United States contributed to approximately $25 billion in retail sales and $17 billion in salaries and wages and employed 575,000 people (IAFWA, 2002). These animals include mule deer, bighorn sheep, moose, elk, and bear, all of which use or thrive within heavily burned habitats.

As described here, a great many mammals benefit from mixed- and high-severity fire and play essential roles in postfire ecosystem dynamics. Land managers rarely weigh these benefits when evaluating the impacts of large fires of mixed- and high-severity, however, thus undervaluing their ecological and economic importance. The vital ecosystem services of mammals in postfire areas should be quantified and carefully considered when planning potentially harmful management activities such as postfire logging and common management activities following postfire logging, such as the application of herbicides and rodenticides.

4.7 CONCLUSIONS

The extraordinary abundance and diversity of mammals using (e.g., American marten, Pacific fisher, grizzly bear) and even thriving (e.g., deer mice, kangaroo rats, bats, mule deer, elk, bighorn sheep, wolves, mountain lions) in severely burned grassland, shrubland, and forested habitats is an important indicator of the high habitat suitability of these areas. Prescribed burning does not provide the expected gains in biological diversity for a range of mammal, reptile, bird, and plant taxa (Pastro et al., 2014). Only large, severe wildfires create significant ecological changes associated with increases in fire-loving species, and, as demonstrated herein, only larger fires can "swamp" the effects of ungulate herbivory on postfire vegetation. Mixed- and high-severity fires globally have unique ecological value that must be weighed against the dominant paradigm that such natural disturbance events are "catastrophic" (Zwolak and Forsman, 2008; also see Chapters 1. 2 and 12). Mammals and other wildlife using intensely burned forests provide myriad ecological services that benefit people and ecosystems alike.

APPENDIX 4.1 THE NUMBER OF STUDIES BY TAXA SHOWING DIRECTIONAL RESPONSE (NEGATIVE, NEUTRAL, OR POSITIVE) TO SEVERE WILDFIRE OVER THREE TIME PERIODS FOLLOWING FIRE. STUDIES CITED INCLUDE UNBURNED AREAS COMPARED TO SEVERELY BURNED AREAS WITH NO POSTFIRE LOGGING, AND EXCLUDED PRESCRIBED BURNS. FOR SMALL MAMMALS, ONLY SPECIES WITH ENOUGH DETECTIONS TO DETERMINE DIRECTIONAL RESPONSE WERE REPORTED

	1–5 year postfire			6–10 year postfire			>10 year postfire		
	Negative	Neutral	Positive	Negative	Neutral	Positive	Negative	Neutral	Positive
Bats[a]		2	4				1		
Small Mammals[b]									
Masked shrew	1								
White-toothed shrew			1						1
Tamias spp.		4							
Pacific kangaroo rat		1	2			1			
Dulzura kangaroo rat			2						
Merriam's kangaroo rat			1			1			1
California pocket mouse	1	1	1						
San Diego pocket mouse	1	2							
Bush rat	1								
Long-haired rat	1								
Pale rat		1							
Red-backed vole	2								
Bank vole	1								
California vole		1	2						
Arizona montane vole		1							
Canyon mouse	1			1			1		
Brush mouse		1	1						

Species								
Deer mouse	3		3		5			1
California mouse	3	1			1			
Cactus mouse	1		1		1			
Pinyon mouse	1			1	1			
Harvest mouse	2		1		1			
Desert woodrat	2		1					
Big-eared woodrat	1		1		1			
Snowshoe hare					1			
Antechinus	1							
Potoroo			1					1
Bandicoot			1		1	1	1	
Wombat			1			1		
Greater glider	1			1				
Macrocarps (3 spp)			1	1				
Carnivores[c]								
American marten							1	1
Stone marten								1
Pacific fisher								1
Gray fox		2						
Red fox								1

Continued

—cont'd

	1–5 year postfire			6–10 year postfire			>10 year postfire		
	Negative	Neutral	Positive	Negative	Neutral	Positive	Negative	Neutral	Positive
Black bear	1	1				1			1
Grizzly bear									1
Gray wolf						1			
Mountain lion						1			
Ungulates[d]									
Caribou	3			3					
Moose			1						
Bighorn sheep			3			2	3		2
Elk			1			1			
Mule deer			4			2			1

[a]Bat citations: Schwab (2006), Malison and Baxter (2010), Buchalski et al. (2013), Law et al. (2018), Jung (2020). Bats are categorized by phonic groups: of six phonic groups in three studies, five phonic groups showed positive response, and one showed neutral response.
[b]Small mammal citations: Keith and Surrendi (1971), Cook (1959), Wirtz (1977), West (1982), Price and Waser (1984), Torre and Díaz (2004), Converse et al. (2006), Zwolak and Forsman (2007), Zwolak and Forsman (2008), Vamstad and Rotenberry (2010), Brehme et al. (2011), Banks et al. (2011), Arthur et al. (2012), Borchert and Borchert (2013), Letnic et al., (2013), Borchert et al. (2014), Ecke et al. (2019), Horncastle et al. (2019), Shaw et al. (2021), Campbell-Jones et al. (2022).
[c]Carnivore citations: Paragi et al. (1996), McLellan and Hovey (2001), Cunningham and Ballard (2004), Cunningham et al. (2006), Soyumert et al. (2010), Borchert (2012), Hanson (2013), Crabb et al. (2022), Lewis et al. (2022).
[d]Ungulate citations: Taber and Dasmann (1957), McCulloch (1969), Bendell (1974), Peek (1974), Davis (1977), Keay and Peek (1980), Smith (2000), Holl et al. (2004), Dalerum et al. (2007), Bleich et al. (2008), Biggs et al. (2010), Joly et al. (2010), Greene et al. (2012), Palm et al. (2022), Lewis et al. (2022).

Studies cited include unburned areas compared with severely burned areas with no postfire logging; they exclude prescribed burns. For small mammals, only species with enough detections to determine directional response are reported.

REFERENCES

Arthur, A.D., Catling, P.C., Reid, A., 2012. Relative influence of habitat structure, species interactions and rainfall on the post-fire population dynamics of ground-dwelling vertebrates. Austral Ecol. 37, 958–970.

Banks, S.C., Dujardin, M., McBurney, L., Blair, D., Barker, M., Lindenmayer, D.B., 2011. Starting points for small mammal population recovery after wildfire: recolonisation or residual populations? Oikos 120, 26–37.

Bendell, J.F., 1974. Effects of fire on birds and mammals. In: Kozlowski, T.T., Ahlgren, C.E. (Eds.), Fire and Ecosystems. Academic Press, New York, USA, pp. 73–138.

Biggs, J.R., VanLeeuwen, D.M., Holechek, J.L., Valdez, R., 2010. Multi-scale analyses of habitat use by elk following wildfire. Northwest Sci. 84, 20–32.

Birtsas, P., Sokos, C., Exadacylos, S., 2012. Carnivores in burned and adjacent unburned areas in a Mediterranean ecosystem. Mammalia 76, 407–415.

Blakey, R.V., Webb, E.B., Kesler, D.C., Siegel, R.B., Corcoran, D., Johnson, M., 2019. Bats in a changing landscape: linking occupancy and traits of a diverse montane bat community to fire regime. Ecol. Evol. 9, 5324–5337.

Bleich, V.C., Johnson, H.E., Holl, S.A., Konde, L., Torres, S.G., Krausman, P.R., 2008. Fire history in a chaparral ecosystem: implications for conservation of a native ungulate. Rangel. Ecol. Manage. 61, 571–579.

Bock, C.E., Jones, Z.F., Kennedy, L.J., Block, J.H., 2011. Response of rodents to wildfire and livestock grazing in an Arizona desert grassland. Am. Midl. Nat. 166, 126–138.

Borchert, M.I., 2012. Mammalian carnivore use of a high-severity burn in conifer forests in the San Bernardino Mountains of southern California, USA. Hystrix Ital. J. Mammal. 23, 50–56.

Borchert, M.I., Borchert, S., 2013. Small mammal use of the burn perimeter following a chaparral wildfire in southern California. Bull. South. Calif. Acad. Sci. 112, 63–73.

Borchert, M.I.D., Farr, P., Rimbenieks-Negrete, M.A., Pawlowski, M.N., 2014. Responses of small mammals to wildfire in a mixed conifer forest in the San Bernardino Mountains, California. Bull. South. Calif. Acad. Sci. 113, 81–95.

Boyles, J.G., Aubrey, D.P., 2006. Managing forests with prescribed fire: implications for a cavity-dwelling bat species. For. Ecol. Manag. 222, 108–115.

Boyles, J.G., Cryan, P.M., McCracken, G.F., Kunz, T.H., 2011. Economic importance of bats in agriculture. Science 332, 41–42.

Brehme, C.S., Clark, D.R., Rochester, C.J., Fisher, R.N., 2011. Wildfires alter rodent community structure across four vegetation types in southern California, USA. Fire Ecol. 7, 81–98.

Brigham, R.M., Vonhof, M.J., Barclay, R.M.R., Gwilliam, J.C., 1997. Roosting behavior and roost-site preferences of forest-dwelling California bats (*Myotis californicus*). J. Mammal. 78, 1231–1239.

Brooks, M.L., 2000. Competition between alien annual grasses and native annual plants in the Mojave Desert. Am. Midl. Nat. 144, 92–108.

Buchalski, M.R., Fontaine, J.B., Heady III, P.A., Hayes, J.P., Frick, W.F., 2013. Bat response to differing fire severity in mixed-conifer forest California, USA. PLoS One 8 (3), e57884. https://doi.org/10.1371/journal.pone.0057884.

Burbidge, A.A., McKenzie, N.L., 1989. Patterns in the modern decline of Western Australia's vertebrate fauna: causes and conservation implications. Biol. Conserv. 50, 143–198.

Buskirk, S.W., Zielinski, W.J., 2003. Small and mid-sized carnivores. In: Zabel, C.J., Anthony, R.G. (Eds.), Mammal Community Dynamics: Management and Conservation in the Coniferous Forests of Western North America. Cambridge University Press, New York, USA, pp. 207–249.

Campbell, J.L., Donato, D.C., 2014. Trait-based approaches to linking vegetation and food webs in early-seral forests of the Pacific Northwest. For. Ecol. Manag. 324, 172–178.

Campbell-Jones, M.M., Bassett, M., Bennett, A., Chia, E.K., Leonard, S., Collins, L., 2022. Fire severity has lasting effects on the distribution of arboreal mammals in a resprouting forest. Austral Ecol. 47, 1456–1469.

Canon, S.K., Urness, P.J., DeByle, N.V., 1987. Habitat selection, foraging behavior, and dietary nutrition of elk in burned aspen forest. J. Range Manag. 40, 433–438.

Carey, A.B., 2000. Effects of new forest management strategies on squirrel populations. Ecol. Apps. 10, 248–257.

Carey, A.B., Horton, S.P., Biswell, B.L., 1992. Northern spotted owls: influence of prey base and landscape character. Ecol. Monogr. 62, 223–250.

Carlson, P.C., Tanner, G.W., Wood, J.M., Humphrey, S.R., 1993. Fire in key deer habitat improves browse, prevents succession, and preserves endemic herbs. J. Wildl. Manage. 57, 914–928.

Chew, R.M., Butterworth, B.B., Grechman, R., 1959. The effects of fire on the small mammal populations of the chaparral. J. Mammal. 40, 253.

Converse, S.J., White, G.C., Block, W.M., 2006. Small mammal responses to thinning and wildfire in ponderosa pine-dominated forests of the southwestern United States. J. Wildl. Manag. 70, 1711–1722.

Cook, S.F., 1959. The effects of fire on a population of small rodents. Ecology 40, 102–108.

Coppeto, S.A., Kelt, D.A., Van Vuren, D.H., Wilson, J.A., Bigelow, S., 2006. Habitat associations of small mammals at two spatial scales in the northern Sierra Nevada. J. Mammal. 87, 402–413.

Crabb, M.L., Clement, M.J., Jones, A.S., Bristow, K.D., Harding, L.E., 2022. Black bear spatial responses to the Wallow Wildfire in Arizona. J. Wild. Manage. 8, e22182.

Crooks, K.R., Soulé, M.E., 1999. Mesopredator release and avifaunal extinctions in a fragmented system. Nature 400, 563–566.

Cunningham, S.C., Ballard, W.B., 2004. Effects of wildfire on black bear demographics in central Arizona. Wildl. Soc. Bull. 32, 928–937.

Cunningham, S.C., Monroe, L.M., Kirkendall, L., Ticer, C.L., 2001. Effects of the Catastrophic Lone Fire on Low, Medium, and High Mobility Wildlife Species. Technical Guidance Bulletin No. 5. Arizona Game and Fish Department, Phoenix, Arizona, USA.

Cunningham, S.C., Kirkendall, L., Ballard, W., 2006. Gray fox and coyote abundance and diet responses after a wildfire in central Arizona. Western N. Am. Nat 66, 169–180.

Dalerum, F., Boutin, S., Dunford, J.S., 2007. Wildfire effects on home range size and fidelity of boreal caribou in Alberta, Canada. Can. J. Zool. 85, 26–32.

Davis, P.R., 1977. Cervid response to forest fire and clearcutting in southeastern Wyoming. J. Wildl. Manag. 41, 785–788.

DeCesare, N.J., Pletscher, D.H., 2006. Movements, connectivity, and resource selection of Rocky Mountain Bighorn sheep. J. Mammal. 87, 531–538.

Diffendorfer, J., Fleming, G.M., Tremor, S., Spencer, W., Beyers, J.L., 2012. The role of fire severity, distance from fire perimeter and vegetation on post-fire recovery of small- mammal communities in chaparral. Int. J. Wildlife Fire 21, 436–448. https://doi.org/10.1071/WF10060.

Doty, A.C., Stawski, C., Law, B.S., Geiser, F., 2016. Post-wildfire physiological ecology of an Australian mirobat. J. Comp. Physiol. B. 186, 937–946.

Ecke, F., Nematollahi Mahani, S.S., Evander, M., Hörnfeldt, B., Khalil, H., 2019. Wildfire-induced short-term changes in a small mammal community increase prevalence of a zoonotic pathogen? Ecol. Evol. 9, 12459–12470.

Fellers, G.M., Pierson, E.D., 2002. Habitat use and foraging behavior of Townsend's big- eared bat (*Corynorhinus townsendii*) in coastal California. J. Mammal. 83, 167–177.

Fontaine, J.B., Kennedy, P.L., 2012. Meta-analysis of avian and small-mammal response to fire severity and fire surrogate treatments in U.S. fire-prone forests. Ecol. Appl. 22, 1547–1561.

Gellman, S.T., Zielinski, W.J., 1996. Use of bats of old-growth redwood hollows on the north coast of California. J. Mammal. 77, 255–265.

Gosse, J.W., Cox, R., Avery, S.W., 2005. Home-range characteristics and habitat use by American martens in eastern Newfoundland. J. Mammal. 86, 1156–1163.

Greene, L., Hebblewhite, M., Stephenson, T.R., 2012. Short-term vegetation response to wildfire in the eastern Sierra Nevada: implications for recovering an endangered ungulate. J. Arid Environ. 87, 118–128.

Gruver, J.C., Keinath, D.A., 2006. Townsend's Big-Eared Bat (*Corynorhinus townsendii*): A Technical Conservation Assessment. USDA Forest Service, Rocky Mountain Region. https://www.fs.usda.gov/Internet/FSE_DOCUMENTS/stelprdb5181908.pdf.

Haim, A., Izhaki, I., 1994. Changes in rodent community during recovery from fire—relevance to conservation. Biodivers. Conserv. 3, 573–585.

Hale, S., Mendoza, L., Yeatman, T., Cooke, R., Doherty, T., Nimmo, D., White, J.G., 2021. Evidence that post-fire recovery of small mammals occurs primarily via in situ survival. Div. Distr. 28, 404–416.

Hanson, C.T., 2013. Habitat use of Pacific fishers in a heterogeneous post-fire and unburned forest landscape on the Kern Plateau, Sierra Nevada, California. Open For. Sci. J. 6, 24–30.

Herrera, C.M., 1989. Frugivory and seed dispersal by carnivorous mammals, and associated fruit characteristics, in undisturbed Mediterranean habitats. Oikos 55, 250–262.

Hobbs, N.T., 1996. Modification of ecosystems by ungulates. J. Wildl. Manag. 60, 695–713.

Holl, S.A., Bleich, V.C., Torres, S.T., 2004. Population dynamics of bighorn sheep in the San Gabriel Mountains, California, 1967-2002. Wildl. Soc. Bull. 32, 412–426.

Howard, J., 1995. *Lepus californicus* (black-tailed jackrabbit). In: Fire Effects Information System. U.S. Department of Agriculture, Forest Service, Rocky Mountain Research Station. https://www.fs.usda.gov/database/feis/animals/mammal/leca/all.html.

Horncastle, V.J., Chambers, C.L., Dickson, B.G., 2019. Grazing and wildfire effects on small mammals inhabiting montane meadows. J. Wildl. Manage. 83, 534–543.

Hutchen, J., Hodges, K.E., 2019. Impact of wildfire size on snowshoe hare relative abundance in southern British Columbia, Canada. Fire Ecol. 15, 37.

International Association of Fish and Wildlife Agencies (IAFWA), 2002. Economic Importance of Hunting in America. IAFWA, Washington, DC.

Johnson, J.B., Edwards, J.W., Ford, W.M., Gates, J.E., 2009. Roost tree selection by northern myotis (Myotis septentrionalis) maternity colonies following prescribed fire in a Central Appalachian Mountains hardwood forest. For. Ecol. Manag. 258, 233–242.

Jolly, C.J., Dickman, C.R., Doherty, T.S., van Eeden, L.M., Geary, W.L., Legge, S.M., Woinarski, J.C.Z., Nimmo, D.G., 2022. Animal mortality during fire. Glob. Change Biol. 28, 2053–2065.

Joly, K., Chapin III, F.S., Klein, D.R., 2010. Winter habitat selection by caribou in relation to lichen abundance, wildfires, grazing, and landscape characteristics in northwest Alaska. Ecoscience 17, 321–333.

Jung, T.S., 2020. Bats in the changing boreal forest: response to a megafire by endangered little brown bats (*Myotis lucifugus*). Ecoscience 27, 59–60.

Keay, J.A., Peek, J.M., 1980. Relationships between fires and winter habitat of deer in Idaho. J. Wildl. Manag. 44, 372–380.

Keith, L.B., Surrendi, D.C., 1971. Effects of fire on a snowshoe hare population. J. Wildl. Manag. 35, 16–26.

Kelly, L.T., Nimmo, D.G., Spence-Bailey, L.M., Taylor, R.S., Watson, S.J., Clarke, M.F., Bennett, A.F., 2012. Managing fire mosaics for small mammal conservation: a landscape perspective. J. Appl. Ecol. 49, 412–421.

Killgore, A., Jackson, E., Whitford, W.G., 2009. Fire in Chihuahuan Desert grassland: short-term effects on vegetation, small mammal populations, and faunal pedoturbation. J. Arid Environ. 73, 1029–1034.

Klinger, R.C.M., Kutilek, J., Shellhammer, H.S., 1989. Population responses of black-tailed deer to prescribed burning. J. Wildl. Manag. 53, 863–871.

Kunkel, K.E., 2003. Ecology, conservation, and restoration f large carnivores in western North America. In: Zabel, C.J., Anthony, R.G. (Eds.), Mammal Community Dynamics: Management and Conservation in the Coniferous Forests of Western North America. Cambridge University Press, New York, USA, pp. 250–295.

Lawler, T.E., 2003. Faunal composition and distribution of mammals in western coniferous forests. In: Zabel, C.J., Anthony, R.G. (Eds.), Mammal Community Dynamics: Management and Conservation in the Coniferous Forests of Western North America. Cambridge University Press, New York, USA, pp. 41–80.

Law, B., Doty, A., Chidel, M., Brassil, T., 2018. Bat activity before and after a severe wildfire in Pilliga forests: resilience influenced by fire extent and landscape mobility? Austral Ecol. 43, 706–718.

Lawrence, G.E., 1966. Ecology of vertebrate animals in relation to chaparral fire in the Sierra Nevada foothills. Ecology 47, 278–291.

Lee, D.E., Tietje, W.D., 2005. Dusky-footed woodrat demography and prescribed fire in a California oak woodland. J. Wildl. Manag. 69, 760–769.

Letnic, M., Tischler, M., Gordon, C., 2013. Desert small mammal responses to wildfire and predation in the aftermath of a La Nina driven resource pulse. Austral Ecol. 38, 841–849.

Lewis, J.S., LeSueur, L., Oakleaf, J., Rubin, E.S., 2022. Mixed-severity wildfire shapes habitat use of large herbivores and carnivores. For. Ecol. Manage. 506, 119933.

Lewis, S.A., Robichaud, P.R., Hudak, A.T., Austin, B., Liebermann, R.J., 2012. Utility of remotely sensed imagery for assessing the impact of salvage logging after forest fires. Remote Sens. 4, 2112–2132.

Lillywhite, H.B., 1977. Effects of chaparral conversion on small vertebrates in southern California. Biol. Conserv. 11, 171–184.

Malison, R.L., Baxter, C.V., 2010. The fire pulse: wildfire stimulates flux of aquatic prey to terrestrial habitats driving increases in riparian consumers. Can. J. Fish. Aquat. Sci. 67, 570–579.

Maser, C., Trappe, J.M., Nussbaum, R.A., 1978. Fungal-small mammal interrelationships with emphasis on Oregon coniferous forests. Ecology 59, 799–809.

Matthews, J.K., Stawski, C., Kortner, G., Parker, C.A., Geiser, F., 2017. Torpor and basking after a severe wildfire: mammalian survival strategies in a scorched landscape. J. Comp. Physiol. B. 187, 383–393.

Mazurek, M.J., 2004. A maternity roost of Townsend's big-eared bats (*Corynorhinus townsendii*) in coast redwood basal hollows in northwestern California. Northwest. Nat. 85, 60–62.

McCulloch, C.Y., 1969. Some effects of wildfire on deer habitat in pinyon-juniper woodland. J. Wildl. Manag. 33, 778–784.

McLellan, B.N., Hovey, F.W., 2001. Habitats selected by grizzly bears in a multiple use landscape. J. Wildl. Manag. 65, 92–99.

Merrick, M.J., Morandini, M., Greer, V.L., Koprowski, J.L., 2021. Endemic population response to increasingly severe fire: a cascade of endangerment for the Mt. Graham red squirrel. Bioscience 71, 161–173.

Mikac, K.M., Knipler, M.L., Gracanin, A., Newbery, M.S., 2023. Ground dwelling mammal response to fire: a case study from Monga National Park after the 2019/2020 Clyde Mountain fire. Austral Ecol. 48, 19–23.

Minshall, G.W., 2003. Responses of stream benthic macroinvertebrates to fire. For. Ecol. Manag. 178, 155–161.

Molina-Freaner, F., Eguiarte, L.E., 2003. The pollination biology of two paniculate agaves (Agavaceae) from northwestern Mexico: contrasting roles of bats as pollinators. Am. J. Bot. 90, 1016–1024.

Nowak, J., Cooper, C.E., Geiser, F., 2016. Cool echidnas survive the fire. Proc. R. Soc. A B 283, 20160382.

Ogen-Odoi, A.A., Dilworth, T.G., 1984. Effects of grassland burning on the savanna hare- predator relationship in Uganda. Afr. J. Ecol. 22, 101–106.
Palm, E.C., Suitor, M.J., Joly, K., Herriges, J.D., Kelly, A.P., Hervieux, D., Russell, K.L.M., Bentzen, T.W., Larter, N.C., Hebblewhite, M., 2022. Increasing fire frequency and severity will increase habitat loss for a boreal forest indicator species. Ecol. Appl. 32, e2549.
Paragi, T.F., Johnson, W.N., Katnik, D.D., Magoun, A.J., 1996. Marten selection of postfire seres in the Alaskan taiga. Can. J. Zool. 74, 2226–2237.
Pastro, L.A., Dickman, C.R., Letnic, M., 2011. Burning for biodiversity or burning biodiversity? Prescribed burn vs. wildfire impacts on plants, lizards, and mammals. Ecol. Appl. 21, 3238–3253.
Pastro, L.A., Dickman, C.R., Letnic, M., 2014. Fire type and hemisphere determine the effects of fire on the alpha and beta diversity of vertebrates: a global meta-analysis. Glob. Ecol. Biogeogr. 23, 1146–1156.
Peek, J.M., 1974. Initial response of moose to a forest fire in northeastern Minnesota. Am. Midl. Nat. 91, 435–438.
Powell, R.A., Zielinski, W.J., 1994. Fisher. In: Ruggiero, L.F., Aubry, K.B., Buskirk, S.W., Zielinski, W.J. (Eds.), The Scientific Basis for Conserving Forest Carnivores: American Marten, Fisher, Lynx, and Wolverine. USDA Forest Service General Technical Report RM-254. Rocky Mountain Forest and Range Experimental Station, Fort Collins, CO.
Price, M.V., Waser, N.M., 1984. On the relative abundance of species: postfire changes in a coastal sage scrub rodent community. Ecology 65, 1161–1169.
Quinn, R.D., 1979. Effects of fire on small mammals in the chaparral. In: Koch, D.L. (Ed.), Cal-Neva Wildlife Transactions. Western Section of the Wildlife Society, Smartsville, California.
Ream, C.H., 1981. The effects of fire and other disturbances on small mammals and their predators: an annotated bibliography. In: USDA Forest Service, Intermountain Forest and Range Experiment Station General Technical Report INT-106.
Reiskind, M.H., Wund, M.A., 2009. Experimental assessment of the impacts of northern long-eared bats on ovipositing Culex (Diptera: Culcidae) mosquitoes. J. Med. Entomol. 46, 1037–1044.
Roberts, T.A., Tiller, R.L., 1985. Mule deer and cattle responses to a prescribed burn. Wildl. Soc. Bull. 13, 248–252.
Rost, J., Pons, P., Bas, J.M., 2012. Seed dispersal by carnivorous mammals into burnt forests: an opportunity for non-indigenous and cultivated plant species. Basic Appl. Ecol. 13, 623–630.
Saint-Germain, M., Drapeau, P., Hébert, C., 2004. Comparison of Coleoptera assemblages from a recently burned and unburned black spruce forests of northeastern North America. Biol. Conserv. 118, 583–592.
Schwab, N.A., 2006. Roost-site Selection and Potential Prey Sources after Wildland Fire for Two Insectivorous Bat Species (*Myotis Evotis and Myotis Lucifugus*) in Mid-Elevation Forests of Western Montana. M.S. thesis. University of Montana.
Schwartz, C.C., Franzmann, A.W., 1991. Interrelationship of black bears to moose and forest succession in the northern coniferous forest. Wildl. Monogr. 113, 1–58.
Schwilk, D.W., Keeley, J.E., 1998. Rodent populations after a large wildfire in California chaparral and coastal sage scrub. Southwest. Nat. 43, 480–483.
Shaffer, K.E., Laudenslayer Jr., W.F., 2006. Fire and animal interactions. In: Sugihara, N.G., van Wagtendonk, J., Shaffer, K.E., Fites-Kaufman, J., Thode, A.E. (Eds.), Fire in California's Ecosystems. University of California Press, Berkeley and Los Angeles, CA, pp. 118–143.
Shaw, R.E., James, A.I., Tuft, K., Legge, S., Cary, G.J., Peakall, R., Banks, S.C., 2021. Unburnt habitat patches are critical for survival and in situ population recovery in a small mammal after fire. J. Appl. Ecol. 58, 1325–1335.
Singer, F.J., Schreier, W., Oppenheim, J., Garton, E.O., 1989. Fire impact on Yellowstone. Bioscience 39, 716–722.

Singer, F.J., Wang, G., Hobbs, N.T., 2003. The role of ungulates and large predators on plant communities and ecosystem processes in western national parks. In: Zabel, C.J., Anthony, R.G. (Eds.), Mammal Community Dynamics: Management and Conservation in the Coniferous Forests of Western North America. Cambridge University Press, New York, USA, pp. 444–486.

Smith, J.K., 2000. Wildland Fire in Ecosystems: Effects of Fire on Fauna. USDA Forest Service, Rocky Mountain Research Station General Technical Report RMRS-GTR-42-vol 1, Ogden, UT.

Soyumert, A., Tavsanoglu, C., Macar, O., Kaynas, B.Y., Gürkan, B., 2010. Presence of large and medium-sized mammals in a burned pine forest in southwestern Turkey. Hystrix Ital. J. Mammal. 21, 97–102.

Spies, T.A., Turner, M.G., 1999. Dynamic forest mosaics. In: Hunter Jr., M.L. (Ed.), Maintaining Biodiversity in Forest Ecosystems. Cambridge University Press, Cambridge, UK, pp. 95–160.

Taber, R.D., Dasmann, R.F., 1957. The dynamics of three natural populations of the deer Odocoileus hemionus columbianus. Ecology 38, 233–246.

Tarbill, G.L., 2010. Nest Site Selection and Influence of Woodpeckers on Recovery in a Burned Forest of the Sierra Nevada. Master's Thesis. California State University, Sacramento.

Terborgh, J., Lopez, L., Nunez, P., Rao, M., Shahabuddin, G., Orihuela, G., Riveros, M., Ascanio, R., Adler, G.H., Lambert, T.D., Balbas, L., 2001. Ecological meltdown in predator-free forest fragments. Science 294, 1923–1925.

Torre, I., Díaz, M., 2004. Small mammal abundance in Mediterranean post-fire habitats: a role for predators? Acta Oecol. 25, 137–143.

Vamstad, M.S., Rotenberry, J.T., 2010. Effects of fire on vegetation and small mammal communities in a Mojave Desert Joshua tree woodland. J. Arid Environ. 74, 1309–1318.

Vanbianchi, C.M., Murphy, M.A., Hodges, K.E., 2017. Canada lynx use of burned areas: conservation implications of changing fire regimes. Ecol. Evol. 7, 2382–2394.

Wan, H.Y., Cushman, S.A., Ganey, J.L., 2020. The effect of scale in quantifying fire impacts on species habitats. Fire Ecol. 16, 9.

Wan, H.Y., Olson, A.C., Muncey, K.D., Clair, S.B.S., 2014. Legacy effects of fire size and severity on forest regeneration, recruitment, and wildlife activity in aspen forests. For. Ecol. Manag. 329, 59–68.

Ward Jr., J.P., Gutiérrez, R.J., Noon, B.R., 1998. Habitat selection by northern spotted owls: the consequences of prey selection and distribution. Condor 100, 79–92.

West, S.D., 1982. Dynamics of colonization and abundance in Central Alaskan populations of the northern red-backed vole, *Clethrionomys rutilus*. J. Mammal. 63, 128–143.

Wirtz, W.O., 1977. Vertebrate post-fire succession. In: Mooney, H.A., Conrad, C.E. (Eds.), Proceedings of the Symposium on Environmental Consequences of Fire and Fuel Management in Mediterranean Ecosystems. USDA Forest Service General. Technical Report WO-3.

Wirtz, W.O., 1995. Responses of rodent populations to wildfire and prescribed fire in southern California chaparral. In: Keeley, J.E., Scott, T. (Eds.), Brushfires in California: Ecology and Resource Management. International Association of Wildlife Fire, Fairfield, Washington, USA, pp. 63–67.

Wisdom, M.J., Vavra, M., Boyd, J.M., Hemstrom, M.A., Ager, A.A., Johnson, B.K., 2006. Understanding ungulate herbivory-episodic disturbance effects on vegetation dynamics: knowledge gaps and management needs. Wildl. Soc. Bull. 34, 283–292.

Zielinski, W.J., Gellman, S.T., 1999. Bat use of remnant old-growth redwood stands. Conserv. Biol. 13, 160–167.

Zimmerman, T.J., Jenks, J.A., Leslie Jr., D.M., 2006. Gastrointestinal morphology of female white-tailed and mule deer: effects of fire, reproduction, and feeding type. J. Mammal. 87, 598–605.

Zwolak, R., Forsman, K.R., 2007. Effects of a stand-replacing fire on small-mammal communities in montane forest. Can. J. Zool. 85, 815–822.

Zwolak, R., Forsman, K.R., 2008. Deer mouse demography in burned and unburned forest: no evidence for source-sink dynamics. Can. J. Zool. 86, 83–91.

Section II

Global Perspectives on Mixed- and High-Severity Fires

Chapter 5

Bark Beetles and High-Severity Fires in Rocky Mountain Subalpine Forests

Dominik Kulakowski[1] and Thomas T. Veblen[2]
[1]*Graduate School of Geography, Clark University, Worcester, MA, United States;* [2]*Department of Geography, University of Colorado-Boulder, Boulder, CO, United States*

5.1 FIRE, BEETLES, AND THEIR INTERACTIONS

For millennia the forests of the Rocky Mountains in the western United States have been shaped by wildfires (Romme and Despain, 1989; Sibold et al., 2006), outbreaks of insects (Eisenhart and Veblen, 2000; Jarvis and Kulakowski, 2015), and the potential interactions between these disturbances (Veblen et al., 1994) (Figure 5.1). The subalpine forests in this region are dominated by lodgepole pine (*Pinus contorta*) at lower elevations and Engelmann spruce (*Picea*

FIGURE 5.1 The forests of the Rocky Mountains in the western United States are shaped by wildfires (background), outbreaks of insects (foreground), and the potential interactions between these disturbances. *Photo: D. Kulakowski.*

engelmannii) and subalpine fir (*Abies lasiocarpa*) at higher elevations. Specifically, these ecosystems include large, high-severity fires; outbreaks of mountain pine beetle (*Dendroctonus ponderosae*) in lodgepole pine forests and spruce beetle (*Dendroctonus rufipennis*) in spruce-fir forests; and their interactions. The potential role of modern fire suppression in altering stand conditions of these high-elevation forests is debatable, but the fact that these forests were historically shaped by fires that killed most canopy trees over extensive areas (i.e., patches > 1000 ha) at long intervals (>100 years) is not disputed (Romme and Despain, 1989; Sibold et al., 2006). The focus of this chapter is subalpine forests characterized by lodgepole pine, mixed spruce, and subalpine fir, but we also touch on upper montane forests where Douglas-fir (*Pseudotsuga menziesii*) either form relatively pure stands or can be found mixed with the tree species typical of the subalpine zone. In these upper montane forests high-severity fires are characteristic, as are outbreaks of Douglas-fir bark beetle (*Dendroctonus pseudotsugae*) (Harvey et al., 2013; Sherriff et al., 2014). This chapter deals specifically with interactions between fire and *Dendroctonus* bark beetles, but we also briefly mention other biotic and abiotic causes of massive tree mortality to place the main theme of the chapter into an appropriate context.

Disturbances such as wildfires and insect outbreaks interact with underlying environmental variability to determine the spatial and temporal heterogeneity of forest landscapes (White and Pickett, 1985) (Figure 5.2). Even large and severe disturbances do not homogenize the landscape but rather promote spatial heterogeneity as a result of variability of disturbance severity and surviving residuals (Turner, 2010). The resulting patterns affect subsequent ecological processes including postdisturbance regeneration and susceptibility to subsequent disturbances (Turner, 2010), as well as biodiversity (Box 5.1).

FIGURE 5.2 Wildfires and insect outbreaks interact with underlying environmental variability to determine the spatial and temporal heterogeneity of forest landscapes. *Photo: D. Kulakowski.*

BOX 5.1 Bark Beetle Outbreaks and Biodiversity

Endemic and epidemic bark beetle outbreaks are important sources of structural heterogeneity and biodiversity in the conifer forests of western North America. Bark beetles are parts of many forest food webs and can be associated with a large number of organisms (Dahlsten, 1982). They can be hosts for parasites and food for a variety of animals, including spiders, birds, and other beetles (Koplin and Baldwin, 1970). The actual effect of any particular bark beetle outbreak on subsequent biodiversity depends on the initial forest conditions, the intensity of the outbreak, and the types of organisms considered.

Bark beetles can have far-reaching effects on ecological structures and biodiversity, which, when considered across scales from individual trees to entire landscapes, reveal their important roles as ecosystem engineers. At the scale of individual beetle galleries, they establish and maintain a microflora of fungi and bacteria that create a complex web of biosynthetic interactions affecting tree resistance and success of beetle attack. By reducing tree resistance, beetle attack creates opportunities for a wide diversity of saprogenic competitors (Raffa et al., 2008). Bark beetles themselves are an important food source for a diverse group of arthropods and vertebrates, including woodpeckers that are highly adapted to digging out larvae of wood-boring insects. In general, a bark beetle outbreak initializes a release of resources that, in the short term, promotes the growth of insectivorous bird populations (Saab et al., 2014; Kelly et al., 2019). Overall, approximately twice as many bird species have increased in forests with bark beetle outbreaks (Saab et al., 2014). The longer-term impact of large outbreaks on avian diversity has not been widely studied but is likely to depend on the amount of tree mortality and the rate of recovery of unattacked host conifers as well as nonhost trees.

For individual forest stands, tree mortality caused by bark beetles increases structural heterogeneity through the creation of canopy gaps and enhanced growth of understory plants, which is likely to create a favorable habitat for many invertebrates and vertebrates. Outbreaks create snags that may be used by various birds and mammals, including woodpeckers, owls, hawks, wrens, warblers, bats, squirrels, American martens (*Martes americana*), and lynx (*Lynx canadensis*). Populations of cavity-nesting birds often increase following bark beetle outbreaks (Saab et al., 2014; Hanson and Chi, 2020).

Across landscapes, beetle outbreaks are likely to alter biodiversity through the creation of more diverse patch configurations and edge effects favoring some wildlife species. Wildlife associated with early seral habitats, such as deer and elk, are expected to be favorably influenced by an outbreak once there has been enough time for understory resources to respond to the creation of canopy openings (Saab et al., 2014). The consequences of a beetle outbreak for biodiversity at the scale of large stands and landscapes depend both on the intensity of the outbreak and on the preoutbreak forest landscape structure. For example, large areas of monotypic lodgepole pine stands that originated after late 19th- or early 20th-century fires are typically low in structural diversity compared with landscapes with patches of beetle-killed trees. By contrast, a spruce beetle outbreak affecting old stands of Engelmann spruce and subalpine fir, which are already

Continued

> **BOX 5.1 Bark Beetle Outbreaks and Biodiversity—cont'd**
>
> characterized by abundant standing dead and fallen large trees, would result in a less extreme alteration of stand conditions that may have a smaller impact on biodiversity.
>
> Another particularly important effect of beetle outbreaks on ecosystem structure and biodiversity is evident in riparian habitats of mountain streams (Jackson and Wohl, 2015). Beetle-killed trees contribute to the recruitment of large, coarse, woody debris into riparian areas and stream systems, which exerts important beneficial influences on the storage of sediment and organic matter and on river and floodplain habitat for numerous animal species, including trout. Compared with timber harvesting, which can remove all riparian wood and severely deplete subsequent in-stream wood recruitment, beetle outbreaks provide a source of in-stream wood loads for decades following a beetle outbreak.

However, the importance of predisturbance conditions on susceptibility to disturbances decreases as the intensity of disturbance increases. Therefore, how disturbances and their interactions determine landscape heterogeneity is likely to change as climatically driven disturbances become more intense under climate change.

Two-way interactions between fires and bark beetle outbreaks occur, whereby outbreaks can affect subsequent fires and fires can affect subsequent outbreaks. Fire regimes are fundamentally a function of fuels, weather, and topography. Because outbreaks can result in visible and abundant dead trees, expecting this change in fuels to affect fire regimes is intuitive. Conversely, because the time since the last severe fire strongly influences stand structure and tree size, which affect susceptibility to bark beetles, expecting fire regimes to affect susceptibility to outbreaks is also intuitive. These apparently simple relationships are made complex and nuanced by various sources of spatial heterogeneity that affect probabilities of fire and outbreaks, as well as rates of forest development following both of these disturbances. In addition to the major changes in stand structure that determine interactions between outbreaks and fires, microclimatic changes following each of these disturbances may also affect the interactions of these disturbances (Carter et al., 2022). The topic of disturbance interactions includes questions of (1) linked disturbances (how the occurrence or severity of one disturbance affects the occurrence or severity of a subsequent disturbance) and (2) compounded disturbances (how two or more disturbances that occur in relatively short succession affect the overall disturbance intensity and postdisturbance development). In this chapter, we focus primarily on linked beetle-fire disturbances but briefly address compounded effects because they are important in how disturbance interactions create spatial heterogeneity. We mainly address the ecological interactions between wildfire and bark beetle outbreaks, but it is also worth noting that these interactions are also central to questions of forest management in the Rocky Mountains and elsewhere. Management related to forest disturbances often aims to promote either resistance (i.e., reduced probability or magnitude of disturbance) or resilience (i.e., capacity of a forest to develop after a disturbance without tipping into an alternate state) (DeRose and Long, 2014). Understanding the ecological relationships between outbreaks and fires is fundamental to developing management strategies for both increased resistance and resilience.

The potential influences of bark beetle outbreaks on subsequent fire hazard have been the subject of a lively public, political, and scientific debate; an emerging body of scientific research; and several substantive literature reviews (Parker et al., 2006; Romme et al., 1986; Jenkins et al., 2008, 2012, 2014; Kaufmann et al., 2008; Simard et al., 2008; Hicke et al., 2012a; Black et al., 2013; Six et al., 2014; Meigs et al., 2015; Carter et al., 2022). A lack of consensus on potential relationships between bark beetle outbreaks and actual or potential wildfire activity in the published literature has led to confusion among scientists, resource managers, and the general public. Some of this confusion and debate may be attributable to the specific research questions posed, the parameters selected to measure potential effects on fire regimes, and the type and initial conditions of ecosystems considered. In this chapter, we examine the current state of knowledge about interactions between bark beetle outbreaks and fires in lodgepole pine and spruce-fir forests of the Rocky Mountains, with particular focus on high-severity fires and the role of disturbance interactions in creating landscape heterogeneity.

5.2 HOW DO OUTBREAKS AFFECT SUBSEQUENT HIGH-SEVERITY FIRES?

Methodological Considerations

Studies that address the question of how bark beetle outbreaks potentially affect subsequent wildfires typically focus on fire hazard (the fuel complex, including the type, volume, and arrangement of fuels that determine the ease of ignition and resistance to control regardless of the fuel type's weather-influenced moisture content); fire risk (the chance that a fire might start based on all causative agents, including fuel hazard, ignition source, and weather); or fire behavior (including flame length, rate of spread, or other measures of the fire) (Hardy, 2005). Most studies have used research designs that can be broadly classified as (1) field experiments; (2) fire behavior modeling; and (3) retrospective case studies of actual fire events in beetle-affected forests. To date, field experiments in which fire behavior is intensely monitored in stands with actual or simulated bark beetle outbreaks have been rare and limited to a few locations in western Canada

(Schroeder and Mooney, 2012). While these studies have yielded useful comparative observations of fire behavior in stands with differing fuel properties, the broader implications of such experimental results are severely limited by the weather conditions under which the burns are implemented and the narrow range of initial conditions (e.g., severity of beetle kill, time since beetle kill, and stand conditions before beetle kill) included in experimental burns. Although experimental burns comparing simulated beetle kill with control stands provide insights for improving fire behavior models, they have not contributed significantly to the more general question of whether outbreaks significantly affect fire regime parameters at broad spatial scales and over a range of time periods.

Given the paucity of experimental studies, this chapter necessarily focuses on studies using the other two methodological approaches: wildfire behavior modeling and retrospective case studies. Wildfire behavior modeling is applied across a range of spatial and temporal scales primarily for planning the management of a wildfire incident. In predicting wildland fire behavior, managers are particularly in need of a basis for predicting ignitions, rate of fire spread, energy released and associated flame-front dimensions, perimeter and area of fire growth, and closely related phenomena such as torching, crowning, and spotting (Alexander and Cruz, 2013). Wildland fire behavior models are commonly classified as empirical versus physics-based models. The former are applied in an operational decision-making context, whereas the latter are developed primarily for enhancing theoretical understanding of fire propagation.

Several empirical modeling systems have coupled the surface fire spread model by Rothermel (1972) with criteria and models of crown fire initiation and rate of crown fire propagation. Operational fire behavior models have been widely used to assess the effectiveness of fuel treatments. As pointed out by Cruz and Alexander (2010), however, simulations of the onset of crowning and rate of spread of active crown fire in conifer forests in the US West using these modeling systems exhibit significant underprediction bias. For example, standard fire behavior models such as NEXUS (Scott and Reinhardt, 2001) underpredict crown fire behavior because unrealistically high wind speeds are required for the onset and spread of crown fires in comparison with observed fire behavior. Despite recognition of the shortcomings of operational fire models, they continue to be applied to assess the effectiveness of thinning treatments and to gauge the effects of forest insect infestations on crown fire potential. When applying operational fire behavior models to the latter objective, a near void of empirical evidence relating fuel characteristics and foliage flammability representative of different stages of bark beetle attack conditions has been widely recognized (Jolly et al., 2012). Thus, nearly all studies have been forced to use unverified models of foliage (live and dead) to address the effects of bark beetle outbreaks on potential fire behavior. Because stand-scale models assign a single average set of conditions to the entire stand, they cannot fully consider the significant amount of fine-scale heterogeneity in fuel moisture resulting from neighboring living and beetle-killed trees (Jolly et al., 2012). Although physics-based fire models can accommodate fine-scale variability in tree mortality, they are computationally demanding and have been applied to the assessment of the effects of bark beetles on potential fire behavior in only a few instances (Hoffman et al., 2012).

A series of studies using operational fire behavior models has led to a generalized expectation of the nature of changes in stand flammability following a bark beetle outbreak (Figure 5.3a). Operational fire behavior models have been used in the major research focus on how outbreaks may affect subsequent high-severity fires by altering fuel quantity and quality. Empirical research at scales from individual needles to stands has consistently shown major changes in the quality and arrangement of fuels following outbreaks. Each of these changes their magnitude, and their timing can potentially affect various aspects of fire regimes—sometimes in a contradictory manner. As a result, the effects of outbreaks on subsequent fires are contingent on a

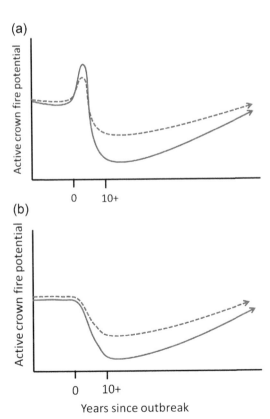

FIGURE 5.3 Active crown fire potential in lodgepole pine (solid lines) and spruce-fir stands (dashed lines) expected from modeling (a) and empirical (b) studies. Because of the structural differences between typical lodgepole pine and spruce-fir stands in the Rocky Mountains, the magnitude of structural changes caused by spruce beetle outbreaks in spruce-fir forests is likely to be less than that associated with mountain pine beetle outbreaks in lodgepole pine forests.

number of factors, including the intensity of the outbreak and the time since the outbreak.

Although changes in fuels following outbreaks are feasible to quantify, correctly understanding and accurately modeling their implications for actual fire regimes has proved to be much more elusive, in part because of important limitations in available fire models that compromise their effectiveness in characterizing fire behavior in beetle-affected forests (Jenkins et al., 2012; Page et al., 2014a, 2014b). Consequently, much of what is known about the cumulative effects of outbreaks on fires is based on retrospective studies that have examined the occurrence of actual fires as they are related to preceding outbreaks. One major advantage of such studies is that they are free of the conceptual limitations of existing fire models and instead depict the actual relationships between outbreaks and fires. These studies also come with important limitations, however, including a limited number and variety of initial forest conditions, variable and sometimes unknown fire weather, and covariation among key variables—all of which can make disentangling complex ecological relationships and identifying underlying mechanisms difficult. As a result, how outbreaks affect fire behavior (e.g., fire line intensity, rate of spread) is poorly understood, whereas the effect of outbreaks on some fire regime parameters such as fire extent, frequency, probability of occurrence, and severity at landscape or broader spatial scales can be addressed more feasibly by retrospective studies.

Retrospective studies have generally relied on either field or remote sensing methods to examine how outbreaks affect fire attributes (Table 5.1). While field methods yield much more accurate data on disturbance severity, the required labor intensity results in the actual sampling of only relatively small areas, even though sample points can be distributed over much larger landscapes. By contrast, remote sensing provides contiguous data on disturbance extent and severity, but it is less accurate because of the relatively low spatial resolution of data and difficulties in remotely detecting signals of disturbances. Both methodological approaches are also vulnerable to unknown preoutbreak variability of stand structure, differences in burning conditions resulting from changing local fire weather during the event, as well as other variability that is not incorporated into a research design but that can affect fire behavior.

We stress that fire behavior modeling and retrospective approaches used to gauge effects of bark beetle outbreaks on subsequent wildfire activity differ fundamentally in the types of questions they are appropriate for addressing. Clearly, understanding fire behavior is important because of its ecological effects and its relevance for resistance of fire to control (Page et al., 2013b). That notwithstanding, in the context of understanding high-severity fire regimes, the central question is whether, and how, outbreaks may fundamentally or cumulatively alter those fire regimes.

TABLE 5.1 Characteristics of Retrospective Studies that Have Examined How Outbreaks Affect Various Fire Parameters in the US Rocky Mountain Subalpine Forests

Authors	Location	Forest	Scale[a] (km²)	Main methods	Fire parameters
Turner et al. (1999)	Yellowstone	Lodgepole	1 s	Field	Extent and severity
Kulakowski and Jarvis (2011)	Colorado	Lodgepole	1 s	Field	Occurrence
Harvey et al. (2013)	Yellowstone	Douglas fir	1 s	Field	Severity
Harvey et al. (2014a)	Northern Rockies	Lodgepole	1 s	Field	Severity
Harvey et al. (2014b)	Yellowstone	Lodgepole	1 s	Field	Severity
Kulakowski et al. (2003)	Colorado	Spruce	10 s	Field	Extent
Bigler et al. (2005)	Colorado	Spruce	10 s	Remote sensing	Extent and severity
Lynch et al. (2006)	Yellowstone	Lodgepole	100 s	Remote sensing	Occurrence
Kulakowski and Veblen (2007)	Colorado	Spruce and lodgepole	100 s	Remote sensing	Extent and severity
Renkin and Despain (1992)	Yellowstone	Lodgepole	1000 s	Documentary	Occurrence and severity
Bebi et al. (2003)	Colorado	Spruce	1000 s	Remote sensing	Occurrence
Andrus et al. (2016)	Colorado	Spruce	1000 s	Field, remote sensing	Severity
Mietkiewicz and Kulakowski (2016)	Western US	Lodgepole	1000 s	Remote sensing	Occurrence

[a]For field-based studies the scale indicates the total area sampled, even if samples were distributed across a much larger area.

Lodgepole Pine Forests

Outbreaks of mountain pine beetle (MPB) clearly alter the fuel structures of lodgepole pine forests (Page and Jenkins, 2007; Klutsch et al., 2009; Simard et al., 2011). Immediately following an MPB attack, important changes in foliar moisture, starch, sugar levels, fiber, and crude fat affect the flammability of lodge-pole pine needles and can shorten ignition times for red needles as opposed to live green needles on unattacked trees (Jolly et al., 2012; Page et al., 2012). But how these changes scale up to affect high-severity fires at stand and landscape scales is complex and continues to be an active topic of research. Modeling studies have suggested that the potential of active crown fire may increase immediately after an outbreak, depending on the fire intensity generated by surface fuels (Hoffman et al., 2013). However, important differences in how foliar moisture content (FMC) on recently killed trees versus other fine fuels respond to changing environmental conditions can hamper accurate modeling of fire behavior (Page et al., 2013a, 2013b, 2014a, 2014b).

Limitations of fire behavior models mean that corresponding results are tentative and suggestive, particularly given the paucity of field experimentation to validate fire behavior models (Cruz and Alexander, 2010). Nevertheless, fire modeling studies of beetle-affected stands are useful for gaining insight into and developing hypotheses about possible consequences of outbreaks on fires. Schoennagel et al. (2012) predicted some effects of fuel moisture on fire behavior under certain weather scenarios. For example, active crown fire was modeled to be more probable at lower wind speeds and less extremely dry conditions assuming lower canopy fuel moisture in red and gray stages of MPB compared with the green stage (Schoennagel et al., 2012). Later in the outbreak more open canopies and high loads of large surface fuels resulting from tree fall were suggested to increase surface fireline intensities, possibly facilitating active crown fire at lower wind speeds (Schoennagel et al., 2012). However, Schoennagel et al. (2012) also suggested that if transition to crown fire occurs (outside the stand or within the stand via ladder fuels or wind gusts), active crown fire can be sustained at similar wind speeds, implying observed fire behavior may not be qualitatively different among MPB stages under extreme burning conditions. In sum, the probability of crown fire is likely to be similar across MPB stages and is characteristic of lodgepole pine forests where extremely dry, gusty weather conditions are key factors in determining fire behavior.

While much has been learned about how outbreaks affect fuels, how quickly canopy bulk density (CBD) decreases following tree mortality and how this varies with biophysical setting is perhaps the most important issue that remains poorly understood. CBD is important to fire regimes and fire behavior because it is an indicator of the amount and continuity of canopy fuels that are available to burn and carry a fire. Some conceptual frameworks suggest that CBD remains unchanged initially following an outbreak (Jenkins et al., 2012; Hicke et al., 2012a), but the limited empirical data that directly address this issue indicate that CBD decreases shortly after tree mortality (Simard et al., 2011). The issue of how quickly CBD decreases following outbreaks necessarily hinges on the synchrony of beetle attack and tree death within a stand, as well as site conditions. In other words, the question can be expressed as whether, and to what degree, the so-called red phase of outbreaks (characterized by beetle-killed trees with red needles) is homogeneous in timing. After a substantial reduction of stand-scale FMC, stand-level CBD would be most likely to remain unchanged following outbreaks if 100% of trees in a stand were killed during the initial year of the outbreak and if site moisture, temperature, and wind conditions promoted retention of dead needles. Retention of needles would be expected to be especially short at relatively dry, warm, and windy sites. If an outbreak lasts several years within a given stand (as it normally does; Schmid and Amman 1992), then it becomes increasingly likely that some trees would lose their needles before other trees are killed, effectively reducing CBD before major reductions in stand-scale FMC. Similarly, lower-severity outbreaks would affect fuels less than higher-severity outbreaks.

Empirical field observations support the notion that FMC and CBD decrease approximately simultaneously. After an MPB outbreak in Yellowstone National Park, reduced canopy moisture content was coupled with reduced CBD (Simard et al., 2011). Based on these data, simulation models of fire behavior predicted that under intermediate wind conditions (40–60 km/h), the probability of active crown fire in stands recently affected by beetles would be lower than in stands not affected by beetles (Simard et al., 2011). In addition, if winds were below 40 or above 60 km/h, stand structure would have little effect on fire behavior. Although the canopy is drier immediately after an outbreak, this does not translate to an increase in fire risk, likely because of the overriding effect of reduced CBD. Decreased CBD is predicted to continue being important in latter stages of outbreaks (5–60 years after) and lead to a consequential lower risk of active crown fire (Jenkins et al., 2008).

Given important uncertainties with modeling predictions of fire behavior after outbreaks, studies that have empirically examined actual high-severity fires have generally found that wildfire activity does not substantially increase, even immediately following MPB outbreaks (Figure 5.3b). For example, ongoing outbreaks of MPB (and spruce beetle) had no detectable effect on the extent or the severity of fire in a large complex of two major fires that burned over 10,000 ha in 2002 in northwestern Colorado, possibly because changes in fuels resulting from the outbreaks may have been overridden by climatic conditions (Kulakowski and Veblen, 2007). Similarly, an experimental burn in jack pine (*Pinus banksiana*) forests in Alberta indicated little difference in rates of canopy fire spread and weather threshold for crown involvement between stands with simulated effects of MPB versus control stands (Schroeder and Mooney, 2012). The effects of MPB on fire severity in the Northern Rockies have been shown to be contingent on burning conditions (Harvey et al., 2014a). Outbreaks in the red stage did not affect any measure of fire severity, except that under extreme burning conditions MPB outbreaks were associated with increased charring on trees that were presumably dead before the fire (Harvey et al., 2014a).

Studies of interactions among disturbances can be hampered by colinearity among stand attributes and a lack of constancy in the many variables expected to affect fire behavior in beetle-affected versus beetle-unaffected stands. For example, compared with stands with lower MPB-caused mortality, more lodge-pole stands in which >50% of susceptible trees were killed in the preceding 5−15 years burned at high severity in 1988 in Yellowstone (Turner et al., 1999). Relatively old stands, however, which because of differences in fuel structures were inherently more likely to burn at a high severity than younger stands, were also more affected by MPB (Renkin and Despain, 1992). Therefore, whether the differences in the spatial patterns of the severe fires in Yellowstone were primarily the consequences of the outbreak or of prefire stand structure that was unrelated to the outbreak is not clear (Simard et al., 2012).

Other studies have reported that MPB outbreaks have negligible or mixed results on subsequent fire regimes. In Yellowstone National Park, stands affected by MPB outbreaks burned at a lower severity in the Robinson Fire in 1994 compared with adjacent stands that were not affected by MPBs before that fire (Omi, 1997). But stands affected by outbreak 13−16 years before the 1988 Yellowstone fires were slightly (about 11%) more likely to burn compared with stands unaffected by beetles (Lynch et al., 2006). By contrast, stands that were affected by outbreak 5−8 years before the 1988 fires were no more likely to burn compared with unaffected stands (Lynch et al., 2006). Over longer time periods in Colorado, the occurrence of fires in lodgepole pine forests has been shown to be unrelated to preceding MPB outbreaks but strongly associated with drought (Kulakowski and Jarvis, 2011). Likewise, outbreaks of MPB in lodgepole pine forests across the western US have been less important than climatic variability for the occurrence of large fires over the past decades (Mietkiewicz and Kulakowski, 2016). Across this broad region, occurrence of large fires was determined primarily by current and antecedent high temperatures and low precipitation but was unaffected by preceding outbreaks, implying that trends of increasing co-occurrence of wildfires and outbreaks are due to a common climatic driver rather than interactions between these disturbances.

Field-based research has examined prefire stand conditions attributable to beetle kill and numerous field-based measures of fire severity in lodgepole pine and mixed conifer forests (Harvey et al., 2013, 2014a, 2014b). These studies have shown that field-based measurements of fire severity (i.e., fire-killed basal area, number of trees killed by fire, char height, and percentage of bole scorched) were primarily driven by burning conditions and topographic position. In lodgepole pine stands in the gray stage of MPB outbreak, fire severity was not related to outbreak severity under moderate burning conditions, except increased charring on trees that were presumably dead before the fire (Harvey et al., 2014a). But under extreme burning conditions, several measures of fire severity, including fire-caused mortality, increased with outbreak severity, possibly because of increased fireline intensity (Harvey et al., 2014a). In this study, however, outbreak severity in stands in the gray stage was relatively low (0%—56% beetle-killed basal area). In fact, it was lower than the severity in stands in the red stage of outbreak in this study and also lower than other studies that found fire severity to be unrelated to outbreak in the gray stage (Simard et al., 2011; Schoennagel et al., 2012; Harvey et al., 2014b). Further research is needed in stands in the gray stage following low-to moderate-severity outbreaks. Given the characteristic high-severity fires in lodgepole pine, however, fire behavior may not be qualitatively different among stages of MPB outbreak under extreme burning conditions (Schoennagel et al., 2012).

In addition to changing fuels, outbreaks also have been suggested to affect fire regimes by increasing lightning strikes because the number of standing dead trees increases in the years to decades following outbreaks. Although some evidence 1 and 6 years following an MPB outbreak in British Columbia has been presented to support this hypothesis, the influence of MPBs on ignitions has been reported to be less important than that of temperature and precipitation (Bourbonnais et al., 2014). That outbreaks could affect fire ignitions by indirectly affecting climate is also theoretically possible. Following an MPB outbreak in British Columbia, outgoing sensible and radiative heat fluxes increased enough to potentially modify local atmospheric processes, cloud cover, and precipitation (Maness et al., 2012). To date, however, no published study has found fire regimes actually to be influenced by these mechanisms. Although important questions remain, available studies indicate that outbreaks of MPB in lodgepole pine forests have not resulted in observable increases in fire risk, extent, or severity.

Spruce-Fir Forests

The spruce-fir forests of the Rocky Mountains are similar to those dominated by lodgepole pine in that they dominate the subalpine landscape, are relatively dense, and are characterized by infrequent high-severity fires and outbreaks of bark beetles. There are also a number of important differences between these two forest types that influence interactions between outbreaks and wildfires. Key differences are that spruce-fir forests are more mesic and include a large component of tree species not susceptible to the most important bark beetle. In addition, spruce-fir forests are represented by a much higher proportion of older (e.g., >200 years old) stands relative to lodgepole pine forests that are represented overwhelmingly by younger (e.g., 100−200 years old) stands in the Rocky Mountain region (Veblen, 1986; Veblen and Donnegan 2006). This difference in stand age distributions reflects the longer mean fire return intervals typical of the higher-elevation spruce-fir forests (Sibold et al., 2006). The result is that, compared with younger postfire stands of lodgepole pine, old stands of spruce-fir have a high component of standing dead large trees as a result of cumulative mortality from many causes over long time periods, even in the absence of a spruce beetle outbreak (Veblen, 1986; Veblen et al., 1991). Old spruce-fir stands also typically have much more biomass in the understory compared with younger lodgepole pine stands. The net effect of these structural differences is that the magnitude of structural changes caused by spruce beetle outbreak in old spruce-fir forests is likely to be less than that associated with MPB in younger postfire lodgepole pine forests (Figure 5.3a and b).

Nevertheless, similar to MPB outbreaks in lodgepole pine forests, important changes in foliar moisture and chemistry occur in Engelmann spruce following attack by spruce beetles. Immediately after a spruce beetle attack, FMC can be lower, proportions of lignin and cellulose can be higher, and proportions of carbohydrate-based compounds can be lower compared with that in green needles on unattacked trees, each of which can result in increased flammability of the foliage (Page et al., 2014a, 2014b), which may lead to expectations of increased risk of active crown fire immediately following outbreaks (Figure 5.3a). However, any increase in crown flammability is short-lived because foliage on killed trees drops soon after a mass attack. Furthermore, as with MPB outbreaks, any increase in fire hazard would be contingent on the intensity and within-stand synchrony of an outbreak. For decades after high-severity spruce beetle outbreaks, models predict reduced probability of active crown fire because of persistent decreased CBD (Jenkins et al., 2008; DeRose and Long, 2009). In addition to altering foliar properties, outbreaks of spruce beetle also can affect the microclimate, often resulting in cooler and more mesic stand conditions (Reid, 1989; Carlson et al., 2021).

How changes in microclimatic conditions as well as foliar moisture and chemistry relate to the actual risk of active crown fire and fire behavior has not yet been definitively established, in part because of the limitations of existing fire models and uncertainty about retention of dead needles on killed trees. Given these limitations and knowledge gaps, the results of retrospective studies of actual fires following outbreaks are important to understanding the consequences of beetle activity for fire risk and severity. After a 1940s spruce beetle outbreak that killed most canopy spruce over thousands of hectares in western Colorado, there was no increase in the numbers of fires over the period from 1950 to 1990 compared with unaffected subalpine forests (Bebi et al., 2003). Likewise, stands affected by beetles were unaffected by a low-severity fire that spread through adjacent forests several years after the outbreak subsided (Kulakowski et al., 2003), possibly because of increased moisture on the forest floor following the outbreak, which may have contributed to a proliferation of mesic understory plants (Reid, 1989). In fact, these beetle-affected stands did not burn until the extreme drought of 2002, during which large severe fires affected extensive areas of Colorado, including some spruce-fir stands that had been affected by the 1940s outbreak. The high-severity fires during that extreme drought were substantially affected by neither the 1940s outbreak (Bigler et al., 2005) nor by ongoing spruce beetle outbreaks (Kulakowski and Veblen, 2007). Similarly, in spruce-fir forests in southwestern Colorado where high-severity fires in 2012−13 occurred in tens of thousands of hectares of gray stage (<5 years since severe outbreak), both field-based and remote-sensing metrics of fire severity were primarily driven by topography, preoutbreak basal area, and fire weather conditions rather than beetle infestation (Andrus et al., 2016). In sum, outbreaks of spruce beetle seem to have little or no effect on the occurrence or severity of fires in spruce-fir forests, primarily because high-severity fires in these forests depend on infrequent, severe droughts (e.g., Schoennagel et al., 2007). Under such extreme weather conditions, changes in fuels resulting from bark beetle outbreaks have only a minor, if any, effect on fire risk.

Why the Apparent Conflict Between Modeling and Observational Results?

As noted previously, fire behavior modeling and retrospective approaches are appropriate for addressing different sets of questions about how bark beetle outbreaks may affect subsequent wildfire activity. Fire behavior models are particularly useful in revealing insights about how beetle-killed fuels might result in uncharacteristic fire behavior at a stand scale that is, in turn, of fundamental importance to firefighter safety. Taken as a whole and extrapolated to larger landscapes, however, modeling-based studies lead to the expectation that, in general, beetle-killed forests should exhibit altered fire behavior (Figure 5.3a). Nevertheless, the pattern emerging from a growing number of observational (retrospective) studies of wildfire activity in beetle-affected forests does not support that expectation (Figure 5.3b). The apparent discrepancy between expectations derived from fire behavior models and observational studies may have numerous, nonmutually exclusive explanations. Most important, these explanations include preoutbreak differences in forest attributes affecting postoutbreak fuel characteristics (Mietkiewicz et al., 2018), fine-scale heterogeneity in infestation severity and synchrony (Simard et al., 2011), or overriding effects of topography and fire weather on patterns of burning. For example, fire weather, especially in high-elevation subalpine forests, strongly influences fire severity (Harvey et al., 2014a, 2014b), fire occurrence (Turner et al., 1999), fire intensity (Bessie and Johnson, 1995), fire spread (Coen, 2005), fire ignition (Bourbonnais et al., 2014), and crown fire behavior (Simard et al., 2011; Schoennagel et al., 2012). During extreme fire weather that promotes high fire activity in subalpine forests, which is predicted to become increasingly common under current climate trends (Coop et al., 2022), fuels are likely dry enough to promote extensive burning regardless of alterations to fuels as a result of bark beetle infestation. While fire behavior models are essential in conceptualizing potentially important driving factors of wildfire activity, particularly at a stand scale, their lack of validation by field experiments (Cruz and Alexander, 2010) limits their suitability for addressing questions of how bark beetle outbreaks may affect fire occurrence, extent, and severity across larger landscapes.

5.3 HOW DO HIGH-SEVERITY FIRES AFFECT SUBSEQUENT OUTBREAKS?

Fires and other disturbances can affect susceptibility to subsequent outbreaks as well as other disturbances by creating long-lasting legacies of forest structure. The effect of severe stand-replacing fires on subsequent outbreaks of bark beetles has been particularly clear as beetles preferentially attack larger trees and stands in advanced stages of development (Schmid and Frye, 1977).

Lodgepole Pine Forests

Tree size is a major determinant of susceptibility to MPBs. Small-diameter trees provide less phloem to support beetle populations, and greater subcortical cooling of small trees contributes to higher mortality of beetles during winter (Safranyik, 2004). Therefore, small-diameter trees have historically been less susceptible to MPB attack, even during outbreaks that may kill most nearby canopy-size trees.

Because the high-severity fires characteristic of lodgepole pine forests in the US Rocky Mountains (Sibold et al., 2006; Veblen and Donnegan, 2006) result in postfire cohorts of small, young trees with thin bark, they can reduce stand susceptibility to MPBs (Kulakowski et al., 2012, 2016). However, reduced susceptibility of younger postfire stands was most pronounced for a 1940s/1950s outbreak, less so for a 1980s outbreak, and did not hold true for a 2000s/2010s outbreak. There are alternate but not mutually exclusive explanations for the varying relationship between severe fires and susceptibility to MPBs over the past century.

One possible explanation is that stand age no longer affects susceptibility to outbreak after stands reach a threshold age of >100–150 years (Taylor and Carroll, 2004). Another possible explanation is related to the theoretical expectation that tree and stand attributes before a disturbance become less important in determining susceptibility to disturbance as the intensity of that disturbance increases. The warm and dry climate of the 2000s contributed to a high-intensity outbreak and likely stressed host trees and thereby reduced tree resistance to beetle attack (Safranyik, 2004). Thus, the susceptibility of younger stands during the 2000s/2010s outbreak may not have been reduced by preceding fires because of the very high intensity of that outbreak (Chapman et al., 2012) compared with the 1980s outbreak, which was much less intense (Smith et al., 2012).

Spruce-Fir Forests

Similar to dynamics in lodgepole pine forests, tree size is important in determining susceptibility of spruce to spruce beetle outbreaks (Bakaj et al., 2016) because beetles preferentially attack larger trees and stands with structures associated with the latter stages of development (Schmid and Frye, 1977). Consequently, young (<80 years) postfire stands of Engelmann spruce in Colorado have been less susceptible to attack by spruce beetle in the 19th (Kulakowski and Veblen, 2006) and 20th centuries (Veblen et al., 1994; Bebi et al., 2003; Kulakowski et al., 2003, 2016), even when landscape-level outbreaks killed most large spruce in surrounding stands.

In contrast to the definitive influence of preceding high-severity fires on spruce beetle outbreaks over the past centuries, high-severity fires that occurred >100 years ago did not strongly influence stand structural traits linked to the 2002−12 bark beetle outbreak (Hart et al., 2014b). This apparent decoupling may be associated with a threshold in stand age, beyond which stand structure is no longer critical in determining susceptibility to outbreak. Alternatively, climate-driven increases in outbreak intensity may be diminishing the importance of tree and stand attributes for outbreaks (Hart et al., 2014b) because warm and dry conditions promote larger beetle populations (Bentz et al., 2010) and decrease tree resistance to beetle attack (Mattson and Haack, 1987; Hart et al., 2014a). This hypothesis is supported by numerous small-diameter and suppressed trees that were affected by spruce beetle in the 21st century before any eventual host saturation, suggesting that tree-level constraints have been relaxed in comparison with previous outbreaks (Hart et al., 2014b).

Nonbeetle Causes of Mortality

While outbreaks *of Dendroctonus* bark beetles in the Rocky Mountains have received most of the attention in research and management over the past 20 years, recognizing that there is widespread evidence of increasing tree mortality that may be attributed to a variety of biotic and abiotic factors is important. Late 20th- and early 21st-century increases in tree mortality, manifested either as gradual increases in background tree mortality rates or pulses of forest die-off, have been widely documented across the western United States and associated with rising regional temperatures (van Mantgem et al., 2009; Williams et al., 2013). Some, but not all, of these increases in tree mortality are associated with lethal insects. Since the mid-1990s the forests of Colorado have experienced profound pulses of tree mortality coincident with warmer temperatures and episodes of reduced precipitation that have affected all the common tree species of the subalpine forests (Bigler et al., 2007;

Worrall et al., 2010; Colorado State Forest Service, 2012; Hanna and Kulakowski, 2012; Smith et al., 2015; Andrus et al., 2021a, 2021b). Sudden and massive mortality of conifers since the mid-1990s in Colorado is well documented in relation to outbreaks of bark beetles: primarily MPB affecting lodgepole pine, limber pine (*Pinus flexilis*), and ponderosa pine (*P. ponderosa*) and spruce bark beetle affecting Engelmann spruce (Chapman et al., 2012; Colorado State Forest Service, 2012). Less well documented is the extensive mortality of subalpine fir attributed to western balsam bark beetle (*Dryocoetes confusus*) and fungal pathogens (Negron and Popp, 2009; Colorado State Forest Service, 2012; Harvey et al., 2021). However, increases in background tree mortality in the absence of evidence of bark beetle infestation are also evident in subalpine forests in Colorado (Bigler et al., 2007; van Mantgem et al., 2009; Anderegg et al., 2014; Smith et al., 2015; Andrus et al., 2021a, 2021b). Thus, increased tree mortality related to broad-scale warming but directly mediated by factors other than *Dendroctonus* beetles seems to be altering forest conditions in general across the Rocky Mountain region and focuses our attention on the need to better understand the complex factors that will affect forest resilience under continued climate warming. The potential influence of this mortality on fire regimes is informed by research on how outbreaks affect fire regimes, but it remains a priority area of research in and of itself.

5.4 HOW ARE INTERACTING FIRES AND BARK BEETLES AFFECTING FOREST RESILIENCE IN THE CONTEXT OF CLIMATE CHANGE?

In the absence of postoutbreak fire, recovery of a tree cover following severe bark beetle outbreaks has long been recognized to result in shifts in species dominance through a combination of growth releases of understory trees, new conifer seedling establishment, and enhanced growth of aspen (Schmid and Frye, 1977; Veblen et al., 1991). Extensive data on forest structure and tree regeneration following widespread bark beetle outbreaks in the Southern Rocky Mountains from the late 1990s to 2017 confirm that despite outbreak-induced shifts in tree species dominance, forest cover is not being replaced by nonforest vegetation (shrublands or grasslands; Rodman et al., 2022). The consistent pattern of trajectories toward recovery of forest cover following bark beetle outbreaks, contrasts with some models suggesting that following large, severe fires poor conifer regeneration due to seed source limitations and inability to tolerate warming and drying climate conditions could result in a decline in forest cover in places (Davis et al., 2023); although the modeled scenarios in Davis et al. (2023) appear to be very uncommon spatially (DellaSala and Hanson, 2019). Nonetheless, outbreaks are not increasing fire frequency or severity, and the fact that both disturbances are increasing simultaneously as a result of the common driver of climatic warming (Raffa et al., 2008; Hart et al., 2014a; Dennison et al., 2014) leads to an increasing probability that forests will be affected by both of these disturbances in short succession. The occurrence of multiple disturbances in relatively short succession combined with a warming climate potentially may overcome the resilience of particular forest ecosystems and usher in transitions to alternative stable states (Buma and Wessman, 2011). The implications of such compounded disturbances on future forest conditions and resilience hinge on important differences in initial forest conditions, spatial heterogeneity in a biophysical setting, magnitude of the disturbance, and details of climate change (see Box 5.2).

BOX 5.2 Bark Beetle Outbreaks and the Carbon Cycle

By killing many trees over large areas, bark beetle outbreaks can affect carbon uptake, sequestration, and release. Initially, killed trees stop taking up carbon as photosynthesis ceases; over the following decades, as trees decompose, carbon is released into the atmosphere and the soil (Hicke et al., 2012b). The rate of carbon uptake during postoutbreak stand regeneration depends on the stand conditions at the time of the outbreak, the postdisturbance environment, and the effect of disturbance interactions.

Kurz et al. (2008) modeled extensive and severe bark beetle outbreaks in Canada and hypothesized the effects of outbreaks on the carbon balance to be large and approximately 75% of the average carbon emitted as a result of fire in the largest outbreak year, but they did not measure carbon uptake from stand regeneration. Stinson et al. (2011) found that these forests were a carbon sink (net positive carbon sequestration) in 1990−2008, despite widespread beetle outbreaks and fires. Studies that have focused on postdisturbance regeneration over smaller areas have found that beetle-disturbed stands are carbon sources in the early years (Brown et al., 2010) but often recover to predisturbance levels over 10 −25 years (Romme et al., 1986; Pfeifer et al., 2011) and in as little as 3−4 years as a result of increased productivity and growth of surviving trees and understory vegetation (Brown et al., 2010, 2012). Ghimire et al. (2015) recently combined postdisturbance forest regrowth trajectories derived from forest inventory data, a process-based carbon cycle model tracking decomposition, and aerial detection survey data to quantify the impact of outbreaks across the western United States. This study reported modeling results that predicted the amount of net carbon release to be large, but somewhat lower than that reported in Canada, likely because of differences in the assumptions of underlying models as well as in outbreaks dynamics between the two regions.

Bark beetle outbreaks clearly have, and will continue to have, important effects on the carbon cycle. However, important challenges remain in understanding the magnitude and duration of these effects, including how tree physiology controls ecosystem carbon fluxes (Frank et al., 2014). Specifically, accurately depicting carbon dynamics following current and future outbreaks depends on an improved understanding of forest regeneration following compounded disturbances and under potentially unfavorable climatic conditions.

Studies that have examined the potential effects of compounded disturbances by outbreaks and fires have found variable results probably reflecting differences in the severity of prefire bark beetle outbreaks as well as differences in preoutbreak forest attributes, abiotic site factors, postfire weather conditions, and how regeneration was quantified. Some studies have reported no major differences in regeneration compared with regeneration following only fire. For example, regeneration after severe fires in lodgepole pine in the northern Rockies was not affected by prefire MPB outbreaks (Harvey et al., 2014a). Yet, in the southern Rockies low seed viability of long dead lodgepole pine cones affected by mountain pine beetle outbreak is expected to result in postfire recovery to stands of lower density than the preoutbreak forest (Rhoades et al., 2022).

High-severity fires that follow spruce beetle outbreaks may affect some dimensions of postfire vegetation development (Carlson et al., 2017). However, following spruce beetle outbreaks from the mid-1990—2010s across northwestern Colorado tree regeneration after fires several years later did not reflect compound or additive effects (Schapira et al., 2021). Likewise, over a longer time period following the severe 1940s spruce beetle outbreak in northwestern Colorado, regeneration after severe fire in 2002 in spruce-fir forests was not affected by that spruce beetle outbreak (Kulakowski et al., 2013; Gill et al., 2017). Regeneration in most spruce-fir forests was unusually low, however, possibly because of generally unfavorable climatic conditions in the 21st century. This general scarcity of regeneration made investigating the consequences of compounded disturbances difficult, and it may be another indicator of overarching climatic influences on disturbance interactions. In another example, regeneration following high-severity fires in Douglas-fir forests was generally low in contrast to that in stands burned by low-severity fires, except where those low-severity fires were preceded by Douglas-fir beetles (Harvey et al., 2013).

Tree species vary in how they are affected by compounded disturbances (Buma and Wessman, 2012; Kulakowski et al., 2013). Therefore, compounded disturbances have the potential to fundamentally change forest composition and future forest trajectories. Compounded disturbances can alter postdisturbance regeneration by either reducing seed source (Rhoades et al., 2022) or by increasing the intensity of the secondary disturbance (Kulakowski and Veblen, 2007), which in turn may negatively influence soil and other microenvironmental conditions (Fonturbel et al., 2011). These two influences may be of minimal consequence for species that reproduce vegetatively, such as quaking aspen (*Populus tremuloides*), which has been shown to dominate initial regeneration following compounded disturbances, sometimes in stands dominated by conifers before the initial disturbance (Kulakowski et al., 2013; Andrus et al., 2021a, 2021b). If compounded disturbances become more common under future climate scenarios, quaking aspen and other species that reproduce vegetatively may be favored over those that reproduce exclusively from seed. However, the susceptibility of aspen to climate-induced die-off (Anderegg et al., 2012; Hanna and Kulakowski, 2012), as well as predicted reductions in its area of suitable habitat under scenarios of climate warming (Rehfeldt et al., 2009), lead to considerable uncertainty about how compounded disturbances under a varying climate will affect future forest trajectories.

5.5 CONCLUSIONS

In lodgepole pine and spruce-fir forests of the Rocky Mountains, effects of bark beetle outbreaks on fuels are complex. The magnitude and heterogeneity of changes in fuels following outbreaks vary with stand structure before the outbreak, the severity of the outbreak, and the rate of the outbreak (Jenkins et al., 2014). Important questions about how biophysical setting may affect the timing and duration of postoutbreak stages and about the importance of within-stand outbreak synchrony on fuel configurations remain. Still, much has been learned about how outbreaks affect fuels. Bark beetle outbreaks reduce FMC, modify foliar chemistry, and increase the volume of dead wood, which can promote several aspects of wildfires. But importantly, outbreaks also reduce canopy density, which lowers the amount of available fuel and can thereby decrease the probability of active crown fires and the likelihood of large, severe fires (Simard et al., 2011; Schoennagel et al., 2012). The relative importance of these contradictory effects during different phases of outbreaks continues to be an active area of research. Empirical studies of fuels immediately following outbreaks (Simard et al., 2011) and retrospective studies of fires in forests recently affected by outbreaks (Kulakowski and Veblen, 2007; Harvey et al., 2014a, 2014b, Andrus et al., 2016) suggest that outbreaks do not substantially increase—and may actually decrease—the risk of high-severity fires, even during and immediately following outbreaks. Likewise, there is general scientific agreement that the risk of active, high-severity crown fires decreases in the years to decades following outbreaks because of reduced CBD (Romme et al., 2006; Jenkins et al., 2008, 2012, 2014; Kaufmann et al., 2008; Simard et al., 2008, Hicke et al., 2012a, Black et al., 2013; Six et al., 2014; Hart et al., 2015a, 2015b; Meigs et al., 2016) (Figure 5.3b). Although this chapter has mainly focused on lodgepole pine and spruce forests, we note that similar conclusions are emerging for other forest types. For example, in upper montane forests dominated by Douglas-fir, prefire outbreak severity did not increase any measure of fire severity (Harvey et al., 2013).

Given the primacy of the importance of weather and climate in overall fire risk across broad regions and forest types (Dennison et al., 2014; Coop et al., 2022), any effects of bark beetle outbreaks on fire regimes should be considered in the context of climatic variability and the relative importance of climate versus changes in fuels associated with outbreaks. Research conducted thus far has consistently indicated that weather and climate are more important than the effects of outbreaks in determining fire risk and behavior in Rocky Mountain subalpine forests (Kulakowski and Jarvis, 2011, Schoennagel et al., 2012; Harvey et al., 2013, 2014a, 2014b; Hart et al. 2015a, 2015b; Andrus et al., 2016; Mietkiewicz and Kulakowski, 2016).

Severe wildfires, such as those that have been burning across western North America over recent decades, have been shown to decrease subsequent susceptibility to bark beetle outbreaks in lodgepole pine (Kulakowski et al., 2012) and spruce-fir forests (Veblen et al., 1994; Kulakowski et al., 2003; Bebi et al., 2003). This implies that climate-driven increases in wildfires have the potential to buffer ecosystems against the risk or severity of future bark beetle outbreaks. However, recent research also suggests that the modulating effect of fires on susceptibility to outbreaks may be contingent on current and future climate influences on beetle populations and tree resistances (Kulakowski et al., 2012; Hart et al., 2014b). Furthermore, prior occurrence of a large severe bark beetle outbreak may sufficiently deplete host populations to reduce the probability of the occurrence of a subsequent outbreak for periods of 70 years or more (Hart et al., 2015a,2015b). The consequences of these interactions under future climate scenarios can be examined through simulation modeling. For example, under a weather scenario of sufficiently dry climate in the future, reduction of host tree populations may lower the probability of bark beetle outbreaks in some habitats (Temperli et al., 2015).

As various forest disturbances become more frequent and extensive, understanding how multiple disturbances interact is of increasing importance. While additional questions remain, the best available science indicates that outbreaks of bark beetles do not increase the risk of high-severity fires in lodgepole pine and spruce-fir forests of the Rocky Mountains. Furthermore, the effects of outbreaks are much less important to fire risk than are weather and climate. By contrast, severe wildfires have been shown to reduce subsequent susceptibility to outbreaks in both forest types. The current state of knowledge does not support the common assumption that increases in bark beetle activity have resulted in increased wildfire activity. Therefore, policy discussions should focus on societal adaptation to the effects of the underlying driving factor of increased tree mortality from insects and from burning: climate warming.

ACKNOWLEDGMENTS

Some of the research included in this review was supported by National Science Foundation Awards (no. 1262687 and 1262691). The authors thank R. Andrus, N. Gill, and N. Mietkiewicz for helpful comments on this work. This work was supported by the Joint Fire Science Program.

REFERENCES

Alexander, M.E., Cruz, M.G., 2013. Are the applications of wildland fire behaviour models getting ahead of their evaluation again? Environ. Model. Softw. 41, 65–71.

Anderegg, W.R.L., Berry, J.A., Smith, D.D., Sperry, J.S., Anderegg, L.D.L., Field, C.B., 2012. The roles of hydraulic and carbon stress in a widespread climate-induced forest die-off. Proc. Natl. Acad. Sci. USA. 109, 233–237.

Anderegg, W.R.L., Anderegg, L.D.L., Berry, J.A., Field, C.B., 2014. Loss of whole-tree hydraulic conductance during severe drought and multi-year forest die-off. Oecologia 175, 11–23.

Andrus, R.A., Veblen, T.T., Harvey, B.J., Hart, S.J., 2016. Fire severity unaffected by spruce beetle outbreak in spruce-fir forests in southwestern Colorado. Ecol. Appl. 26 (3), 700–711.

Andrus, R.A., Hart, S.J., Tutland, N., Veblen, T.T., 2021a. Future dominance by quaking aspen expected following short-interval, compounded disturbance interaction. Ecosphere 12 (1), 1.

Andrus, R.A., Chai, R.K., Harvey, B.J., Rodman, K.C., Veblen, T.T., 2021b. Increasing rates of subalpine tree mortality linked to warmer and drier summers. J. Ecol. 109 (5), 2203–2218.

Bakaj, F., Mietkiewicz, N., Veblen, T.T., Kulakowski, D., 2016. The relative importance of tree and stand properties in predicting susceptibility to spruce beetle outbreak in the mid-20th century. Ecosphere 7 (10), e01485. https://doi.org/10.1002/ecs2.1485.

Bebi, P., Kulakowski, D., Veblen, T.T., 2003. Interactions between fire and spruce beetles in a subalpine Rocky Mountain forest landscape. Ecology 84, 362–371.

Bentz, B.J., Régnière, J., Fettig, C.J., Hansen, E.M., Hayes, J.L., Hicke, J.A., Kelsey, R.G., Negrón, J.F., Seybold, S.J., 2010. Climate change and bark beetles of the Western United States and Canada: direct and indirect effects. BioScience 60, 602–613.

Bessie, W.C., Johnson, E.A., 1995. The relative importance of fuels and weather on fire behavior in subalpine forests. Ecology 76, 747.

Bigler, C., Kulakowski, D., Veblen, T.T., 2005. Multiple disturbance interactions and drought influence fire severity in Rocky Mountain subalpine forests. Ecology 86, 3018–3029.

Bigler, C., Gavin, D., Gunning, C., Veblen, T.T., 2007. Drought induces lagged tree mortality in a subalpine forest in the Rocky Mountains. Oikos 116, 1983–1994.

Black, S.H., Kulakowski, D., Noon, B.R., DellaSala, D.A., 2013. Do bark beetle outbreaks increase wildfire risks in the central U.S. Rocky Mountains? Implications from recent research. Nat. Area J. 33, 59–65.

Bourbonnais, M.L., Nelson, T.A., Wulder, M.A., 2014. Geographic analysis of the impacts of mountain pine beetle infestation on forest fire ignition. Can. Geogr. 58, 188–202.

Brown, M., Black, T.A., Nesic, Z., Foord, V.N., Spittlehouse, D.L., Fredeen, A.L., Grant, N.J., Burton, P.J., Trofymow, J.A., 2010. Impact of mountain pine beetle on the net ecosystem production of lodgepole pine stands in British Columbia. Agric. For. Meteorol. 150, 254–264.

Brown, M.G., Black, T.A., Nesic, Z., Fredeen, A.L., Foord, V.N., Spittlehouse, D.L., Bowler, R., Burton, P.J., Trofymow, J.A., Grant, N.J., Lessard, D., 2012. The carbon balance of two lodgepole pine stands recovering from mountain pine beetle attack in British Columbia. Agric. For. Meteorol. 153, 82–93.

Buma, B., Wessman, C.A., 2011. Disturbance interactions can impact resilience mechanisms of forests. Ecosphere 2. Article 64.

Buma, B., Wessman, C.A., 2012. Differential species responses to compounded perturbations and implications for landscape heterogeneity and resilience. For. Ecol. Manag. 266, 25–33.

Carlson, A.R., Sibold, J.S., Assal, T.J., Negrón, J.F., 2017. Evidence of compounded disturbance effects on vegetation recovery following high-severity wildfire and spruce beetle outbreak. PLoS ONE 12 (8), e0181778. https://doi.org/10.1371/journal.pone.0181778.

Carlson, A.R., Sibold, J.S., Negrón, J.F., 2021. Wildfire and spruce beetle outbreak have mixed effects on below-canopy temperatures in a Rocky Mountain subalpine forest. J. Biogeogr. 48, 216–230.

Carter, T.A., Hayes, K., Buma, B., 2022. Putting more fuel on the fire... or maybe not? A synthesis of spruce beetle and fire interactions in North American subalpine forests. Landsc. Ecol. 37, 2241–2254.

Chapman, T., Veblen, T.T., Schoennagel, T., 2012. Spatio-temporal patterns of mountain pine beetle activity in the southern Rocky Mountains. Ecology 93, 2175–2185.

Coen, J.L., 2005. Simulation of the Big Elk Fire using coupled atmosphere-fire modeling. Int. J. Wildland Fire 14, 49–59.

Colorado State Forest Service, 2012. Report on the Health of Colorado's Forests: Forest Stewardship through Active Management. https://www.fs.usda.gov/foresthealth/docs/fhh/CO_FHH_2012.pdf.

Coop, J.D., Parks, S.A., Stevens-Rumann, C.S., Ritter, S.M., Hoffman, C.M., 2022. Extreme fire spread events and area burned under recent and future climate in the western USA. Glob. Ecol. Biogeogr. 31 (10), 1949–1959.

Cruz, M.G., Alexander, M.E., 2010. Assessing crown fire potential in coniferous forests of western North America: a critique of current approaches and recent simulation studies. Int. J. Wildland Fire 19, 377–398.

Dahlsten, D.L., 1982. Relationships between bark beetles and their natural enemies. In: Mitton, J.B., Sturgeon, K.B. (Eds.), Bark Beetles in North American Conifers. University of Texas Press, Austin, pp. 140–182.

Davis, K.T., Robles, M., Kemp, K.B., Higuera, P.E., Chapman, T., Metlen, K., Peeler, J.L., Rodman, K.C., Woolley, T., Addington, R.N., et al., 2023. Reduced fire severity offers near-term buffer to climate-driven declines in conifer resilience across the western United States. Proceedings of the National Academy of Sciences of the United States of America 120 (11). https://doi.org/10.1073/pnas.2208120120.

DellaSala, D.A., Hanson, C.T., 2019. Are wildland fires increasing large patches of complex early seral forest habitat? Diversity 11. Article 157. https://doi.org/10.3390/d11090157.

Dennison, P.E., Brewer, S.C., Arnold, J.D., Moritz, M.A., 2014. Large wildfire trends in the western United States, 1984–2011. Geophys. Res. Lett. 41, 2928–2933.

DeRose, R.J., Long, J.N., 2009. Wildfire and spruce beetle outbreak: simulation of interacting disturbances in the central Rocky Mountains. Ecoscience 16, 28–38.

DeRose, R.J., Long, J.N., 2014. Resistance and resilience: a conceptual framework for silviculture. For. Sci 60 (6), 1205–1212.

Eisenhart, K., Veblen, T.T., 2000. Dendrochronological detection of spruce bark beetle outbreaks in northwestern Colorado. Can. J. For. Res. 30, 1788–1798.

Fonturbel, M.T., Vega, J.A., Perez-Gorostiaga, P., Fernandez, C., Alonso, M., Cuinas, P., Jimenez, E., 2011. Effects of soil burn severity on germination and initial establishment of mari-time pine seedlings, under greenhouse conditions, in two contrasting experimentally burned soils. Int. J. Wildland Fire 20, 209–222.

Frank, J.M., Massman, W.L., Ewers, B.E., Huckaby, L.S., Negrón, J.F., 2014. Ecosystem CO_2/H_2O fluxes are explained by hydraulically limited gas exchange during tree mortality from spruce bark beetles. J. Geophys. Res. Biogesci. 119, 1195–1215.

Ghimire, B., Williams, C.A., Collatz, G.J., Vanderhoof, M., Rogan, J., Kulakowski, D., Masek, J.G., 2015. Large carbon release from bark beetle outbreaks across Western United States imposes climate feedback. Glob. Change Biol. 21, 3087–3101.

Gill, N., Jarvis, D., Veblen, T.T., Pickett, S.T.A., Kulakowski, D., 2017. Is initial post-disturbance regeneration indicative of longer-term trajectories? Ecosphere 8 (8), e01924. https://doi.org/10.1002/ecs2.1924.

Hanna, P., Kulakowski, D., 2012. The influences of climate on aspen dieback. For. Ecol. Manag. 274, 91–98.

Hanson, C.T., Chi, T.Y., 2020. Black-backed woodpecker nest density in the Sierra Nevada, California. Diversity 12 (1), 364. https://doi.org/10.3390/d12100364.

Hardy, C.C., 2005. Wildland fire hazard and risk: problems, definitions, and context. For. Ecol. Manag. 211, 73–83.

Hart, S.J., Veblen, T.T., Eisenhart, K.S., Jarvis, D., Kulakowski, D., 2014a. Droughtinducesspruce beetle (*Dendroctonus rufipennis*) outbreaks across northwestern Colorado. Ecology 95, 930–939.

Hart, S.J., Veblen, T.T., Kulakowski, D., 2014b. Do tree and stand-level attributes determine susceptibility of spruce-fir forests to spruce beetle outbreaks? For. Ecol. Manag. 318, 44–53.

Hart, S.J., Schoennagel, T., Veblen, T.T., Chapman, T.B., 2015a. Area burned in the western United States is unaffected by recent mountain pine beetle outbreaks. Proc. Natl. Acad. Sci. 112 (14), 4375–4380.

Hart, S.J., Veblen, T.T., Mietkiewicz, N., Kulakowski, D., 2015b. Negative feedbacks on bark beetle outbreaks: widespread and severe spruce beetle infestation restricts subsequent infestation. PLoS One 10 (5), e0127975.

Harvey, B.J., Donato, D.C., Romme, W.H., Turner, M.G., 2013. Influence of recent bark beetle outbreak on fire severity and postfire tree regeneration in montane Douglas-fir forests. Ecology 94, 2475–2486.

Harvey, B.J., Donato, D.C., Turner, M.G., 2014a. Recent mountain pine beetle outbreaks, wildfire severity, and postfire tree regeneration in the US Northern Rockies. Proc. Natl. Acad. Sci. USA 111, 15120–15125. https://doi.org/10.1073/pnas.1411346111.

Harvey, B.J., Donato, D.C., Romme, W.H., Turner, M.G., 2014b. Fire severity and tree regeneration following bark beetle outbreaks: the role of outbreak stage and burning conditions. Ecol. Appl. 24, 1608–1625.

Harvey, B.J., Andrus, R.A., Battaglia, M.A., Negrón, J.F., Orrego, A., Veblen, T.T., 2021. Drought times in mesic places: factors associated with forest mortality vary by scale in a temperate subalpine region. Ecosphere 12 (1), e03318.

Hicke, J.A., Johnson, M.C., Hayes, J.L., Preisler, H.K., 2012a. Effects of bark beetle-caused tree mortality on wildfire. For. Ecol. Manag. 271, 81–90.

Hicke, J.A., Allen, C.D., Desai, A.R., Dietze, M.C., Hall, R.J., Hogg, E.H., Kashian, D.M., Moore, D., Raffa, K.F., Sturrock, R.N., Vogelmann, J., 2012b. Effects of biotic disturbances on forest carbon cycling in the United States and Canada. Glob. Chang. Biol. 18, 7–34.

Hoffman, C.M., Morgan, P., Mell, W., Parsons, R., Strand, E., Cook, S., 2012. Numerical simulation of crown fire hazard following bark beetle-caused mortality in lodgepole pine forests. For. Sci. 58, 178–188.

Hoffman, C.M., Morgan, P., Mell, W., Parsons, R., Strand, E., Cook, S., 2013. Surface fire intensity influences simulated crown fire behavior in lodgepole pine forests with recent mountain pine beetle-caused tree mortality. For. Sci. 59, 390–399.

Jackson, K.J., Wohl, E., 2015. Instream wood loads in montane forest streams of the Colorado Front Range, USA. Geomorphology 234, 161–170.

Jarvis, D., Kulakowski, D., 2015. Long-term history and synchrony of mountain pine beetle outbreaks. J. Biogeogr 42, 1029–1039. https://doi.org/10.1111/jbi.12489.

Jenkins, M.J., Hebertson, E., Page, W., Jorgensen, C.A., 2008. Bark beetles, fuels, fires and implications for forest management in the Intermountain West. For. Ecol. Manag. 254, 16–34.

Jenkins, M.J., Page, W.G., Hebertson, E.G., Alexander, M.E., 2012. Fuels and fire behavior dynamics in bark beetle-attacked forests in Western North America and implications for fire management. For. Ecol. Manag. 275, 23–34.

Jenkins, M.J., Runyon, J.B., Fettig, C.J., Page, W.G., Bentz, B.J., 2014. Interactions among the mountain pine beetle, fires, and fuels. For. Sci. 60, 489–501.

Jolly, W.M., Parsons, R.A., Hadlow, A.M., Cohn, G.M., McAllister, S.S., Popp, J.B., Hubbard, R.M., Negron, J.F., 2012. Relationships between moisture, chemistry, and ignition of Pinus contorta needles during the early stages of mountain pine beetle attack. For. Ecol. Manag. 269, 52−59.

Kaufmann, M.R., Aplet, G.H., Babler, M., Baker, W.L., Bentz, B., Harrington, M., Hawkes, B.C., Huckaby, L.S., Jenkins, M.J., Kashian, D.M., Keane, R.E., Kulakowski, D., McHugh, C., Negron, J., Popp, J., Romme, W.H., Schoennagel, T., Shepperd, W., Smith, F.W., Sutherland, E.K., Tinker, D., Veblen, T.T., 2008. The Status of Our Scientific Understanding of Lodgepole Pine and Mountain Pine Beetles - A Focus on Forest Ecology and Fire Behavior. The Nature Conservancy, Arlington, VA. GFI technical report 2008−2.

Kelly, J.J., Latif, Q.S., Saab, V.A., Veblen, T.T., 2019. Spruce Beetle outbreaks guide American Three-toed Woodpecker Picoides dorsalis occupancy patterns in subalpine forests. Ibis 161, 172−183.

Klutsch, J.G., Negron, J.F., Costello, S.L., Rhoades, C.C., West, D.R., Popp, J., Caissie, R., 2009. Stand characteristics and downed woody debris accumulations associated with a mountain pine beetle (Dendroctonus ponderosae Hopkins) outbreak in Colorado. For. Ecol. Manag. 258, 641−649.

Koplin, J.R., Baldwin, P.H., 1970. Woodpecker predation on an endemic population of Engelmann spruce beetles. Am. Midl. Nat. 83, 510−515.

Kulakowski, D., Jarvis, D., 2011. The influence of mountain pine beetle outbreaks on severe wildfires in northwestern Colorado and southern Wyoming: a look at the past century. For. Ecol. Manag. 261, 1686−1696.

Kulakowski, D., Veblen, T.T., 2006. The effect of fires on susceptibility of subalpine forests to a 19th century spruce beetle outbreak in western Colorado. Can. J. For. Res. 36, 2974−2982.

Kulakowski, D., Veblen, T.T., 2007. Effect of prior disturbances on the extent and severity of wildfire in Colorado subalpine forests. Ecology 88, 759−769.

Kulakowski, D., Veblen, T.T., Bebi, P., 2003. Effects of fire and spruce beetle outbreak legacies on the disturbance regime of a subalpine forest in Colorado. J. Biogeogr. 30, 1445−1456.

Kulakowski, D., Jarvis, D., Veblen, T.T., Smith, J., 2012. Stand-replacing fires reduce susceptibility to mountain pine beetle outbreaks in Colorado. J. Biogeogr. 39, 2052−2060.

Kulakowski, D., Matthews, C., Jarvis, D., Veblen, T.T., 2013. Compounded disturbances in subalpine forests in western Colorado favor future dominance by quaking aspen (Populus tremuloides). J. Veg. Sci. 24, 168−176.

Kulakowski, D., Veblen, T.T., Bebi, P., 2016. Fire severity controlled susceptibility to a 1940s spruce beetle outbreak in Colorado, USA. PLoS ONE 11 (7), e0158138.

Kurz, W.A., Dymond, C.C., Stinson, G., Rampley, G.J., Neilson, E.T., Carroll, A.L., Ebata, T., Safranyik, L., 2008. Mountain pine beetle and forest carbon feedback to climate change. Nature 452, 987−990.

Lynch, H.J., Renkin, R.A., Crabtree, R.L., Moorcroft, P.R., 2006. The influence of previous mountain pine beetle (Dendroctonus ponderosae) activity on the 1988 Yellowstone fires. Ecosystems 9, 1318−1327.

Maness, H., Kushner, P.J., Fung, I., 2012. Summertime climate response to mountain pine beetle disturbance in British Columbia. Nat. Geosci. 6, 65−70.

Mattson, W.J., Haack, R.A., 1987. The role of drought in outbreaks of plant-eating insects. BioScience 27 (2), 110−119.

Meigs, G.W., Campbell, J.L., Zald, H.S.J., Bailey, J.D., Shaw, D.C., Kennedy, R.E., 2015. Does wildfire likelihood increase following insect outbreaks in conifer forests? Ecosphere 6 (7), 118. https://doi.org/10.1890/ES15-00037.1.

Meigs, G.W., Zald, H.S.J., Campbell, J.L., Keeton, W.S., Kennedy, R.E., 2016. Do insect outbreaks reduce the severity of subsequent forest fires? Environ. Res. Lett. 11, 045008.

Mietkiewicz, N., Kulakowski, D., 2016. Relative importance of climate and mountain pine beetle outbreaks on the occurrence of large wildfires in the western US. Ecol. Appl. 26 (8), 2523—2535.

Mietkiewicz, N., Kulakowski, D., Veblen, T.T., 2018. Pre-outbreak forest conditions mediate the effects of spruce beetle outbreaks on fuels in subalpine forests of Colorado. Ecol. Appl. 28 (2), 457—472.

Negron, J.F., Popp, J.B., 2009. The flight periodicity, attack patterns, and life history of Dryocoetes confusus Swaine (Coleoptera: Curculionida: Scolytinae), the western basalm bark beetle, in north central Colorado. West. N. Am. Nat. 69 (4), 447—458.

Omi, P.N., 1997. Final Report: Fuels Modification to Reduce Large Fire Probability. Submitted to US Department of Interior, Fire Research Committee, Colorado State University, Fort Collins.

Page, W.G., Jenkins, M.J., 2007. Mountain pine beetle-induced changes to selected lodgepole pine fuel complexes within the intermountain region. For. Sci. 53, 507—518.

Page, W.G., Jenkins, M.L., Runyon, J.B., 2012. Mountain pine beetle attack alters the chemistry and flammability of lodgepole pine foliage. Can. J. For. Res. 42, 1631—1647.

Page, W.G., Jenkins, M.J., Alexander, M.E., 2013a. Foliar moisture content variations in lodgepole pine over the diurnal cycle during the red stage of mountain pine beetle attack. Environ. Model. Softw. 49, 98—102.

Page, W.G., Alexander, M.E., Jenkins, M.J., 2013b. Wildfire's resistance to control in mountain pine beetle-attacked lodgepole pine forests. For. Chron. 89, 783—794.

Page, W.G., Jenkins, M.J., Runyon, J.B., 2014a. Spruce beetle-induced changes to Engelmann spruce foliage flammability. For. Sci. 60, 691—702.

Page, W.G., Jenkins, M.J., Alexander, M.E., 2014b. Crown fire potential in lodgepole pine forests during the red stage of mountain pine beetle attack. Forestry 87, 347—361.

Parker, T.J., Clancy, K.M., Mathiasen, R.L., 2006. Interactions among fire, insects and pathogens in coniferous forests of the interior western United States and Canada. Agric. For. Entomol. 8, 167—189.

Pfeifer, E.M., Hicke, J.A., Meddens, A.J.H., 2011. Observations and modeling of 336 aboveground tree carbon stocks and fluxes following a bark beetle outbreak in the western United States. Glob. Chang. Biol. 17, 339—350.

Raffa, K.F., Aukema, B.H., Bentz, B.J., Carroll, A.L., Hicke, J.A., Turner, M.G., Romme, W.H., 2008. Cross-scale drivers of natural disturbances prone to anthropogenic amplification: the dynamics of bark beetle eruptions. Bioscience 58, 501—517.

Rehfeldt, G.E., Ferguson, D.E., Crookston, N.L., 2009. Aspen, climate, and sudden decline in western USA. For. Ecol. Manag. 258, 2353—2364.

Reid, M., 1989. The Response of Understory Vegetation to Major Canopy Disturbance in the Subalpine Forests of Colorado. Masters Thesis,Geography. University of Colorado, Boulder.

Renkin, R.A., Despain, D.G., 1992. Fuel moisture, forest type, and lightning-caused fire in Yellowstone National Park. Can. J. For. Res. 22, 37—45.

Rhoades, C.C., Fegel, T.S., Hubbard, R.M., Chambers, M.E., 2022. Limited seed viability in long-dead serotinous lodgepole pine trees in the Southern Rockies, USA. For. Ecol. Manag. 526, 120565.

Rodman, K.C., Andrus, R.A., Carlson, A.R., Carter, T.A., Chapman, T.B., Coop, J.D., Fornwalt, P.J., Gill, N.S., Harvey, B.J., Hoffman, A.E., et al., 2022. Rocky Mountain forests are poised to recover following bark beetle outbreaks but with altered composition. J. Ecol. 110 (12), 2929.

Romme, W.H., Despain, D.G., 1989. Historical perspective on the Yellowstone fires of 1988. Bioscience 39, 695–699.

Romme, W.H., Knight, D.H., Yavitt, J.B., 1986. Mountain pine beetle outbreaks in the Rocky Mountains: regulators of primary productivity? Am. Nat. 127, 484–494.

Romme, W.H., Clement, J., Hicke, J., Kulakowski, D., MacDonald, L.H., Schoennagel, T.L., Veblen, T.T., 2006. Recent Forest Insect Outbreaks and Fire Risk in Colorado Forests: A Brief Synthesis of Relevant Research. Colorado Forest Restoration Institute, Fort Collins, CO.

Rothermel, R.C., 1972. A mathematical model for predicting fire spread in wildland fuels. In: Res. Pap. INT-115. U.S. Department of Agriculture, Intermountain Forest and Range Experiment Station, Ogden, UT, 40 p.

Saab, V.A., Latif, Q.S., Rowland, M.M., Johnson, T.N., Chalfoun, A.D., Buskirk, S.W., Heyward, J.E., Dresser, M.A., 2014. Ecological consequences of mountain pine beetle outbreaks for wildlife in western North American forests. For. Sci. 60, 539–559.

Safranyik, L., 2004. Mountain pine beetle epidemiology in lodgepole pine. In: Ke lowna, B.C., Shore, T.L., Brooks, J.E., Stone, J.E. (Eds.), Mountain Pine Beetle Symposium: Challenges and Solutions, October 30–31, 2003. Natural Resources Canada, Canadian Forest Service, Pacific Forestry Centre, Victoria, B.C., pp. 33–40. Information Report BC-X-399.

Schapira, Z., Stevens-Rumann, C., Shorrock, D., Hoffman, C., Chambers, A., 2021. Beetlemania: is the bark worse than the bite? Rocky Mountain subalpine forests recover differently after spruce beetle outbreaks and wildfires. For. Ecol. Manag. 482, 118879.

Schmid, J.M., Amman, G.D., 1992. *Dendroctonus* beetles and old-growth forests in the Rockies (technical coordinators). In: Kaufmann, M.R., Moir, W.H., Bassett, R.L. (Eds.), Old-Growth Forests of the Southwest and Rocky Mountain Regions: Proceedings of a Workshop. USDA Forest Service General Technical, pp. 51–59. Report RM-213.

Schmid, J.M., Frye, R.H., 1977. Spruce Beetle in the Rockies. General Technical Report RM-49. USDA Forest Service, Fort Collins, CO, USA.

Schoennagel, T., Veblen, T.T., Kulakowski, D., Holz, A., 2007. Multidecadal climate variability and climate interactions affect subalpine fire occurrence, western Colorado (USA). Ecology 88, 2891–2902.

Schoennagel, T., Veblen, T.T., Negron, J.F., Smith, J.M., 2012. Effects of mountain pine beetle on fuels and expected fire behavior in lodgepole pine forests, Colorado, USA. PLoS One 7, e30002.

Schroeder, D., Mooney, C., 2012. Fire Behavior in Simulated Mountain Pine Beetle-Killed Stands. Final Report. Wildfire Operations Research, FP Innovations, Hinton, Alberta.

Scott, J.H., Reinhardt, E.D., 2001. Assessing crown fire potential by linking models of surface and crown fire behavior. In: Res. Pap. RMRS-RP-29. U.S. Department of Agriculture, Forest Service, Rocky Mountain Research Station, Fort Collins, CO, 59 p.

Sherriff, R.L., Platt, R.V., Veblen, T.T., Schoennagel, T.L., Gartner, M.H., 2014. Historical, observed, and modeled wildfire severity in montane forests of the Colorado Front Range. PLoS One 9, e106971.

Sibold, J.S., Veblen, T.T., Gonzalez, M.E., 2006. Spatial and temporal variation in historic fire regimes in subalpine forests across the Colorado Front Range in Rocky Mountain National Park. J. Biogeogr. 32, 631–647.

Simard, M., Powell, E.N., Griffin, J.W., Raffa, K.F., Turner, M.G., 2008. Annotated Bibliography for Forest Managers on Fire-Bark Beetle Interactions. USFS Western Wildlands Environmental Threats Assessment Center, Prineville, Ore.

Simard, M., Romme, W.H., Griffin, J.M., Turner, M.G., 2011. Do bark beetle outbreaks change the probability of active crown fire in lodgepole pine forests? Ecol. Monogr. 81, 3–24.

Simard, M., Romme, W.H., Griffin, J.M., Turner, M.G., 2012. Do mountain pine beetle outbreaks change the probability of active crown fire in lodgepole pine forests? Ecol. Monogr. 93, 946–950.

Six, D.L., Biber, E., Long, E., 2014. Management for mountain pine beetle outbreak suppression: does relevant science support current policy? Forests 5, 103–133. https://doi.org/10.3390/f5010103.

Smith, J.M., Hart, S.J., Chapman, T.B., Veblen, T.T., Schoennagel, T., 2012. Dendroecological reconstruction of 1980s mountain pine beetle outbreak in lodgepole pine forests in northwestern Colorado. Ecoscience 19, 113–126.

Smith, J.M., Paritsis, J., Veblen, T.T., Chapman, T.B., 2015. Permanent forest plots show accelerating tree mortality in subalpine forests of the Colorado Front Range from 1982 to 2013. For. Ecol. Manag. 341, 8–17.

Stinson, G., Kurz, W.A., Smyth, C.E., Neilson, E.T., Dymond, C.C., Metsaranta, J.M., Boisvenue, C., Rampley, G.J., Li, Q., White, T.M., Blain, D., 2011. An inventory-based analysis of Canada's managed forest carbon dynamics, 1990 to 2008. Glob. Chang. Biol. 17, 2227–2244.

Taylor, S.W., Carroll, A.L., 2004. Disturbance, forest age dynamics, and mountain pine beetle outbreaks in BC: a historical perspective. In: Shore, T.L., Brooks, J.E., Stone, J.E. (Eds.), Challenges and Solutions. Proceedings of the Mountain Pine Beetle Symposium, Kelowna, BC, Canada, October 30–31, 2003, Canadian Forest Service, Pacific Forestry Centre, Information Report BC-X-399. NRC Research Press, pp. 41–51.

Temperli, C., Veblen, T.T., Hart, S.J., Kulakowski, D., Tepley, A.J., 2015. Interactions among spruce beetle disturbance, climate change and forest dynamics captured by a forest landscape model. Ecosphere 6 (11), 1–20.

Turner, M.G., 2010. Disturbance and landscape dynamics in a changing world. Ecology 91, 2833–2849.

Turner, M.G., Gardner, R.H., Romme, W.H., 1999. Prefire heterogeneity, fire severity, and early postfire plant reestablishment in subalpine forests of Yellowstone National Park, Wyoming. Int. J. Wildland Fire 9, 21–36.

van Mantgem, P.J., Stephenson, N.L., Byrne, J.C., Daniels, L.D., Franklin, J.F., Fulé, P.Z., Harmon, M.E., Larson, A.J., Smith, J.M., Taylor, A.H., Veblen, T.T., 2009. Widespread increase of tree mortality rates in the western United States. Science 323, 521–524.

Veblen, T.T., 1986. Age and size structure of subalpine forests in the Colorado Front Range. Bull. Torrey Bot. Club 113, 225–240.

Veblen, T.T., Donnegan, J.A., 2006. Historical Range of Variability of Forest Vegetation of the National Forests of the Colorado Front Range. USDA Forest Service, Rocky Mountain Region and the Colorado Forest Restoration Institute, Fort Collins, 151 pages.

Veblen, T.T., Hadley, K.S., Reid, M.S., Rebertus, A.J., 1991. The response of subalpine forests to spruce beetle outbreak in Colorado. Ecology 72, 213–231.

Veblen, T.T., Hadley, K.S., Nel, E.M., Kitzberger, T., Reid, M., Villalba, R., 1994. Disturbance regime and disturbance interactions in a Rocky Mountain subalpine forest. J. Ecol. 82, 125–135.

White, P.S., Pickett, S.T.A., 1985. Natural disturbance and patch dynamics: an introduction. In: Pickett, S.T.A., White, P.S. (Eds.), The Ecology of Natural Disturbance and Patch Dynamics. Academic, New York, pp. 3–13.

Williams, A.P., Allen, C.D., Macalady, A.K., Griffin, D., Woodhouse, C.A., Meko, D.M., Swetnam, T.W., Rauscher, S.A., Seager, R., Grissino-Mayer, H.D., Dean, J.S., Cook, E.R.,

Gangodagamage, C., Cai, M., McDowell, N.G., 2013. Temperature as a potent driver of regional forest drought stress and tree mortality. Nat. Clim. Chang. 3, 292–297.

Worrall, J.J., Marchetti, S.B., Egeland, L., Mask, R.A., Eager, T., Howell, B., 2010. Effects and etiology of sudden aspen decline in southwestern Colorado, USA. For. Ecol. Manag. 260, 638–648.

Chapter 6

High-Severity Fire in Chaparral: Cognitive Dissonance in the Shrublands

Richard W. Halsey[1] and Alexandra D. Syphard[1,2]
[1]The California Chaparral Institute, Escondido, CA, United States; [2]Conservation Biology Institute, Corvallis, OR, United States

6.1 CHAPARRAL AND THE FIRE SUPPRESSION PARADIGM

The conflict between facts and beliefs concerning fire in California's native shrublands is an example of cognitive dissonance—the psychological discomfort caused when an individual is confronted with new facts or ideas in conflict with currently held opinions (Festinger, 1957).

The most characteristic native shrubland in California is chaparral, a sclerophyllous, drought-hardy plant community composed of such iconic species as manzanita (*Arctostaphylos* sp.) and ceanothus (*Ceanothus* sp.) (Figure 6.1). Once the preferred habitat of the California grizzly bear (*Ursus arctos californicus*), chaparral covers many of the state's hills and mountains with rich biodiversity that reaches its peak on the central coast. Chaparral is also the most extensive plant community characterizing the California Floristic Province, extending into southern Oregon and Baja California with disjunct patches in central and southeastern Arizona (Sonoran) and northern Mexico (Keeley, 2000). There are also interesting Madrean-Oriental chaparral communities in west Texas, and mexical shrublands, scattered throughout semi-arid, rain-shadow portions of Mexico's central and eastern mountain ranges as far south as Oaxaca. Despite the separation and differences in climate, the various sclerophyllous shrublands in North America still share many of the same plant genera. Mexical shares at least 14 genera with California chaparral, 7 of which are also found in the Mediterranean Basin (Valiente-Banuet et al., 1998).

FIGURE 6.1 Chaparral is a unique plant community characterized by large, contiguous stands of drought-hardy shrubs, a Mediterranean-type climate, and infrequent, high-intensity/high-severity crown fires. *Photo R.W. Halsey.*

California chaparral is shaped by a Mediterranean climate, with hot, dry summers and mild, wet winters. Mexical has a climate pattern opposite of California with summer rain and winter drought. Sonoran and Madrean-Oriental chaparral have rain occurring during both winter and summer months.

Although often portrayed as a fire-adapted ecosystem, a more accurate description is one adapted to a particular fire pattern or regime that is characterized by large, infrequent, and high-intensity fires (Keeley et al., 2012). Increase the frequency, reduce the intensity, or change the seasonality of fire and chaparral species can be eliminated, often replaced by nonnative weeds and grasses. As ignitions have increased as a result of human activity, chaparral is being threatened by too much fire in much of its range, particularly in southern California.

The role fire plays in chaparral is often misunderstood by policymakers, land and fire managers, forest scientists, and the public (Keeley et al., 2012). The primary cause of this misunderstanding is a powerful belief system that has formed around what can be characterized as the *fire suppression paradigm*, resulting in cognitive dissonance as new scientific information has emerged. The fire suppression paradigm asserts that a century of fire suppression policy has largely eliminated fires and allowed habitats to accumulate to unnatural levels of plant growth so that today when wildfires begin, they burn uncontrollably, often producing catastrophic effects in human communities (Keeley et al., 1999). However, for many plant communities, especially chaparral, the fire suppression paradigm is irrelevant. For example, data for the past 100 years show that despite a policy of fire suppression, wildfire frequency has not been reduced in most southern California landscapes. In fact, fires are more common today than historically. Because of this misconception about fire suppression, many managers believe that wildfires are primarily "fuel-driven" events and, as a consequence, can be controlled by modifying vegetation.

Deeply embedded in the paradigm is the preconception that small, low-intensity/low-severity surface fires are the natural pattern, whereas high-intensity/high-severity crown fires are not. When a high-severity fire burns more than ~40 ha, it is often considered a direct result of past fire suppression, regardless of the plant community involved. The paradigm was originally developed to describe the fire regime found in lightning-saturated, dry ponderosa pine (*Pinus ponderosa*) forests of the Southwest, where somewhat frequent, lower-intensity surface fires represent a common pattern (Steel et al., 2015). However, large patches of high intensity fire can occur in such systems, especially during extreme fire weather and drought (Sherriff and Veblen, 2007; Baker et al., 2023).

Because the fire suppression paradigm is forest-centric, understory shrubs and small trees are viewed as fuel rather than important components of habitat. This has led to a set of values, facilitated by lumber and ranching interests, that view chaparral as "worthless brush," an "invader" of forests and rangeland, and an "unsightly menace" (Halsey, 2011). The bias has led to other derogatory characterizations such as claiming chaparral plants are so pyrogenic that they are literally "oozing combustible resins" (Shea, 2008). The paradigm has effectively demonized a native ecosystem that supports significant biodiversity.

The key point is that chaparral fires are unlike forest fires, yet dry-forest fire ecology has been misapplied to explain how fire should burn in chaparral. The basic facts about chaparral fires can be summarized as follows:

— Fire suppression has not caused excessive amounts of chaparral to accumulate (Keeley et al., 1999).
— Fire suppression can play a role in protecting chaparral stands from ecological damage resulting from excessive fire during nonwind driven events.
— Infrequent, large, high-intensity crown fires are natural in chaparral (Keeley and Zedler, 2009).
— There are few, if any, justifiable ecological/resource benefits in conducting prescribed burning or other vegetation treatments in chaparral (Keeley et al., 2009a).

Research over the past 2 decades has rejected the fire suppression paradigm when applied to ecosystems subject to crown fires, especially shrublands like chaparral. Not surprisingly, the cognitive dissonance caused by this research (e.g., Conard and Weise, 1998; Mensing et al., 1999; Keeley et al., 1999; Keeley and Zedler, 2009; Lombardo et al., 2009) has fostered resistance by the supporters of the challenged paradigm (e.g., Minnich, 2001). Consequently, it continues to influence public policy and opinion about chaparral specifically, and wildfire in general. But, as the evidence has accumulated, the fire suppression paradigm is slowly shifting to a new model acknowledging that infrequent (intervals of 30–150 years or more), large, high-intensity crown fires do in fact represent the natural fire regime for chaparral and that weather, not fuel type, is the most important variable controlling fire intensity, spread, and size (e.g., Moritz et al., 2004; Keeley and Zedler, 2009).

6.2 THE FACTS ABOUT CHAPARRAL FIRES: THEY BURN INTENSELY AND SEVERELY

The natural, physical structure of chaparral shrubs (contiguous cover, dense accumulation of fine leaves and stems, and retention of dead wood) and the seasonal pattern of drought that includes low humidity, high temperatures, and low live fuel moistures create conditions favoring high-intensity crown fires (Figure 6.2).

Crown fires are those that burn into the canopies of the dominant vegetation. These are opposed to surface or understory fires that burn vegetation close to the ground. *Surface fires* are common components of certain forested ecosystems with mixed-severity fire regimes where there is a distinct separation between understory vegetation and the tree canopy. Chaparral creates a contiguous fuel bed from the ground up that makes high-intensity crown fires inevitable.

Fire intensity represents the energy released during various phases of a fire. High-intensity fires typically consume most of the living, aboveground plant material, leaving behind only charred stems and branches.

FIGURE 6.2 The natural, physical structure of chaparral shrubs (contiguous fuel from the ground to the crown) and the seasonal pattern of drought create conditions favoring high-intensity/high-severity crown fires. *Photo R.W. Halsey.*

Fire-severity is also used to describe wildland fire but in relation to how fire intensity affects ecosystems. It is typically measured by the amount of organic material consumed by the flames (above- and belowground), or plant mortality. The manner in which fire intensity and severity are used interchangeably by different authors sometimes leads to considerable confusion (Keeley, 2009). For chaparral, however, severity measures may not be particularly helpful because high-intensity chaparral fires typically burn all the aboveground living material, leaving behind only dead, charred shrub skeletons. Fire severity has been measured by the twig diameter remaining on the terminal branches of shrub skeletons and has been shown to correlate with one measure of fire intensity (Moreno and Oechel, 1989). Even though the mature, aboveground forms of some plant species are killed, the belowground portions remain alive as lignotubers that resprout vigorously within a few weeks after the fire. In the first year after fire, massive numbers of seeds from fire-killed obligate seeding shrubs and fire-following annuals are stimulated by fire cues to germinate in the postfire environment (Keeley, 1987; Keeley and Keeley, 1987). Obligate seeding shrubs are nonsprouting species, like many Ceanothus and manzanita species, that require a fire cue for seed germination. As long as fire arrives above the lower limit of the natural fire return interval of 30–40 years, the postfire chaparral ecosystem is extraordinarily resilient and vibrant (Figure 6.3).

The size of chaparral fires varies, but the seasonal occurrence of high winds, usually from September through December at the end of California's drought period, nearly guarantees periodic large, high-intensity fire events

FIGURE 6.3 A large variety of chaparral plant species quickly resprout from underground lignotubers after a fire. In addition, the germination of seeds of obligate seeding (nonresprouting) shrubs is stimulated by heat, charred wood, or smoke. Resprouting species shown include chamise (*Adenostoma fasciculatum*; *front right*), two laurel sumac (*Malosma laurina*; *center left*), and three mission manzanitas (*Xylococcus bicolor*). Obligate seeding *Ceanothus tomentosus* seedlings are pictured in the middle of the photo. Note the diameter of the burned stems. The lack of small twigs indicates a high-severity fire. *Photo R.W. Halsey.*

across the shrubland landscape. The historical, natural occurrence of such large crown fires two to three times per century has been confirmed by multiple investigators studying charcoal sediments (Mensing et al., 1999), tree rings of big-cone Douglas-fir (*Pseudotsuga macrocarpa*) that occur in small populations on steep slopes within the chaparral (Lombardo et al., 2009), and historic records (Keeley and Zedler, 2009).

Large crown fires that have historically burned with high intensity characterize all Mediterranean-type climate shrublands around the world (California, central Chile, South Africa, southwestern Australia, and the Mediterranean Basin) (Keeley et al., 2012). In particular, the likely scenario for chaparral-dominated wildfires in California before human settlement was one of large, infrequent fires (once or twice per century) that were ignited by lightning at higher elevations during the moderate summer monsoon period between August and September. Remnants of the fire, such as smoldering logs, likely persisted into the fall. When extreme weather variables coincided (i.e., several years of drought, low humidity, high temperature, and strong winds), the fire could have reignited and rapidly spread. Fires stopped when they reached the coast or when the weather changed. Today fires ignited at higher elevations during monsoonal storms are typically extinguished by firefighters. At lower elevations, fires are vastly more frequent as a result of human-caused ignitions (Keeley, 2001).

Although counterintuitive, chaparral plant communities are much more resilient to infrequent, high-intensity fires than they are to more frequent, lower-intensity fires (Keeley et al., 2008). If chaparral does not have sufficient time to replenish the soil seed bank, accumulate the biomass necessary to produce fires hot enough to successfully germinate fire-cued seeds, or allow resprouting species time to restore starch supplies in underground lignotubers, a cascading series of events begins that can significantly change or eliminate the plant community. If the fire return interval is less than 10−20 years, biodiversity is reduced and nonnative weeds and grasses typically invade, ultimately type-converting native shrubland to nonnative grassland (Brooks et al., 2004). The process of type conversion can often occur in stages, where obligate seeding species are first eliminated, then obligate resprouters, and finally the invasion and dominance of nonnatives. Consequently, the first stages of type conversion can be missed by remote sensing techniques.

Today the average fire rotation interval (time between fires) for wildlands in southern California is 36 years, but this varies widely among different locations. Fire return intervals can vary from fires every few years in some locations to fires every 100 years or more at others (Keeley et al., 1999).

6.3 FIRE MISCONCEPTIONS ARE PERVASIVE

In conflict with ecological facts is the presumption of the fire suppression paradigm that large, high-intensity fires in chaparral are unnatural. Popularized versions of the paradigm as characterized by public opinion, the press, and Congressional testimony claim these fires are so hot that they destroy plant communities and leave behind lifeless moonscapes that are prone to mudslides that occur because of sterilized soils. It is concluded that this is the direct result of 20th-century fire suppression that allowed the chaparral to become overgrown with dense shrubs, creating massive amounts of fuel. Also, the dramatic level of recovery of postfire chaparral communities has likely reinforced the image of a fire-adapted community that "needs" fire to "rejuvenate" itself. These perspectives are not supported by research (reviewed in Keeley et al., 2012).

Following the logic of the fire suppression paradigm, two potentially damaging land management decisions can be rationalized: chaparral fires should be allowed to burn without suppression, and more fire should be added to the landscape through prescribed burning.

The problem with such decisions is twofold. First, fire is suppressed for at least this reason, to protect lives and property. Responsible fire managers will not allow a wildland fire to burn near a community. The much-maligned US Forest Service's "10 a.m. policy," whereby all possible resources are thrown at the fire with the intention of suppression by 10 a.m. the next day, or the California Department of Forestry and Fire Protection's goal of keeping all fires confined to <4 ha, are public safety policy objectives near homes. "We're protecting private lands and public lands where there's many lives at stake and homes at stake, [and] infrastructure," Duane Shintaku, California Department of Forestry and Fire Protection's Deputy Director for Resource Protection said. "… [A]nd you can't tell someone 'You know what? We're just going to see what would happen if we wait to see if it gets big.'" (Goldenstein, 2015).

Secondly, as we discuss later, too much fire—rather than not enough—is threatening many native shrubland ecosystems. The overgeneralization and misapplication of the fire suppression paradigm is the underlying cause of many of the misconceptions about wildland fire in chaparral. Ironically, fire suppression often is criticized by the agencies responsible for suppression and by citizens who have been misled by the fire suppression paradigm, yet whose lives and property are being protected.

Confusing Fire Regimes

In dry forests, the idealized behavior of frequent, low-intensity fire caused by lightning has been characterized as the "good" kind of fire because it is considered controllable, typically burning 40 ha or less and only "pruning plants" rather than "consuming" them (Sneed, 2008; Kaufmann et al., 2005; also see Chapter 13). However, such a fire is physically impossible in vegetation with the characteristics of chaparral (Figure 6.4).

FIGURE 6.4 The 2007 Zaca Fire burned more than 97,200 ha in the Los Padres National Forest, the third-largest recorded fire in California after the 1889 Santiago Canyon Fire and the 2003 Cedar Fire. Although there are unburned patches within the perimeter (note vegetation strips at the lower right, along the central ridge, and the unburned area to the *left*), wherever the flames burned, they did so at high intensity/high severity. The fire burned over the entire scene shown in the photo. *Photo R.W. Halsey.*

Emblematic of the impact caused by the misapplication of the forest-centric fire suppression paradigm is a statement made by the chair of the Santa Barbara County Fish and Game Commission, who criticized a proposal to designate chaparral as a protected, environmentally sensitive habitat: "Fire in our local ecosystems is one of the best ways to achieve the goal of good biodiversity. The local Native Americans burned almost every year. Early Spanish explorer records prove this to be true. There are many lightning-caused fires in our area, but we routinely put them out, creating an unnatural condition of heavy, dense fuel loading and harming our ecosystem in the process" (Giorgi, 2014).

The commission chair's statement would have been closer to supportable, at least in some locations, if it had only referred to the region's few higher-elevation pine forests or the mixed-conifer forests on the western slopes of the Sierra Nevada (see Chapters 1 and 2). Extending it to the chaparral ecosystem that dominates the surrounding Los Padres National Forest, however, is not supported by research. In addition, while lightning is common in high-elevation forests, the south coastal region of southern California does not experience sufficient lightning frequency to sustain the type of fire imagined by the board's chair. In fact, the region has one of the lowest lightning frequencies in North America (Keeley, 2002).

Although the research was provided to the Fish and Game Commission through testimony before and during the hearing, the commission voted to reject the proposal to designate chaparral as a sensitive habitat.

Native American Burning

The burning of landscapes by Native Americans has become an integral part of the fire suppression paradigm because it supports the practice of prescribed burning to reduce fuel loads. While it is true Native Americans burned the landscape, especially along the central coast of California, there is strong evidence that such burning led to the elimination of shrublands near population centers, rather than maintaining them in a healthy condition (Keeley, 2002). The assumption that anthropogenic burning is important to maintain healthy vegetation communities in North America conflicts with the fact that these communities existed as functioning ecosystems for millions of years prior to the arrival of humans.

Native American burning practices were performed to on a localized level near settlements to support a hunter-gatherer existence. Attempting to replicate such practices today in wildland areas would likely cause significant ecological damage. Most shrubland ecosystems already experience more fire than they can tolerate (e.g., Keeley et al., 1999). In addition, Native Americans were not faced with human-caused climate change conditions, the spread of combustible, nonnative weeds, and increased ignitions caused by millions of additional people on the landscape.

Some have also speculated that Native Americans used "controlled burning" to prevent large wildfires and promote "good fire" (Anderson, 2006; SBCFWC, 2008; https://www.universityofcalifornia.edu/news/how-indigenous-practice-good-fire-can-help-our-forests-thrive, accessed August 29, 2023). Evidence of Native American burning shows it was generally conducted within a half-day's walk from villages (Keeley, 2002), not across the landscape in an attempt to reduce the intensity and size of uncontrolled wildfires (Vachula et al., 2019). Indeed, one ethnographic report describes a massive wildfire in San Diego County before European contact that resulted in a significant migration of Native American residents to the desert (Odens, 1971).

Succession Rather Than Destruction

The perspective that high-intensity fires "destroy" the natural environment is a common theme in media stories after nearly every wildland fire (see Chapter 13). The concept is so pervasive it makes its way from public media to professional reports for decision makers. For example, Los Angeles City Council staff reported that the 2007 Griffith Park Fire "... caused significant damage to the vegetation, destroying the majority of the mixed chaparral and mixed shrub plant communities" (LACC, 2007).

As long as fire is within the parameters of the natural fire regime, a more accurate view is that large, high-intensity fires are part of a natural successional process for chaparral. Interestingly, chaparral is "autosuccessional," meaning that after chaparral burns, chaparral returns (Hanes, 1971). The first year or two after a fire, ephemeral fire-following annuals and short-lived perennials dominate, but then begin to be replaced by shrub seedlings and resprouts. The shrubs continue to grow and eventually reform the chaparral canopy within 10–15 years. This confounded early ecologists and foresters who were trained in traditional ecology to value trees over shrubs. Their response during the 1920s was to plant over a million conifers, a substantial share of which were nonnative, in the San Gabriel Mountains in Los Angeles County. Most were soon killed by drought or eventually by fire, convincing most foresters that chaparral, not forest, was the most sustainable plant community in the area (Halsey, 2011).

Although postfire ecological succession stories do sometimes make the news, they are generally overwhelmed by sensationalized reports of flames, destruction, and blackened landscapes. As remarkable as the postfire chaparral environment is—with hills covered with colorful wildflowers, resprouting shrubs, and large clusters of seedlings emerging from the dark soil—the perception that the environment has been destroyed by fire remains a pervasive image.

Decadence, Productivity, and Old-Growth Chaparral

When discussing the impact of fire, one must take care not to fall into the trap of anthropomorphizing a wild ecosystem like chaparral and thinking fire is needed to "refresh" or "clean out" old, "decadent" or "senescent" growth (Hanes, 1971). These characterizations of older chaparral stands have not been supported by subsequent research (see, e.g., Moritz et al., 2004; Keeley, 1992).

Multiple studies have demonstrated the ability of old-growth chaparral, nearly a century old or more, to maintain productive growth and recover with high biodiversity after a fire (Hubbard, 1986; Keeley and Keeley, 1977; Larigauderie et al., 1990). In fact, long fire-free periods are required for many species to properly regenerate (Odion and Tyler, 2002; Odion and Davis, 2000; Keeley, 1992).

With legacy manzanitas having waist-sized trunks, a rich flora of lichens rarely found anywhere else (Lendemer et al., 2008), and a dense canopy forming a protective watershed, old-growth chaparral provides an important habitat for a wide array of species and valuable ecosystem services to surrounding human communities. As such, old-growth chaparral represents a crucial component in the preservation of California's biodiversity (Keeley, 2000; Halsey and Keeley, 2016) (Figure 6.5).

Sometimes, a trailside sign or textbook description of chaparral includes the specter of "undisturbed climax chaparral" eventually becoming so thick that it will either "choke itself," "die out," or be replaced by woodland (Ricciuti, 1996). While trees will overtop and shade out chaparral in areas with higher annual rainfall and richer soil conditions than exist in most chaparral sites, the general belief that chaparral will eventually disappear because of age is not supported by data (Keeley, 1992).

The imagined fate of old-growth chaparral illustrates the common genesis of many misconceptions where anecdotal evidence has replaced scientific

FIGURE 6.5 Old-growth chaparral in San Diego County, California. A big-berry manzanita (*Arctostaphylos glauca*) has wrapped itself around an Engelmann oak (*Quercus engelmannii*). The manzanita is estimated to be over a century old. *Photo R.W. Halsey.*

investigation—observations that may have merit in a limited, specific instance but have been broadly misapplied to support a binary paradigm. The remarkable nuances of nature as revealed by science are often ignored.

Unfortunately, with increasing fire frequency, old-growth stands of chaparral (in excess of 60-year-old) are becoming increasingly rare (Knudsen, 2006). And, while biodiversity does temporarily increase after a fire because of the germination of ephemeral fire-following species, there is no danger that this biodiversity is threatened by long fire-return intervals. The soil seed bank can likely remain viable for a significant amount of time. Shrublands burned after approximately 150 years respond with a rich array of seedlings (Keeley et al., 2005b) (Figure 6.6). Considering the number of human-caused ignitions, there is no need to be concerned over the lack of fire.

FIGURE 6.6 A large number of fire-following annuals and short-lived perennials emerge from the soil seed bank after a high-intensity chaparral fire. In addition, geophytes emerging from underground tubers, like this brodiaea (*Dichelostemma capitatum*), are likely stimulated to flower by additional sunlight provided by the removal of the chaparral canopy by fire. *Photo R.W. Halsey. Tyler and Borchert (2007).*

Allelopathy

Another factor mentioned to support the notion that fire is "needed" in chaparral is allelopathy, the theorized phenomenon of plants releasing chemicals to suppress the growth or germination of neighboring competitors. It has been suggested that such chemical inhibition explained the lack of plant growth under the canopy of mature chaparral stands in southern California (Muller et al., 1968). When the chaparral burned, the theory suggested, flames denatured the toxic substances in the soil, thereby releasing the seeds from inhibition and suggesting the need for fire. One problem with this explanation is that the soil chemicals suspected of suppressing growth actually increase after a fire (Christensen and Muller, 1975).

The seeds of most chaparral plants are innately dormant before they contact the soil because of their dependency on fire cue-stimulated germination. In addition, the presence of herbivores has been demonstrated to be a major factor in eliminating seedlings that do germinate (Bartholomew, 1970). Therefore, the lack of seedlings under the canopy and the postfire seedling response in chaparral can be easily explained without considering chemical inhibition (Halsey, 2004). Despite the research, however, allelopathy in chaparral is still presented as fact in college courses and texts (SBCC, 2002; George et al., 2014).

Fire Suppression Myth

"Fuel build-up," as per the fire suppression paradigm, is invariably blamed in most media stories for the occurrence of large chaparral wildfires, despite being rejected by research.

In analyzing the California Statewide Fire History Database since 1910, Keeley et al. (1999) concluded that for shrub-covered landscapes of southern and central coastal California, "there is no evidence that fire suppression has altered the natural stand replacing fire regime in the manner suggested by others." In fact, fire suppression in California's Pacific south coast has played an important role in *protecting* much of the chaparral from *too* much fire. The authors of a comprehensive summary of the literature about fires in the region concluded the following (Keeley et al., 2009a): "The fire regime in this region is dominated by human caused ignitions, and fire suppression has played a critical role in preventing the ever increasing anthropogenic ignitions from driving the system wildly outside the historical fire return interval. Because the net result has been relatively little change in overall fire regimes, there has not been fuel accumulation in excess of the historical range of variability, and as a result, fuel accumulation or changes in fuel continuity do not explain wildfire patterns."

Unfortunately, fire suppression in shrublands has not been completely successful in protecting chaparral and sage scrub habitats from too much fire. Shrublands in areas surrounding the San Diego, Los Angeles, and Santa Barbara metropolitan areas have some of the most negative fire return interval departures in California, meaning they are experiencing more fire than they have historically, threatening the chaparral's resilience (Safford and Van de Water, 2014). The problem seems to be spreading north into the northern Santa Lucia Range and may likely continue to spread as climate change and population growth increase the potential for ignitions.

Too Much Fire Degrades Chaparral

Chaparral is highly resilient to infrequent fire, within the natural range of variability, and postfire communities are remarkable in their capacity to return to prefire composition within a decade or so, with the community assembly finely balanced with resprouting and seeding species. Nevertheless, given increases in fire frequency, this resiliency can be interrupted. "Type conversion" is the term given to changes in vegetation type caused by changes in the external environment, and one of the most common disturbances is accelerated fire frequency. When keystone, nonresprouting (obligate seeding) shrub species, like most *Ceanothus* species, experience closely spaced fires, their populations often are decimated and effect a type conversion to a less diverse, resprouting-dominated shrubland (Zedler et al., 1983). Such stands become more open and often are subsequently invaded by nonnative herbaceous species. Fire return intervals of less than 6 years have been shown to be highly detrimental to the persistence of nonresprouting chaparral species (Jacobsen et al., 2004); in fact, multiple fires within a 6-year interval have even reduced resprouting species, further opening the chaparral environment (Haidinger and Keeley, 1993).

Type conversion has been an ongoing process since the arrival of humans in California (Wells, 1962). The process is complex, dependent on fire history, community composition, and site factors. The loss of shrub cover and the invasion of combustible nonnative grasses creates a positive feedback loop (Keeley et al., 2005a) whereby the community assembly changes, further increasing fire frequency and causing further type conversion away from the original stand composition. The speed of the type conversion process can be increased dramatically by numerous variables such as drought, cool-season fires (Knapp et al., 2009), livestock grazing, soil type, soil disturbance, and mechanical clearance activities (Bentley, 1967).

During extended periods of drought, seedling success of obligate seeding shrubs, like many *Ceanothus* species, is reduced after fire. In fact, excessive soil temperatures resulting from drought-induced canopy reduction after adult die back between fires has been shown to cause the premature germination of *Ceanothus megacarpus* seedlings just before the seasonal drought period (Burns et al., 2014). Seedling survival under such conditions is questionable, potentially depleting the seed bank.

Record drought conditions after fire also increase the mortality of resprouting chaparral shrubs like chamise (*Adenostoma fasciculatum*) and greenbark (*Ceanothus spinosus*). Resprouting shrub species likely deplete their carbohydrate reserves during the resprouting process, making them particularly vulnerable to drought because of the need to transpire water to acquire carbon dioxide that is used to supply energy to a large, respiring root system (Pratt et al., 2014). An additional fire within a 10-year window adds even more stress to resprouting species.

It is beyond question that type conversions occur and that severe type conversion from evergreen chaparral to alien-dominated grasslands has significantly altered the Californian landscape in the past (Wells, 1962; Keeley, 1990). For example, Talluto and Suding (2008) found that over a 76-year period, 49% of the sage scrub shrublands in one southern California county had been replaced by annual grasses and that a substantial amount of this could be attributed to fire frequency.

In recent years, southern California has experienced some rather extensive reburns at anomalously short intervals (Keeley et al., 2009b), potentially setting the stage for the disruption of natural ecosystem processes and type-converting these shrublands to a mosaic of exotic and native species. This has already been documented for a number of sites where short-interval fires have extirpated some native species and greatly enhanced alien species (Keeley and Brennan, 2012). As discussed above, within the four southern and central/coastal national forests in California, most of the shrublands—the dominant plant communities within these federal preserves—are threatened by excessive fire (Figure 6.7).

Quantifying how much chaparral has been compromised or completely type converted is a challenging research question because much of the damage was likely accomplished before accurate records of plant cover were kept. Based on relic patches of chamise and historical testimony, Cooper (1922) speculated that extensive areas of chaparral have been eliminated and converted to grasslands, including the floor of the Santa Clara Valley, large portions of the Sacramento and San Joaquin Valleys, and many of the grassy regions in the Coast Ranges and the western Sierra foothills. Large areas along Interstate 5 in the Cajon Pass region, the foothills above San Bernardino, and the Chino Hills south of Pomona also appear to be type converted landscapes.

The focus on complete type conversion to grassland has led some to ignore the beginning stages of the process: the simplification of habitat by the loss of biodiversity (Keeley, 2005). For example, in a comment letter on the draft 2010 California Fire Plan, San Diego County claimed that chaparral burned in

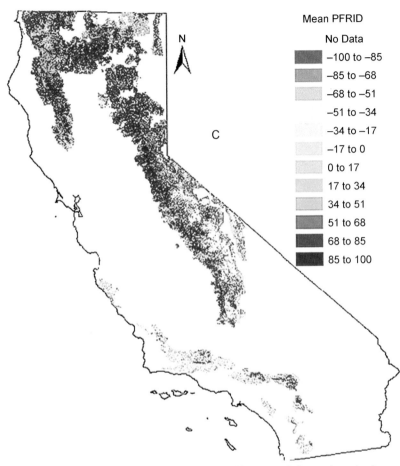

FIGURE 6.7 Most chaparral in California is threatened by too much fire, as shown by the map's color variations representing the fire return interval departure (PFRID) percentages for national forest lands in California. Note the color differences between the southern California national forests, which are dominated by chaparral (*yellows*), and the conifer-dominated forests in the Sierra Nevada (*blues*). The warm colors identify areas where the current fire return interval is shorter than that before European settlement (negative PFRID percentages). Cool colors represent current fire return intervals that are longer than those before European settlement (positive PFRID percentages). *Photo R.W. Halsey. From Safford and Van de Water (2014).*

both the 2003 and 2007 wildland fires "remained chaparral and is recovering" (Steinhoff, 2010). In fact, much of the chaparral in question was not recovering properly because of the loss of several keystone shrub species and was showing significant invasion by nonnative grasses (Keeley and Brennan, 2012) (Figure 6.8). Despite the research, some continued to question the fact that

FIGURE 6.8 The impact of excessive fire on chaparral. The entire area shown was burned in 1970. The *middle/left* area burned again in 2001 and is returning with a full complement of native chaparral species. In the *right portion*, which burned again in 2003, obligate seeding species are absent, the number of resprouting species has been reduced, and nonnative weeds have invaded. The interval between the last two fires was too short, causing a dramatic reduction in biodiversity and leading to type conversion. The location pictured is near Alpine, San Diego County, California. *Photo R.W. Halsey.*

chaparral in San Diego County was being type converted by too frequent fire (Oberbauer, 2018).

Meng et al. (2014) recently raised some skepticism about the ability of repeat fire to effect type conversion by pointing out the difficulty early 20th-century range managers experienced when using fire to "improve" ranges that were supposedly plagued by chaparral. These managers typically relied on herbicides and mechanical destruction for thorough replacement of shrubs to create more useful grazing lands. As pointed out by Keeley and Brennan (2012), however, managers utilize fire only under narrow prescription conditions, which are generally not capable of carrying repeat fires at short fire return intervals—hence their difficulty in meeting their objective. By contrast, wildfires typically burn outside prescription, often with 100 km/h (about 60 mile/h) wind gusts and relative humidity less than 5%.

Using remote sensing, Meng et al. (2014) attempted to answer the question of how extensive type conversion is caused by repeat fires occurring in the past decade. While the technique cannot address changes in diversity and species composition that are known to occur with short-interval fires, it has some potential for viewing more gross changes in functional types such as shrubs and annual plants. Although these authors concluded that widespread type conversion is not an immediate threat in southern California, this conclusion deserves closer scrutiny because documenting fire-related vegetation change across large landscapes over just a 25-year period using remote sensing is fraught with potential errors and cannot serve as an effective proxy for field data.

One reason for error is that numerous spatially and temporally different human and biophysical factors can influence the process of postfire recovery; these factors should be controlled for before attribution can be determined. In the paper by Meng et al. (2014), the control and overlap areas were located on somewhat adjacent, but very different, parts of the landscape that varied by factors such as aspect, terrain, or soil type. The areas also could have experienced different landscape disturbance histories. This is especially possible given the topographic complexity of the region and researchers' use of the California's Fire Resource and Assessment Program's Fire History Database (FRAP) for discerning precise stand ages. This database is broadly useful for management planning but must be used carefully in a research context. For example, Keeley et al. (2008) found that across 250 sites the FRAP database did not accurately portray stand age (as determined by ring counts) for 47% of the sites, presumably because of the scale at which fires are mapped and by generally ignoring fires less than 40 ha.

Another concern is that the method of documenting vegetation change used by Meng et al. (2014) may not be sensitive enough to resolve gradual shifts in composition that would likely occur after only one repeat fire event. They used a vegetation index derived from imagery sensed remotely from a satellite as a way of assessing vegetation "cover," or the "greenness," of each 30-m image pixel. Because different pigments are stimulated by different parts of the light spectrum, this index essentially assesses chlorophyll content, which is correlated with biomass and assumed to represent the relative cover of evergreen shrubs. It does not, however, account for differences among chaparral species, whose composition in the plots was unknown. In addition, different species of chaparral have varying sensitivities to repeat fires, and thus multiple repeat fires of differing intervals might be required to discern enough vegetation change to be detected by this index.

Given that vegetation change is likely a gradual, cumulative process, the results reported by Meng et al. (2014) are actually consistent with a potential for widespread chaparral conversion—contrary to their conclusions. Over half of the area that burned twice in their study did have lower cover, as defined by the index, than the control area. Given enough fire on the landscape over a long enough period, gradual shifts may result in significant change and impact.

Type Conversion and Prescribed Fire

The priorities of land management agencies have led some to deny the existence of chaparral type conversion. For example, in the same comment letter mentioned above, San Diego County wrote that it "strongly disagreed" with the draft 2010 California Fire Plan because it contained the following statement: "… fires have been too frequent in many shrublands, especially those of southern California, which are then at risk of type conversion from native species to invasives that can pose a fire threat every fire season."

The county explained that recognizing the threat of chaparral type conversion in the Fire Plan would impact its ability to obtain funding to carry out vegetation clearance activities.

Prescribed burning—one of the clearance activities that San Diego County was hoping to conduct—has been shown to seriously compromise chaparral plant communities. In a study that simulated the effect of frequent fire on southern California coastal shrublands, Syphard et al. (2006) concluded that, "Due to this potential for vegetation change, caution is advised against the widespread use of prescribed fire in the region."

One of the problems with prescribed burning in chaparral is that there is a narrow window when such burns can occur: in the cool season (late spring). Plants have too much moisture in their tissues in the winter and early spring months to carry a fire. In the summer and fall, the risk of wildfire is too high because of low moisture levels and weather conditions. Consequently, prescribed burns are conducted when the chaparral ecosystem is likely most vulnerable to damage by fire: the plants are growing, the soil is still moist, many animal species are breeding, and some birds are occupying the chaparral during their annual migrations. Thus, significant ecological damage can occur as a result of a prescribed burn (Knapp et al., 2009).

The exact mechanisms are not clearly understood, but cool-season burns likely cause significant damage to plant growth tissues and destroy seeds in the soil as soil moisture turns into steam. A prescribed burn conducted in the 1990s in Pinnacles National Park, California, led to immediate type conversion of chamise chaparral to nonnative grassland (Keeley, 2006). An escaped prescribed burn in 2013 consumed more than 1090 ha of fragile desert habitat in San Felipe Valley, California, much of which was chaparral that was recovering from a fire 11 years before. The fire compromised one of the last old-growth desert chaparral stands in the region (CCI, 2013) (Figure 6.9).

Combustible Resins and Hydrophobia

The loss of vegetation after a fire exposes more soil surface and increases the kinetic force of precipitation on the soil, which can increase the flow of water on the surface. The result can be significant erosion, flash flooding, and large debris flows. However, a factor that seems to get more attention than its proven influence indicates is water repellency, or "hydrophobic soils."

The observation that heat during a fire can change or intensify the water repellency of soil depending on temperature and other factors has been studied extensively (DeBano, 1981; Hubbert et al., 2006) and was first identified after chaparral fires. The hydrophobic soils theory suggests that because of gas released by burning plants and soil litter, hot fires create an impermeable "waxy layer" a few inches below the surface. According to popular accounts, this layer then prevents water from permeating into the ground, causing large chunks of topsoil to break loose during rain storms and slide down the hill

FIGURE 6.9 Photo shows an escaped 40 ha prescribed fire in the San Felipe Valley Wildlife Area, San Diego County, California, that ultimately burned more than 1000 ha, most of which was 11-year-old desert chaparral. Considering the ecological fragility of the area because of its age and the multiple fires that have burned much of the valley over the previous decade, there likely will be a significant reduction of biodiversity in the region. *Photo R.W. Halsey.*

(LAT, 2014). Warnings about the hazards of such waterproof layers are commonly raised by the media after fires.

However, the actual impact hydrophobic soils have on erosion is questionable. Contrary to popular accounts, water repellency is not like a layer of plastic wrap under the surface; instead it is quite patchy and transient, abating once soils are wetted. Water repellency is also a natural condition of many unburned soils. In fact, high-intensity fires have been found to destroy water repellency in the soil (Doerr et al., 2006). In a review of the literature, Busse et al. (2014) concluded the following: "Most studies have only inferred a causal link between water repellency and erosion, and have failed to isolate the erosional impacts of water repellency from the confounding effects of losses in vegetation cover, litter cover, or soil aggregate stability."

The theorized role hydrophobic soils play in erosion is used to justify questionable, and sometimes expensive, land management decisions. The chaparral has been especially targeted.

To justify the clearance of native chaparral habitat, the Arizona Game and Fish Department claimed that "... catastrophic wildfire in the chaparral type can burn intensely enough to create hydrophobic soils, reducing soil productivity, increasing erosion, and causing severe downstream flooding" (AGFD, 2007). The City of Los Angeles spent $2 million to spread mulch after the 2007 Griffith Park Fire in part because "... chaparral vegetation has a natural tendency to develop water repellent or hydrophobic soils due to their natural high wax content. As a result, burned watersheds generally respond to runoff faster than unburned watersheds ..." (LACC, 2007).

FIGURE 6.10 Postfire treatments in chaparral are costly and often of questionable value. Strips of mulch were dropped by aircraft on the side of the Viejas Mountain in San Diego County after the 2003 Cedar Fire. *Photo R.W. Halsey.*

More than $1.25 million was spent laying down strips of mulch on Viejas Mountain in San Diego County after the 113,473 ha, high-intensity 2003 Cedar Fire, ostensibly to control erosion (Figure 6.10). However, Viejas Mountain is composed of gabbro-type soils that are not typically prone to extensive erosion (Halsey, 2008). Hydrophobic soils also have been used to justify postfire "salvage" logging after the 2013 Rim Fire in the Stanislaus National Forest (USFS, 2013).

6.4 REDUCING COGNITIVE DISSONANCE

Despite research that disproves many of the commonly held misconceptions about fire in chaparral that are fostered by the fire suppression paradigm, misconceptions persist. Many have found their way into land management plans that advocate landscape-scale "fuel treatments" or vegetation management projects for the stated purpose of "returning" California's chaparral ecosystem to a more "natural," less dangerous fire regime. How the media, policymakers, and managers have responded to the cognitive dissonance that occurs when their assumptions about fire are challenged by science provides insight into the difficulties encountered when new ideas confront embedded paradigms.

Festinger (1957) suggested there are several ways an individual can reduce the tension caused by facts or ideas that conflict with their own opinions. Individuals can respond with cognitive competence by accepting the new data point or idea and change their opinions accordingly. Alternatively, individuals can respond incompetently by rejecting the new data point or idea either by ignoring or denying it or by justifying their opinion with new information or beliefs, occasionally using logical fallacies in the process.

For example, as mentioned earlier, when the chair of the Santa Barbara County Fish and Game Commission cited Native American burning as an argument for why we should not suppress fires in chaparral, he was using the common logical fallacy of appealing to antiquity. Such an appeal assumes older ideas or practices are better than newer ones because they have been around for a long time.

Local Agency

In an attempt to alter the natural fire regime, San Diego County tried to establish a chaparral clearance program that targeted more than 780 km^2 of back country habitat with "prescribed fire, mechanical or biochemical fuel treatments" (SDCBS, 2009). This effort was based on a report issued earlier by the county. In misapplying the fire suppression paradigm to native shrublands, the report claimed that, "A fire regime of smaller, more frequent fires was being replaced by one of fewer, larger, and more intense fires" because of an unnatural density of "fuel" due to past fire suppression (SDCBS, 2003).

Despite data submitted by reviewing scientists over a period of more than 4 years, indicating the county was basing its policies on incorrect assumptions, the county's Planning and Land Use Department issued several new drafts of its vegetation management plan without correcting the errors (Halsey, 2012).

In a comment letter by the Conservation Biology Institute, scientists wrote (Spencer, 2009): "Although this fourth draft is an improvement over previous drafts, it reflects partial and piece-meal updating based on various submitted comments and the workshop discussions rather than the comprehensive re-write that is necessary. This results in the report being internally inconsistent, confusing, and often self-contradictory. Moreover, despite scientific facts and logic presented to the county by numerous individuals, the report continues to perpetuate disproved myths about fires and fire management in southern California."

In addition to ignoring information contrary to its position, the county misinterpreted the science in a manner that justified its viewpoint. One of the scientists whose work was the subject of the county's misinterpretations in the 2003 report wrote: "We were disturbed by the way our research findings were completely mischaracterized in this report on page 8. Not only are the specific statements about our findings completely false, but also, more generally, our research does not support the claims and recommendations of this section of the report" (Schoenberg and Peng, 2004).

The San Diego County Board of Supervisors eventually adopted a final vegetation management plan in 2009, with most of the inaccurate information removed but with a number of questionable ecological assumptions about chaparral remaining. Within a month of the plan's adoption, the county attempted to implement the report's first clearance project without conducting the appropriate environmental review as required by the California Environmental Quality Act. The county claimed an "emergency exemption."

However, the California Chaparral Institute (CCI), an environmental nonprofit organization based in San Diego County, successfully challenged the project in court. The court rejected the county's position that a 3- to 4-year, $7 million vegetation management project was a "short-term project" addressing an immediate, emergency occurrence. In an attempt to influence the judge, county counsel used the logical fallacy of appealing to emotion by warning of death and destruction during future fires if the court ruled against the county. Since the hearing was considering a point of law, not evaluating emotional pleas, the court was not swayed, ordering the county to follow the proper procedures under the law.

The county ultimately produced a full environmental impact report (EIR) on the project after being challenged again by CCI when it attempted to avoid the review process through a negative declaration. The EIR was certified and the county completed the initial site-specific clearance project in 2012. The county later dropped the larger regional plan that had been so severely criticized.

State Agency

In 2012, the California Board of Forestry and Fire Protection proposed a statewide Vegetation Treatment Plan (VTP) that targeted more than one-third of the state for potential vegetation clearance operations. The VTP stated that large-scale wildland treatments should focus on areas "… up to the watershed scale, or even greater, that are treated to reduce highly flammable or dense fuels, including live brushy plants in some vegetation types (such as chaparral), a buildup of decadent herbaceous vegetation or, dead woody vegetation." One of the rationales for the VTP was that "[p]ast land and fire management practices (fire suppression) have had the effect of increasing the intensity, rate of spread, as well as the annual acreage burned on these lands" (CSBF, 2012).

As with the San Diego County example, the research contradicted this conclusion when applied to chaparral. Commenting on the VTP's stated intent to "reintroduce fire into (natural) communities where fire has been excluded through past fire suppression efforts," the California Department of Fish and Wildlife (CDFW, 2013) wrote: "There is substantial evidence that the frequency of fires continues to increase in coastal southern California (USDI NPS, 2004; Keeley et al., 1999). Fire management of California's shrublands has been heavily influenced by policies designed for coniferous forests; however, fire suppression has not effectively excluded fire from chaparral and coastal sage scrub landscapes and catastrophic wildfires are not the result of unnatural fuel accumulations (Keeley, 2002). There is also considerable evidence that high fire frequency is a very real threat to native shrublands in southern California, sometimes leading to loss of species when fire return intervals are shorter than the time required to reach reproductive maturity (Keeley, 2002)."

In contrast to San Diego County's reluctance to accept new scientific research, the state initially responded with cognitive competence. After the state board received criticism from fire scientists that the VTP did not reflect the most current research, the California State Legislature asked the California Fire Science Consortium, an independent network of fire scientists and managers, to review the proposal. The Consortium recommended that the VTP "undergo major revision if it is to be a contemporary, science-based document" (CFSC, 2014). The board then began the process of rewriting the document in 2014, with assurances they would be modifying their plan by incorporating the new information and offering opportunities for the original reviewers to provide input on the developing draft.

However, in 2019 the Board of Forestry dropped its previous VTP effort, restarting the process with a new EIR. The new document called for the management of thousands of hectares of chaparral despite recognizing that the system was endangered by increasing fire frequencies. Although acknowledging the threat of type conversion, the document turned over the decision of what constituted the process to the proponent of the chaparral clearance project. The EIR was approved by the Board of Forestry, but was immediately challenged with a California Environmental Quality Act lawsuit by the California Chaparral Institute and the Endangered Habitats League. The litigation was scheduled to be heard in court in 2023.

Media

The popular media poses a particular problem because reporters often do not specialize in one topic long enough to become familiar with contrary data that question prevailing paradigms. When confronted with new information, however, the media outlet has several options. It can provide time or space for an editorial response, publish another story on the subject, or make a concerted effort to incorporate the new information into its editing process for future stories.

For example, significant cognitive competence has been demonstrated by one of California's most influential newspapers, the *Los Angeles Times*. The paper has become familiar with the science and has helped its readers understand that too much fire is threatening the chaparral (LAT, 2009), recommended the California Board of Forestry withdraw its original vegetation management plan and produce a new one using the best available science (LAT, 2013), and commonly describes the state's characteristic ecosystem as chaparral rather than using the older, derogatory term "brush."

A San Francisco Bay-area publication, on the other hand, provides an example of how older, inaccurate information was allowed to persist. In an article about fires in the chaparral-dominated Ventana Wilderness area of the Los Padres National Forest, Rowntree (2009) wrote, "Because of fire suppression policies and strategies put into place in 1907, fires became relatively infrequent. But when fires happened, and Marble Cone is a prime example, the immense accumulated fuel led to hotter, more intense fires compared to those associated with a more natural fire regime."

There is no scientific evidence to support the claim that chaparral fires are burning hotter or more intensely than they have historically. The Marble Cone Fire cited in the article burned approximately 72,000 ha in 1997. However, in 1906, before the fire suppression era began, approximately 60,700 ha burned with equal intensity in the same area. Other large, intense fires in the region were recorded even earlier (J. Keeley, unpublished data).

In southern California, the 2003 Cedar Fire in San Diego County, which burned 113,473 ha, is often referred to as southern California's largest fire. But in 1889, the Santiago Canyon Fire burned an estimated 125,000 ha (and possibly as much as 200,000 ha, depending on the estimates used) in San Diego, Orange, and Riverside Counties (Keeley and Zedler, 2009).

The area burned in the 2017 Tubbs Fire in and around Santa Rosa, California, had burned in the equally large Hanly Fire in 1964, and again prior to that in the 1870 Great Fire (Elliot, 2019).

Although the capacity for large fires has not changed, the number of people and homes in the way of the flames certainly has. Over the past century, high-intensity chaparral-related wildfires have continued to be some of the largest and most devastating conflagrations in the United States in terms of property and lives lost (Halsey, 2008).

The author of the aforementioned article on the Marble Cone Fire reinforced the misconception that large, high-intensity fires are unnaturally destructive because they roar across the landscape, "destroying oak, madrone, chamise, manzanita, and all other shrubs and trees in its path." The impression made was that if the Marble Cone Fire had been natural, it would have been a low-intensity surface fire that "smolders as it slowly works its way through grasslands and chaparral." The presumed destructive nature of hydrophobic soil was also cited in the article as being responsible for creating "the slippery foundation for the mud-flows that caused havoc on Highway 1 ...."

After receiving a critique citing the errors, the publishers stood by their story. Although a website-based opportunity was offered for a short critique, the publishers rejected a follow-up article or comment letter because there was not enough room in the magazine for an additional discussion of an issue as complex as fire (D. Loeb, personal correspondence, 2010).

6.5 PARADIGM CHANGE REVISITED

In his seminal work on the structure of scientific revolutions, Thomas Kuhn (1962) wrote "... the proponents of competing paradigms practice their trades in different worlds."

For the proponents of the fire suppression paradigm, wildfire is primarily a fuel-driven event. Thus, controlling fuels controls fires, as the thinking goes, and native vegetation is viewed not as habitat but rather as unwanted fuel.

Alternatively, by framing wildfire in context of the entire environment, whereby other variables such as weather can play more important roles than fuel and where vegetation is viewed as wildlife habitat, a less invasive fire management paradigm can be developed.

The first paradigm is embedded in a controllable world where nature can be tamed, whereas in the second one, nature will ultimately defeat control. One sees nature as fuel; the other sees nature as providing important habitat in both its pre- and postburned conditions (also discussed in the Preface and Chapters 1, 2 and 12). One focuses on manipulating wildlands to control wildfire, the other on community retrofits and planning to make communities more fire-safe (Penman et al., 2014). As Kuhn explains, the two groups, "… see different things when they look from the same point in the same direction. Again, that is not to say that they can see anything they please. Both are looking at the world, and what they look at has not changed. But in some areas they see different things, and they see them in different relations one to the other. That is why a law that cannot even be demonstrated to one group of scientists may occasionally seem intuitively obvious to another.

The feeling one may have during an argument that the other party is operating in another universe can in fact be an accurate description of what is happening."

Although other drivers of fire behavior are sometimes acknowledged, the practical implementation of policies resulting from the fire suppression paradigm is an exclusive focus on fuels (wildland vegetation). In this view, any fuel is too much fuel. Such a viewpoint was offered during congressional testimony after the 2003 chaparral-dominated wildfires in southern California (Bonnicksen, 2003): "Some people believe that horrific brushland fires are wind-driven events. They are wrong. Science and nearly a century of professional experience shows that they are fuel driven events. Wind contributes to the intensity of a fire, but no fire can burn without adequate fuel, no matter how strong the wind."

Besides the logical fallacy of appealing to unnamed authorities, this argument sets up a straw man fallacy. By misrepresenting the science that challenges the fuel-centric position and then refuting it, the congressional witness concludes that the science itself has been refuted. This is a fallacy because the science that is claimed to be refuted is actually being misrepresented.

Clearly, fire needs fuel to burn. Excepting extreme situations, all terrestrial environments have some kind of fuel, be it grass, shrubs, trees, or houses; all can provide adequate fuel for a fire under the right conditions. The science that challenges the fire suppression paradigm does not hold that fire can burn without fuel.

As the ~365,000 ha East Amarillo Complex grassland fire in Texas demonstrated in 2006, fine, grassy fuels also can also cause horrific fires. Twelve people died and 89 structures were destroyed in a fire that moved

72 km in just 9 h and had flame lengths >3.5 m (Zane et al., 2006). The 2021 grass-fueled, 2438 ha Marshall Fire in Colorado killed two people and burned more than 1000 homes (Holmstrom et al., 2022).

Notably, fuel reduction projects have been shown to be ineffective when it matters most: during extreme fire weather. During such conditions, the fire is not controllable because it will burn through, over, or around fuel treatments (Keeley et al., 2004, 2009b). Many fuel breaks never intersect fires, but those that do nearly always require the presence of a fire crew to be effective, demonstrating the importance of a fuel break's strategic location (Syphard et al., 2011). An extensive study of chaparral fires throughout central and southern California showed that there is not a strong relationship between fuel age and fire probabilities (Moritz et al., 2004). Even in fuels-reduced forests, burning under extreme weather conditions can produce large areas of high-severity fire (Lydersen et al., 2014). Extensive fuel treatments in a forest can also fail to prevent extensive damage to a community, such as Lake Arrowhead during the 2008 Grass Valley Fire, if the structures themselves are not fire-safe (Rogers et al., 2008).

Paradigms have a challenging intellectual duality because not only can they guide productive research but they can also blind. Proponents of an older paradigm can ignore overwhelming, contrary evidence or force it to fit their model. As Thomas Chamberlin (1890) wrote in his paper concerning the value of multiple working hypotheses, "There is an unconscious selection and magnifying of the phenomena that fall into harmony with the theory and support it, and an unconscious neglect of those that fail of coincidence."

In addition to the force of paradigm, financial pressure can be involved in propelling an idea beyond its proven effective value. When the 2003 Healthy Forests Restoration Act was passed by Congress, a significant source of money was made available for fuel treatments on public and private land. Shortly after the passing of the act, a US Forest Service supervisor summit was held in Nebraska, where forest supervisors were asked to sign a pledge to meet their forests' hazardous fuel targets. A clear signal was being sent from Washington, DC, that clearing vegetation was going to be a primary goal. The act codified the fire suppression paradigm and encouraged the perspective of habitat as hazardous fuel, regardless of the natural fire regime.

Don G. Despain, one of the original scientists who advocated allowing fire to perform its natural role in ecosystems, met with other wildland fire pioneers like Les Gunzel, Robert Mutch, and Bruce Kilgore in Missoula, Montana in 1972 to discuss ways they could change how fire was viewed. "We were a pretty lonely bunch back then," Despain explained in a 2006 interview (D. Despain, personal communication). But, as time went on and attitudes about fire began to shift, Don began to notice that the impact of past fire suppression was being taken too far. Alternative variables that may have influenced fire behavior in the West were being ignored. "So many assumptions about fire were being made that had never been observed," Don said. "I came to think I was the only person to watch a fire actually burn. People need to get out and observe and apply natural curiosity with what is going on instead of running to the legislature."

6.6 CONCLUSION: MAKING THE PARADIGM SHIFT

In the 1990s, the predominate view of chaparral within the Pacific Southwest Region of the US Forest Service was that the ecosystem represented primarily fuel, needed more fire, and that large chaparral wildfires were a direct product of 20th-century fire suppression. Although there remain Forest Service managers who still hold these views, the agency has demonstrated cognitive competence by accepting new information, rejecting the fire suppression paradigm as it had been applied to chaparral, and adjusting its official policies accordingly.

The shift began in 2000, after three papers that seriously questioned the prevailing views were published (Keeley et al., 1999; Mensing et al., 1999; Zedler and Seiger, 2000). These papers stimulated a significant volume of research, confirming that the fire suppression paradigm was not applicable to California's chaparral ecosystem.

John Tiszler (2000) wrote a white paper questioning the use of prescribed fire in the chaparral-dominated Santa Monica Mountains National Recreation Area (SMMNRA) within the National Park system. After the 2000 Cerro Grande Fire in Colorado, the National Park Service established a moratorium on prescribed fire and began a reexamination of its parks' fire management policies. New fire ecologists reexamined the SMMNRA's approach to fire and rejected the fire suppression paradigm. By 2005, a new fire management plan was formalized for the park (SMMNRA, 2005). The new approach is summarized on the park's website (SMMNRA, 2015): "In the last forty years fire managers have promoted the idea that prescribed fire is necessary to protect ecosystems and communities by restoring fire's natural role in the environment to thin forest stands and to reduce hazardous fuels. This is true for western forests where the natural fire regime was frequent, low intensity surface fires started by lightning ... However, this is not true for the shrubland dominated ecosystems of southern California and the Santa Monica Mountains."

After the 2003 Cedar Fire in San Diego County, California, organizations such as the California Chaparral Institute and the Endangered Habitats League raised awareness about the value of native shrublands. Through publications, public outreach, and occasional legal challenges, the organizations helped to communicate the new science to both the public and government agencies.

By 2013, the paradigm shift occurred and the US Forest Service published a guiding document that redefined their view of chaparral and recognized how excessive fires were threatening the ecosystem (USFS, 2013): "There is an additional crisis taking place in our Southern California Forests as an unprecedented number of human-caused fires have increased fire frequency to the extent that fire-adapted chaparral can no longer survive and is being replaced with non-native annual grasses at an alarming rate ... Only an environmental restoration program of unprecedented scale can alter the direction of current trends."

On June 18, 2013, during an important US Forest Service symposium at the headquarters of the Angeles National Forest, Martin Dumpis, the coordinator for a new Forest Service initiative focusing on the protection and restoration of chaparral, summarized the new approach well. "Chaparral should be seen as a natural resource, rather than a fire hazard."

A complete paradigm shift will require a growing appreciation for not only the "ecosystem services" provided by chaparral and the intrinsic value of the plant community itself (Halsey et al., 2018) but also the elimination of the financial incentives to clear chaparral habitat based on fuel-centric notions about wildfire.

Attesting to the psychological component of adhering to older paradigms, psychologist Carl Jung (1992) provided a way out that is especially poignant to the subject of this chapter: "Nature is an incomparable guide if you know how to follow her."

REFERENCES

AGFD, 2007. South Mowry Habitat Improvement Project. Arizona Game and Fish Department. Habitat Partnership Program. Habitat Enhancement and Wildlife Management Proposal. Project N. 07-521. Tucson, Arizona.

Anderson, M.K., 2006. Tending the Wild. University of California Press, Berkeley, p. 256 p.

Baker, W.L., Hanson, C.T., Williams, M.A., DellaSala, D.A., 2023. Countering omitted evidence of variable historical forests and fire regime in western USA dry forests: the low-severity-fire model rejected. Fire 6, 146. https://doi.org/10.3390/fire6040146.

Bartholomew, B., 1970. Bare zone between California shrub and grassland communities: the role of animals. Science 170, 1210−1212.

Bentley, J.R., 1967. Conversion of Chaparral Areas to Grassland: Techniques Used in California. Agriculture Handbook No. 328. US Department of Agriculture, Washington, DC, p. 35 p.

Bonnicksen, T., 2003. Testimony to the subcommittee on forests and forest health of the committee on resources, U.S. House of representatives. In: One Hundred Eighth Congress, First Session. Lake Arrowhead, California.

Brooks, M.L., D'Antonio, C.M., Richardson, D.M., DiTomaso, J.M., Grace, J.B., Hobbs, R.J., Keeley, J.E., Pellant, M., Pyke, D., 2004. Effects of invasive alien plants on fire regimes. Bioscience 54, 677−688.

Burns, A., Homlund, H.L., Lekson, V.M., Davis, S., 2014. Seedling survival after novel drought-induced germination in Ceanothus megacarpus. In: Abstract. Berea College Research Symposium, Kentucky.

Busse, M.D., Hubbert, K.R., Moghaddas, E.E.Y., 2014. Fuel Reduction Practices and Their Effects on Soil Quality. Gen. Tech. Rep. PSW-GTR-241, U.S. Department of Agriculture, Forest Service, Pacific Southwest Research Station, Albany, CA, p. 156 p.

CCI, July 8, 2013. Escaped Cal Fire Prescribed Burn, San Felipe Valley Wildlife Area in a Letter From the California Chaparral Institute to the South Coast Region. Department of Fish and Wildlife, California.

CDFW, 2013. California Department of Fish and Wildlife Memorandum on the California Board of Forestry and Fire Protection Draft Vegetation Treatment Program Environmental Impact Report. Attachment A.

CFSC, August 2014. Panel Review Report of Vegetation Treatment Program Environmental Impact Report by California Board of Forestry and Fire Protection in Association With CAL FIRE Agency. California Fire Science Consortium. Coordinated by.

Chamberlin, T.C., 1890. The method of multiple working hypotheses. Science 148, 754–759.

Christensen, N.L., Muller, C.H., 1975. Effects of fire on factors controlling plant growth in Adenostoma chaparral. Ecol. Monogr. 45, 29–55.

Conard, S.G., Weise, D.R., 1998. Management of fire regimes, fuels, and fire effects in southern California chaparral: lessons from the past and thoughts for the future. Tall Timbers Fire Ecol. Conf. Proc. 20, 342–350.

Cooper, W.S., 1922. The broad-sclerophyll vegetation of California. In: An Ecological Study of the Chaparral and its Related Communities. Carnegie Institution of Washington, Washington, DC, p. 124 p. Publication No. 319.

CSBF, October 30, 2012. Draft Programmatic Environmental Impact Report for the Vegetation Treatment Program of the California State Board of Forestry and Fire Protection. The California Department of Forestry, Sacramento, CA.

DeBano, L.F., 1981. Water Repellent Soils: A State-Of-The-Art. United States Department of Agriculture Forestry Service General Technical Report, PSW-46, Berkley, California.

Doerr, S.H., Shakesby, R.A., Blake, W.H., Chafer, C.J., Humphreys, G.S., Wallbrink, P.J., 2006. Effects of differing wildfire severities on soil wettability and implications for hydrological response. J. Hydrol. 319, 295–311.

Elliot, J., 2019. The 1964 Hanly Fire. Santa Rosa History. http://santarosahistory.com/wordpress/2019/09/the-1964-hanly-fire/.

Festinger, L., 1957. A Theory of Cognitive Dissonance. Stanford University Press, Stanford, CA.

George, M.R., Roche, L.M., Eastburn, D.J., 2014. Ecology. Annual Rangeland Handbook, Division of Agriculture and Natural Resources. University of California, Davis, CA.

Giorgi, W.T., November 20, 2014. Testimony at the Santa Barbara County Fish and Game Commission.

Goldenstein, T., January 19, 2015. Water Stress Takes Toll on California's Large Trees, Study Says. Los Angeles Times, Los Angeles, CA.

Haidinger, T.L., Keeley, J.E., 1993. Role of high fire frequency in destruction of mixed chaparral. Madrono 40, 141–147.

Halsey, R.W., 2004. In search of allelopathy: an eco-historical view of the investigation of chemical inhibition in California coastal sage scrub and chamise chaparral. J. Torrey Bot. Soc. 131, 343–367.

Halsey, R.W., 2008. Fire, Chaparral, and Survival in Southern California. Sunbelt Publications, San Diego, CA.

Halsey, R.W., 2011. Chaparral as a natural resource. In: Proceedings of the California Native Plant Society Conservation Conference, January 17-19, 2009, pp. 82–86.

Halsey, R.W., May 1, 2012. The politics of fire, the struggle between science & ideology in San Diego County. Serialized Chaparralian 40 (3/4), 6–20. Independent Voter Network, May 15, 2013. https://ivn.us/2012/05/15/the-politics-of-fire-the-struggle-between-science-ideology-in-san-diego-county.

Halsey, R.W., Keeley, J.E., 2016. Conservation issues: California chaparral. In: Reference Module in Earth Systems and Environmental Sciences. Elsevier Publications, Inc.

Halsey, R.W., Halsey, V.W., Gaudette, R., 2018. Connecting Californians with the chaparral. In: Underwood, E., Safford, H., Molinari, N., Keeley, J.E. (Eds.), Valuing Chaparral, Economic, Socio-Economic, and Management Perspectives. Springer International Publishing.

Hanes, T.L., 1971. Succession after fire in the chaparral of southern California. Ecol. Monogr. 41, 27–52.

Holmstrom, M., Orient, S., Gordon, J., Johnson, R., Rodeffer, S., Money, L., Rickert, I., Pietruszka, B., Duarte, P., 2022. Marshall Fire. Facilitated Learning Analysis. Colorado Division of Fire Prevention & Control. https://storymaps.arcgis.com/stories/83af63bd549b4b8ea7d42661531de512.

Hubbard, R.F., 1986. Stand Age and Growth Dynamics in Chamise Chaparral. Master's Thesis. San Diego State University, San Diego, CA.

Hubbert, K.R., Preisler, H.K., Wohlgemuth, P.M., Graham, R.C., Narog, M.G., 2006. Prescribed burning effects on soil physical properties and soil water repellency in a steep chaparral watershed, southern California, USA. Geoderma 130, 284–298.

Jacobsen, A.L., Davis, S.D., Fabritius, S.L., 2004. Fire frequency impacts non-sprouting chaparral shrubs in the Santa Monica Mountains of southern California. In: Arianoutsou, M., Papanastasis, V.P. (Eds.), Ecology, Conservation and Management of Mediterranean Climate Ecosystems. Millpress, Rotterdam, Netherlands.

Jung, C.G., 1992. C.G. Jung Letters, AdlerJaffe, A.R. (Ed.), vol. 1. R.F.C. Hull. Princeton University.

Kaufmann, M.R., Shlisky, A., Marchand, P., 2005. Good Fire, Bad Fire: How to Think About Forest Land Management and Ecological Processes. U.S. Department of Agriculture, Forest Service, Rocky Mountain Research Station, Fort Collins, CO.

Keeley, J.E., 1987. Role of fire in seed germination of woody taxa in California chaparral. Ecology 68 (2), 434–443.

Keeley, J.E., 1990. The California valley grassland. In: Schoenherr, A.A. (Ed.), Endangered Plant Communities of Southern California. Southern California Botanists, Fullerton, California, pp. 2–23.

Keeley, J.E., 1992. Demographic structure of California chaparral in the long-term absence of fire. J. Veg. Sci. 3, 79–90.

Keeley, J.E., 2000. Chaparral. In: Barbour, M.G., Billings, W.D. (Eds.), North American Terrestrial Vegetation, second ed. Cambridge University Press, Cambridge, UK, pp. 203–253.

Keeley, J.E., 2001. We still need Smokey bear! Fire Manage.Today 61 (1), 21–22.

Keeley, J.E., 2002. American Indian influence on fire regimes in California's coastal ranges. J. Biogeogr. 29, 303–320.

Keeley, J.E., 2005. Fire as a Threat to Biodiversity in Fire-Type Shrublands. Gen. Tech. Rep. PSW-GTR-195: 97-106, USDA Forest Service.

Keeley, J.E., 2006. Fire management impacts on invasive plant species in the western United States. Conserv. Biol. 20, 375–384.

Keeley, J.E., 2009. Fire intensity, fire severity and burn severity: a brief review and suggested usage. Int. J. Wildland Fire 18, 116–126.

Keeley, J.E., Brennan, T.J., 2012. Fire-driven alien invasion in a fire-adapted ecosystem. Oecologia 169, 1043–1052.

Keeley, J.E., Keeley, S.C., 1977. Energy allocation patterns of a sprouting and nonsprouting species of *Arctostaphylos* in the California chaparral. Am. Midl. Nat. 98, 1–10.

Keeley, J.E., Keeley, S.C., 1987. Role of fire in the germination of chaparral herbs and suffrutescents. Madrono 34, 240–249.
Keeley, J.E., Zedler, P.H., 2009. Large, high-intensity fire events in southern California shrublands: debunking the fine-grain age patch model. Ecol. Appl. 19, 69–94.
Keeley, J.E., Fotheringham, C.J., Morais, M., 1999. Reexamining fire suppression impacts on brushland fire regimes. Science 284, 1829–1832.
Keeley, J.E., Fotheringham, C.J., Moritz, M., 2004. Lessons from the 2003 wildfires in southern California. J. For. 102, 26–31.
Keeley, J.E., Keeley, M., Fotheringham, C.J., 2005a. Alien plant dynamics following fire in Mediterranean-climate California shrublands. Ecol. Appl. 15, 2109–2125.
Keeley, J.E., Pfaff, A.H., Safford, H.D., 2005b. Fire suppression impacts on postfire recovery of Sierra Nevada chaparral shrublands. Int. J. Wildland Fire 14, 255–265.
Keeley, J.E., Brennan, T.J., Pfaff, A.H., 2008. Fire severity and ecosystem responses from crown fires in California shrublands. Ecol. Appl 18 (6), 1530–1546.
Keeley, J.E., Aplet, G.H., Christensen, N.L., Conard, S.C., Johnson, E.A., Omi, P.N., Peterson, D.L., Swetnam, T.W., 2009a. Ecological Foundations for Fire Management in North American Forest and Shrubland Ecosystems. Gen. Tech. Report PNW-GTR- 779. USDA, USFS PNW Research Station, Portland, OR, p. 92 p.
Keeley, J.E., Safford, H., Fotheringham, C.J., Franklin, J., Moritz, M., 2009b. The 2007 southern California wildfires: lessons in complexity. J. For. 107, 287–296.
Keeley, J.E., Bond, W.J., Bradstock, R.A., Pausas, J.G., Rundel, W., 2012. Fire in Mediterranean Climate Ecosystems: Ecology, Evolution and Management. Cambridge University Press, Cambridge, UK, p. 528 p.
Knapp, E.E., Estes, B.L., Skinner, C.N., 2009. Ecological Effects of Prescribed Fire Season: A Literature Review and Synthesis for Managers. Gen. Tech. Report PSW-GTR-224, USDA, Forest Service. PSW Research Station, p. 80 p.
Knudsen, K., 2006. Notes on the lichen flora of California # 2. Bull. Calif. Lichen Soc. 13 (1), 10–13.
Kuhn, T.S., 1962. The Structure of Scientific Revolutions. The University of Chicago Press, Chicago, IL.
LACC, August 1, 2007. Griffith Park Fire Recovery Plan/expenses. Council File Number 07-0600-S38. Los Angeles City Council, Los Angeles, CA.
LAT, April 22, 2009. A Burning Problem. Los Angeles Times editorial, Los Angeles, CA.
LAT, March 11, 2013. Cal Fire's Flawed Fire Plan. Critics Say It's Outdated, Contains Many Inaccuracies and Could Cause Major Environmental Damage. Los Angeles Times editorial, Los Angeles, CA.
LAT, December 18, 2014. Flash Floods After Fires. Los Angeles Times, Los Angeles, CA.
Larigauderie, A., Hubbard, T.W., Kummerow, J., 1990. Growth dynamics of two chaparral shrub species with time after fire. Madrono 37, 225–236.
Lendemer, J.C., Kocourkov, J., Knudsen, K., 2008. Studies in lichens and lichenicolous fungi: notes on some taxa from North America. Mycotaxon 105, 379–386.
Lombardo, K.J., Swetnam, T.W., Baisan, C.H., Borchert, M.I., 2009. Using bigcone Douglas-fir fire scars and tree rings to reconstruct interior chaparral fire history. Fire Ecol. 5, 32–53.
Lydersen, J.M., North, M.P., Collins, B.M., 2014. Severity of an uncharacteristically large wildfire, the Rim Fire, in forests with relatively restored frequent fire regimes. For. Ecol. Manag. 328, 326–334.
Meng, R., Dennison, P.E., D'Antonio, C.M., Moritz, M.A., 2014. Remote sensing analysis of vegetation recovery following short-interval fires in southern California shrublands. PLoS One 9, e110637.

Mensing, S.A., Michaelsen, J., Byrne, R., 1999. A 560 year record of Santa Ana fires reconstructed from charcoal deposited in the Santa Barbara Basin, California. Quat. Res. 51, 295–305.

Minnich, R.A., 2001. An integrated model of two fire regimes. Conservat. Biol. 15, 1549–1553.

Moreno, J.M., Oechel, W.C., 1989. A simple method for estimating fire intensity after a burn in California chaparral. Acta Oecol 10, 57–68.

Moritz, M.A., Keeley, J.E., Johnson, E.A., Schaffner, A.A., 2004. Testing a basic assumption of shrubland fire management: how important is fuel age? Front. Ecol. Environ. 2, 67–72.

Muller, C.H., Hanawalt, R.B., McPherson, J.K., 1968. Allelopathic control of herb growth in the fire cycle of California chaparral. Bull. Torrey Bot. Soc. 95, 225–231.

Oberbauer, T., 2018. Botany in San Diego Before European Contact. California Native Plant Society, San Diego Chapter Newsletter.

Odens, P., 1971. The Indians and I. Imperial Printers, El Centro, CA, p. 80 p.

Odion, D.C., Davis, F.W., 2000. Fire, soil heating, and the formation of vegetation patterns in chaparral. Ecol. Monogr. 70, 149–169.

Odion, D., Tyler, C., 2002. Are long fire-free periods needed to maintain the endangered fire-recruiting shrub *Arctostaphylos morroensis* (Ericaceae)? Conserv. Ecol. 6, 4.

Penman, T.D., Collins, L., Syphard, A.D., Keeley, J.E., Bradstock, R.A., 2014. Influence of fuels, weather and the built environment on the exposure of property to wildfire. PLoS One 9 (10), e111414.

Pratt, R.B., Jacobsen, A.L., Ramirez, A.R., Helms, A.M., Traugh, C.A., Tobin, M.F., Heffner, M.S., Davis, S.D., 2014. Mortality of resprouting chaparral shrubs after a fire and during a record drought: physiological mechanisms and demographic consequences. Global Change Biol. 20, 893–907.

Ricciuti, E.R., 1996. Chaparral. Biomes of the WorldBenchmark Books. Marshall Cavendish, New York, p. 64 p.

Rogers, G., Hann, W., Martin, C., Nicolet, T., Pence, M., 2008. Fuel Treatment Effects on Fire Behavior, Suppression Effectiveness, and Structure Ignition. US Department of Agriculture, Forest Service, Vallejo, CA, p. 35 p.

Rowntree, L., 2009. Forged by fire. Bay Nat 9 (4), 24–29.

Safford, H.D., Van de Water, K.M., 2014. Using Fire Return Interval Departure (FRID) Analysis to Map Spatial and Temporal Changes in Fire Frequency on National Forest Lands in California. Research Paper PSW-RP-266. U.S. Department of Agriculture, Forest Service, Pacific Southwest Research Station, Albany, CA, p. 59 p.

SBCC, 2002. Biology 100. Concepts in Biology. Introduction to Chaparral. Santa Barbara City College, Santa Barbara, CA. https://catalog.sbcc.edu/course-descriptions/biol/. (Accessed 18 December 2014).

SBCFWC, February 26, 2008. Draft Resolution to the Santa Barbara County Board of Supervisors From the Santa Barbara County Fish and Wildlife Commission.

Schoenberg, F.P., Peng, R.D., January 26, 2004. Letter to the San Diego Board of Supervisors.

SDCBS, August 13, 2003. Mitigation Strategies for Reducing Wildland Fire Risks. San Diego County Wildland Fire Task Force Findings and Recommendations. San Diego County Board of Supervisors, San Diego, CA.

SDCBS, 2009. County of San Diego Vegetation Management Report. Final Draft, 2/11/09. San Diego County Board of Supervisors.

Shea, N., July 2008. Under Fire. National Geographic Magazine, Washington DC.

SMMNRA, 2005. Final Environmental Impact Statement for a Fire Management Plan, Santa Monica Mountains National Recreation Area. US Department of the Interior, National Park Service, Washington, DC.

Sherriff, R.L., Veblen, T.T., 2007. A spatially-explicit reconstrtuction of historical fire occurrence in the ponderosa pine zone of the Colorado front range. Ecosystems 10, 311−323.

SMMNRA, 2015. Why This Park Does Not Use Prescribed Fire. US Department of the Interior, National Park Service, Washington, DC. Santa Monica Mountains National Recreation Area website: http://www.nps.gov/samo/parkmgmt/prescribedfires.htm.

Sneed, D., July 12, 2008. Fires in California Devastate Wildlife, Sensitive Habitats. The Tribune. San Luis Obispo.

Spencer, W.D., January 5, 2009. Letter to the County of San Diego Planning Commission. Conservation Biology Institute, San Diego, CA.

Steel, Z.L., Safford, H.D., Viers, J.H., 2015. The fire frequency-severity relationship and the legacy of fire suppression in California forests. Ecosphere 6 (1), 8.

Steinhoff, 2010. San Diego County Comment Letter on the Draft California Fire Plan.

Syphard, A.D., Franklin, J., Keeley, J.E., 2006. Simulating the effects of frequent fire on southern California coastal shrublands. Ecol. Appl. 16 (5), 1744−1756.

Syphard, A.D., Keeley, J.E., Brennan, T.J., 2011. Comparing the role of fuel breaks across southern California national forests. For. Ecol. Manag. 261, 2038−2048.

Talluto, M.V., Suding, K.N., 2008. Historical change in coastal sage scrub in southern California, USA in relation to fire frequency and air pollution. Landsc. Ecol. 23, 803−815.

Tiszler, J., May 2000. Fire Regime, Fire Management, and the Preservation of Biological Diversity in the Santa Monica Mountains NRA. Report for the Santa Monica National Recreation Area, Thousand Oaks, CA.

Tyler, C.M., Borchert, M.I., 2007. Chaparral geophytes: fire and flowers. Fremontia 35 (4), 22−24.

USFS, 2013. Ecological Restoration Implementation Plan. R5-MB-249. US Department of Agriculture. Forest Service, Pacific Southwest Region, Vallejo, CA, p. 154 p.

USDI NPS, 2004. Draft Environmental Impact Statement Fire Management Plan, Santa Monica Mountains National Recreation Area. United States Department of the Interior, National Park Service, Washington, DC.

Vachula, R.S., Russell, J.M., Huang, Y., 2019. Climate exceeded human management as the dominant control of fire at the regional scale in California's Sierra Nevada. Environ. Res. Lett. 14, 104011. https://doi.org/10.1088/1748-9326/ab4669.

Valiente-Banuet, A., Flores-Hernandez, N., Verdu, M., Davila, P., 1998. The chaparral vegetation in Mexico under non-mediterranean climate: the convergence and Madrean-Tethyan hypothesis reconsidered. Am. J. Bot. 85, 1398−1408.

Wells, P.V., 1962. Vegetation in relation to geological substratum and fire in the San Luis Obispo quadrangle, California. Ecol. Monogr. 32, 79−103.

Zane, D., Henry, J.H., Lindley, C., Pedergrass, P.W., Galloway, D., Spencer, T., Stanford, M., 2006. Surveillance of Mortality During the Texas Panhandle Wildfires (March 2006). Regional and Community Coordination Branch. Public Health Preparedness Unit, Texas Department of State Health Services.

Zedler, P.H., Seiger, L.A., 2000. Age mosaics and fire size in chaparral: a simulation study. In: 2nd Interface Between Ecology and Land Development in California, pp. 9−18. USGS Open-File Report 00-02.

Zedler, P.H., Gautier, C.R., McMaster, G.S., 1983. Vegetation change in response to extreme events: the effect of a short interval between fires in California chaparral and coastal scrub. Ecology 64, 809−818.

Chapter 7

Regional Case Studies: Southeast Australia, Sub-Saharan Africa, Central Europe, and Boreal Canada

Case Study: The Ecology of Mixed-severity Fire in Mountain Ash Forests

Laurence E. Berry[1] and Holly Sitters[2,3]
[1]Department of Energy, Environment and Climate Action, Melbourne, VIC, Australia; [2]Australian Wildlife Conservancy, Subiaco East, WA, Australia; [3]School of Ecosystem and Forest Sciences, University of Melbourne, Creswick, VIC, Australia

7.1 THE SETTING

The eucalypt forests of southeast Australia are among the most flammable ecosystems worldwide (Pyne, 1992). Most of the forest in the region is dominated by a single overstory eucalypt species. Here we describe stands of mountain ash (*Eucalyptus regnans*), which is the tallest flowering plant species in the world, at heights approaching 100 m (Beale, 2007; Lindenmayer and Bowd, 2022). Mountain ash forest occurs in the states of Victoria and Tasmania, and our focus is on a 121,000-ha region in the state of Victoria's Central Highlands, where undulating landscapes merge with mountainous terrain at the foothills of the Great Dividing Range.

The region lies at altitudes of between 200 and 1100 m above sea level and experiences mild, humid winters and warm summers; mean annual rainfall increases from approximately 1200 mm in the west to 1800 mm in the east (Bureau of Meteorology, 2023). The productive wet forest comprises a layer of understory trees that reach heights of more than 20 m and include myrtle beach (*Nothofagus cunninghamii*), southern sassafras (*Atherosperma moschatum*), and silver wattle (*Acacia dealbata*). A shrub layer of 2–15 m supports tree ferns (*Dicksonia antarctica* and *Cyathea australis*), hazel pomaderris (*Pomaderris aspera*), musk daisy bush (*Olearia argophylla*), and blanket leaf (*Bedfordia salicina*). The well-developed understory and shrub layers of mountain ash forest accumulate high fuel loads that are too wet to burn except following periods of drought or during extremely hot and dry conditions (Ashton, 1981). Consequently, large fires in mountain ash forest are infrequent but intense (Jackson, 1968).

Unlike many eucalypt species, mountain ash is considered fire-sensitive because it does not reproduce vegetatively and is killed by severe fire (Lindenmayer, 2009). As an obligate seeder, however, its regeneration is dependent on high-intensity fire, which desiccates seed capsules and releases up to 14 million seeds per hectare (Attiwill, 1994). Seedlings are shade intolerant and rarely establish in mature forest; instead, they thrive under the high light levels of the nitrogen-rich ash bed that characterizes the postfire landscape (Ashton and Martin, 1996). For several decades the prevailing paradigm was that regenerating stands were primarily even-aged (Ashton, 1976; Griffiths, 1992; Loyn, 1985).

It is increasingly recognized that large, intense fires create mosaics of fire severity and generate multi-aged forest (Simkin and Baker, 2008; Lindenmayer et al., 2000). Because of the substantial topographic reliefs (up to 1000 m) and resulting gradients in wetness, large fires in the Victorian Central Highlands (VCH) rarely burn homogeneously (Figures 7.1 and 7.2). Patterns of fire severity in the region are a function of interplay among topography, vegetation, and weather (Berry et al., 2015a,b); however, when fire weather is severe (high temperatures, low relative humidity, strong winds), unburnt patches rarely occur irrespective of drought severity, topography, or vegetation (Collins et al., 2019).

In general, high-severity crown fire occurs most predictably on ridgetops because of wind exposure and associated decreases in fuel moisture, whereas unburnt patches tend to occur in sheltered valleys where fuel moisture levels are elevated (Wood et al., 2011; Bradstock et al., 2010). Eighty-four unburnt patches of >1 ha were present within the perimeter ($\sim$250,000 ha) of the 2009 wildfire (average patch size was 27 ha; Leonard et al., 2014). Collectively, they covered only 1% of the fire-affected area, but they probably provided sufficient refuge habitat to allow animals and fire-intolerant plants to survive, persist, and recolonize following the fire (Robinson et al., 2013; Banks et al., 2017). Further, a spectrum of moderate- to high-severity fire causes only partial mortality of overstory trees (mountain ash survival averages 39% in areas of moderate crown scorch and is over 90% in areas affected by low-severity fire; Benyon and Lane (2013)), providing additional refuge habitat for fauna and giving rise to multi-aged stands (Simkin and Baker, 2008). We discuss the implications of mixed- and high-severity fire for the flora and fauna of mountain ash forest before considering the consequences of increasingly frequent mega-fires, such as the unprecedented fires of 2019–20 that burned more than seven million hectares across eastern Australia (Bradstock et al., 2014; Bowman et al., 2021).

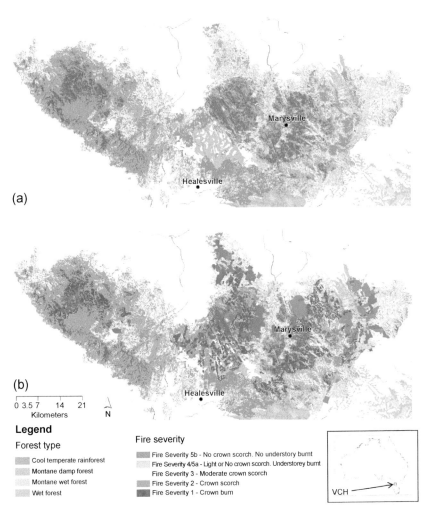

FIGURE 7.1 (a) Map of the 2009 Kilmore-Murrindindi fire complex in the Victorian Central Highlands (VCH), Australia. The map displays the extent of forest types containing mountain ash and major roads and towns. (b) The map is overlayed with the extent of logging per decade beginning in 1900 (light blue) to 2010 (dark blue).

7.2 MOUNTAIN ASH LIFE CYCLE

As an obligate seeder, mountain ash is susceptible to rapid decline under altered fire regimes (McCarthy et al., 1999) (Figure 7.3). Crucially, the fire return interval must be longer than the species' maturation age (15–20 years) and shorter than its life span (350–500 years) (Wood et al., 2010). If successive stand-replacing fires occur <20 years apart, mountain ash is succeeded by other species more tolerant of shorter fire intervals, such as silver wattle

FIGURE 7.2 A topographically sheltered, unburnt gully embedded within a forest that burned at high severity. These areas may act as fire refuges within the burn extent, enabling the persistence of fauna otherwise vulnerable to the effects of fire on habitat. *Photo credit: Laurence Berry.*

FIGURE 7.3 Mountain ash trees displaying fine-scale variability in canopy consumption and a high density of new saplings. This photograph was taken approximately 5 years after fire. *Photo credit: Laurence Berry.*

(Lindenmayer, 2009; Bowd et al., 2023). The spatial and temporal attributes of fire regimes interact with climate, topography, and other disturbances to potentially alter species' distributions.

It is plausible, however, that the direct effects of climate change, such as increased temperature or rainfall, pose a greater threat to obligate seeders than altered fire regimes per se (Lawson et al., 2010). Smith et al. (2014) sought to disentangle the relative importance of fire-related factors (return interval and severity) and environmental factors (climate and topography) on seedling establishment in mountain ash. Regenerating mountain ash stands feature several million seedlings per hectare, and rapid self-thinning reduces sapling density to around 400 per hectare 40 years after fire. Only 40–80 mature trees per hectare remain after 150–200 years (Ashton and Attiwill, 1994). Competition causes the death of many growing seedlings, and the collapse of small, suppressed pole and sapling trees compounds mortality (Lindenmayer et al., 2023). Smith et al. (2014) focused their investigation on the critical stage of seedling establishment (within a year of fire) to identify factors that drive successful rejuvenation.

Climatic variables, topographic position, and fire return interval all were identified as important determinants of eucalypt seedling establishment (Smith et al., 2014). Seedling abundance was greater in flat, elevated areas of comparatively high rainfall and low temperature. Moreover, seedlings were more abundant at sites with a longer fire return interval, indicating that seed storage potential and rejuvenation capacity increase with forest age (Ashton, 1975). There was no detectable difference in seedling establishment between sites affected by moderate- and high-severity fire. Conceivably, opposing forces hid fire-severity effects; for example, light levels at high-severity sites may be optimal for seedling growth, but this positive effect could have been counteracted by greater seed mortality during fire (Smith et al., 2014). The authors suggest that both climate change and altered fire regimes influence the regeneration capacity of mountain ash, the distribution of which will potentially shift as the climate warms.

7.3 INFLUENCE OF STAND AGE ON FIRE SEVERITY

A warmer, drier climate is linked to increases in the frequency of large, intense fires in southeast Australia (Cary et al., 2012; Bowman et al., 2021). Taylor et al. (2014) investigated the influence of stand age on fire severity in mountain ash forest across an array of growth stages and disturbance histories with a view toward better understanding the implications of increasingly frequent fire for species' capacity to regenerate. Mountain ash forests have been a key source of pulp wood and sawlogs since the 1930s (Lutze et al., 1999), so stand age classes were derived from multiple data sets relating to both fire and logging history, and fire severity data were sampled at 100 m intervals across a grid comprising nearly 10,000 sites.

Strong nonlinear relationships between forest age and fire severity were identified (Taylor et al., 2014). Severity was highest in stands that were 7–36 years old, which sustained canopy consumption and scorching. By contrast, canopy consumption rarely occurred in very young stands (<7 years old), and was uncommon in stands >40 years old. Regenerating stands support a highly flammable fuel layer of densely spaced seedlings, and self-thinning yields large volumes of fire-prone fine fuels. Fine fuel loads are thought to peak at 35 years after fire, when stand height and floristic composition can also compound flammability. A lack of dead fine fuel in stands <7 years old may explain the scarcity of crown fire in very young stands (Taylor et al., 2014). Correspondingly, the flammability of older forest is reduced by the establishment of rainforest species such as myrtle beach, which lessen light penetration to the ground layer and foster the development of a cool, moist microclimate (Wood et al., 2014). High-severity fire is intrinsic to mountain ash forest, but old stands, where moisture levels are higher and the probability of tree survival is greater, play a crucial role in reducing fire intensity and rates of spread (Taylor et al., 2014; Lindenmayer and Bowd, 2022).

7.4 DISTRIBUTION OF OLD-GROWTH FORESTS

Large, old trees in mountain ash forests provide biological legacies that perform myriad ecological roles; they store carbon, provide faunal habitat, and promote stand rejuvenation (Bowman et al., 2014). Sixty to 80% of mountain ash forest in VCH was historically considered old-growth 100–150 years ago (Lindenmayer, 2009). Wildfire and logging during the past century have reduced old-growth forests to only 1.16% of the area within the extent of current mountain ash forests in the VCH (Lindenmayer and Bowd, 2022). Old-growth stands naturally occur in flat plateaus and valley bottoms, where fire is less intense; clearcutting often targets such areas, precipitating the replacement of less flammable forest with dense regenerating stands of self-thinning trees and potentially altering patterns of spatial propagation of fire behavior (Lindenmayer et al., 2011). Rapid advances in fire severity mapping and modeling have benefitted the study of the spatial characteristics of large, intense fires and are informing management strategies that seek to promote the distribution and conservation of older forest stands (Berry et al., 2015a,b; Lindenmayer and Bowd, 2022).

7.5 MIXED-SEVERITY FIRE AND FAUNA OF MOUNTAIN ASH FORESTS

Wildfires are a major form of disturbance in mountain ash forests. Extensive, high-intensity, stand-replacing fires dramatically alter habitat structure and resource distribution across large spatial scales (Bowman et al., 2009; Bradstock et al., 2005). How fauna respond to these fires is dependent on the interacting properties of the fire regime and the behavioral ecology of each species. The underlying history of fire in the landscape and the immediate effects of wildfire influence the availability and distribution of key wildlife habitat resources. Mountain ash forests often exhibit complex early seral mosaics resulting from the layering of historic and contemporary disturbances.

7.6 FAUNA AND FIRE-AFFECTED HABITAT STRUCTURES

Hollow-bearing trees are an essential habitat feature of tall mountain ash forests. Tree hollows begin to form in mature trees, known as "stags," after 120 years (Figure 7.4). Stags provide essential denning habitat for a diverse fauna of arboreal marsupials, birds, bats, and lizards. The density of hollow-bearing trees per hectare is the primary determinant of the presence of the globally endangered Leadbeater's possum (*Gymnobelideus leadbeateri*), a small,

FIGURE 7.4 A large, hollow-bearing tree, known as a "stag," in an area of long unburnt, late successional forest. This particular example may have taken over 400 years to form. *Photo credit: Laurence Berry.*

colonial, nocturnal, arboreal marsupial endemic to the VCH. Recent declines in the distribution and numbers of the possum are directly related to the fire- and logging-influenced decline in the number and distribution of hollow-bearing trees (Lindenmayer et al., 2013a,b, 2017; Nitschke et al., 2020).

Large, high-severity wildfires radically alter vegetation structure in mountain ash forests. Fires that burn the understory remove important habitat features such as silver wattle, which provides a major food resource (sap) for Leadbeater's possum and also provides a dense canopy that facilitates the movement of fauna through the forest. Leadbeater's possum has a complex social structure, where many individuals form part of a large matriarchal colony. These colonies need to gather food resources over large areas. The impact of fire on food resource availability may reduce the total foraging area available to support these large colonies. Fires that consume the canopy lead to a decline in the abundance of foliage feeders such as the southern greater glider (*Petauroides volans*) (Lindenmayer et al., 2013b, 2022b). While occupancy of most common birds decreased after a large, high-severity wildfire, most species recovered to prefire levels within 6 years (Lindenmayer et al., 2022a). One species, the flame robin (*Petroica phoenicea*), had the highest occupancy in high-severity fire areas. By the end of the study period (7−10 years postfire), the occupancy probability of some species, such as the yellow-faced honeyeater (*Lichenostomus chrysops*), was higher in high-severity fire areas than in the same forests prior to the fire (Lindenmayer et al., 2022a). Some birds that feed on open ground, such as the buff-rumped thornbill (*Acanthiza reguloides*), scarlet robin (*Petroica boodang*), and flame robin (*Petroica pheonicea*) become more abundant immediately after understory fires (Loyn, 1997).

Following the infamous 2009 "Black Saturday" wildfires that burned under extreme weather (high temperatures, drought, high winds) through ~ 250,000 ha of mountain ash forest in Victoria, Australia, the mountain brushtail possum (*Trichosurus cunninghami*) was the only arboreal marsupial occupying severely burnt sites 1−3 years postfire (Lindenmayer et al., 2013b). This possum is a large (2−4 kg), nocturnal, arboreal marsupial with generalized foraging habits and diet (Seebeck et al., 1984). A study that fitted proximity measuring radio transmitters to possums during the 2009 fires found no evidence of fire-associated possum mortality (Banks et al., 2011). However, 80% of the dead, hollow-bearing trees within the range of the collared animals were consumed during the fire. Individuals adapted to this loss in critical resources by displaying flexibility in den site selection. Banks et al. (2011) concluded that although the adaptability in resource use displayed by the mountain brushtail possum may buffer populations against decline in the short-term, increased frequency of stand-replacing fires is likely to cause major limitations on shelter availability and pose significant conservation challenges for hollow-dependent fauna.

7.7 FAUNAL RESPONSE TO THE SPATIAL OUTCOMES OF FIRE

A growing body of research has highlighted the importance of the spatial outcomes of fire on biodiversity responses (e.g., Robinson et al., 2014; Sitters et al., 2014; Berry et al., 2015a; Banks et al., 2017; Collins et al., 2019). Species able to persist within recently burned forest may benefit from mixed-severity fire because unburnt patches of forest embedded within large burns may provide essential habitat for species whose life cycles are dependent on resources reduced by fire. The relevance of these fire refuges to species conservation following large fires may be dependent on their size and the amount of additional unburnt resources available in the surrounding landscape. For example, Robinson et al. (2014) compared bird response to fire severity in patches of mountain ash forest with varying fire intervals: short (<3 years) and long (>20 years). They found that unburnt patches that persisted within the extent of large fires may act as refuges for birds associated with late-successional conditions. In particular, unburnt patches of forest with a long time since fire contained more individuals (56%) and higher species richness (20–40) than severely burned forest <3 years after fire. Similarly, unburnt patches of forest located in mesic gullies at the edge of extensively burned landscapes may facilitate the presence of the mountain brushtail possum (Berry et al., 2015a; Lindenmayer et al., 2017). In contrast, southern greater glider required large, intact, unburnt areas of forest; the species is known for its distinct habitat requirements, which render it particularly vulnerable to climate change and disturbance (Wagner et al., 2021).

The southern greater glider, sugar glider, and Leadbeater's possum declined as the amount of burned forest increased in the surrounding landscape (Lindenmayer et al., 2013b). The landscape context effects of fires on fauna are expected to be transitory as rapid postfire recovery of vegetation occurs. However, the temporal period in which landscape context effects should be considered is related to the rate of recovery of limiting resources, the scale and extent of the fire, and the time since previous fire.

7.8 CONSERVATION CHALLENGES AND FUTURE FIRE

Large fires in the mountain ash forests of Victoria coupled with short fire intervals and extensive clearcutting and postfire "salvage" logging operations have reduced the total area of available habitat for a range of late-successional specialists. Clearcutting removes keystone habitat structures before fire and may alter the spread of fire throughout the landscape by providing homogenous fuels across large areas. Postfire logging removes important biological legacies that may remain after fire and further fragments habitat after fire (see Chapter 9). These multilayered disturbances may interact to cause irreparable shifts in ecosystem state. For nearly a century, extensive postfire logging operations have followed large fires in the mountain ash forests of Victoria (Lindenmayer and Ough, 2006; Lindenmayer et al., 2019). The abundance of hollow-bearing trees is greatly reduced in such logged forests. Lindenmayer and Noss (2006) outlined key measures to improve biodiversity retention following large wildfires, including the preservation of unburnt or partially burned forest patches and excluding postfire logging from nature reserves and water catchments, extensive areas of old-growth forest, and areas with few roads.

A recent risk assessment of the mountain ash forests of the VCH ranked the ecosystem as critically endangered, with a predicted 92% chance of ecosystem collapse by 2067 (Burns et al., 2015). The authors of the study recommended immediate protection of undisturbed areas of mountain ash, substantial restorative action in degraded stands, and the cessation of widespread industrial logging. Spies et al. (2012) outlined a list of common conservation objectives in fire-prone, temperate, tall forests in southeast Australia and the Pacific Northwest of the United States, including identifying the socioeconomic context of appropriate fire management, identifying desirable disturbance for biodiversity needs, moving beyond fuel treatment as an end point in fire management by basing management goals on species ecology, and planning for the spatial outcomes of future large fire events.

The future integrity of mountain ash ecosystems is in doubt. The increasing frequency of large, unplanned fires coupled with ongoing extensive logging operations is decreasing the availability of suitable habitat for a range of late-successional specialist species (Lindenmayer et al., 2012; Lindenmayer and Bowd, 2022). In some areas the interaction of fire and logging disturbances are creating irreversible shifts in ecosystem state (Lindenmayer et al., 2011; Bowd et al., 2023). Studies show that recently logged forest burns at higher severity than old-growth forests (Taylor et al., 2014). Large areas of homogenously aged young forest regenerating from logging will likely increase the severity of a fire passing through the landscape. This process may catalyze an increase in fire severity in adjacent areas of forest that, because of topography, stand age, and forest type, may otherwise have burned at lower severity.

Contemporary land uses have fragmented the remaining 1887 ha of old-growth mountain ash forest, which is dispersed across 147 different patches (Taylor and Lindenmayer, 2020). Land-use changes influence the spatiotemporal mosaic of fire regimes, which may contribute to the continued decline of already endangered species such as Leadbeater's possum, the yellow-bellied glider, and the southern greater glider.

The authors of a major long-term research project in the VCH concluded that the only way to counter downward species trajectories is to remove the principal cause of the decline in keystone habitat structures and ecologically important biological legacies: widespread industrial logging (Lindenmayer et al., 2013a; Lindenmayer et al., 2019). At present, the majority of logged forest areas are regrowth from the massive 1939 "Black Friday" wildfires. Within these areas, most trees are felled approximately 40 years from developing the key ecological features required by fauna associated with late-successional conditions. The lack of accurate fire severity maps from the 1939 fires also leads to the felling of old trees embedded within regrowth forest that survived the fire.

The Government of Victoria has indicated that it will end the native forest logging industry by 2030; in the meantime, restoration of the ecosystem will benefit from strategic decisions to avoid logging in wetter and more sheltered areas, and new targets for the proportion of old-growth forest (Lindenmayer and Bowd, 2022). Given that 60% of VCH mountain ash forest may have been old-growth before large-scale logging commenced, a target of "substantially more than 60%" old-growth has been recommended based on the likelihood that large amounts of existing forest will be affected by future frequent fire (Lindenmayer and Bowd, 2022).

The unprecedented Australian mega-fires of 2019–20 predominantly affected eastern Victoria and New South Wales; however, such massive fires will become more frequent in the VCH under a rapidly changing climate (Bowman et al., 2021). The protection of large areas of ash forest increases the likelihood that a greater area of old-growth forest will remain unburnt following future large wildfires. It would also ensure that regrowth forests across large spatial extents will reach the necessary level of maturity needed to provide essential habitat structures for endangered native, late-successional fauna in the region. Further, it would allow for ecologically appropriate levels of mixed- and high-severity fire to provide unmanaged complex early successional conditions for wildlife that benefits from higher-severity fire at some postfire time period.

Acknowledgments

Candice Power helped with preparation of figures and fire data; Ken Baldwin assisted with obtaining the map of the boreal distribution and the map of the Canadian ecozones; James Brandt and Roger Brett helped obtaining the distribution of the North American map. John Little assisted us in obtaining access to the large fire database. Christian Hébert, Brad Hawkes, John Parminter, Brian Simpson kindly provided excellent pictures. Brian Hearn, Christian Hébert, and Caroline Simpson conducted official internal review of the manuscript and provided valuable comments. Additional comments were provided by Candice Power and Patricia Baines. NRC Press provided permissions to use figures published in two papers published in Environmental Reviews.

REFERENCES

Ashton, D.H., 1975. Studies of flowering behaviour in *Eucalyptus regnans* F. Muell. in central Victoria. Aust. J. Bot. 23, 399–411.

Ashton, D.H., 1976. Development of even-aged stands of *Eucalyptus regnans* F. Muell. in Central Victoria. Aust. J. Bot. 24, 397–414.

Ashton, D.H., 1981. Fire in tall open forests (wet sclerophyll forests). In: Gill, A.M., Groves, R.H., Noble, I.R. (Eds.), Fire and the Australian Biota. Australian Academy of Science, Canberra.

Ashton, D.H., Attiwill, P.M., 1994. Tall open forests. In: Groves, R.H. (Ed.), Australian Vegetation. Cambridge University Press, Melbourne, pp. 157–196.

Ashton, D.H., Martin, D.G., 1996. Regeneration in a pole-stage forest of eucalyptus regnans subjected to different fire intensities in 1982. Aust. J. Bot. 44, 393–410.

Attiwill, P.M., 1994. Ecological disturbance and the conservative management of eucalypt forests in Australia. For. Ecol. Manag. 63, 301–346.

Banks, S.C., Knight, E.J., Mcburney, L., Blair, D., Lindenmayer, D.B., 2011. The effects of wildfire on mortality and resources for an arboreal marsupial: resilience to fire events but susceptibility to fire regime change. PLoS One 6, e22952.

Banks, S.C., McBurney, L., Blair, D., Davies, I.D., Lindenmayer, D.B., 2017. Where do animals come from during post-fire population recovery? Implications for ecological and genetic patterns in post-fire landscapes. Ecography 40, 1325–1338.

Beale, R., 2007. If Trees Could Speak. Stories of Australia's Greatest Trees. Allen & Unwin, Sydney.

Benyon, R.G., Lane, P.N.G., 2013. Ground and satellite-based assessments of wet eucalypt forest survival and regeneration for predicting long-term hydrological responses to a large wildfire. For. Ecol. Manag. 294, 187–207.

Berry, Laurence E., Driscoll, Don A., Banks, Samuel C., Lindenmayer, David B., 2015a. The use of topographic fire refuges by the greater glider (*Petauroides volans*) and the mountain brushtail possum (*Trichosurus cunninghami*) following a landscape-scale fire. Aust. Mammal. 37, 39–45.

Berry, L.E., Driscoll, D.A., Stein, J., Blanchard, W., Banks, S., Bradstock, R., Lindenmayer, D.B., 2015b. Identifying the location of fire refuges in wet forest ecosystems. Ecol. Appl. 25 (8), 2337–2348.

Bowd, E.J., McBurney, L., Lindenmayer, D.B., 2023. The characteristics of regeneration failure and their potential to shift wet temperate forests into alternate stable states. For. Ecol. Manag. 529, 120673.

Bowman, D., Balch, J.K., Artaxo, P., Bond, W.J., Carlson, J.M., Cochrane, M.A., D'antonio, C.M., Defries, R.S., Doyle, J.C., Harrison, S.P., Johnston, F.H., Keeley, J.E., Krawchuk, M.A., Kull, C.A., Marston, J.B., Moritz, M.A., Prentice, I.C., Roos, C.I., Scott, A.C., Swetnam, T.W., Van Der Werf, G.R., Pyne, S.J., 2009. Fire in the earth system. Science 324, 481–484.

Bowman, D., Murphy, B.P., Neyland, D.L.J., Williamson, G.J., Prior, L.D., 2014. Abrupt fire regime change may cause landscape-wide loss of mature obligate seeder forests. Glob. Chang. Biol. 20, 1008–1015.

Bowman, D.M.J.S., Williamson, G.J., Gibson, R.K., Bradstock, R.A., Keenan, R.J., 2021. The severity and extent of the Australia 2019–20 Eucalyptus forest fires are not the legacy of forest management. Nat. Ecol. Evol. 5 (7), 1003–1010.

Bradstock, R.A., Bedward, M., Gill, A.M., Cohn, J.S., 2005. Which mosaic? A landscape ecological approach for evaluating interactions between fire regimes, habitat and animals. Wildl. Res. 32, 409–423.

Bradstock, R.A., Hammill, K.A., Collins, L., Price, O., 2010. Effects of weather, fuel and terrain on fire severity in topographically diverse landscapes of south-eastern Australia. Landsc. Ecol. 25, 607–619.

Bradstock, R., Penman, T., Boer, M., Price, O., Clarke, H., 2014. Divergent responses of fire to recent warming and drying across south-eastern Australia. Glob. Chang. Biol. 20, 1412–1428.

Bureau of Meteorology, 2023. Climate Data Online. Bureau of Meteorology. Available: http://www.bom.gov.au/climate/data. (Accessed April 2023).

Burns, E.L., Lindenmayer, D.B., Stein, J., Blanchard, W., Mcburney, L., Blair, D., Banks, S.C., 2015. Ecosystem assessment of mountain ash forest in the Central Highlands of Victoria, south-eastern Australia. Aust. Ecol. 40 (4), 386–399.

Cary, G.J., Bradstock, R.A., Gill, A.M., Williams, R.J., 2012. Global change and fire regimes in Australia. In: Bradstock, R.A., Gill, A.M., Williams, R.J. (Eds.), Flammable Australia: Fire Regimes, Biodiversity and Ecosystems in a Changing World. CSIRO Publishing, Melbourne.

Collins, L., Bennett, A.F., Leonard, S.W.J., Penman, T.D., 2019. Wildfire refugia in forests: severe fire weather and drought mute the influence of topography and fuel age. Glob. Chang. Biol. 25, 3829–3843.

Griffiths, T., 1992. Secrets of the Forest. Allen & Unwin, Melbourne.

Jackson, W., 1968. Fire, air, water and earth—an elemental ecology of Tasmania. Proc. Ecol. Soc. Aust. 3, 9–16.

Lawson, D.M., Regan, H.M., Zedler, P.H., Franklin, A., 2010. Cumulative effects of land use, altered fire regime and climate change on persistence of *Ceanothus verrucosus*, a rare, fire-dependent plant species. Glob. Chang. Biol. 16, 2518–2529.

Leonard, S.W.J., Bennett, A.F., Clarke, M.F., 2014. Determinants of the occurrence of unburnt forest patches: potential biotic refuges within a large, intense wildfire in south-eastern Australia. For. Ecol. Manag. 314, 85–93.

Lindenmayer, D., 2009. Forest Pattern and Ecological Process: A Synthesis of 25 Years of Research. CSIRO Melbourne, Melbourne.

Lindenmayer, D., Noss, R., 2006. Salvage logging, ecosystem processes, and biodiversity conservation. Conserv. Biol. 20, 949–958.

Lindenmayer, D., Ough, K., 2006. Salvage logging in the montane ash eucalypt forests of the Central Highlands of Victoria and its potential impacts on biodiversity. Conserv. Biol. 20, 1005–1015.

Lindenmayer, D.B., Cunningham, R.B., Donnelly, C.F., Franklin, J.F., 2000. Structural features of old-growth Australian montane ash forests. For. Ecol. Manag. 134, 189–204.

Lindenmayer, D.B., Hobbs, R.J., Likens, G.E., Krebs, C.J., Banks, S.C., 2011. Newly discovered landscape traps produce regime shifts in wet forests. Proc. Natl. Acad. Sci. U. S. A. 108, 15887–15891.

Lindenmayer, D.B., Blanchard, W., Mcburney, L., Blair, D., Banks, S., Likens, G.E., Franklin, J.F., Laurance, W.F., Stein, J.A., Gibbons, P., 2012. Interacting factors driving a major loss of large trees with cavities in a forest ecosystem. PLoS One 7, e41864.

Lindenmayer, D.B., Blair, D., Mcburney, L., Banks, S.C., Stein, J.A., Hobbs, R.J., Likens, G.E., Franklin, J.F., 2013a. Principles and practices for biodiversity conservation and restoration forestry: a 30 year case study on the Victorian montane ash forests and the critically endangered Leadbeater's possum. Aust. Zoo. 36, 441–460.

Lindenmayer, D.B., Blanchard, W., Mcburney, L., Blair, D., Banks, S.C., Driscoll, D., Smith, A.L., Gill, A.M., 2013b. Fire severity and landscape context effects on arboreal marsupials. Biol. Conserv. 167, 137–148.

Lindenmayer, D.B., Blanchard, W., Blair, D., McBurney, L., Banks, S.C., 2017. Relationships between tree size and occupancy by cavity-dependent arboreal marsupials. For. Ecol. Manag. 391, 221–229.

Lindenmayer, D.B., Blanchard, W., Bowd, E., Scheele, B.C., Foster, C., Lavery, T., McBurney, L., Blair, D., 2022a. Rapid bird species recovery following high-severity wildfire but in the absence of early successional specialists. Divers. Distrib. 28, 2110–2123.

Lindenmayer, D.B., McBurney, L., Blanchard, W., Marsh, K., Bowd, E., Watchorn, D., Taylor, C., Youngentob, K., 2022b. Elevation, disturbance, and forest type drive the occurrence of a specialist arboreal folivore. PLoS ONE 17, e0265963.

Lindenmayer, D., Bowd, E., 2022. Critical ecological roles, structural attributes and conservation of old growth forest: lessons from a case study of Australian Mountain Ash Forests. Front. For. Glob. Change 5, 878570.

Lindenmayer, D., McBurney, L., Blanchard, W., 2023. Drivers of collapse of fire-killed trees. Aust. Ecol. 48, 134–142.

Loyn, R.H., 1985. Bird populations in successional forests of mountain ash *Eucalyptus regnans* in Central Victoria. EMU 85, 213–230.

Loyn, R.H., 1997. Effects of an extensive wildfire on birds in far eastern Victoria. Pac. Conserv. Biol. 3, 221.

Lutze, M.T., Campbell, R.G., Fagg, P.C., 1999. Development of silviculture in the native state forests of Victoria. Aust. For. 62, 236−244.

Mccarthy, M.A., Gill, A.M., Lindenmayer, D.B., 1999. Fire regimes in mountain ash forest: evidence from forest age structure, extinction models and wildlife habitat. For. Ecol. Manag. 124, 193−203.

Nitschke, C.R., Trouvé, R., Lumsden, L.F., Bennett, L.T., Fedrigo, M., Robinson, A.P., Baker, P.J., 2020. Spatial and temporal dynamics of habitat availability and stability for a critically endangered arboreal marsupial: implications for conservation planning in a fire-prone landscape. Landsc. Ecol. 35, 1553−1570.

Pyne, S.J., 1992. Burning Bush: A Fire History of Australia. Allen & Unwin, Sydney.

Robinson, N., Leonard, S., Ritchie, E., Bassett, M., Chia, E., Buckingham, S., Gibb, H., Bennett, A.F., Clarke, M.F., 2013. Refuges for fauna in fire-prone landscapes: their ecological function and importance. J. Appl. Ecol. 50, 1321−1329.

Robinson, N.M., Leonard, S.W.J., Bennett, A.F., Clarke, M.F., 2014. Refuges for birds in fire-prone landscapes: the influence of fire severity and fire history on the distribution of forest birds. For. Ecol. Manag. 318, 110−121.

Seebeck, J., Warneke, R., Baxter, B., 1984. Diet of the bobuck, *Trichosurus caninus* (Ogilby) (Marsupialia: Phalangeridae) in a mountain forest in Victoria. Possums and Gliders. In: Smith, A.P., Hume, I.D. (Eds.), Possums and Gliders. Surrey Beatty and Sons, Sydney, pp. 145−154.

Simkin, R., Baker, P.J., 2008. Disturbance history and stand dynamics in tall open forest and riparian rainforest in the Central Highlands of Victoria. Aust. Ecol. 33, 747−760.

Sitters, H., Christie, F.J., Di Stefano, J., Swan, M., Penman, T., Collins, P.C., York, A., 2014. Avian responses to the diversity and configuration of fire age classes and vegetation types across a rainfall gradient. For. Ecol. Manag. 318, 13−20.

Smith, A.L., Blair, D., Mcburney, L., Banks, S.C., Barton, P.S., Blanchard, W., Driscoll, D.A., Gill, A.M., Lindenmayer, D.B., 2014. Dominant drivers of seedling establishment in a fire-dependent obligate seeder: climate or fire regimes? Ecosystems 17, 258−270.

Spies, T.A., Lindenmayer, D.B., Gill, A.M., Stephens, S.L., Agee, J.K., 2012. Challenges and a checklist for biodiversity conservation in fire-prone forests: perspectives from the Pacific Northwest of USA and Southeastern Australia. Biol. Conserv. 145, 5−14.

Taylor, C., McCarthy, M.A., Lindenmayer, D.B., 2014. Nonlinear effects of stand age on fire severity. Conserv. Lett. 7, 355−370.

Taylor, C., Lindenmayer, D.B., 2020. Temporal fragmentation of a critically endangered forest ecosystem. Aust. Ecol. 45, 340−354.

Wagner, B., Baker, P.J., Nitschke, C.R., 2021. The influence of spatial patterns in foraging habitat on the abundance and home range size of a vulnerable arboreal marsupial in southeast Australia. Conserv. Sci. and Prac. 3, e566.

Wood, S.W., Hua, Q., Allen, K.J., Bowman, D., 2010. Age and growth of a fire prone Tasmanian temperate old-growth forest stand dominated by *Eucalyptus regnans*, the world's tallest angiosperm. For. Ecol. Manag. 260, 438−447.

Wood, S.W., Murphy, B.P., Bowman, D., 2011. Firescape ecology: how topography determines the contrasting distribution of fire and rain forest in the south-west of the Tasmanian Wilderness World Heritage Area. J. Biogeogr. 38, 1807−1820.

Wood, S., Bowman, D., Prior, L.D., Lindenmayer, D.B., Wardlaw, T., Robinson, R., 2014. Tall eucalypt forests. In: Lindenmayer, D.B., Burns, E., Thurgate, N.Y., Lowe, A.J. (Eds.), Biodiversity and Environmental Change: Monitoring, Challenges and Direction. CSIRO Publishing, Melbourne.

Case Study: The Importance of Mixed- and High-severity Fires in Sub-Saharan Africa

Ronald W. Abrams[1,2]

[1]*Dru Associates, Inc., Glen Cove, NY, United States;* [2]*ADA Pty. Ltd, Chintsa, South Africa*

7.9 THE BIG PICTURE

Biodiversity in sub-Saharan Africa evolved and persists in this complex region in part because of the various mechanisms of habitat change initiated by mixed- and high-severity fires. Fischer (2021) reports that up to 70% of all global forest fires occur in Africa, mostly anthropogenic. The greatest expanse of vegetated habitat in Africa is savannah, where large, severe fires are uncommon and anthropogenic burns are widespread (Figure 7.5). Lehmann et al. (2014) indicate that savannah ecology sustains as much as one-fifth of the

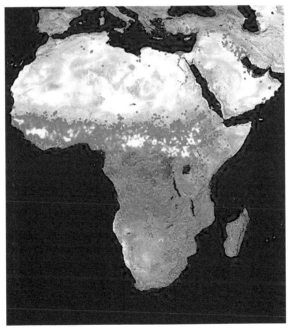

FIGURE 7.5 MODIS Rapid Response System Global Fire Maps showing regions burning (red) as of November 29, 2008. The image has been cropped to show sub-Saharan Africa. http://modis.gsfc.nasa.gov/gallery/individual.php?db_date=2008-11-29; accessed May 15, 2024.

global human population and—in Africa more than anywhere—hosts "most of the remaining megafauna." Across Africa where conservation is practiced, resource managers have embraced the importance of mixed- and high-severity fire, particularly where national parks and preserves serve ecotourism. A reigning leader in fire management is South Africa's Kruger National Park (see various articles in Du Toit et al., 2003; Biggs and Rogers, 2003). In sub-Saharan African savannah key drivers of the balance between grasses and woody vegetation include both natural wildfire (rare) and prescribed burns (Du Toit et al., 2003; Govender et al., 2006). Prescribed burns in the grasslands of African savannah burn fast and hot, spread quickly, and create habitat mosaics that in turn are associated with physiographic diversity. In this discussion, all fires that have biodiversity implications are mixed with portions that are high severity (termed "mixed severity").

In contrast to the story told by most of this book, Africa is comparatively closer to a "historic" condition for both ecological and anthropological reasons. The preface to this book sets a framework that seeks to overturn unreasonable "fear" of fire, as well as a suite of misunderstandings about the importance of mixed- and high-severity fire to various biomes. Pyne (2004) called for a new attitude that embraces the reality of wildfire, changes in land use, and prevention of continuing habitat loss, and optimistically expected such progress in 5—6 years. Two decades later, we tackle the issue, and this section discusses how people and natural resource managers have perceived mixed-severity fire in sub-Saharan Africa—a very different perspective than that of the developed regions of the world. Surprisingly, in the past 5 years, there is some evidence in an analysis by FAO that forest loss in Africa has slowed significantly (Shapiro et al., 2022, but see further discussion of this topic below).

7.10 WHERE IS FIRE IMPORTANT IN SUB-SAHARAN AFRICA?

Large tracts of Congo, Lakes Tanganyika and Victoria, tropical montane forests, and coastal rainforests are generally too wet for naturally occurring fires, and while anthropogenic fires occur, they are largely associated with pastoral agriculture (Cochrane, 2003; Thonicke et al., 2001; Brncic et al., 2007; Krawchuk and Moritz, 2011). Bowman et al. (2009) showed from a global perspective that fire frequency is relatively high in central Africa, but because the region of the Congo Basin is too wet for natural fires at high frequency, this is clearly anthropogenically induced fire frequency. The Congo Basin and its margins of tropical forests have lightning frequencies that are among the world's highest (Christian et al., 2003), but the high degree of moisture limits not only spread but also the intensity of fire, and thus severity is not high. Yet Laurance (2003) explained that the subtle nature of fire effects on tropical forests and their biota is widespread but poorly inventoried. Brncic et al. (2007) noted that lightning strikes in tropical forests tend to burn one spot (individual trees), which may contribute on a very small scale to mosaics of habitat but do not create the effects of large or intense fires.

In rain forest and other habitats, however, the potential for fire to create ubiquitous ecological change does exist; this is underscored by Bardgett and van der Putten (2014) in their discussion of the high sensitivity of microbial biota to microhabitat changes: fire can influence such changes, especially with new evidence of subtle range shifts as microbiota respond to climate change. Govender et al. (2006) described how variable fire intensity differentially heats the ground, which would influence desiccation and soil exposure to both wind and water erosion and thus can substantially alter the substrate and fauna. Microbiota may be the elephant in the room with respect to the future of biodiversity in Africa (see below for more on microbes and fire).

In sub-Saharan Africa fire-induced habitat change promotes biodiversity in both tropical forest and savannah, although most of the research in Africa is focused on savannah, because it is so dominant on the subcontinent (Thonicke et al., 2001; van Wilgen et al., 2003, 2004; Brncic et al., 2007, Vermeire et al., 2021). Some research on montane habitat also confirms the contribution of mixed- and high-severity fire to biodiversity, but at smaller scales that are bounded by physical habitat barriers (Geldenhuys, 2004), a different scenario than the wildfires of interest in the northern hemisphere. These are often intense (severe), but measures of this factor are sparse for Africa. As explained in the preface of this book, wildfire has an important role in shaping vegetation communities, and we need to help managers to understand where the "free fire" should occur (with much of the "new forest" in savannahs due to woody encroachment, should we resist such changes by active management of such growth?). This book is focused on a conservation-oriented understanding of wildfire, but much of what is discussed here about fire's contribution to biodiversity in Africa comes from studies of controlled burns or long-term observations that infer fire patterns from landscape history.

7.11 WHAT ABOUT PEOPLE AND FIRE?

Many people in Africa have a daily relationship with fire (cooking, heating). Though occasionally situations become dire because of uncontrolled, human-induced burns or the rare, naturally occurring wildfire, for many rural Africans fire is not something to be feared as a threat. This is in contrast to the perception of fire in developed nations, where the fear of wildfire limits its potential to perform a fundamental role in biodiversity maintenance (Bowman et al., 2009). Some people in Africa respond with suppression when fires threaten sacred woodlands or specimen trees, so there is an influence of religion or culture on habitat protection that may or may not support biodiversity (Hens, 2006; Ormsby, 2012). Criminal fires in Africa from war (Dudley et al., 2002) or civil conflicts (Hamilton et al., 2000) are clearly political and beyond the scope of this book. Nevertheless, criminal disturbance of habitats is a high priority in Africa, with a potentially positive effect on biodiversity as part of the overall relationship between the people and fire (as opposed to poaching or irresponsible waste disposal). Successful conservation strategies in Africa must consider the benefits that fire provides to a local human population (heat and fuel), just as conservation research now needs to consider the agenda of people living on the land (Abrams et al., 2009).

But is it possible for rural people to combine judicious use of fire that is both productive for them and coincidentally for the ecosystem? In South America studies of cycads (plants that are abundant in Africa) show that fires are used to encourage seed production for human use, and at the same time cycad reproduction increases in postfire recovery (Norstog and Nichols, 1997). The cycad's coralloid root adaptation seems to enable the plants to persist and even proliferate after either brush fires or anthropogenic burns (Norstog and Nichols, 1997). The only negative effect cited is when a fire is intense (hot) enough to kill the invertebrate pollinators residing in the surrounding leaf litter (which can result in shifting of the reproductive periodicity while the litter is recolonized). Of course, there is the perspective that the extra cycads (and other species encouraged by human-set fire) lead to ecological imbalance, but for a group of plants as resilient as cycads, their dominance in the ecosystem may occur in a wide range of conditions. In fact, this group of plants is among the most resilient of organisms, with species that change their phenotype when transplanted between latitudes (Heenan, 1977, David Heenan, personal communication).

7.12 COEVOLUTION OF SAVANNAH, HERBIVORES, AND FIRE

Fire has influenced African ecological systems over evolutionary timescales, for example, in the historical shift of southwestern African grasslands from C3 (3-carbon molecules) to C4 (4-carbon molecules) dependent photosynthesis over millennia (Hoetzl et al., 2013; Lehmann et al., 2014). A similar historical trend has been debated for North America, however, with a more specific look at the mechanisms at work (see Chapters 1, 2 and 8 in this book). Heisler et al. (2003) described how fire may not always control domination by woody cover; shrub cover increased despite annual fire in a long-term study of North American grasslands. Johnson (2009) explained the decline of North American megafauna as a cause of the loss of prehistoric grasslands, an illustration from history that concurs with observations about Africa's herbivores as integral to the maintenance of Africa's grasslands. As discussed more below, there is going to be a rising debate if, indeed, savannah turned woodland continues its spread in sub-Saharan Africa (Higgins et al., 2023; Sagang et al., 2022).

In contrast to North America, sub-Saharan Africa is still dominated by savannah, where heterogeneity is maintained by two chief mechanisms. First, mixed-severity fires and herbivores (moderated by substrates that contain limited moisture for at least part of the year) directly suppress woody vegetation and favor grass-dominated habitat (Westbroek et al., 1993; Sankaran et al., 2005; Archibald, 2008). Second, and perhaps more important, is the way fire influences structural features that effect biodiversity, such as heating/desiccation of soils that alters chemistry and microbiota, which in turn influences the germination of species through control of nutrient availability in soils (Venter et al., 2003; Bardgett and van der Putten, 2014). In tropical peatlands (perhaps found in marshlands of the Congo Basin) fire poses a threat to biodiversity when apparently low-intensity smoldering burns occur because of the long burnout times (again, Laurance's (2003) insidious fire characterization) and the deep and wide reach of smoldering peat layers (subject to desiccation by climate change) that result in adverse effects to soils reaching beyond the surfaces affected (Turetsky et al., 2015). In other situations fire in marshland may create habitat mosaics that can enhance structural habitat diversity (Turetsky et al., 2015), but the subject needs new research in Africa.

The mid-20th century reasoning for biodiversity conservation through controlled burns had support because of the rising fear that Africa, through the loss of naturally balanced herbivore populations (i.e., the keystone herbivore hypothesis), would follow the path of other continents that lost grasslands and native forest diversity (Owen-Smith, 1989). As early as the 1980s, Owen-Smith (1989) confirmed that in Africa the natural cycling of fire effects was disrupted by fire suppression and elephant (*Loxodonta africana*) population declines, which together have led to more wooded savannah habitat, and a loss of grassland, in the Tsavo area of Kenya (Owen-Smith, 1989).

7.13 HERBIVORES AND FIRE

Herbivore consumption and excretion are critical processes supporting biodiversity in African savannahs. Similarly, Bond and Keeley (2005) described fire as a global herbivore with significant influence on biodiversity through its capacity to alter habitats. Habitat alteration by grazing, browsing, and burning release plant species held in check by other plants that have a competitive advantage without disturbance, altering species assemblages and ecosystem dynamics, even if only for a short time. The resulting vegetative diversity, in height, form, and percentage of cover at different layers of woodland and savannah habitat, combine to support food diversity and thus faunal diversity, making the role of fire integral to the ecology and sociology of sub-Saharan Africa.

Sankaran et al. (2005) modeled the tree—grass balance in African savannah, showing that variations in rainfall, soils, and fire/herbivory drive and maintain diversity. Infrequent or low-intensity fire (>10.5-year return cycle) promotes woody cover, and conversely, frequent mixed-severity fires deter woody cover (except on sandy soils) (Sankaran et al., 2005). Of course, a savannah region that is drier accumulates greater fuel loads, leading to increased fire frequency and intensity, further suppressing woody habitat in favor of grasslands (Higgins et al., 2000) upon which many megafaunal species depend. Sankaran et al. (2005) found that most savannahs maintain woody cover well below the resource-limited upper bound, not generally reaching their climatic bounds, suggesting that drivers other than simply climate prevail (e.g., mixed severity fire, herbivores, soils), leading to the conclusion "... that water limits the maximum cover of woody species in many African savannah systems, but that disturbance dynamics control savannah structure below the maximum."

The long-term research in the Serengeti ecosystem of East Africa led by Sinclair et al. (2007) gives a useful picture of the relationship between fire and savannah biodiversity. Population size and dispersion of wildebeest (*Connochaetes taurinus*) and African buffalo (*Syncerus caffer*) have a negative correlation with fire return frequency and intensity (more or less accumulated fuel), and standing elephant population levels have a negative correlation with woodland recovery (i.e., a positive correlation with mixed-severity fire that causes direct tree mortality or weakening, which makes individuals more susceptible to lower-intensity fires); both mechanisms show strong biofeedback with the frequency and intensity (areal coverage = severity) of fires. As reduced fire activity continued over the years since the high fire activity of the 1960s, however, wildebeest populations later declined as woodlands expanded into grasslands, providing lions (*Panthera leo*) with better hiding cover and hunting success (Sinclair et al., 2007). In the Serengeti-Mara system there have been no known naturally occurring fires in recent decades—all were anthropogenic (Sinclair et al., 2007). Records of the extent of burns during the dry season in East Africa, documented by aerial reconnaissance, show a steady decline in the area burned from the early 1960s to the 1980s. More recent records confirm this trend (unpublished data cited by Sinclair et al. (2007)). These trends reflect an increase in herbivore reduction of grass biomass caused by grazing, meaning fires cannot progress as mixed-high severity burns, which can affect the grassland–woodland balance, leading to further shifts from grassland savannah to woodland (see the discussion below about forest increases).

In contrast to East Africa, in the Kruger Park of South Africa, where up to 20% of savannah fires are caused by lightning (van Wilgen et al., 2003), and in deference to the protection of villages and homes, most fires are subject to some form of control because the majority of them are set by humans. South African National Parks, however, have a long history in Kruger Park of maintaining grasslands through a combination of mechanisms (i.e., manipulation of vegetation by elephants, herbivore grazing, and frequent fire return). In 1992, Kruger Park managers decided to use a "lightning fire" approach, but within 10 years of monitoring they recognized that almost 80% of fires in the park were anthropogenic, so the experiment could not be continued, and they reverted to patch-mosaic burns and allowed natural fires to reach high intensity and to burn out, using controlled burns where needed and monitoring biodiversity with fire as a parameter (Biggs and Rogers, 2003; van Wilgen et al., 2004).

In modeling the triggers that switch habitats between grass- or tree-dominated savannahs, Higgins et al. (2000) confirmed Warner and Chesson's (1985) view that the "storage effect" hypothesis explains observations that varying fire intensities produce different species releases and thus coexistence of potentially competitive species (Warner and Chesson, 1985). Howe (2014) modernized this concept, especially for tropical tree diversity maintenance, terming it the diversity storage hypothesis. Higgins et al. (2000) hypothesized that grass—tree habitat-sharing is strongly influenced by the fortune of tree seedlings in surviving drought or avoiding the flame zone of fires intense enough to kill young trees, wherein tree recruitment is controlled by rainfall, which limits seed establishment, and by mixed- to high-severity fire that prevents recruitment to adult sizes. Thus, variations in fire intensity are a factor controlling the ability of trees to escape the flame zone, where mortality is greater. Variable fire intensity also influences species "selection" as the more fire-resistant size classes are adapted to recolonizing more quickly or, where enough individuals escape severe fires, to maintaining the local population (van Wilgen et al., 2003). In other words, variance in fire intensity produces the variance in recruitment rates that is necessary for the storage effect to operate (Higgins et al., 2000) as a mechanism contributing to biodiversity.

A pair of studies reviewed by Mayer and Khalyani (2011) showed fire frequency drives shifts between forest, savannah, and grassland biomes. Staver et al. (2011) found that, globally, fire is a strong indicator of savannah distribution, where the absence of fire in sub-Saharan Africa leads to bimodal tree cover in terms of frequency when rainfall is intermediate (1000—2500 mm Mean Annual Rainfall (MAR)). Hirota et al. (2011) similarly found that fire (interacting with rainfall regimes) has the potential to create three modes: forest, savannah, and grasslands. Mayer and Khalyani (2011), concurring with Sinclair et al. (2007), summarized their interpretation of these data by writing: "Both reports identify an unstable state at 50%—60% tree cover; either trees take hold and promote their own growth hydrologically (and suppress fire), or grasses take hold and promote their expansion through fire".

Staver et al. (2011) and Hirota et al. (2011) identified the transitions among biomes using global data sets showing that areas remain forested (>60% tree cover) with rainfall routinely >2500 mm/year. Habitats that receive a middle range of precipitation (1000—2500 mm/year) were forest or savannah depending on the strength of the fire—grass feedback. Habitats are unstable when tree cover is between 50% and 60%, a condition of transition between forest and savannah. When habitats are subject to strong seasonal rainfall (i.e., between ~ 750 and 1500 mm/year) they were either savannah or grassland, depending on fire frequency.

One outcome of the recent literature on drivers of biodiversity management shows clearly that climate, of which fire regimes are an integral associate, do in fact dominate the shifting between African savannah—forest ecosystems (Higgins et al., 2023). In their statistical review of herbaria data they showed that, by citing to a long-term study by Sagang et al. (2022), that the 'shifting' is not a stable or predictable scenario, much more dependent on extrinsic factors than previously thought. The future of Africa's savannah habitats is not at all clear and requires dedicated interdisciplinary research.

7.14 BEYOND AFRICA'S SAVANNAH HABITAT

Geldenhuys (2004) described how forests in South Africa once might have covered 7% of the terrain based on suitable microclimate, but now cover only 0.1% of their potential range. Geldenhuys attributed this to anthropogenic activities, with fire setting the boundaries of many habitats. In the coastal and certain inland regions of southern Africa, mountain range physiography creates wind patterns that drive the dispersion of fire (and intensity), and these locations actually serve as "fire refugia" that, based on Geldenhuys' description, also provide biodiversity storage depots (again, the storage effect or Howe's diversity storage hypothesis). Latimer et al. (2005) presented data showing that speciation in South Africa's fynbos biome (a Mediterranean-type habitat) is more rapid than in other comparable regimes, and release by fire is a leading management option.

Du Toit et al. (2003), Sinclair, and Geldenhuys interpret the research on fire ecology in sub-Saharan Africa such that land use managers must seek a balance between fire suppression and preserving its role in biodiversity maintenance. But they face a daunting challenge. Conservationists are wary of any nonnatural alteration of natural cycles (e.g., climate-related and anthropogenic loss of the natural fire cycles) as the potential downfall of entire food webs, including in sub-Saharan Africa, where the bird and mammal populations have a prevailing role in biodiversity maintenance (Laurance and Useche, 2009; Mayer and Khalyani, 2011; Howe, 2014). Nimmo et al. (2013) showed how fire-created habitat mosaics contribute to faunal diversity in Australia and echoed the "patch-mosaic burning" strategy used globally. Gregory et al. (2010) confirmed, by studies in East African savannahs, that avifaunal diversity and species persistence in a given area are dependent on both herbivore grazing and fire return cycles. The adaptive management strategy used in Kruger Park in South Africa has derived the concept of "thresholds of potential concern" (TPCs), which takes a holistic approach to habitat, and thus wildlife management, by following ecological responses to TPCs and adapting their controlled burns schedule accordingly (Biggs and Rogers, 2003).

7.15 HABITAT CHANGES FOREST LOSS/GAIN AND OTHER CONSIDERATIONS

As the African human population grows and agriculture changes, including pastoral practices, the forests of the region respond. The extent and spread of forests in Africa, including where savannah grows in, can add substantial areas prone to fire, but favored by far-ranging reforestation policies. Some reports recently suggest that forest loss has slowed, so more 'new' forest becomes a challenge to planning fire management. It will be important to track such a tendency. Even though the current literature leaves the question open, there are enough reports of forest gain to merit discussion here. Intriguing questions arise in the face of increasing forest area in Africa. Are these areas restored, or colonized anew associated with gains in biodiversity? Or should such "encroachments" into grasslands be controlled, perhaps by managed burning or cessation of fire suppression (i.e., allowing "wild fire" to go unchecked).

Pelletier et al. (2020) report that in Zambia small holdings that generate useful income result in less overall forest loss than was happening from commercial scale development, showing a mechanism by which forest loss reduction can be encouraged. Research by the UN FAO published a surprising report that, based on data from 2015 to 2020 in the Congo Basin, deforestation has substantially declined between 2017 and 2020 (Shapiro et al., 2022). Sagang et al. (2022) demonstrate a significant increase in forest cover due to woody encroachment into savannah habitats. If indeed there is a trend toward reduced forest cover in sub-Saharan Africa, management of fire and other resources will be significantly affected. In order to better understand such findings a brief look at other deforestation reports is warranted here.

Olagunju (2015) presented data showing little change in forest loss rates of sub-Saharan Africa between 1990 and 2010. Ordway et al. (2019) disagree with many of the reports of limited forest loss to plantations (e.g., oil palm in Cameroon), citing statistics of considerable loss between 2000—10, due to the creation of more plantations. Masayi et al. (2021) reports an 18% decline in forest cover between 1977 and 2019 in a region of western Kenya. Kogo et al. (2019) using remote sensing reported 43% net gain between 1995 and 2001, and then a decline up to 2017 that resulted in a net loss of 12.5%. In Zimbabwe Zvobgo and Tsoka (2021) looked at the Upper Manyame Sub Catchment 1990; 2020, where losses ranged from 29.48% to 20.10% of the region, with pronounced loss of about 6% between 2000 and 2010. Cropland increased by 10% and development by over 25%.

But in Malawi, Bone et al. (2017) reported a wide range of forest loss statistics, including over half of the districts sampled showing no forest loss, with some considerable gains in certain areas. Forest gains, up to 8%, are caused by woody plant encroachment (Vermeire et al., 2021), where herbivore distribution has changed (human impact) and fire frequency has been reduced. While the actual cellular mechanisms remain poorly studied still, work by this group has showed that removal of natural change dynamics in favor of anthropomorphic impacts do not substantially alter bacterial and fungal community ecology. This surprising finding means that the restoration of savanna habitats can be promoted in concert with carefully managed human activity. Hence, rangeland management in Africa involves knowledge of site-specific biomes, soil fertility, fire regimes, and grazing patterns, so at least recent studies offer us a way forward, albeit cumbersome because each area of concern needs its own intensive background study.

A regional site-specific study by Coetsee et al. (2023) in Kenya documented how differently grazers and herbivores effect soil fertility. On Mt Kenya, in the moorland fire of 2012, the soils were not substantially impacted by the fire in analyses completed 3 years after the fire (Downing et al., 2017). A modeling study of fire history on Mt. Kilimanjaro found that impacts to the carbon-based resources would require up to 150 years to fully recover to prefire levels (Fischer, 2021). The conversion of vegetation and associated wildlife habitat to pioneer habitats or successional forest implies a substantial adverse impact to soil microbes and their role in woodland ecology. This translates to the difficulty forest managers have in the choices associated with fire suppression, habitat restoration, or the use of prescribed fire. A topic now receiving attention is the geochemistry of fires. While this topic is beyond the scope of this chapter, the findings by Bauters et al. (2021) describe a high input of phosphorus to soils by fire in tropical African forests. The source of phosphorus represented here may be attractive to forest managers looking to enhance plant growth but could also represent a liability if such deposits occur near surface waters susceptible to eutrophication.

Snyman (2015) reports that even after 2 years, the effects of fire that remove organic content from the soils have not recovered in terms of hydrology, an important impact when water resources are limited. Maquia et al. (2020) found that bacteria assemblages recover so quickly after a fire despite the heat and burn effects that kill most organisms in the surface soils. A study in Nigeria found that the electrochemistry character of soils under burned habitat were significantly altered (Abdulcadir, 2021). Bottom line on geochemistry and hydrogeology in tropical Africa: much more research is needed!

7.16 HABITAT MANAGEMENT THROUGH CONTROLLED BURNS

Venter et al. (2003) described how the interaction of fire with patch dynamics and soil biogeochemistry are important drivers of physiographic and, consequently, overall ecological heterogeneity, including the wildlife of Kruger Park. In considering the size or intensity of two historic episodes of severe and intense fires in Kruger (>20% of the park's area), Venter et al. (2003) concluded that, while these fires resulted in temporary shifts in vegetation in any given location, the broader ecosystem remained stable, and they argued that prescribed burns can serve as scale-independent disturbances. In addition, while Latimer et al. (2005) agree that wildlife assemblages are structured in response to fire-driven cycles, they also found that species release follows, to some extent, Hubbell's neutral theory of ecology in the fynbos habitat because all of the many species are present and their appearance follows the patterns of disturbance (i.e., drought and fire) that cause microhabitat variations. Over evolutionary time scales, the system remains stable, with similar biodiversity metrics, but results in dynamic, shifting patches to which wildlife respond as the plants respond to mixed- and high-severity fire (Gregory et al., 2010). Thus, what would sub-Saharan Africa look like without anthropogenic fire?

But there are additional questions about our collective knowledge of the results of controlled burns for biodiversity conservation. Ironically, Romme et al. (1998) demonstrated that, at least in certain systems, adaptation to long-standing natural fire return cycles can actually suppress species release. The practitioners who manage Kruger National Park manipulate fire to prevent such suppression, using patch-mosaic strategies as a basis to ensure varying stages of succession (Biggs and Rogers, 2003). Parr and Andersen (2006) question whether sufficient monitoring of patch-mosaic burning has been used to parlay results in the field into adaptive management strategies, although they credit South African National Parks with a comprehensive approach to the issue, and they do not doubt that pyrodiversity contributes to biodiversity. It seems that when one considers the combined effects on savannah of large herbivores and fire, they amount to biodiversity enhancement because herbivores maintain long-term, consistent pressure on grasses that is comparable to that of large, intense wildfire (widespread in its vegetative effects), which also mediates the interactions between forest, woodland, and grassland. In Kruger Park the herbivore contribution to grassland maintenance is supplemented with rotated patch-mosaic burns, making the combined outcome comparable to a circumstance in which large, mixed-severity fires were common and allowed to burn out, all adaptively managed by the park's TPC approach (Biggs and Rogers, 2003).

The TPC approach has achieved the overall objective of using an understanding of mixed- and high-severity fire dynamics to optimize support of the natural food web. Govender et al. (2006) provided some of the limited measures of fire intensity and severity for sub-Saharan Africa, looking at the savannah of Kruger Park. They reported results of a 21-year experiment designed at the time to test ideas about eradication and recovery, and post hoc analysis using fire intensity measurements and an array of technological and evidentiary methods showed how the occurrence and role of mixed-severity fire relate to a set of variables. Van Wilgen et al. (2004) described fire cycling in the Kruger Park as dependent on grass biomass (influenced by wildlife and rainfall regimes) but also observed that it could be managed by season and fire intensity.

The practices used now, for example, at Tygerberg Preserve in the Western Cape and at Kruger integrate the effects of variable fire intensity in selection of time, size, and target species. In the Western Cape there is less grass and more woody material to burn than in Kruger's savannah. So, in Kruger, wildfire spread rates (which control the ultimate intensity of a controlled burn) are a management choice (Govender et al., 2006). In the Western Cape (see Box 7.1

> **BOX 7.1 Southwestern Cape Renosterveld Management**
>
> Habitat type: Swartland Shale Renosterveld (critically endangered) is restricted to the Cape floristic kingdom within a Mediterranean climate. Renosterveld vegetation occurs in nutrient-rich (clay-based) soils. Structurally, Renosterveld is much less complex than fynbos, consisting mostly of a single layer of shrubs. The most dominant species is *Elytropappus rhinocerotis* (Renosterbos); hence, the name of the vegetation type. The grassy understory can be extremely rich in geophytes and annual plant species, which are particularly conspicuous after fire. Renosterveld is a late-successional vegetation type. If the vegetation is not burned, however, it is invaded (encroached upon) by forest precursor species such as *Olea europaea* (wild olive) and *Kiggelaria africana* (wild peach) (adapted from Trinder-Smith et al., 2006).
>
> *Species being suppressed:*
> *Olea europaea* (wild olive)—tree
> *Kiggelaria africana* (wild peach)—tree
> *Searsia angustifolia* (Wilgerkorentebos)—large shrub
> *Post-fire recovery species:*
> *Drimia capensis* (Maerman)—perennial geophyte
> *Watsonia marginata*—perennial geophyte
> *Haemanthus sanguineus* (April fool)—perennial geophyte
> *Oxalis* species (e.g., *Oxalis eckloniana* and *Oxalis livida*)—perennial herb
> *Otholobium hirtum*—shrub (pioneer species)
> *Helichrysum cymosum*—woody shrub
> *Felicia fruticosa* (wild aster)—shrub
> *Chrysanthemoides monilifera* (tick berry/Bietou)—shrub

below), and in areas where uncontrolled spread is less of a concern, like the self-bounded mountain areas of South Africa discussed by Geldenhuys (2004), mixed- and high-severity burns are used to remove woody species or alien species such as black wattle (*Acacia mearnsii*).

However, Govender et al. (2006) showed that in the Kruger Park mixed- and high-severity fire serves to suppress the recruitment of large trees because they must outgrow the flame zone, so such fires are a tool that are being used by managers to maintain the tree islands amid the grasslands. Govender et al. (2006) used a 3000 kW/m threshold to define high intensity (and thus potential severity) from van Wilgen et al. (1990), citing the example that saplings shorter than 1 m would be top-killed, whereas those taller than 2 m could persist.

An interesting finding by Govender et al. (2006) is that for 4—5 years after a fire, grass growth is robust and fuel accumulates, but into the sixth year and beyond, growth levels off and grazing pressure reduces the vitality of the stand. Does this mean that an optimal, natural fire periodicity would be 4—5 years? The current periodicity in Kruger is 3—6 years, and this seems to be the management target for South Africa's savannah. These fires are often large in area (>5000 ha) and burn at a range of intensities, with the front or back fire lines being of very high intensity. As reported in Du Toit et al. (2003), Govender et al. (2006) found that such fires progress to high severity effects through an array of burn scenarios, including variations in intensity associated with time of day or night, slopes, and wind direction/force, so that by the time the fire burns out the result is a "fine-scale mosaic of varying fire intensities" that drive postfire habitat conditions.

7.17 SOUTHWESTERN CAPE RENOSTERVELD MANAGEMENT

For Africa, there are not as many studies specific to fire regimes as for other regions. I close with an example of habitat management that does parallel other parts of the world. Syphard et al. (2009) included South Africa's Cape floristic region as one of the Mediterranean ecotypes, where fire periodicity is critical to the maintenance of biodiversity through seed release mechanisms. Latimer et al. (2005) provided evidence that this region is among the most biodiverse on earth. As an example of current biodiversity management with fire in this region, a case study using the Renosterveld habitat of South Africa's Western Cape is presented. Here, mixed-severity fire (Figure 7.6) is used to remove three invasive plants, and the patches are allowed to burn out as if they were natural (but with a professional crew on site to protect against the fire running to an unwanted area); this results in an immediate release of eight

FIGURE 7.6 The ignition of a patch burn in March 2012 at Tygerberg Preserve in South Africa's Cape Floristic Region, a World Heritage region. *Photo: Penelope Glanville.*

native species essential to conserving this critically endangered Mediterranean type habitat (Box 7.1).

At Tygerberg Preserve, the pattern of controlled burn implementation follows the patch-mosaic concept, aimed at creating a pattern of a succession that mirrors that of a series of natural fires to release the "stored species" (Figure 7.7). Although this burn (Figure 7.8) was done at the end of summer

FIGURE 7.7 The patch burn is allowed to burn hot and expend itself in a prescribed area at Tygerberg Preserve, attended by a professional fire ecology management crew. Different stages of succession can be seen in distant mosaic areas. Note the proximity of residential areas. *Photo: Penelope Glanville.*

FIGURE 7.8 Patch mosaic recovery at Tygerberg Preserve, July 2012, where the burn occurred merely 3 months prior (i.e., fall to winter period). *Photo: Neal Schachat.*

(for fire protection reasons), the fast recovery of the native vegetation is evident even as winter begins in June/July.

The controlled burns at Tygerberg are supplemented by onsite inventories and monitoring of soil and hydrology, maintaining a flexibility that is responsive to local residents, season, and stages of succession, much as described by Biggs and Rogers (2003) for Kruger Park. Although they are no longer present, the Western Cape Renosterveld once supported rhinoceros (*Diceros bicornis* or *Ceratotherium simum simum*) and still supports bontebok (*Damaliscus pygargus pygarus*) and many smaller mammals, as well as some herpetofauna (N. Schachat, personal communication). The Cape floristic region is extremely sensitive to disturbance but is so valued that even municipalities expend considerable effort at biodiversity management.

7.18 CONCLUSION

An evaluation of the benefits of fires in Africa and other emerging nations is complicated because people living off the land have different priorities than do conservation scientists seeking an academic understanding of the issues surrounding wildfire. In this context, is considering the values of agri-environmental approaches to biodiversity conservation necessary (Laurance and Useche, 2009; Gregory et al., 2010)? Depending on the circumstance, anthropogenic fires can be considered by some communities to be damaging to local ecosystems, and to other groups acceptable. Accordingly, the term benefits must be understood as the degree to which the ecologically "desired outcome" is achieved and collateral damage is recognized as unintended. Therefore, a simple consensus model in sub-Saharan Africa cannot properly measure or evaluate pyrogenic benefits, so I prefer to discuss "outcomes" of fire management policy and research. In the literature and media coverage of fire in sub-Saharan Africa, there is no parallel to the "panic and fear" that occur in, for example, the western region of the United States. There is, of course, less political focus on subtle issues of environmental policy: much of the continent is justifiably absorbed in basic human survival.

The outcomes of either wildfire or anthropogenic fires are a concern shared by those planning the protection of an ecosystem's biodiversity. For a simplified example, if the return periodicity of fire in a savannah ecosystem is suppressed and too much tinder allowed to accumulate, then the result would be an excess of high-intensity fire that could go beyond reducing the ground and shrub layers (a desired outcome) to damaging the canopy and disrupting the balance of shade and nutrient use that maintains the grass-tree mosaic (van Wilgen et al., 2004; Staver et al., 2011; Lehmann et al., 2014). This suggests an ongoing and integral role for anthropogenic burning in African savannah.

The nexus between people and habitat management makes it difficult to get a clear broad picture of the benefits of mixed- and high-severity fires in sub-Saharan Africa. Controlled burns in Kruger Park's rigorous long-term "experimentation" have taught us much. Many people making their living from the land would consider fire suppression as a benefit that protects their holdings from unwanted damage, possibly including outcomes that are protective of crops or forage for livestock. Even in regions where wildlife is raised for game farming, ecotourism, or even hunting, fire suppression is sometimes seen as a benefit to those endeavors, whereas the natural return cycle of wildfires would produce a different, and alternatively beneficial, outcome.

Nevertheless, the history of fire and habitat management in sub-Saharan Africa teaches us two main lessons: (1) mixed- and high-severity fire is a critical mechanism in the maintenance of biodiversity; and (2) people can and do live with severe fires as part of their local ecosystem. There is a serious need to inventory and assess the role of mixed- and high-severity wildfire in Africa, but between politics, war, and basic human needs, this sort of research has not been done at the same levels at which it is occurring in, for example, North America. For sub-Saharan Africa, where savannah ecology has gripped researchers for 5 decades or more, results show that the combined effects of fire on savannah, as well as on elephants, rhinos, wildebeest, buffalo, and many more ungulates, is to maintain a balance of woody versus grassy vegetation. When current fire management practices are combined with herbivore population monitoring (as in Kruger Park), the outcome is similar to what must have been the long-term evolution of that balance as influenced by natural, mixed-severity fire. In those parts of Africa where conservation of biodiversity is a priority, and where there is a commitment of human resources in this regard (skilled labor led by science), fire is an important tool and is given its deserved respect.

REFERENCES

Abdulcadir, A., 2021. Effect of bush burning on the cation exchange capacity of the soil in bali local government, Bali, Taraba State, Nigeria. Afr. Schol. J. Agric. Agric. Tech. 22 (1).

Abrams, R.W., Anwana, E.D., Ormsby, A., Dovie, D.B.K., Ajagbe, A., Abrams, A., 2009. Integrating top-down with bottom-up conservation policy in Africa. Conserv. Biol. 23, 799–804.

Archibald, S., 2008. African grazing lawns: how fire, rainfall, and grazer numbers interact to affect grass community states. J. Wildl. Manag. 72, 492–501.

Bardgett, R.D., van der Putten, W.H., 2014. Belowground biodiversity and ecosystem functioning. Nature 515, 505–511. http://www.nature.com/doifinder/10.1038/nature13855.

Bauters, M., Drake, T.W., Wagner, S., et al., 2021. Fire-derived phosphorus fertilization of African tropical forests. Nat. Commun. 12, 5129. https://doi.org/10.1038/s41467-021-25428-3.

Biggs, H.C., Rogers, K.H., 2003. An adaptive system to link science, monitoring and management in practice. In: Toit, D., et al. (Eds.), The Kruger Experience: Ecology and Management of Savannah Heterogeneity. Island Press, Washington, DC.

Bone, R.A., Parks, K.E., Hudson, M.D., Tsirinzeni, M., Willcock, S., 2017. Deforestation since independence: a quantitative assessment of four decades of land-cover change in Malawi. South. For. J. For. Sci. 79 (4), 269–275. https://doi.org/10.2989/20702620.2016.1233777.

Bond, W.J., Keeley, J.E., 2005. Fire as a global 'herbivore': the ecology and evolution of flammable ecosystems. Trends Ecol. Evol. 20, 387–394.

Bowman, D.M.J.S., Balch, J.K., Artaxo, P., Bond, W.J., Carlson, J.M., Cochrane, M.A., D'Antonio, C.M., Defries, R.S., Doyle, J.C., Harrison, S.P., Johnston, F.H., Keeley, J.E., Krawchuk, M.A., Kull, C.A., Marston, J.B., Moritz, M.A., Prentice, I.C., Roos, C.I., Scott, A.C., Swetnam, T.W., Van Der Werf, G.R., Pyne, S.J., 2009. Fire in the Earth system. Science 324, 481–484.

Brncic, T.M., Willis, K.J., Harris, D.J., Washington, R., 2007. Culture or climate? The relative influences of past processes on the composition of the Lowland Congo Rainforest. Philos. Trans. R. Soc. Lond. Ser. B Biol. Sci. 362, 229–242.

Christian, H.J., Blakeslee, R.J., Boccippio, D.J., Boeck, W.L., Buechler, D.E., Driscoll, K.T., Goodman, S.J., Hall, J.M., Koshak, W.J., Mach, D.M., Stewart, M.F., 2003. Global frequency and distribution of lightning as observed from space by the optical transient detector. J. Geophys. Res. 108, 4-1–4-15.

Coetsee, C., Wigley, B.J., Sankaran, M., Ratbam, J., Augustine, D.J., 2023. Contrasting effects of grazing vs browsing herbivores determine changes in soil fertility in an East African Savanna. Ecosystems 26, 161–173.

Cochrane, M.A., 2003. Fire science for rainforests. Nature 421, 913–919.

Downing, T.A., Moses, I., Johnstone, K., Nekesa, O.A., 2017. Effects of wildland fire on the tropical alpine moorlands of Mount Kenya. Catena 149 (1), 300–308.

Du Toit, J.T., Rogers, K.H., Biggs, H.C. (Eds.), 2003. The Kruger Experience: Ecology and Management of Savannah Heterogeneity. Island Press, Washington, DC.

Dudley, J.P., Ginsberg, J.R., Plumptre, A.J., Hart, J.A., Campos, L.C., 2002. Effects of war and civil strife on wildlife and wildlife habitats. Conserv. Biol. 16, 319–329.

Fischer, R., 2021. The long-term consequences of forest fires on the carbon fluxes of a tropical forest in Africa. Appl. Sci. 11, 4696. https://doi.org/10.3390/app11104069.

Geldenhuys, C.J., 2004. Concepts and process to control invader plants in and around natural evergreen forest in South Africa. Weed Technol. 18, 1386–1391.

Govender, N., Trollope, W.S.W., van Wilgen, B.W., 2006. The effect of fire season, fire frequency, rainfall and management on fire intensity in savannah vegetation in South Africa. J. Appl. Ecol. 43, 748–758.

Gregory, N.C., Sensenig, R.L., Wilcove, D.S., 2010. Effects of controlled fire and livestock grazing on bird communities in East African Savannahs. Conserv. Biol. 24, 1606–1616.

Hamilton, A., Cunningham, A., Byarugaba, D., Kayanja, F., 2000. Conservation in a region of political instability: Bwindi Impenetrable Forest, Uganda. Conserv. Biol. 14, 1722–1725.

Heenan, D., 1977. Some observations on the cycads of Central Africa. Bot. J. Linn. Soc. 74, 279–288.

Heisler, J.L., Briggs, J.M., Knapp, A.K., 2003. Long-term patterns of shrub expansion in a C4-dominated grassland: fire frequency and the dynamics of shrub cover and abundance. Am. J. Bot. 90, 423–428.

Hens, L., 2006. Indigenous knowledge and biodiversity conservation and management in Ghana. J. Hum. Ecol. 20, 21–30.

Higgins, S.I., Bond, W.J., Trollope, W.S.W., 2000. Fire, resprouting and variability: a recipe for grass-tree coexistence in Savannah. J. Ecol. 88, 213–229.

Higgins, S.I., Conradie, T., Kruger, L.M., O'Hara, R.B., Slingsby, J.A., 2023. Limited climatic space for alternative ecosystem states in Africa. Science 380, 1038.

Hirota, M., Holmgren, M., Van Nes, E.H., Scheffer, M., 2011. Global resilience of tropical forest and savannah to critical transitions. Science 334, 232−235.

Hoetzl, S., Dupont, L., Schefuß, E., Rommershirchen, F., Weber, G., 2013. The role of fire in miocene to pliocene C_4 grassland and ecosystem evolution. Nat. Geosci. 6, 1027−1030.

Howe, H.F., 2014. Diversity storage: implications for tropical conservation and restoration. Glob. Ecol. Conserv. 2, 349−358. https://doi.org/10.1016/j.gecco.2014.10.004.

Johnson, C., 2009. Paleontology, megafaunal decline and fall. Science 326, 1072−1073.

Kogo, B.K., Kumar, L., Koech, R., 2019. Forest cover dynamics and underlying driver forces affecting ecosystems services in western Kenya. Rem. Sens. Appl. Soc. Environ. 14, 75−83.

Krawchuk, M.A., Moritz, M.A., 2011. Constraints on global fire activity vary across a resource gradient. Ecology 92, 121−132.

Latimer, A.M., Silander, J.A., Cowling, R.M., 2005. Neutral ecological theory reveals isolation and rapid speciation in a biodiversity hot spot. Science 309, 1722−1725.

Laurance, W.F., 2003. Slow burn: the insidious effects of surface fires on tropical forests. Trends Ecol. Evol. 18, 209−212.

Laurance, W.F., Useche, D.C., 2009. Environmental synergisms and extinctions of tropical species. Conserv. Biol. 23, 1427−1437.

Lehmann, C.E.R., Anderson, T.M., Sankaran, M., Higgins, S.I., Archibald, S., Hoffmann, W.A., Hanan, N.P., Williams, R.J., Fensham, R.J., Felfili, J., Hutley, L.B., Ratnam, J., San Jose, J., Montes, R., Franklin, D., Russell-Smith, J., Ryan, C.M., Durigan, G., Hiernaux, P., Haidar, R., Bowman, D.M.J.S., Bond, W.J., 2014. Savannah vegetation-fire-climate relationships differ among continents. Science 343, 548−552.

Maquia, I.S., Chauque, A., Ferriera-Pinto, M.M., Lumini, E., Berruti, A., Ribeiro, N.S., Marques, I., Ribeiro-Barros, A.I., 2020. Mining the microbiome of key species from African Savanna Woodlands: potential for soil health improvement and plant growth promotion. Microrganisms 8 (9), 129.

Masayi, N.N., Omondi, P., Tsingalia, M., 2021. Assessment of land use and land cover changes in Kenya's Mt. Elgon For. Ecosyst. Afr. J. Ecol. 59 (4), 988−1003.

Mayer, A.L., Khalyani, A.H., 2011. Grass trumps trees with fire. Science 334, 188−189.

Nimmo, D.G., Kelly, L.T., Spence-Bailey, L.M., Watson, S.J., Taylor, R.S., Clarke, M.F., Bennett, A.F., 2013. Fire mosaics and reptile conservation in a fire-prone region. Conserv. Biol. 27, 345−353.

Norstog, K.J., Nichols, T.J., 1997. The Biology of Cycads. Cornell University Press, Ithaca.

Olagunju, T.E., 2015. Impacts of human-induced deforestation, forest degradation and fragmentation on food security. New York Sci. J. 8 (1). http://www.sciencepub.net/newyork.

Ordway, E.M., Naylor, R.L., Nkongho, R.N., Lambin, E.F., 2019. Oil palm expansion and deforestation in Southwest Cameroon associated with proliferation of informal mills. Nat. Commun. 10, 114. https://doi.org/10.1038/s41467-018-07915-2.

Ormsby, A., 2012. Cultural and conservation values of sacred forests in Ghana. In: Pungetti, G., Ovideo, G., Hooke, D. (Eds.), Sacred Species and Sites: Advances in Biocultural Conservation. Cambridge University Press, Cambridge.

Owen-Smith, N., 1989. Megafaunal extinctions: the conservation message from 11,000 years BP. Conserv. Biol. 3, 405−412.

Parr, C.L., Andersen, A.N., 2006. Patch mosaic burning for biodiversity conservation: a critique of the pyrodiversity paradigm. Conserv. Biol. 20, 1610−1619.

Pelletier, J., Mason, N.M., Barrett, C.B., 2020. Does smallholder maize intensification reduce deforestation? Evidence from Zambia. In: Version of Record. https://www.sciencedirect.com/science/article/pii/S095937802030710X_Manuscript_0d7ac6ba9f8abfdeed4b89c9ba1135c7.

Pyne, S.J., 2004. Pyromancy: reading stories in the flames. Conserv. Biol. 18, 874−877.

Romme, W.H., Everham, E.H., Frelich, L.E., Moritz, M.A., Sparks, R.E., 1998. Are large, infrequent disturbances qualitatively different from small, frequent disturbances? Ecosystems 1, 524−534.

Sagang, L.B.T., Ploton, P., Viennois, G., Féret, J.-B., Sonké, B., Couteron, P., as Barbier, N., 2022. Monitoring vegetation dynamics with open earth observation tools: the case of fire-modulated savanna to forest transitions in Central Africa. ISPRS J. Photogr Rem. Sens. 188, 142−156.

Sankaran, M., Hanan, N.P., Scholes, R.J., Ratnam, J., Augustine, D.J., Cade, B.S., Gignoux, J., Higgins, S.I., Le Roux, X., Ludwig, F., Ardo, J., Banyikwa, F., Bronn, A., Bucini, G., Caylor, K.K., Coughenour, M.B., Diouf, A., Ekaya, W., Feral, C.J., February, E.C., Frost, P.G.H., Hiernaux, P., Hrabar, H., Metzger, K.L., Prins, H.H.T., Ringrose, S., Sea, W., Tews, J., Worden, J., Zambatis, N., 2005. Determinants of woody cover in African savannahs. Nature 438, 846−849.

Shapiro, A., d'Annunzio, R., Jungers, Q., Desclée, B., Kondjo, H., Mbulito, J., Fancis, I., Rambaud, G.P., Sonwa, D., Mertens, B., Tchana, E., Obame, C., Khasa, D., Bougouin, C., Nana, T., Ouissika, H., Kipute, D., 2022. Are Deforestation and Degradation in the Congo Basin on the Rise? An Analysis of Recent Trends and Associated Direct Drivers. https://doi.org/10.21203/rs.3.rs-2018689/v1.

Sinclair, A.R.E., Mduma, S.A.R., Hopcraft, J.G.C., Fryxell, J.M., Hilborn, R., Thirgood, S., 2007. Long-term ecosystem dynamics in the Serengeti: lessons for conservation. Conserv. Biol. 21, 580−590.

Snyman, H.A., 2015. Short-term responses of Southern African semi-arid rangelands to fire: A review of impact on soils. Arid Land Res. Manag. 29 (2), 222−236.

Staver, A.C., Archibald, S., Levin, S.A., 2011. The global extent and determinants of savannah and forest as alternative biome states. Science 334, 230−232.

Syphard, A.D., Radeloff, V.C., Hawbaker, T.J., Stewart, S.I., 2009. Conservation threats due to human-caused increases in fire frequency in Mediterranean-climate ecosystems. Conserv. Biol. 23, 758−769.

Thonicke, K., Venevsky, S., Sitch, S., Cramer, W., 2001. The role of fire disturbance for global vegetation dynamics: coupling fire into a dynamic global vegetation model. Glob. Ecol. Biogeogr. 10, 661−677.

Trinder-Smith, T.H., Kidd, M.M., Anderson, F., 2006. Wild Flowers of the Table Mountain National Park. Botanical Society of South Africa, Cape Town, South Africa.

Turetsky, M.R., Benscoter, B., Page, S., Rein, G., van der Werf, G.R., Watts, A., 2015. Global vulnerability of peatlands to fire and carbon loss. Nat. Geosci. 8, 11−14.

Van Wilgen, B.W., Everson, C.S., Trollope, W.S.W., 1990. Fire management in southern Africa: some examples of current objectives, practices and problems. In: Goldammer, G.J. (Ed.), Fire in the Tropical Biota: Ecosystem Processes and Global Challenges. Springer Verlag, Berlin, Germany.

Van Wilgen, B.W., Trollope, W.S.W., Biggs, H.C., Potgieter, A.L.F., Brockett, B.H., 2003. Fire as a driver of ecosystem variability. In: Du Toit, J.T., Rogers, K.H., Biggs, H.C. (Eds.), The Kruger Experience: Ecology and Management of Savannah Heterogeneity. Island Press, Washington, DC.

Van Wilgen, B.W., Govender, N., Biggs, H.C., Ntsala, D., Funda, X.N., 2004. Response of savannah fire regimes to changing fire-management policies in a large African national park. Conserv. Biol. 18, 1533−1540.

Venter, F.J., Scholes, R.J., Eckhardt, H.C., 2003. The abiotic template and its associated vegetation pattern. In: Du Toit, J.T., Rogers, K.H., Biggs, H.C. (Eds.), The Kruger Experience: Ecology and Management of Savannah Heterogeneity. Island Press, Washington, DC.

Vermeire, M.L., Thoresen, J., Lennard, K., Vikram, S., Kirkman, K., Swemmer, A.M., Te Beest, M., Siebert, F., Gordijn, P., Venter, Z., Brunel, C., Wolfaard, G., Krumins, J.A., Cramer, M.D., Hawkins, H.J., 2021. Fire and herbivory drive fungal and bacterial communities through distinct above- and belowground mechanisms. Sci. Total Environ. 785 (2021), 147−189.

Warner, R.R., Chesson, P.L., 1985. Coexistence mediated by recruitment fluctuations: a field guide to the storage effect. Am. Nat. 125, 769−787.

Westbroek, P., Collins, M.J., Jansen, J.H.F., Talbot, L.M., 1993. World archaeology and global change: did our ancestors ignite the ice age? World Archaeol. 25, 122−133.

Zvobgo, L., Tsoka, J., 2021. Deforestation rate and causes in upper manyame catchment, Zimbabwe: Implications on achieving national climate change mitigation targets. Trees For. People 5.

Case Study: Response of Invertebrates to Mixed- and High-severity Fires in Central Europe

Petr Heneberg
Charles University, Prague, Czech Republic

7.19 THE SETTING

Forest fire is considered a marginal phenomenon in traditional central European forestry. According to the European Forest Fire Information System, the Czech Republic (78,866 km^2, forest cover of over 33.9%) experienced forest fires of over 296 ± 136 ha annually in 2004–08.[1] The situation in other central European countries is similar. Forest fires are considered socially and economically unwelcome, and the burned areas are mandatorily managed within just 2 years of their formation.[2] Management consists of removing partially or entirely burned trees and nearly immediately replanting new tree seedlings (Kunt, 1967). Forest fires concentrate prevalently in regions with pine stands growing on gravel/sand sediments or sandstone rocks. Data from 2013 suggest that large-scale (>40 ha) fires affected 288,169 ha across Europe (data available for 24 countries excluding Russia), with forest fires representing over half of the entire burned area recorded during the year (ranging from 6% in Ireland and 8% in the United Kingdom to 96% in Kosovo and 97% in Sweden). Interestingly, 72,008 ha (29% of the affected area) were protected within the "Natura 2000" network containing habitats of high interest for nature conservation (Evans, 2012). The percentage of the Natura 2000 area affected by fire ranged between 0.002% in Germany (a single fire affecting 133 ha) and 2.137% in Portugal (100 fires affecting 40,837 ha).[3] However sites cover not only forested areas but also meadows, fields, and wetlands as long as they are of conservation interest. Thus, the percentage of Natura 2000-protected forests affected annually by fires is much higher and could be of conservation interest as sites where postfire succession occurs in a relatively unmanaged state.

---

1. https://effis.jrc.ec.europa.eu/effis/applications/data-and-services/report_2021.xlsx; cited March 23, 2024.
2. Czech forestry law no. 289/2005 Coll., §31.
3. https://effis-gwis-cms.s3.eu-west-1.amazonaws.com/effis/reports-and-publications/annual-fire-reports/FireReport2013_final2pdf_2.pdf; cited March 23, 2024.

7.20 AEOLIAN SANDS SPECIALISTS ALONGSIDE THE RAILWAY TRACK NEAR BZENEC-PŘÍVOZ

Public awareness of the importance of mixed- and high-severity fires for biodiversity in central European forests is negligible. In 1990, however, the nature reserve "Váté písky" (the name is literally translated as "Aeolian sands") was established; it consists of a narrow (up to 60 m-wide and 5.5 km-long) deforested strip alongside the railway track between Rohatec and Bzenec-Přívoz in the southeastern Czech Republic (48.91°N, 17.24°E). In the 18th century, native acidophilous and Pannonian oak forests with admixed pine (*Festuco ovinae-Quercetum roboris, Carici fritschii-Quercetum roboris*, locally in wet depressions also *Carpinion* and *Carici elongace-Alnetum*) were removed, and the whole region was subject to desertification; numerous active aeolian sand dunes formed, and associated biota occurred. At the beginning of the nineteenth century, nearly the whole area was planted with pines. However, in 1841, a railway track between Vienna (Austria) and Cracow (Poland) was built across the newly planted forest. Along the railway, the deforested strip was formed and maintained by manual tree cutting and frequent fires induced by sparks generated by the steam engines used by the railway company. The strip was maintained in a strictly deforested state from the 1840s to the 1970s when the operation of steam engines terminated. Since the 1970s, the self-seeded Scots pine (*Pinus sylvestris*) and black locust (*Robinia pseudoacacia*) have been occasionally removed; however, solitary trees have remained present since then, and fires still occur, although less often than in the times of steam engine operations. Pine forests surrounding the railway strip still belong to the two Czech regions with the highest frequency of forest fires; however, all the burned areas are quickly replanted. Removal of dead wood, followed by plowing and replanting trees, reduces the occurrence of numerous rare invertebrates, only part of which can survive at sites subject to other disturbances, such as military training ranges and sand quarries. These replacement habitats provide patches of bare sandy ground (Cizek et al., 2013; Heneberg et al., 2013) but are devoid of standing fire-killed trees and downed logs.

In May 2012, the surrounding forest was affected by a 200-ha fire (184 ha of forest). The affected trees were removed, and 1.2 million tree seedlings were replanted, mostly Scots pines, with only a few oaks, lime trees, and beeches. The dead wood is absent at this site, and the sandy patches are still present, but the whole area has become shaded. The planted trees suffered from an outbreak of forest cockchafer *Melolontha hippocastani*. Local forest authorities reported the loss of 90% of planted seedlings in the first years after the fire, with a gradual decrease in the outbreak intensity in later years. Chemical treatment of the tree seedlings affected by forest cockchafers was banned due to the designation of the affected area as a special protection area under the European Union Directive on the Conservation of Wild Birds.

The area is characterized by a 10 to 30 m-thick stratum of aeolian sands overlaid by arsenic campisols (inceptisol) or arenic regosols (entisol, orthent). Although the strip is limited in its extent, it hosts a rich assemblage of fungi, vascular plants, and animals. Typical plant species are *Corynephorus canescens*, *Festuca vaginata* subsp. *dominii*, *Stipa borysthenica*, *Verbascum phoeniceum*, *Spergula morisonii*, *Helichrysum arenarium*, *Gypsophila paniculata*, and *Hierochloë repens*. Typical invertebrates are *Melolontha hippocastani* and *Polyphylla fullo*; hundreds of aculeate hymenopteran species, including *Bembix rostrata* and *Bombus cryptarum*; butterflies such as *Zerynthia polyxena*, *Hipparchia statilinus*, *Chamaesphecia leocopsiformis*, and *Synansphecia muscaeformis*; and spiders such as *Eresus kollari*.

Lowered frequency of fires in the past 4 decades decreased the availability of bare soil and reduced the diversity and abundance of bare soil specialists. For many of these species, the above-mentioned narrow strip and similar nearby habitats represent the only known sites of their occurrence in the Czech Republic. There are several other similar areas in the surrounding pine forest and its vicinity, which include two military training ranges and two large sandpits. When the military ceased operations at one of its two local training ranges, however, the area was quickly overgrown by dense vegetation. Most bare soil specialists decreased in abundance by over one order of magnitude. Some of them, such as *Bembix rostrata*, are now associated only with the illegal off-road motorcycle tracks that are present at some parts of the former military range. The former military training range is now protected as a nature reserve. Numerous attempts have been made to mimic the disturbance-induced removal of vegetation cover and prevent sod formation, but the effects were limited. In this area, allowing more fires to occur and bare ground to persist through the period of natural succession (i.e., without artificially planting trees) would increase the habitat for now-rare invertebrates that depend on sandy openings created by fire.

7.21 POSTFIRE SUCCESSION NEAR JETŘICHOVICE: A CHANCE FOR DEAD WOOD SPECIALISTS

Following 2 months of drought, on June 22, 2006, a week-long mixed-severity forest fire affected 18 ha of Scots pine and white pine (*Pinus strobus*) forest mixed with sessile oak *Quercus petraea*, European beech *Fagus sylvatica*, Norway spruce *Picea abies*, and silver birch *Betula pendula*. The terrain was rugged, consisting of sandstone rocks (mean slope inclination of 35 degrees; range, 0–90 degrees), located near Jetřichovice (50.86°N, 14.40°E, northern Czech Republic). Because the affected area was located within a national park's borders, the park administration was authorized to demand the burned forest remain intact, allowing no logging or replanting efforts (Marková et al., 2011; Trochta et al., 2012). Extensive additional mortality of the remaining live trees occurred over the first 2 years after the fire. The coverage of green tree crowns decreased from 83% before the fire to 39% within 2 months after the fire and to 15% 1 year later (Trochta et al., 2012).

Within 3 weeks after the fire, the fire-affected area was occupied by antracophilous fungi such as *Pyronema omphalodes*. Four months after the fire, 13 fungal species were recorded; the most common ones were *Rhizina undulata*, *Pyronema omphalodes*, *Geopyxis carbonaria*, and *Pholiota highlandensis*. Some rare fungal species also occurred, represented by, for example, *Rutstroemia carbonicola*. Several bryophytes and ferns (*Pteridium aquilinum*) also appeared during this very early succession phase.

By the first spring after the fire, the burned area was covered by the common liverwort *Marchantia polymorpha*, common mosses *Ceratodon purpureus* and *Funaria hygrometrica*, and seedlings of *Betula pendula*, *Pinus* spp., *Fagus sylviatica*, and *Acer* spp. In autumn, most of the burned area was covered by *B. pendula*, *Populus tremula*, *Salix caprea*, and young *Pinus* spp. There were recorded 37 species of fungi, with antracophilous fungi representing one-quarter of the species found.

In the next 5 years the burned area was dominated by ferns (*Pteridium aquilinum* and *Dryopteris carhusiana agg.*) and vascular plants such as *Epilobium angustifolium* (attracting particularly abundant assemblages of aculeate hymenopterans), *Erythronium montanum*, *Exacum tetragonum agg.*, *Spergula morisonii*, *Cerastium holosteoides* subsp. *triviale*, *Digitalis purpurea*; invasive grasses *Calamagrostis epigejos* and *Agrostis capillaris*; and numerous ragworts and groundsels, including *Senecio sylvaticus*, *Senecio vulgaris*, *Senecio viscosus*, *Senecio vernalis*, and *Senecio ovatus* (Hadinec and Lustyk, 2011; Marková et al., 2011). The family Asteraceae of flowering plants was particularly diverse at the burned site and included *Taraxacum sect. Ruderalia*, *Cirsium vulgare*, *Conyza canadensis*, *Hypochaeris radicata*, *Mycelis muralis*, *Tussilago farfara*, and *Crepis capillaries*. Mosses such as *Polytrichastrum formosum*, *Polytrichum juniperinum*, *Polytrichum piliferum*, *Ceratodon purpureus*, and *Pohlia nutans* also occurred. About 50 fungal species were found annually in the burned area, with a markedly decreasing share of antracophilous fungi, which decreased to 11% in the third, 7% in the fourth, and <2% (one species) in the fifth year after fire (Marková et al., 2011).

The burned area was thoroughly examined for changes in aculeate hymenopteran assemblages. Of the 12 red-listed species (threatened according to the Czech Red List of invertebrate species (Farkač et al., 2005) and according to the International Union for the Conservation of Nature) occurring in the surrounding intact pine forest that served as a control (unburned) area, 10 also were present in the burned forest 1—7 years following the forest fire. More important, the burned forest stands also attracted another 30 red-listed species of bees and wasps, which were absent in the surrounding unburned pine forest. Among them were two species that were considered regionally extinct (*Dipogon vechti* and *Chrysis iris*), two critically endangered species (*Miscophus niger* and *Passaloecus monilicornis*), and numerous endangered and vulnerable species. The burned forest stands hosted 252 bee and wasp species 1—7 years following the forest fire, representing 19% of the total 1343 species reported so far from the Czech Republic (Bogusch et al., 2007, 2015). Importantly, the species spectrum associated with the burned forest stands differed not only from those of the unburned forest but also from the species associated with bare sand and heather patches with retained solitary pine trees at a site 12 km away (Blažej and Straka, 2010; Bogusch et al., 2015). The species absent at the unburned site with bare sand and heather patches consisted mainly of those requiring dead wood, such as standing fire-killed trees (snags), which is rarely available in the current commercially exploited forests of the Czech Republic as well as surrounding countries. "Cavity adopters" represented 43% of the red-listed species recorded in the burned forest studied. The forest fire caused a rapid but very temporary decrease in both the abundance and diversity of bees and wasps. By the first year after fire, however, several species were found in the burned forest that were absent in the surrounding unburned forest.

These were mainly the polylectic species of the open countryside and several broadly distributed cavity nesters (*Mellinus arvensis, Lasioglossum pauxillum, Trypoxylon minus*). After 2—3 years, however, the burned forest attracted a highly diverse and abundant assemblage of very specialized bee and wasp species, the abundance and diversity of which was several times higher than in the surrounding forest. The species occurring temporarily at this stage of succession were represented by, for example, *Lasioglossum nitidiusculum* and *Andrena lapponica*. At 5—7 years following the fire, the diversity and abundance of aculeate hymenopterans decreased to the levels experienced in the surrounding unburned forest, and the species spectrum changed. Even in these later phases of forest succession, the burned site hosted numerous species that were absent or rare in the unburned forest and absent or rare during the earlier phases of forest succession (*Ammophila sabulosa, Lasioglossum punctatissimum, Arachnospila hedickei,* and *Crossocerus exiguus*) (Bogusch et al., 2015).

Though this section is focused on invertebrates and postfire habitat, it is worth noting that avian and mammal research also was conducted in this fire area. While birds associated with dense, mature forests declined after the mixed-severity fire, numerous bird species increased or appeared on the site for the first time (Marková et al., 2011). Regarding small mammals, the unburned pine forests adjacent to the study area are characterized by species-poor assemblages, represented by only *Clethrionomys glareolus* (bank vole) and *Apodemus flavicollis* (yellow-necked mouse), both of which occur at very low densities (Bárta, 1986; Marková et al., 2011). During the first year after the fire, the burned site hosted only these two species; however, their abundance was relatively high. The onset of natural postfire succession in the follow-up years was associated with newly appearing species *Sorex araneus* (Eurasian shrew) and *Sorex minutus* (Eurasian pygmy shrew), which otherwise occur in this region only in the river floodplains and in deciduous forests at basaltic rocks, as well as *Crocidura suaveolens* (lesser white-toothed shrew) and *Apodemus sylvaticus* (wood mouse) (Marková et al., 2011).

Fires in the study area are common. The vegetation is dominated by Scots pine and Norway spruce forests on rugged sandstone bedrock, with an extensive accumulation of dry pine-needle litter and extensive tree dieback caused by drought episodes and an outbreak of European spruce bark beetle *Ips typographus*. In the summer of 2022, a large part of this region was affected by a 27-day forest fire that affected 1850 ha and represented the largest forest fire in the Czech Republic and Sachsen (Germany). When writing this essay, there are still disputes on how large a part of the affected area will be allowed for spontaneous vegetation succession. The fire-affected forests are protected as national parks (in both countries). The national park management would like to extrapolate the experience from the forest fire site near Jetřichovice, where spontaneous vegetation succession was allowed, to the newly emerged forest fire site. However, the park management is afraid that the larger forest fire area may make the extrapolation of this experience challenging, and an identical course of development may not be guaranteed.

7.22 CONCLUSIONS

Combined data support the benefits of a mosaic of postfire successional habitats of various ages. Conservation management that focuses on the formation of bare soil patches by fire (as in the first of the two above-presented case studies) supports a very diverse spectrum of organisms but is not sufficient to provide habitat for hymenopteran cavity adopters, which are associated with dead wood. These species would benefit from greater public acceptance of mixed-severity fires in forests and greater retention of snags after a fire (Bogusch et al., 2015). Results similar to those described above have been found for various invertebrate taxa after mixed-severity fire, including increases in diversity of saproxylic (dependent on decaying wood) beetles in the Swiss Alps (Moretti et al., 2010) and enhanced species richness of ground beetles, hoverflies, bees, wasps, and spiders in the forests of the southern Alps (Moretti et al., 2004). Further support for these conclusions regarding benefits to arthropods from mixed-severity fire is also found in studies regarding bark beetles. Similar to postfire forest stands, forest gaps (snag patches) formed by spruce bark beetles are associated with the increased density and diversity of many red-listed species across numerous taxa in this region (Beudert et al., 2015). These include, for example, dead wood aculeate hymenopteran specialists (Müller et al., 2008; Lehnert et al., 2013). These emerging data indicate a need for more natural disturbances and natural succession and increased protection of fire-prone areas in central Europe as nature reserves. These measures would support arthropods and numerous vertebrate species, including many red-listed species, which thrive in unmanaged postfire habitats (Beudert et al., 2015).

REFERENCES

Bárta, Z., 1986. Drobní zemní savci skalních plošin Jetřichovických stěn (Děčínské mezihoří, CHKO Labské pískovce). In: Závěreěná zpráva úkolu ě.: IU-2/Bá-1985-86. Správa CHKO Labské pískovce, Děčín.

Beudert, B., Bässler, C., Thorn, S., Noss, R., Schroder, B., Dieffenbach-Fries, H., Foullois, N., Muller, J., 2015. Bark beetles increase biodiversity while maintaining drinking water quality. Conserv. Lett. https://doi.org/10.1111/conl.12153.

Blažej, L., Straka, J., 2010. Výsledky monitoring vybraných skupin hmyzu (Coleoptera: Carabidae, Hymenoptera: Aculeata) v bývalé lesní školce u Bynovce (CHKO Labské pískovce). Sborník Oblastního muzea v Mostě, řada přírodovědná 32, 23−42.

Bogusch, P., Straka, J., Kment, P., 2007. Annotated checklist of the Aculeata (Hymenoptera) of the Czech Republic and Slovakia. Acta Entomologica Musei Nationalis Pragae 11 (supplementum), 1−300.

Bogusch, P., Blažej, L., Trýzna, M., Heneberg, P., 2015. Forgotten role of fires in Central European forests: critical importance of early post-fire successional stages for bees and wasps (Hymenoptera: Aculeata). Eur. J. For. Res. 134, 153−166.

Cizek, O., Vrba, P., Benes, J., Hrazsky, Z., Koptik, J., Kucera, T., Marhoul, P., Zamecnik, J., Konvicka, J., 2013. Conservation potential of abandoned military areas matches that of established reserves: plants and butterflies in the Czech Republic. PLoS One 8, e53124.

Evans, D., 2012. Building the European Union's Natura 2000 network. Nat. Conserv. 1, 11−26.

Farkač, J., Král, D., Škorpík, M. (Eds.), 2005. Red List of Threatened Species in the Czech Republic—Invertebrates. AOPK ČR, Prague.

Hadinec, J., Lustyk, P. (Eds.), 2011, Additamenta ad floram Reipublicae Bohemicae, IX, vol. 46. Zprávy České botanické společnosti, Praha, pp. 51−160.

Heneberg, P., Bogusch, P., Řehounek, J., 2013. Sandpits provide critical refuge for bees and wasps (Hymenoptera: Apocrita). J. Insect Conserv. 17, 473−490.

Kunt, A., 1967. Lesní požáry. Knižnice požární ochrany No. 28. Československý svaz požární ochrany, Praha.

Lehnert, L.W., Bässler, C., Brandl, R., Burton, P.J., Muller, J., 2013. Conservation value of forests attached by bark beetles: highest number of indicator species is found in early successional stages. J. Nat. Conserv. 21, 97−104.

Marková, I., Adámek, M., Antonín, V., Benda, P., Jurek, V., Trochta, J., Švejnohová, A., Šteflová, D., 2011. Havraní skála u Jetřichovic v národním parku České Švýcarsko: vývoj flóry a fauny na ploše zasažené požárem. Ochr. přír. 66, 18−21.

Moretti, M., Obrist, M.K., Duelli, P., 2004. Arthropod biodiversity after forest fires: winners and losers in the winter fire regime of the southern Alps. Ecography 27, 173−186.

Moretti, M., De Cáceres, M., Pradella, C., Obrist, M.K., Wermelinger, B., Legendre, P., Duelli, P., 2010. Fire-induced taxonomic and functional changes in saproxylic beetle communities in fire sensitive regions. Ecography 33, 760−771.

Muller, J., Bußler, H., Goßner, M., Rettelbach, T., Duelli, P., 2008. The European spruce bark beetle *Ips typographus* in a national park: from pest to keystone species. Biodivers. Conserv. 17, 2979−3001.

Trochta, J., Král, K., Šamonil, P., 2012. Effects of wildfire on a pine stand in the Bohemian Switzerland National Park. J. For. Sci. 58, 299−307.

The Role of Large Fires in the Canadian Boreal Ecosystem

André Arsenault[1,2,3]

[1]*Memorial University at Grenfell, Corner Brook, NL, Canada;* [2]*Université du Québec en Abitibi-Témiscamingue, Rouyn-Noranda, QC, Canada;* [3]*Natural Resources Canada, Canadian Forest Service—Atlantic Forestry Centre, Corner Brook, NL, Canada*

7.23 THE GREEN HALO

The green halo, visible from space and forming a distinct ring around the northern hemisphere (Figure 7.9), is one of the most striking ecological features on earth (Gawthrop, 1999; Talbot and Meades, 2011). The circumboreal forest biome, encompassing 11% of the earth's land surface and one-fourth of the world's forests, is one of the largest floristic regions of the world, wrapping around the northern parts of the globe including North America, Europe, and Asia. The boreal forest, named after the Greek god Boreas (north wind), is characterized by a cold, snowy climate, a short growing season, and relatively cool summers. Large wildfires tend to be more abundant in western Quebec in

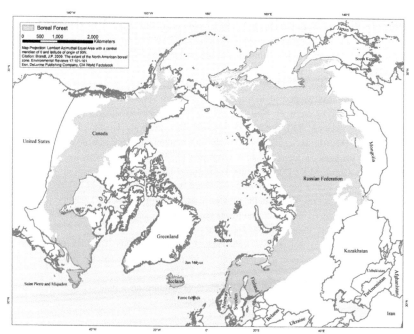

FIGURE 7.9 Extent of boreal forest and woodlands around the globe. *Adapted from Brandt et al. (2013) with permission from James Brandt and NRC Press.*

Crown Copyright © 2024, Published by Elsevier Inc. All rights are reserved, including those for text and data mining, AI training, and similar technologies.

eastern Canada, whereas Newfoundland and Labrador, and Quebec's North shore, have very long fire cycles (Bergeron and Fenton, 2012; Arsenault et al., 2016). Fire activity tends to increase westward from central Canada with some of the largest fires occurring in the west. Although fire activity is increasing in Canada as a result of climate change, predicted rates of increase are believed to be within the upper limits of the natural range of variation (Chavardes et al., 2022). Large fires are important in transforming forests into habitat for specialized ecological communities that require charred wood, by enabling a pulse of dead wood, which is important for many species and ecological processes over the course of natural succession, and by leaving refuges that allow species to persist within burns and recolonize burned areas.

Although some have described the boreal forest as a relatively simple ecosystem with low species diversity and frequent large crown fires, a new paradigm is emerging, that reveals a much more complex and interesting natural history. The vast range of fire cycles implies that old growth and late seral features were historically abundant in certain parts of the boreal forests, more than was appreciated in the past. In fact, new evidence suggests that old-growth forests were a much more predominant feature of the preindustrial landscape in North America (Cyr et al., 2009; Bergeron and Fenton, 2012), Europe (Östlund et al., 1997), and Asia (Eichhorn, 2010) than previously thought. This does not lessen the role of large fires in the boreal, which produce vital habitat for numerous species that have evolved strategies to respond to fire (Rowe, 1983), but it does suggest that a more intricate biogeographical lens is required to understand this ecosystem.

7.24 LAND OF EXTREMES

The boreal forest is a land of extremes; fire cycles can vary from 40 years to more than 900 years, and some trees grow very fast following disturbance, whereas others seem to live in suspended animation as a result of a harsh environment. The spatial scale of ecological processes is also impressive, with annual area burned in Canada measured in the millions of hectares, and even more land affected by defoliating insects and bark beetles. These forests also stand out because of the extraordinary amount of carbon they store (Kasischke et al., 1995) and the fact that they represent one of the last frontier forests on Earth, containing abundant natural resources (Burton et al., 2010). These extreme characteristics of the boreal present great opportunities and challenges (Burton et al., 2010), which are further complicated by the global distribution of this ecosystem across many jurisdictions and emerging issues of conservation of biodiversity and carbon in a changing climate (Bradshaw et al., 2009; Moen et al., 2014). Understanding the natural disturbance regime of the boreal forest and its effect on the environment is critical to ensure that the integrity and resilience of this ecosystem are maintained for future generations. This essay focuses on the role of large wildfires in the Canadian boreal forest ecosystem.

7.25 VEGETATION

Long, cold winters and short, cool summers are key features influencing the vegetation of the boreal. The distribution of the boreal forest in North America is vast, extending from the Pacific to the Atlantic (Figure 7.10). It crosses the state of Alaska and all the Canadian provinces and territories with the exception of the Maritime Provinces (considered hemiboreal), making up 53% of Canada's forests (Brandt, 2009). Although the boreal forests of Canada share similar floristic and ecological characteristics across their distribution, they also exhibit significant differences. At a coarse scale, they are made up of seven different ecozones (Figure 7.11), representing formidable ecological gradients of elevation, longitude, latitude, and continentality. From the productive white spruce (*Picea glauca*) stands in the south, to the krumholtz forests shaped by the harsh climate of the northwestern mountains of the Yukon, to the fierce winds off the coast of Newfoundland to the east, this is a land of beauty and surprising ecological diversity.

Extensive coniferous forests, sometimes mixed with broadleaved trees, interspersed by bogs, barrens, fens, rivers, and lakes cover the landscape (Figure 7.12). Black spruce (*Picea marina*), white spruce, tamarack (*Larix laricina*), white birch (*Betula papyrifera*), trembling aspen (*Populus tremuloides*), and balsam poplar (*Populus balsamifera*) are distributed throughout

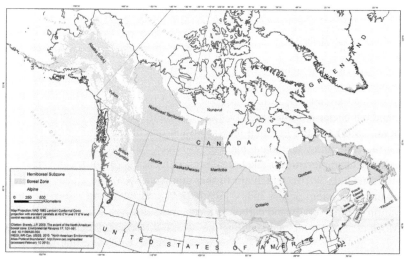

FIGURE 7.10 Distribution of the Canadian boreal forest and Alaska boreal interior. *Adapted from Brandt (2009) with permission from James Brandt and NRC Press.*

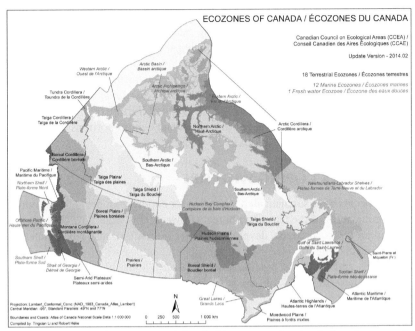

FIGURE 7.11 Map of the ecozones of Canada. The ecozones forming the boreal zone are the boreal shield, boreal plains, taiga plains, taiga shield, Hudson plains, taiga cordillera, and boreal cordillera. *Canadian Council on Ecological Areas (CCEA).*

FIGURE 7.12 A boreal landscape in Terra Nova National Park, Newfoundland and Labrador. *Picture courtesy of André Arsenault, Canadian Forest Service.*

most the Canadian boreal and in interior Alaska. Several tree species are mostly restricted to the east, including balsam fir (*Abies balsamea*), yellow birch (*Betula alleghaniensis*), sugar maple (*Acer saccharum*), red maple (*Acer rubrum*), eastern white pine (*Pinus strobus*), red pine (*Pinus resinosa*), black ash (*Fraxinus nigra*), and red spruce (*Pinus rubens*), whereas others are mostly restricted to the west, including lodgepole pine (*Pinus contorta*), subalpine fir (*Abies lasiocarpa*), and whitebark pine (*Pinus albicaulis*). Jack pine (*Pinus banksiana*) is widely distributed in eastern and central portions of the boreal (Bourgeau-Chavez et al., 2000a).

7.26 PLANTS COPING WITH FIRE

Plant species have sophisticated strategies for coping with fire and surviving in fire-prone areas of the Canadian boreal (Rowe, 1983). Stan Rowe, a renowned Canadian forest ecologist, developed a clever classification of plant functional groups according to how they cope with fire, inspired by the work on vital attributes and the plant regeneration niche developed by Noble and Slatyer (1980). These functional groups are

1. Invaders: examples include white birch, fire moss (*Ceratodon purpureus*), and fireweed (*Chamerion angustifolium*), which depend on copious amounts of wind-dispersed propagules.
2. Evaders: species that store seed in the canopy (e.g., lodgepole and jack pine), in the humus, and in mineral soil.
3. Avoiders: these species, such as balsam fir, have little adaption to fire.
4. Resisters: thick-bark old pine species.
5. Endurers: plants able to resprout after the passage of fire (i.e., the "phoenix" species), such as the trembling aspen (*Populus tremuloides*) and saskatoon (*Amelanchier alnifolia*).

Notably, some species such as fireweed can behave like both invaders and endurers. In general, this visionary framework links well with contemporary integrated approaches aimed toward understanding the role of plant functional traits in response to disturbance.

7.27 FIRE REGIME OF THE CANADIAN BOREAL FOREST

Large fires are considered a key controlling process in boreal forests (Payette, 1992) (Figure 7.13). Although the growing season is relatively short, persistent high-pressure systems in mid-summer can lead to drought conditions, thereby drying out forest fuels and favoring the occurrence of large wildfires (Johnson, 1992). Unlike mixed-severity fires (Odion et al., 2014) or low-severity fires (Agee, 1993), crown fires in the boreal forest usually kill most trees within the burned areas (Johnson, 1992), with the exception of fire skips, also known as remnants or refuges (Perera and Buse, 2014). A classic description of the fire

FIGURE 7.13 Examples of large fires in the boreal forest of Canada. *Pictures courtesy of the Canadian Forest Service.*

regime for the boreal is perhaps best illustrated by Cogbill's (1985) fire history study of the Laurentian Highlands of Quebec, in which he described frequent crown fires producing even-aged stands with fire cycles varying between 70 and 140 years. An important implication of such a fire regime is that a minor component of old stands would be embedded in a matrix of young forests. Bergeron and Fenton (2012) have argued that this view of the dynamics of the boreal forest has been used by some to justify clear-cut harvesting, which also results in a matrix of young stands under a sustained yield regime.

In addition to the profound ecological distinction between a clearcut and complex early seral forest habitat created by crown fire (Chapter 9), this classic interpretation of the fire regime has recently been challenged, suggesting that the fire cycle was variable and included intervals long enough to create an abundance of old-growth forests in northern Quebec (Cyr et al., 2009; Bergeron and Fenton, 2012). The fire regime also varies significantly across the spatial extent of the Canadian boreal. A map of large fires (Figure 7.14) from the Canadian large-fire database (Stocks et al., 2003; Canadian Forest Service, 2015), reveals interesting patterns. The spatial extent of wildfire increases toward western Canada and decreases significantly in eastern Canada. This pattern is closely linked to fire cycles estimated from various sources (Table 7.1). Ironically, the largest burned areas are in ecozones least affected by anthropogenic influences but where fire weather is considered to be severe (Krezek-Hanes et al., 2011).

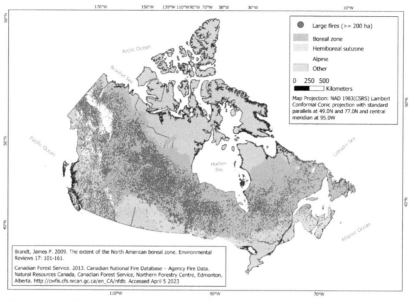

FIGURE 7.14 Distribution of large fires (>200 ha) in Canada and the distribution of the boreal region of Canada from 1959–2020. *Natural Resources Canada, Canadian Forest Service (2023).*

The estimates of fire cycles among the three studies provided should not be compared because they used different methods. They do, however, provide a useful comparison of trends per ecozone. The first two studies, Krezek-Hanes et al. (2011) and Bourgeau-Chavez et al. (2000b), are based on an analysis of the Canadian large-fire database, whereas Bergeron and Fenton (2012) uses a combination of field-based reconstruction of time-since-fire maps and the large-fire database. Unfortunately, fire history reconstruction studies do not yet cover the entire range of the Canadian boreal, and more work is needed in developing methods for interpretations across spatial and temporal scales. The combination of tree ring studies with paleoecological reconstruction of fire history is a promising area to explore, especially in areas where fire intervals are very long, as in Newfoundland and Labrador (Foster, 1983; Arsenault, unpublished data). The temporal variability of the fire cycle over centuries seems to be strongly linked with climate drivers such as the Little Ice Age (Bergeron, 1991) and has changed over the past centuries. This suggests that no single fire cycle can be used as a reference for natural variability (Johnson et al., 1998). Furthermore, most fire cycle calculations using the negative exponential assume that the rate of burning was constant over time, an assumption that is rarely completely met (Huggard and Arsenault, 1999; Bergeron and Fenton, 2012).

The spatial pattern of fire occurrence operates at multiple scales. At a coarse scale, fire is less common in Newfoundland and Labrador, and Quebec's north shore, compared with western Quebec. A similar pattern occurs in the western boreal shield, where fire occurrence increases from western Ontario westward, with very little fire activity in the Hudson plains, which is a

TABLE 7.1 Fire Cycle Estimates for the Canadian Boreal Ecozones and Interior Alaska From Three Studies[a]

	Fire Cycle (Mean Forest Age) Years			
Ecozone	Study 1	Study 2	Study 3	RC
Boreal shield	204	498	122	36.9
Boreal shield east		905	166	
Boreal shield west		91	78	
Newfoundland boreal	769			0.8
Boreal plains	213	181	82	11.4
Taiga plains	141	179	142	13.8
Taiga shield	130	242	122	17.7
Taiga shield east		324	166	
Taiga shield west		160	78	
Hudson plains	588	506	813	3.3
Taiga cordillera	213		202	4.5
Boreal cordillera	263	396	255	7.8
Alaska boreal interior		184		

RC, relative contribution of the ecozone to the total area burned in Canada (percentage), per Krezek-Hanes et al. (2011).
[a]Study 1: Krezek-Hanes et al. (2011); Study 2: Bourgeau-Chavez et al. (2000b); Study 3: Bergeron and Fenton (2012).

very wet ecozone. At a finer scale, fire in Newfoundland is more prevalent in the central ecoregion and virtually absent in western Newfoundland (Arsenault, unpublished data). Interestingly, this pattern is associated with a higher proportion of black spruce in the central ecoregion and a higher proportion of balsam fir in the western ecoregion of Newfoundland. This could reflect at least in part a feedback mechanism between species composition and fire occurrence because black spruce stands are considered more flammable (Furyaev et al., 1983). At an even finer scale, Bergeron (1991) observed that differences in fire severity and frequency were influenced by location on an island and lakeshore landscape. He concluded that the boreal fire regime was controlled by long-term changes in climate at a regional scale and by strong interaction with landscape features at a local scale. The spatial pattern of ignition type also varies greatly in Canada; not surprisingly, most anthropogenic ignitions occur in the southern portion of the boreal (Stocks et al., 2003). All ecozones have larger area burned by lightning fires as opposed to fires of anthropogenic origin, with a notable exception in Newfoundland and Labrador. Interestingly, the ratio of lightning fires to human-set fires is much higher in Labrador than it is on the island of Newfoundland. Virtually all of the large fires on the island since 1959 were of anthropogenic origin. Lightning ignition is relatively rare on the island, and many large fires resulted from railway and forestry operations.

7.28 TEMPORAL PATTERNS OF FIRE AND OTHER CHANGES IN THE BOREAL

The area burned by large fires in Canada's boreal forests is highly variable from year to year (Figure 7.15a) (Stocks et al., 2003) but an increase is detectable at the decadal scale (Figure 7.16a), although this varies among ecoregions, with those in western Canada showing higher increases than in the east (Figure 7.16b). Initially, the large increase in the 1980s was attributed to a possible change in detection accuracy and higher usage of the forests (Stocks et al., 2003). However, Krezek-Hanes et al. (2011) and Hanes et al. (2019) argue that several studies have shown a clear link between the increase in fire activity and temperature over the past 40 years (i.e., Girardin, 2007). This increase in fire activity, likely linked to the lengthening of the fire season, is expected to increase not only in the Canadian boreal but also in boreal forests globally (Flannigan et al., 2013). In addition, more droughts during the fire season are expected to further increase fire occurrence despite an overall increase in precipitation (Christensen et al., 2007).

A comprehensive meta-analysis of fire history studies in Canada clearly show a global decrease in fire activity from the preindustrial to the current era–a decrease linked to long-term changes in climatic patterns, especially in eastern Canada (Girardin et al., 2013), confirming that predicted increases in fire activity are still expected to be within the natural range of variability (Bergeron et al., 2004a, 2004b). However, these predicted increases in fire activity are moving incrementally closer to the upper end of the natural range of variability that may start to create extreme situations seen 7000 years ago, resulting in profound effects on natural communities (Girardin et al., 2024). The 2023 wildfire season is likely a harbinger to this scenario and clearly follows predicted trends of the effects of climate change on wildfire occurrence in Canada associated with prolonged drought and warm conditions, low snowpack, extended fire season, and severe thunderstorms with abundant lightning strikes associated with little precipitation. In addition, some wildfires in the Maritimes and in western Canada resulted from anthropogenic ignitions especially before green-up of the vegetation in the spring. This recent wildfire season was unprecedented (within the context of human record keeping) in area burned (Figure 7.15a,b), spatial and temporal scale (large fires from coast to coast starting very early and some still burning in spring 2024), and in severity of its impact on communities, including the evacuation of over 200,000 people (Jain et al., 2024; Boulanger et al., 2024).

Furthermore, it should be noted that fire frequency in the Yukon flats of interior Alaska may have surpassed the natural range of variation estimated as a >10,000-year period, based on Kelly et al. (2013), though other data presented in this study suggest that fire activity was somewhat higher 500 to 1000

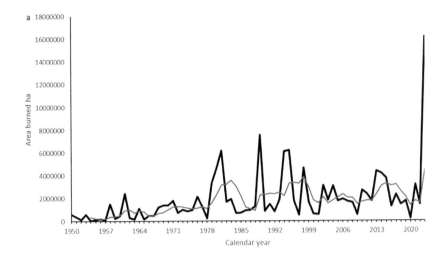

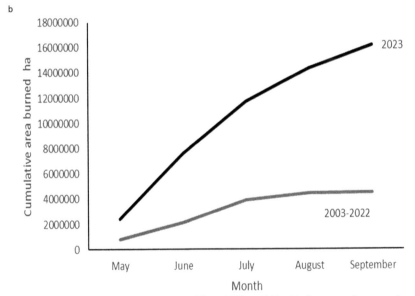

FIGURE 7.15 (a) Total annual area burned from 1950 to 2023 with 5 year running mean (grey line) for the Canadian boreal forest region. (b) Annual cumulative area burned by month. The black line represents values for the 2023 wildfire season and the grey line represents the maximum values between 2003 and 2022. Note that values for 2022 and 2023 are approximate. *Natural Resources Canada, Canadian Forest Service (2024).*

years ago, during the Medieval Climate Anomaly. The predicted proportion of area burned in the Eastern boreal forest of Canada is forecasted to be at the upper end of its historical range, going back 6000 years, by the end of this century (Bergeron et al., 2010; Girardin et al., 2013). Other factors could, however, complicate the forecast of forest changes resulting from rising temperature and increased fire activity (Hessl, 2011). Additional sources of

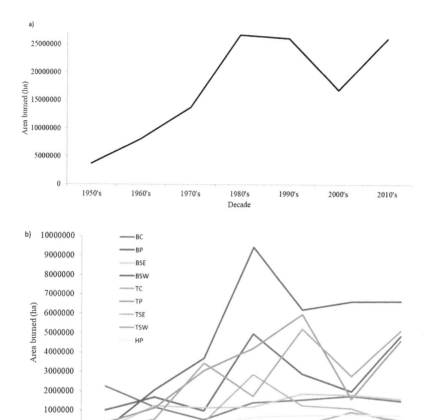

FIGURE 7.16 (a) Total area burned by decade in the Canadian boreal forest, and (b) total area burned by decade in Canadian ecozones. *BC*, boreal cordillera; *BP*, boreal plains; *BSE*, boreal shield east; *BSW*, boreal shield west; *HP*, Hudson plains; *TC*, taiga cordillera; *TP*, taiga plains; *TSE*, taiga shield east; *TSW*, taiga shield west. *Natural Resources Canada, Canadian Forest Service (2024).*

uncertainty include the interaction of disturbances such as drought, fire, pathogens, and insects and their effect on vegetation, especially tree regeneration. An example of this is an interaction between fire severity and climate in Interior Alaska that triggers a change in successional trajectory favoring broadleaf species over black spruce (Johnstone et al., 2010; Beck et al., 2011) and potentially negatively affecting regeneration and growth of black spruce that is a keystone tree species in the boreal forests of North America (Baltzer et al., 2021; Lesven et al., 2024).

7.29 BIODIVERSITY

The biodiversity of the Canadian boreal forest is not fully understood. However, it is clear that the diversity of physical habitat and the many ecotones among forests, bogs, fens, barrens, lakes, streams, and rivers are obviously key.

FIGURE 7.17 Freshly killed trees (a) are prime habitat for the spruce sawyer beetle (b). *Pictures courtesy of Sébastien Bélanger.*

The diversity of forest habitat resulting from a wide range of disturbance types and frequencies is also intimately linked to the distribution and abundance of organisms. For these reasons it is imperative that the boreal forest not be treated as a simple species-poor, fire-maintained ecosystem. It is much more complex. Large fires are important drivers of biodiversity in three key respects.

Firstly, they result in the arrival of an ephemeral but key pyro-community of organisms that are tightly connected to freshly fire-killed trees (Figure 7.17a). Saproxylic insects, such as the spruce sawyer beetle (*Monochamus scutellatus*) (Figure 7.17b), are usually first on the scene after fire (Boucher et al., 2012), soon followed by primary cavity-nesting birds such as the black-backed woodpecker (*Picoides arcticus*) which feed on the wood-boring beetle larvae (Hutto, 1995; Drapeau et al., 2002). This phenomenon is repeated over many different coniferous forests in North America and is a clear indication of the important role of large fires in biodiversity (Hutto,

2008). Recent studies have shown that this food web is linked to nutrient cycling in the boreal forest (Cobb et al., 2010). Several studies suggest the conservation of a mosaic of different burn severity patches is necessary to satisfy requirements of dead wood-associated species in burned forests (Nappi et al., 2010; Azeria et al., 2011, 2012).

Secondly, large fires generate important pulses of dead wood that will be used over time by a variety of communities, including many fire-dependent species and other species associated with rotten wood such as saproxylic insects, lichens, bryophytes, cavity nesters, and fungi (Arsenault and Chapman, 2014).

Thirdly, large fires are often heterogeneous, creating patches of unburned forests and residual structures (Figure 7.18a,b). These refuges can be of different sizes, ages, and permanence (DeLong and Kessler, 2000). They are instrumental in the survival of biota during a fire event, persistence of biota within large burns, and as habitat for source populations for the recolonization of burned areas over time (Robinson et al., 2013; Perera and Buse, 2014).

The large range of fire intervals in the boreal also translates into a range of forest age classes, including old-growth forests (McCarthy and Weetman, 2006; Bergeron and Fenton, 2012). The conservation of old-growth forests and late seral conditions is essential for maintaining biodiversity in boreal forests (Desponts et al., 2004; Harper et al., 2003; Thompson et al., 2003; Bergeron and Fenton, 2012). This is especially true for areas with very long fire intervals, or where insects are the primary drivers of natural disturbance, creating gaps of various sizes and heterogeneous forest structure (McCarthy, 2001) that lead toward the development of some late seral characteristics (McCarthy and Weetman, 2007).

7.30 CONCLUSION

Notwithstanding the benefits of wildfires for biodiversity, some organisms will perish because they cannot avoid the flames. This will lead to a re-organization of the biota and reset succession. Some species that will likely suffer significantly after the large and widespread fires of 2023 are the ones that are already at risk from different types of threats, especially old-growth associated taxa. For example, boreal caribou *(Ranger trandus caribou)* is already at risk resulting from multiple threats, including widespread logging. The cumulative effects of anthropogenic and natural disturbances are increasingly problematic for the survival of this iconic species (Boulanger et al., 2024). Finally, the post-wildfire forest management strategies have the potential to exacerbate impacts on forest biodiversity by removing key forest structures such as deadwood by postfire "salvage logging," introducing non-native species by grass-seeding agronomic plant species, and increasing the overall disturbance severity footprint. The temporary change in forest cover resulting from wildfire is a natural phase of ecological succession. However, the loss of significant

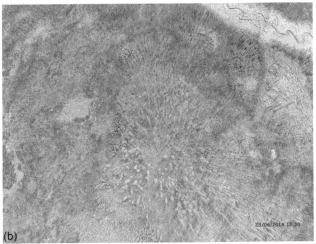

FIGURE 7.18 Examples of residual structure and remnants left after a large fire in Wood Buffalo National Park. *Pictures courtesy of Brian Simpson, Canadian Forest Service.*

amounts of forest habitat due to logging (Gauthier et al., 2015), in combination with climate change, could signal that we are moving to a tipping point towards an uncertain future.

Despite such excellent research and writing on the boreal forest, new paradigms are still emerging. It is a sober reminder that critical thinking is an essential ingredient for keeping bias and assumptions in check. This is especially important when science is applied for managing forests. For a long time, forest management was driven by a strong focus on timber extraction and developed a jargon that infiltrated the dialect of forestry with words such as "decadent" for old-growth forests, "waste wood" for trees that had been killed by natural disturbances, and "salvage" as the practice used to recover that "wasted" timber. Today, forest management in the boreal is strongly driven by themes like ecosystem-based management and sustainable development. The new era will require conservation of boreal forests at different ends of the disturbance spectrum, from newly created burn habitat to multicentury old-growth forests. The new information uncovered regarding the Canadian boreal is also an important reminder that investments in long-term ecological research pay off for the scientific community, decision makers, and for the people who live in, use, play in, and make decisions about this wonderful ecosystem, a biome surely with many more treasures of knowledge to be found under her mossy carpet.

REFERENCES

Agee, J.K., 1993. Fire Ecology of Pacific Northwest Forests. Island Press, Washington, DC.

Arsenault, A., Chapman, B., 2014. How much dead wood is enough in Canadian forests? A review of the science and policy. Page 181. In: Parrrota, J.A., Moser, C.F., Scherzer, A.J., Koerth, N.E., Lederle, D.R. (Eds.), Sustaining Forests, Sustaining People: The Role of Research. XXIV IUFRO World Congress, 5-11 October, 2014, Salt Lake City, vol. 16. USAIn: The International Forestry Review, p. 578.

Arsenault, A., LeBlanc, R., Earle, E., Brooks, D., Clarke, B., Lavigne, D., Royer, L., 2016. Unravelling the past to manage Newfoundland's forests for the future. For. Chron. 92, 487–502.

Azeria, E.T., Ibarzabal, J., Boucher, J., Hébert, C., 2011. Towards a better understanding of beta diversity: deconstructing composition patterns of saproxylic beetles breeding in recently burnt boreal forests. In: Pavlinov, I.Y. (Ed.), Research in Biodiversity: Models and Applications. InTech Open Access Publisher, Rijeka, pp. 75–94.

Azeria, Ermias T., et al., 2012. Differential effects of post-fire habitat legacies on beta diversity patterns of saproxylic beetles in the boreal forest. Ecoscience 19, 316–327.

Baltzer, J.L., Day, N.J., Walker, X.J., Greene, D., Mack, M.C., Alexander, H.D., et al., 2021. Increasing fire and the decline of fire adapted black spruce in the boreal forest. Proc. Natl. Acad. Sci. U. S. A. 118 (45) e2024872118.

Beck, P.S.A., Goetz, S.J., Mack, M.C., Alexander, H.D., Jin, Y., Randerson, J.T., Loranty, M.M., 2011. The impacts and implications of an intensifying fire regime on Alaskan boreal forest composition and albedo. Global Change Biol 17, 2853–2866.

Bergeron, Y., 1991. The influence of island and mainland lakeshore landscapes on boreal forest fire regimes. Ecology 72, 1980–1992.

Bergeron, Y., Fenton, N., 2012. Boreal forests of eastern Canada revisited: old growth, non- fire disturbances, forest succession, and biodiversity. Botany 90, 509–523.

Bergeron, Y., Cyr, D., Girardin, M.P., Carcaillet, C., 2010. Will climate change drive 21st century burn rates in Canadian boreal forest outside of its natural variability: collating global climate model experiments with sedimentary charcoal data. Int. J. Wildland Fire 19, 1127–1139.

Bergeron, Y., Gauthier, S., Flannigan, M., Kafka, V., 2004a. Fire regimes at the transition between mixedwood and coniferous boreal forest in northwestern Quebec. Ecology 85, 1916–1932.

Bergeron, Y., Flannigan, M., Gauthier, S., Leduc, A., Lefort, P., 2004b. Past, current and future fire frequency in the canadian boreal forest: implications for sustainable forest management. Ambio 33, 356–360.

Boucher, J., Azeria, E.T., Ibarzabal, J., Hébert, C., 2012. Saproxylic beetles in disturbed boreal forests: temporal dynamics, habitat associations, and community structure. Ecoscience 19, 328–343.

Boulanger, Y., Arseneault, D., Belisle, A.C., Bergeron, Y., Boucher, J., Boucher, Y., et al., 2024. The 2023 wildfire season in Quebec: an overview of extreme conditions, impacts, lessons learned and considerations for the future. bioRxiv, 2024-02.

Bourgeau-Chavez, L.L., Alexander, M.E., Stocks, B.J., Kasischke, E.S., 2000a. Distribution of forest ecosystems and the role of fire in the North American boreal region. In: Kasischke, E.S., Stocks, B.J. (Eds.), Fire, Climate Change, and Carbon Cycling in the Boreal Forest, Ecological Studies, vol. 138, pp. 111–131.

Bourgeau-Chavez, L.L., Kasischke, E.S., Mudd, J.P., French, N.H.F., 2000b. Characteristics of forest ecozones in the North American Boreal Region. In: Kasischke, E.S., Stocks, B.J. (Eds.), Fire, Climate Change, and Carbon Cycling in the Boreal Forest, Ecological Studies, vol. 138, pp. 258–273.

Bradshaw, C.J.A., Warkentin, I.G., Sodhi, N.S., 2009. Urgent preservation of boreal carbon stocks and biodiversity. Trends Ecol. Evol. 24, 541–548.

Brandt, J.P., 2009. The extent of the North American boreal zone. Environ. Rev. 17, 101–161.

Brandt, J.P., Flannigan, M.D., Maynard, D.G., Thompson, I.D., Volney, W.J.A., 2013. An introduction to Canada's boreal zone: ecosystem processes, health, sustainability, and environmental issues. Environ. Rev. 21, 207–226.

Burton, P.J., Bergeron, Y., Bogdanski, B.E.C., Juday, G.P., Kuuluvainen, T., McAfee, B.J., Ogden, A., Teplyakov, V.K., Alfaro, R.I., Francis, D.A., Gauthier, S., Hantula, J., 2010. Sustainability of boreal forests and forestry in a changing environment. In: Mery, G., Katila, P., Galloway, G., Alfaro, R.I., Kanninen, M., Lobovikov, M., Varjo, J. (Eds.), Forests and Society—Responding to Global Drivers of Change, IUFRO World Series, vol. 25. International Union of Forest Research Organizations (IUFRO) Vienna, pp. 247–282.

Canadian Forest Service, 2023. Canadian National Fire Database – Agency Fire Data. Natural Resources Canada, Canadian Forest Service, Northern Forestry Centre, Edmonton, Alberta. http://cwfis.cfs.nrcan.gc.ca/en_CA/nfdb. Accessed April 5, 2023.

Canadian Forest Service, 2015. National Fire Database – Agency Fire Data. Natural Resources Canada, Canadian Forest Service, Northern Forestry Centre, Edmonton, Alberta. http://cwfis.cfs.nrcan.gc.ca/en_CA/nfdb.

Canadian Forest Service, 2023. Canadian National Fire Database – Agency Fire Data. Natural Resources Canada, Canadian Forest Service, Northern Forestry Centre, Edmonton, Alberta. http://cwfis.cfs.nrcan.gc.ca/en_CA/nfdb. Accessed April 5, 2023.

Chavardes, R.D., Danneyrolles, V., Portier, J., Girardin, M.P., Gaboriau, D.M., Gauthier, Sl, Drobyshev, I., Cyr, D., Wallenius, T., Bergeron, Y., 2022. Converging and diverging burn rates in North American boreal forests from the Little Ice Age to the present. Int. J. Wildland Fire 31, 1184–1193.

Christensen, J.H., Hewitson, B., Busuioc, A., Chen, A., Gao, X., Held, I., Jones, R., Kolli, R.K., Kwon, W.T., Laprise, R., Magana Rueda, W., Mearns, L., Menéndez, C.G., Räisänen, J., Rinke, A., Sarr, A., Whetton, P., 2007. Regional climate projections. In: Solomon, S., Qin, D.,

Hanning, M., Chen, Z., Marquis, M., Averyt, K.B., Tignor, H., Miller, H.L. (Eds.), Climate Change 2007: The Physical Science Basis. Contribution of Working Group I to the Fourth Assessment Report of the Intergovernmental Panel on Climate Change. Cambridge University Press, Cambridge, UK, pp. 847−940.

Cobb, T.P., Hannam, K.D., Kischuk, B.E., Langor, D.W., Quideau, S.A., Spence, J.R., 2010. Wood-feeding beetles and soil nutrient cycling in burned forests: implications of post-fire salvage logging. Agric. For. Entomol. 12, 9−18.

Cogbill, C.V., 1985. Dynamics of the boreal forests of the Laurentian Highlands, Canada. Can. J. For. Res. 15, 252−261.

Cyr, D., Gauthier, S., Bergeron, Y., Carcaillet, C., 2009. Forest management is driving the eastern North American boreal forest outside of its natural range of variability. Front. Ecol. Environ. 10, 519−534.

DeLong, S.C., Kessler, W.B., 2000. Ecological characteristics of mature forest remnants left by wildfire. For. Ecol. Manag. 131, 93−106.

Desponts, M., Brunet, G., Bélanger, L., Bouchard, M., 2004. The eastern boreal old-growth balsam fir forest: a distinct ecosystem. Can. J. Bot. 82, 830−849.

Drapeau, P., Nappi, A., Giroux, J.F., Leduc, A., Savard, J.P., 2002. Distribution patterns of birds associated with snags in natural and managed eastern boreal forests. In: Laudenslayer, B., Valentine, B. (Eds.), Ecology and Management of Dead Wood in Western forests. USDA Forest Service Pacific Southwest Research Station, Albany, Calif. USDA Forest Service General Technical Report PSW-GTR 181.

Eichhorn, M.P., 2010. Boreal forests of Kamchatka: structure and composition. Forests 1, 154−176.

Flannigan, M., Cantin, A.S., de Groot, W.J., Wotton, M., Newbery, A., Gowman, L.M., 2013. Global wildland fire season severity in the 21st century. For. Ecol. Manag. 294, 54−61.

Foster, D.R., 1983. The history and pattern of fire in the boreal forest of southeastern Labrador. Can. J. Bot. 61, 2459−2471.

Furyaev, V.V., Wein, R.W., MacLean, D.A., 1983. Fire influences in Abies-dominated forests. In: Wein, R.W., MacLean, D.A. (Eds.), The Role of Fire in Northern Circumpolar Ecosystems. John Wiley & Sons Ltd, New York, pp. 221−234.

Gauthier, S., Bernier, P., Kuuluvainen, T., Shvidenko, A.Z., Schepaschenko, D.G., 2015. Boreal forest health and global change. Science 349 (6250), 819−822.

Gawthrop, D., 1999. Vanishing halo: saving the boreal forest. Greystone Books, Vancouver, BC.

Girardin, M.P., 2007. Interannual to decadal changes in area burned in Canada from 1781 to 1982 and the relationship to Northern Hemisphere land temperatures. Glob. Ecol. Biogeogr. 16, 557−566.

Girardin, M.P., Ali, A.A., Carcaillet, C., Gauthier, S., Hely, C., Le Goff, H., Terrier, A., Bergeron, Y., 2013. Fire in managed forests of eastern Canada: risks and options. For. Ecol. Manag. 294, 238−249.

Girardin, M.P., Gaboriau, D.M., Ali, A.A., Gajewski, K., Briere, M.D., Bergeron, Y., et al., 2024. Boreal forest cover was reduced in the mid-Holocene with warming and recurring wildfires. Commun. Earth Environ. 5 (1), 176.

Hanes, C.C., Wang, X., Jain, P., Parisien, M.-A., Little, J.M., Flannigan, 2019. Fire regime changes in Canada over the last half century. Can. J. For. Res. 49, 256−269.

Harper, K., Boudreault, C., De Grandpré, L., Drapeau, P., Gauthier, S., Bergeron, Y., 2003. Structure, composition and diversity of old-growth black spruce boreal forest of the Clay Belt region in Québec and Ontario. Environ. Rev. 11 (Suppl. S1), S79−S98.

Hessl, A.E., 2011. Pathways for climate change effects on fire: models, data, and uncertainties. Prog. Phys. Geogr. 35, 393−407.

Huggard, D.J., Arsenault, A., 1999. Reverse cumulative standing age distribution in fire Frequency analysis. Can. J. For. Res. 29, 1449−1456.

Hutto, R.L., 1995. Composition of bird communities following stand-replacement fires in Northern Rocky Mountain (U.S.A.) Conifer forests. Conserv. Biol. 9, 1041−1058.

Hutto, R.L., 2008. The ecological importance of severe wildfires: some like it hot. Ecol. Appl. 18, 1827−1834.

Jain, P., et al., 2024. Canada under fire-drivers and impacts of the record-breaking 2023 wildfire season. ESS Open Archive. https://doi.org/10.22541/essoar.170914412.27504349/v1.

Johnson, E.A., 1992. Fire and Vegetation dynamics: Studies From the North American Boreal Forest. Cambridge University Press, Cambridge, UK, 129 p.

Johnson, E.A., Miyanishi, K., Weir, J.M.H., 1998. Wildfires in the Western Canadian boreal forest: landscape patterns and ecosystem management. J. Veg. Sci. 9, 603−610.

Johnstone, J.F., Hollingsworth, T.N., Chapin III, F.S., Mack, M.C., 2010. Changes in fire regime break the legacy lock on successional trajectories in Alaskan boreal forest. Global Change Biol. 16, 1281−1295.

Kasischke, E.S., Christensen, N.L., Stocks, B.J., 1995. Fire, global warming, and the carbon balance of boreal forests. Ecol. Appl. 5, 437−451.

Kelly, R., Chipman, M.L., Higuera, P.E., Stefanova, I., Brubaker, L.B., Hu, F.S., 2013. Recent burning of boreal forests exceeds fire regime limits of the past 10,000 years. Proc. Natl. Acad. Sci. U. S. A. 110, 13055−13060.

Krezek-Hanes, C.C., Ahern, F., Cantin, A., Flannigan, M.D., 2011. Trends in Large Fires in Canada, 1959-2007. Canadian Biodiversity: Ecosystem Status and Trends 2010. Canadian Councils of Resource Ministers, Ottawa, ON. Technical Thematic Report No. 6. http://www.biodivcanada.ca/default.asp?lang=En&n=137E1147-0. v +48 p.

Lesven, J.A., Druguet Dayras, M., Cazabonne, J., Gillet, F., Arsenault, A., Rius, D., Bergeron, Y., 2024. Future impacts of climate change on black spruce growth and mortality: review and challenges. Environ. Rev.

Lindenmayer, D.B., Blanchard, W., Blair, D., Westgate, M.J., Scheele, B.C., 2019. Spatiotemporal effects of logging and fire on tall, wet temperate eucalypt forest birds. Ecol. Appl. 29(8) e01999.

McCarthy, J., 2001. Gap dynamics of forest trees: a review with particular attention to boreal forests. Environ. Rev. 9, 1−59.

McCarthy, J.W., Weetman, G., 2006. Age and size structure of gap-dynamic, old-growth boreal forest stands in Newfoundland. Silva Fennica 40, 209.

McCarthy, J.W., Weetman, G., 2007. Self-thinning dynamics in a balsam fir (Abies balsamea (L.) Mill.) insect-mediated boreal forest chronosequence. For. Ecol. Manag. 241, 295−309.

Moen, J., Rist, L., Bishop, K., Chapin III, F.S., Ellison, D., Kuuluvainen, T., Petersson, H., Puettmann, K.J., Rayner, J., Warkentin, I.G., Bradshaw, C.J.A., 2014. Eye on the taiga: removing global policy impediments to safeguard the boreal forest. Conserv. Lett. 7, 408−418.

Nappi, A., Drapeau, P., Saint-Germain, M., Angers, V.A., 2010. Effect of fire severity on long-term occupancy of burned boreal conifer forests by saproxylic insects and wood-foraging birds. Int. J. Wildland Fire 19, 500−511.

Noble, I.R., Slatyer, R.O., 1980. The use of vital attributes to predict successional changes in plant communities subject to recurrent disturbances. Vegetatio 43, 5−21.

Odion, D.C., Hanson, C.T., Arsenault, A., Baker, W.L., DellaSala, D.A., Hutto, R.L., Klenner, W., Moritz, M.A., Sherriff, R.L., Veblen, T.T., Williams, M.A., 2014. Examining historical and current mixed-severity fire regimes in ponderosa pine and mixed- conifer forests of western North America. PLoS One 9, e87852.

Ostlund, L., Zackrisson, O., Axelsson, A.L., 1997. The history and transformation of a Scandinavian boreal forest landscape since the 19th century. Can. J. For. Res. 27, 1198–1206.

Payette, S., 1992. Fire as a controlling process in the North American boreal forest. In: Shugart, H.H., Leemans, R., Bonan, G.B. (Eds.), A Systems Analysis of the Global Boreal Forest. Cambridge University Press, New York, pp. 144–169.

Perera, A., Buse, L., 2014. Ecology of Wildfire Residuals in Boreal Forests. John Wiley & Sons, Chichester, West Sussex, UK.

Robinson, N.M., Leonard, S.W.J., Ritchie, E.G., Bassett, M., Chia, E.K., Buckingham, S., Gibb, H., Bennett, A.F., Clarke, M.F., 2013. Refuges for fauna in fire-prone landscapes: their ecological function and importance. J. Appl. Ecol. 50, 1321–1329.

Rowe, J.S., 1983. Concepts of fire effects on plant individuals and species. In: Wein, R.W., MacLean, D.A. (Eds.), The Role of Fire in Northern Circumpolar Ecosystems, 18. J. Wiley, New York, pp. 135–154. SCOPE.

Stocks, B.J., Mason, J.A., Todd, J.B., Bosch, E.M., Wotton, B.M., Amiro, B.D., Flannigan, M.D., Hirsch, K.G., Logan, K.A., Martell, D.L., Skinner, W.R., 2003. Large forest fires in Canada. J. Geophys. Res. 108, 1959–1997. FFR 5-1 through 5-12.

Talbot, S.S., Meades, W.J., 2011. Circumboreal Vegetation Map (CBVM): Mapping the Concept Paper. CAFF Strategy Series Report No. 3. CAFF Flora Group (CFG), CAFF International Secretariat, Akureyri, Iceland.

Thompson, I., Larson, D., Montevecchi, W., 2003. Characterization of old "wet boreal" forests, with an example from balsam fir forests of western Newfoundland. Environ. Rev. 11, S23–S46.

Chapter 8

What's Driving the Recent Increases in Wildfires?

Dominick A. DellaSala[1] and Chad T. Hanson[2]
[1]Wild Heritage, A Project of Earth Island Institute, Berkeley, CA, United States; [2]John Muir Project of Earth Island Institute, Berkeley, CA, United States

8.1 UNDERSTANDING THE PAST, PRESENT, AND FUTURE OF WILDFIRES

Knowing something about the past allows educated guesses (hypotheses) to be made about whether what's happening today is "normal," novel (unprecedented), likely to remain stable or worsen over time. Taking a peek at history can give us clues of how fire-adapted species and ecosystems adapt to fire of all intensities and frequencies. In doing so, the past helps us define a set of reference conditions bounded by the highs and lows of a variable of interest (e.g., wildfire extent) over time in what has been called "the historic range of variability" (HRV, see below).

Regarding the future, global climate change models and scenario planning tools give us important clues regarding what might happen to fire-adapted ecosystems and people based on the accumulation of greenhouse gas emissions (GHGs) and degradation of the land carbon sink by logging and other developments. Importantly, as global overheating dangerously pushes past the 1.5°C threshold (our current trajectory), this will very likely bring accelerating climate disasters, including wildfires spilling over into the built environment (as in Chapter 10). Notably, at the time of this writing (March 19, 2024), the United Nations Secretary General António Guterres issued a planetary distress call (red alert) that humanity will soon cross the 1.5°C threshold with unprecedented biosphere—atmosphere disruptions on the horizon.

To fully assess the causes and consequences of fire across sufficiently representative timescales (past, present, and future), researchers use a combination of backcasting and forecasting techniques as discussed in this chapter. We present exciting new developments in climate attribute science that can help evaluate if contemporary events are exceeding HRV and how to prepare for the novel climate in the coming decades. We organized the climate change section of this chapter using a series of evidentiary statements that make clear associations between extreme fire-weather behind large, uncontrollable wildfires due to anthropogenic climate change (ACC). ACC is used throughout this chapter to distinguish human-caused change to the biosphere—atmosphere feedback that are far in excess of what would be happening without us in the picture.

The era of ACC is upon us and is impacting people and nature in alarming ways, marked by both the buildup of GHGs that we spew daily into the atmosphere and heightened destructive forces of nature that are responding in-kind (Romero-Lankao et al., 2014; IPCC, 2023). The combination of rising temperatures and changes in seasonal and annual precipitation is influencing the size, severity, and occurrence of fires around the world (e.g., Krawchuk et al., 2009; Bowman et al., 2009; Flannigan et al., 2009, Westerling, 2016, also see citations below). Because climate will increasingly dominate fire occurrence and behavior into the foreseeable future (Figure 8.1), it is important to draw on as broad a base of knowledge as possible to understand fire—climate—human interactions and to identify appropriate conservation strategies for nature and risk-reduction strategies for the built environment (Chapter 10 as well).

8.2 LOOKING BACK OVER THE PALEO-RECORD (BACK CASTING)

Historical data make us mindful of the need for a long time span to serve as an adequate reference for forest management decisions, as well as the potential role of fire in the future. However, effectively utilizing historical fire information requires some level of understanding of the available datasets, as well as the time domains at which they describe fire. It also requires an appreciation of human influences on fire, including the degree to which people have altered fire regimes and ecosystems through deliberate burning, fire suppression, and fire elimination, as well as the introduction of invasive species (Whitlock et al., 2010). Fire history should be viewed not as irrelevant storytelling, but rather as vital information that describes the range of possible fire conditions under a broader array of spatial and temporal scales than can be observed at present.

Multiple datasets are available to describe fire activity at different spatial and temporal scales (Gavin et al., 2007; Kehrwald et al., 2013) (Figure 8.2). On time scales of days to decades, remotely sensed data and historical documents register fire occurrence and are used to estimate global area burned. On longer time scales of decades to centuries, tree-ring records, both fire scars on living trees and forest stand structures, provide information on prehistoric fire occurrence, fire frequency, and fire severity. Studies of tree rings in the western United States have been instrumental in describing low and mixed-severity fire

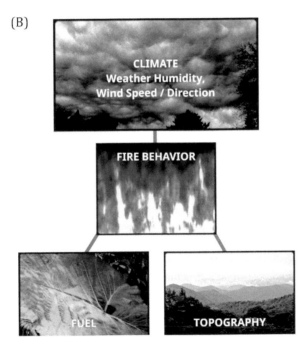

FIGURE 8.1 (a) Fuel-limited fire regimes depicting the interaction of climate, vegetation/fuels, and topography as generally equivalent influences of fire behavior. (b) Climate-limited fire regime depicting the top-down influence of climate on fire behavior. Many fire regimes are shifting from a to b as climate increasingly becomes the limiting factor of fire behavior. Also see Littell et al. (2009, 2018).

regimes (e.g., Brown et al., 1999; Baker et al., 2023a), the character of postfire vegetation development following high-severity fires (Romme, 1982; Sherriff et al., 2001; Odion et al., 2014), and modes of climate variability that lead to

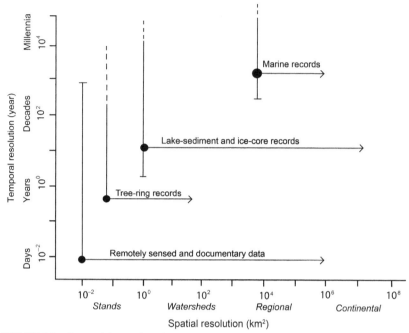

FIGURE 8.2 Types of data and models used to reconstruct past fire on different temporal and spatial scales. *After Gavin et al. (2007).*

years and decades of large fires (Swetnam and Betancourt, 1990; Heyerdahl et al., 2008; Trouet et al., 2010). Fire-scar tree-ring records can produce a reconstruction of fire history with yearly and sometimes seasonal precision that extend our knowledge of past fires back centuries and in some cases millennia. However, they are less useful in understanding the history of forests that experience high-severity stand-replacing fires. In these settings, analysis of stand ages and postfire age structure provides information on past fire events as well as postfire vegetation development. In mixed severity fire regimes, stand age and fire scars have been used to reveal the mosaic of burned and unburned vegetation patterns (Taylor and Skinner, 2003; Schoennagel et al., 2011; Odion et al., 2014).

Sedimentary Charcoal Analysis

On time scales of centuries to millennia, sedimentary records from lakes and natural wetlands provide information about fire history, and particulate charcoal is often a primary proxy. Charcoal records are less spatially resolved and temporally less precise in comparison to tree-ring and stand age data, but they have the advantage of examining fire response to a broader range of climate conditions and vegetation types than exist in the recent past. Evidence of fire in the form of black carbon, charcoal particles, and chemical signatures also is available in marine and ice cores that span several millennia (Daniau et al., 2013; Kehrwald et al., 2013; Brugger et al., 2022).

Fire-history research based on sedimentary charcoal records has undergone a renaissance in recent decades, in part motivated by interest in understanding the historical precedence of recent large, severe fires (e.g., Higuera et al., 2021; Clark-Wolf et al., 2023). Whether such conflagrations occurred in the more distant past and the extent to which they were caused by unusual climate conditions or human activities are topics of both scientific and public concern. Charcoal analysis is based on the premise that charred particles are carried aloft during the fire and travel some distance in the atmosphere before settling on the ground and lake surface. The charcoal particles that fall on lakes and wetlands eventually become sequestered in sediment, and changes in particle abundance at different depths in sediment cores provide a proxy of past fire activity (Whitlock and Larsen, 2001). A suite of radiocarbon dates or other chronologic markers establish an independent chronology in most fire-history studies. The primary data are presented as charcoal accumulation rates (particles per cm^2 per year), although several metrics have been used (Conedera et al., 2009). High-resolution charcoal investigations from lake sediments often focus on the abundance of large charcoal particles (>125 microns in diameter) in contiguous thin slices of the core. Because large particles are transported relatively short distances, they tend to provide a record of local stand-replacing fires, and continuous sampling allows reconstructions often with decadal precision. Changes in overall fire activity, represented by the overall trend and variability in charcoal accumulation through time, are a function of shifts in fuel composition and distribution (vegetation), as well as overall fire frequency, size, and severity.

Calibration studies and process-based models of charcoal transport suggest that charcoal trends are a good proxy of area burned within a <30 km radius of small lakes being studied. This charcoal source area is similar to that of pollen, which is analyzed from the same core to reconstruct vegetation history (Higuera et al., 2011; Kelly et al., 2013; Vachula et al., 2018). Statistically significant charcoal peaks above a prescribed threshold are attributed to individual fire episodes (i.e., one or more fires occurring in the time span of the sample). Peak detection is used to identify fire episodes and describe variations in the frequency as well as the magnitude of fire episodes (Higuera et al., 2009). Some studies make efforts to identify and separate grass and wood charcoal at each stratigraphic level as a tool to discriminate between surface and crown fires (Whitlock et al., 2008). Additional precision also comes when the charcoal particles are themselves identified, a technique that comes from archeology (Carcaillet and Thinon, 1996; Mustaphi and Pisaric, 2014; Marguerie et al., 2020).

Fire History Across a Moisture Gradient

The goals of fire-history research are to distinguish the drivers of fire activity, be they climate, vegetation, or anthropogenic factors; understand the extent and nature of past fire activity; and assess fire's long-term ecological effects. These objectives require: (1) examining multiple charcoal records to separate local from regional patterns; (2) discerning climate−fire−ecosystem feedbacks recorded in paleodata and modeling; and (3) understanding past human influences in shaping fire regimes (Henne et al., 2011; Marlon et al., 2013; Pfeiffer et al., 2013; Iglesias et al., 2015a; Whitlock et al., 2018).

One way to understand fire's role in different ecosystems is to examine its importance across a moisture gradient (Figure 8.3) (Whitlock et al., 2015). At the dry end, deserts may experience frequent ignitions and low vegetation and soil moisture, but discontinuous vegetation often prohibits fire spread, and fires of any significant size are infrequent. At the wet end, vegetation is abundant in rainforests, but the dry season is short and natural ignitions are infrequent or do not coincide with the period of dry, flammable vegetation. In such wet settings, fires are also infrequent, although they can be severe when ignition and drought coincide. At the intermediate scale, temperate dry forests and savanna meet both requirements of sufficient amount and dryness of vegetation and frequency of ignition, and these vegetation types support frequent low- and mixed-severity fires. Thus, fires are generally less frequent at the wet and dry ends of the moisture spectrum than in the intermediate seasonally dry and more pyrogenic landscapes (Whitlock et al., 2010; Archibald et al., 2013; McWethy et al., 2013). Of course, these relationships can be altered by human activities.

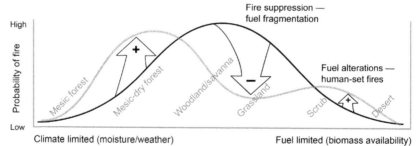

FIGURE 8.3 The magnitude of human influences on natural fire regimes varies along a broad moisture gradient of vegetation types. Climate exerts strong control over fire activity at the extreme wet and dry ends of the moisture gradient as a result of low combustion potential of fuels in mesic settings and the scarcity and disconnected arrangement of fuels in arid regions. Humans have the potential to alter fire regimes (shown by positive and negative arrows) by changing ignition frequency, fuel composition, and pattern as well as by suppressing fires (dashed line). *After Whitlock et al. (2015).*

A comparison of the long-term vegetation trajectories in southern Europe, New Zealand, Patagonia, and the US Pacific Northwest, for example, shows the importance of understanding land-use history and its role in shaping present landscapes and fire regimes (Whitlock et al., 2018; Nanavati et al., 2022). In Europe, for example, natural vegetation—climate—fire linkages were broken 6000—8000 years ago with the onset of Neolithic farming. In New Zealand, such linkages were first lost about 700 years ago with the arrival of Maori people. In the US Pacific Northwest and Patagonia, the greatest landscape alteration came 150 years ago with European colonization that variously increased or decreased fire activity, changed the season of burning, introduced nonnative species that affect flammability (Brooks et al., 2004), disrupted Native American cultural burning, and fragmented natural landscape patterns of vegetation and fire spread (e.g., through logging and fire suppression; Odion et al., 2014; Nanavati et al., 2022; Johnston et al., 2023) (see also Chapter 6 regarding chaparral-fire disruptions).

8.3 WESTERN USA FIRE HISTORY CASE STUDIES

Tree-ring and charcoal data from middle- and high-elevation forests in the western United States indicate that past variations in fire activity are strongly linked to a changing climate. On long time scales, a primary driver of past fire activity has been the slow variations in the seasonal cycle of solar radiation (referred to as insolation, generally the degree of sun exposure). In the early Holocene (ca. 11,000—6000 calendar years before present (cal year BP), with "present" set at 1950 AD), summer insolation was up to 8% higher than at present, and winter insolation was lower by a similar amount. Higher summer insolation led directly to higher-than-present summer temperatures and effectively decreased moisture; it indirectly produced a strengthened northeastern Pacific subtropical high-pressure system, which further suppressed summer moisture in the northwestern United States.

Many parts of the US Northwest show higher fire activity in the early Holocene compared with the late Holocene (Whitlock et al., 2008; Iglesias et al., 2018; Alt et al., 2018). At the same time, stronger-than-present monsoonal circulation, also driven by the summer insolation maximum, may have led to wet summer conditions and fewer fires in the southwestern United States (Bartlein et al., 1998; Anderson et al., 2008; Brunelle, 2022). On interannual to century time scales, ocean atmosphere interactions (El Niño Southern Oscillation, Pacific Decadal Oscillation, American Multidecadal Oscillation) contribute to fire occurrence and severity through atmospheric configurations that create persistent drought (Kitzberger et al., 2007; Morgan et al., 2008; Trouet et al., 2010), although the strength of these short-duration relationships varies greatly from region to region.

Greater Yellowstone Region—In the greater Yellowstone ecosystem, regional analysis of charcoal records describes broad trends in climate, fire, and vegetation change over the past 15,000 years (Iglesias et al., 2015b, 2018). These data indicate that the highest fire activity in the region occurred between 12,000 and 10,000 cal year BP, when summers were warmer than today, winters were colder, and winter precipitation was generally high. The high-fire period was associated with a decline in fire-vulnerable Engelmann spruce (*Picea engelmannii*) and an increase in whitebark pine (*Pinus albicaulis*) at all elevations.

On the rhyolite (a type of silica-rich volcanic rock) plateaus of central Yellowstone, charcoal data highlight the direct connections between fire and climate through time (Millspaugh et al., 2000; Schiller et al., 2020). This area has supported lodgepole pine (*Pinus contorta*) forest for the past 11,000 years because of the strong edaphic (relating to the soil) controls on vegetation composition. By contrast, past fire activity was more dynamic than the vegetation history, showing the highest occurrence between 11,000 and 7000 cal year BP during the summer insolation maximum and decreasing frequencies to the present day. Most Holocene fires were likely mixed- or high-severity events, given the persistence of lodgepole pine. Other studies of Yellowstone show the occurrence of infrequent large fires during the Little Ice Age (1600–1900 AD), and fewer and likely small fire events during the Medieval Climate Anomaly (800–1200 AD) (Meyer et al., 1995; Pierce et al., 2004; Whitlock et al., 2012). By contrast, an analysis of postfire sediment deposits in alluvial fans in ponderosa pine (*Pinus ponderosa*) forests in southern Idaho revealed large, severe-fire events well above recent levels during a warm period from 1050 to 650 cal year BP (Pierce et al., 2004).

Western Cascades—The fire history west of the Cascade crest in Washington and Oregon also was strongly influenced by shifts in the duration and severity of summer drought and the composition of the forest. Between 9500 and 5000 cal year BP, drier-than-present summers supported forests with abundant Douglas-fir (*Pseudotsuga menziesii*), red alder (*Alnus rubra*), and bracken fern (*Pteridium* spp.) (Long et al., 2007; Whitlock et al., 2008; Gavin et al., 2013). This forest composition resembled complex early seral forest stages, and not surprisingly fires were more frequent than today. In valley floors, woodland, prairie, and savanna habitats were expanded in the early Holocene compared with their present distribution, again in association with more fires. As summer insolation decreased in the late Holocene, summers became cooler and wetter than before, and forests of mesophytic (referring to plants adapted to moderate levels of moisture) conifers (e.g., western hemlock (*Tsuga heterophylla*), western red cedar (*Thuja plicata*), fir (*Abies* spp.), and Sitka spruce (*Picea sitchensis*)) prevailed. In association with this cooling trend, fires were less frequent, but, given the vegetation composition, they were likely more severe than in earlier times.

People lived throughout the region by the beginning of the Holocene and population densities at the time of European colonization were high (Boyd, 1990). A compilation of charcoal records across the Washington, Oregon, and British Columbia shows a regional rise in fire activity in the last 6000 years, which is ascribed to more anthropogenic burning with rising human populations and increased climate variability. Despite this regional trend, site-to-site variability in the occurrence of charcoal peaks suggests that fires were mostly local and probably mixed severity (Walsh et al., 2010). The dominant tree species in wet temperate forests, Douglas-fir, has evolved with fire and displays several life-history traits that allow it to persist across a wide range of fire frequencies and severities (Tepley et al., 2013). Its rapid establishment and growth of seedlings after mixed-severity fires and its ability to establish beneath and above competing shrubs promote rapid recovery of Douglas-fir canopy, often within decades after fire (Tepley et al., 2014). The presence of partially intact forest within most burned areas also enables Douglas-fir to rapidly colonize adjacent high-severity patches. Given these factors, it seems highly likely that a targeted ignition strategy by Indigenous peoples in the Pacific Northwest did not lead to large-scale forest conversion, as occurred, for example, in the temperate wet forests of New Zealand, where fire-adapted species are not prevalent (Whitlock et al., 2015).

The fire history is more complex in terms of spatial and temporal variability in the Klamath-Siskiyou region of southwestern Oregon and northern California (Taylor and Skinner, 2003; Colombaroli and Gavin, 2010; Odion et al., 2010; Briles et al., 2011). A study of Bolan Lake showed infrequent fires in the early postglacial period (17,000–14,500 cal year BP), when the climate was cooler than present and subalpine parklands of lodgepole pine, spruce, and mountain hemlock (*Tsuga mertensiana*) were present (Briles et al., 2005). Warming after 14,500 cal year BP was associated with forest closure and increased fire activity. After 11,000 cal year BP, open xerothermic (pertaining to plants adapted to relatively hotter, drier conditions) forests of pine, oak (*Quercus* spp.), incense cedar (*Calocedrus decurrens*), and *Ceanothus* developed, and fires became more frequent than during the late-glacial period. During the middle Holocene (7000–4500 cal year BP), a closed forest of fir, Douglas-fir, red alder, and oak became established, and the frequency of fire episodes reached its highest levels. In the past 4000 years, fir-dominated forests have developed at middle elevations, and mountain hemlock has expanded at high elevations. At most sites, fire frequency has declined in the late Holocene, with the exception of elevated fire activity during the Medieval Climate Anomaly (Briles et al., 2011).

Colorado Rocky Mountains—The fire history of subalpine forests in the Colorado Rocky Mountains shows the importance of changes in forest composition and density on fire behavior (Higuera et al., 2014). Tree-ring records indicate that subalpine forests of Engelmann spruce, subalpine fir

(*Abies lasiocarpa*), and lodgepole pine have supported low-frequency, stand-replacing fires in recent centuries (Buechling and Baker, 2004; Sibold and Veblen, 2006). The vegetation and fire frequency have shown little variation over the past 6000 years, despite long-term trends toward lower summer temperatures and less effective moisture (where effective moisture, precipitation evaporation) (Figure 8.4). Mean fire return intervals have ranged

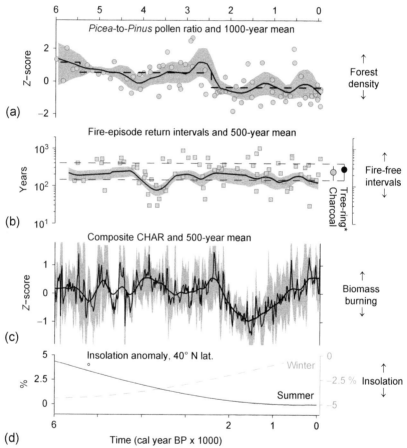

FIGURE 8.4 Millennial-scale vegetation, fire, and climate history from subalpine forests in Rocky Mountain National Park. (a) The *Picea*-to-*Pinus* pollen ratio (points) is a composite from three pollen records (black curves and gray envelopes represent the 95% confidence intervals). The ratio shifts to lower values (more pine) at 5400 and 2400 cal yr BP (calendar years before present). (b) Individual fire return intervals (FRIs) from each site (squares) and the composite mean FRI averaged and smoothed over 500-year periods (gray shading). Mean FRI from the charcoal record for the past 300 years (c.1650–1950) compares well with estimated fire rotations from tree-ring records over the same period (Buechling and Baker, 2004; Sibold and Veblen, 2006) (gray and black circles on right). (c) Composite CHAR record at 15-year intervals (with 95% confidence intervals) and smoothed to 500 years. (d) Holocene insolation for the summer and winter solstice at 40 degrees N latitude. *From Higuera et al. (2014).*

between 150 and 250 years during the past 6000 years, although the variability around the long-term fire return interval mean correlated well with shifts in summer moisture (i.e., more fires during drier summers). Levels of biomass burned (inferred from charcoal trends) decreased significantly at 2400 cal year BP, despite little change in vegetation or fire frequency. This shift is interpreted as evidence of less biomass burned per fire and a decrease in crown fire severity. In the past 1500 years fire severity has steadily increased in these forests. Higuera et al. (2014, 2021) suggest that in central Rocky Mountain subalpine forests (1) fire severity is likely more responsive to climate change than is fire frequency; (2) the indirect influence of climate on vegetation (availability of "fuels") is as important as the direct effects of climate on fire activity; and (3) the large area burned from 2000 to 2020, particularly in 2020, appears to exceed the historical range over the past two millennia.

8.4 HISTORICAL RANGE OF VARIATION

Historical range of variation (HRV) in the fire context generally refers to the variability in spatial and temporal dynamics of wildfires over some historical timeline. The HRV fire concept is typically applied in land management circles at the level of individual species and ecosystem responses and other changes to help set management priorities (Keane and Loehman, 2019). However, many studies describing current and projected trends in fire activity draw on an inappropriately short baseline of historical data for comparison. This is often referred to as the "shifting baseline perspective," whereby the baseline is shifted to a more recent "historical" timeline. Reliance on inadequate baselines can result in researchers, land managers, and policymakers assuming that current fires exceed levels typical of that ecosystem than what the more distant past might otherwise reveal.

Consider that variability in nature is the "spice of life" that creates the environmental stimulus for species to adapt to their surroundings. Thus, the greater the genetic variants within a population, the more likely adaptive traits will emerge and spread throughout the population, allowing a species to persist in the face of environmental stimulus. If the stimulus are extreme (outside HRV), response thresholds are crossed and species extinction risks go up, particularly in the case of populations with limited dispersal capabilities that cannot simply move to more suitable locations. Historical conditions can therefore serve as a proxy for comparing species and ecosystem responses to contemporary conditions as well as projecting forward under a range of climate and emission scenarios, as in the case of climate change downscaling (see below).

The HRV of fire regimes can serve as important "guideposts" for interpreting current and projected wildfire activity by integrating the role of climate and humans in shaping long-term landscape and fire dynamics. This nuanced perspective allows us to better gauge what responses might be possible under a novel climate. Veering too far and too quickly from HRV as a result of land-use stressors synergistically acting with rapid climate change, for example, is likely to cross thresholds—or tipping points—in upping the ante on extinctions and novel ecosystem conditions.

Command and Control versus Conservation Science Approaches

Interpreting HRV and wildfire effects in general in land management decisions has been fraught with debate between those who practice command-and-control approaches (CAC) to nature versus those that strive for conservation science approaches (CSA) that can include both active and passive restoration (Hanson et al., 2009); although both sides acknowledge climate as an important driver of recent fire increases. A major difference in approaches is that CAC is reliant on various forms of logging/thinning, building fire breaks, building and maintaining roads, suppressing "bad" fires, and introducing prescribed fires to ameliorate wildfire effects (e.g., calls for more "good fire" to thwart "bad fire," Chapter 10). CAC proponents also reject adherence to the precautionary principle and, in doing so, typically underestimate collateral ecosystem and climate damages from their approach (e.g., Hessburg et al., 2021 vs. DellaSala et al., 2022, Chapter 10).

Differences between CAC versus CSA can at least be partially attributed to perspectives in how wildfire is viewed, as in the case of cognitive dissonance explained in Chapter 6 (i.e., there is a tendency to see what one wants to see *a priori*). For example, a high severity fire that impacts a mature forest is either viewed as a negative because of loss of the older forest stage, or as a positive in generating complex early seral forests. While there is general agreement in the importance of "some" high severity fire patches, disagreements remain over how much, how big should the patches be, and how frequent should they occur on the landscape—leading to even more CAC in attempts to contain high severity (e.g., megafires, Hessburg et al., 2021). Methodological approaches to high severity fire monitoring also differ in tree mortality estimates from severe fire in areas logged before a fire without determining the contribution of logging to tree mortality postfire (Hanson, 2022; Baker and Hanson, 2022). Further differences arise in how carbon emissions from fire versus logging (especially of large trees) are estimated and taken into account in management (Law et al., 2018; Mildrexler et al., 2023), historical timelines in trend analyses (e.g., shifting baseline as discussed), the type of data and degree of sampling intensity (tree ring fire scars alone vs. multiple data sources, Odion et al., 2014), inappropriate statistical analyses (e.g., the median fire return interval that compresses variation in fire return intervals leading to questionable HRV determinations, Baker 2017), spatial scale differences (regional vs. subregional) (DellaSala and Hanson, 2019; Parks and Abatzoglou, 2020), whether data and studies were omitted (Baker et al., 2023a), and the degree of impacts of CAC on high fire severity dependent ecosystems (DellaSala et al., 2022).

8.5 LINKING WILDFIRE TO ANTHROPOGENIC CLIMATE CHANGE

Because ACC remains controversial among some in the public and many decision makers, we present this section as a series of evidentiary statements around key questions with confidence levels assigned to the evidence using criteria developed by the Intergovernmental Panel on Climate Change (https://archive.ipcc.ch/publications_and_data/ar4/wg1/en/ch1s1-6.html; accessed April 3, 2024). Confidence interval statements herein are based on our knowledge of the literature and our conservation science approach to wildfire as a natural disturbance that cannot be fully controlled in a changing climate. In the evidentiary sense, confidence in the assumptions made about fire and the climate range from very low to very high, depending on the reliability of the evidence presented.

Additionally, much of our focus in this section is based on the so-called 2020 Labor Day fires in Oregon and Washington that destroyed thousands of structures during a climate driven event also noted in Chapter 10. Importantly, while there remains a fire deficit overall throughout the western USA in forests, that deficit is beginning to close in certain landscapes because of ACC and landscapes made vulnerable to extreme fire by logging (see Bradley et al., 2016; Zald and Dunn 2018). This section explores mostly the link between ACC and wildfire increases.

Extreme fire-weather such as the Oregon 2020 Labor Day fires was unlike anything in the state's recorded history (see Chapter 10). Those fires occurred during excessive heat (heat domes parked over the region), drought, and high winds. Such events have been increasing due to ACC caused primarily by the burning of fossil fuels and land-use emissions. This section attributes specific ACC factors to wildfire increases based on robust, peer-reviewed science related to the growing field of attribute-specific climate-change studies. Twenty-one facts and supporting materials are presented that link wildfire activity to ACC factors that have caused smoke-induced human health issues, loss of life, substantial economic damages, and urban-wildfire disasters across the West with an emphasis on Oregon as consistent with Chapter 10.

How Might the Climate Tipping Point Affect Wildfires and People?

Tipping points are scary places on which to be teetering, but that is where we find ourselves. Multiple warnings have been issued by scientists and the United Nations that humanity is on a dangerous collision course that is disrupting the relatively stable biosphere—atmosphere connection that allowed humans to prosper over millennia that will not end well without major changes in population growth deceleration and the cumulative effects of our unprecedented ecological footprint. For example, at the United Nations Earth Summit in 1992, some 1700 scientists issued a warning about humanity's increasing impact to the planet's life-giving systems (https://www.ucsusa.org/resources/1992-world-scientists-warning-humanity; accessed April 1, 2024). There have been two recent warnings since then signed by 27,000 scientists from 180 countries (https://scientistswarning.forestry.oregonstate.edu/; accessed April 1, 2024). We note the following supporting facts that sound alarm bells on further disruptions.

FACT 1. The World Meteorological Organization (2023) indicates there is now a 66% chance that global temperature increases will breach the critical 1.5°C threshold as soon as 2027 with increasing catastrophic consequences for all of society. Unprecedented oceanic temperature increases (Cheng et al., 2023) and annual heat records (each year is a new record) are indicative of the speed at which climate change is proceeding. Many extreme events draw excess energy from oceanic temperature increases in particular.

FACT 2. Overshooting the 1.5°C threshold and failing to reduce GHG emissions would substantially increase extinctions of native species and impact ecosystems, according to the IPCC (2023) report. The 1.5°C threshold is a well-documented global "safety net" beyond which impacts from ACC will increasingly be catastrophic to all of society as noted from the report:

> *In scenarios with increasing CO_2 emissions, the land and ocean carbon sinks are projected to be less effective at slowing the accumulation of CO_2 in the atmosphere (high confidence). While natural land and ocean carbon sinks are projected to take up, in absolute terms, a progressively larger amount of CO_2 under higher compared to lower CO_2 emissions scenarios, they become less effective, that is, the proportion of emissions taken up by land and ocean decreases with increasing cumulative net CO_2 emissions (high confidence).*

Notably, for at least older forests, nearly all of the carbon in live pools that was present before severe fire is simply transferred laterally to dead carbon pools post fire, remaining mainly in standing large trees and burned soils if not logged (Harmon et al., 2022). This is followed by natural forest regeneration enabling carbon uptake at different rates depending on site productivity and fire severity.

FACT 3. The global climate assessment report (IPCC, 2023) concluded that "human influence has likely increased the chance of compound extreme events since the 1950s. Concurrent and repeated climate hazards have occurred in all regions, increasing impacts and risks to health, ecosystems, infrastructure, livelihoods and food (high confidence)." Additionally, as ACC proceeds, "compound extreme events include increases in the frequency of concurrent heatwaves and droughts (high confidence); fire weather in some regions (medium confidence); and compound flooding in some locations (medium confidence)." The IPCC report also notes that carbon flux from ecosystems due to extreme wildfires will, in turn, increase carbon emissions (high confidence). However, we note that so far emissions from logging at least in the United States exceed that of wildfires, insects, and wind storms combined by 5—10 times depending on the region (Harris et al., 2016; Law et al., 2018). Nonetheless, there is precious little time remaining to change this trajectory to avert disastrous consequences for humanity and nature.

FACT 4. The United Nations Environment Programme Report (2022), authored by 52 international scientists, linked global spread of landscape-scale wildfires to planet-wide overheating that is "turning landscapes into tinderboxes, while more extreme weather means stronger, hotter, drier winds to fan the flames." We note, however, that the tinderboxes are mostly concentrated in heavily logged landscapes (Bradley et al., 2016; Zald and Dunn 2018) that contrast with fire refugia provided by older forests and intact areas (Lesmeister et al., 2021).

FACT 5. The Oregon Department of Energy (2023) in their sixth climate assessment indicates total area burned each year in Oregon increased in the past 35 years. Similar increases have occurred throughout the western US and Canada (Westerling 2016, Chavardes et al., 2022). The number of days with extreme wildfire danger has also more than doubled in most regions along with frequent droughts, reductions in humidity, and declining snowpack, all of which are attributed to the buildup of GHGs (greenhouse gases). The Oregon report concluded that particulate pollution from smoke could double or triple by the end of the century, increasing human health and socioeconomic impacts. Similarly, the National Climate Assessment (2023) projected declining air quality in the western US by the end of the century due to an increase in wildfire smoke.

FACT 6. Excessive smoke plumes and unhealthy air quality during extreme wildfires are linked to a growing number of human health problems (Reid and Maestas 2019), especially in vulnerable populations such as children, the elderly, pregnant woman, people of color, and the economically disadvantaged. For example, the Oregon Health Authority (2021) documented a nearly one-third increase in asthma-related emergency room visits in the Portland-Metro region in a 4 week span during and after the Labor Day fires of 2020.

FACT 7. The Oregon Global Warmin Commission (2023) is responsible for tracking and evaluating impacts of climate change stating that:

Climate change is already having a measurable impact on Oregon's landscape, communities and economy. Oregon is experiencing increased temperatures, changing precipitation patterns, reduced snowpack, drier summers, and more frequent and damaging wildfires (note to the reader - we interpret "damaging" as referring to the built environment) ... .These impacts are projected to become more frequent and severe as temperatures increase and global climate conditions become more extreme and unpredictable. Oregon's average annual temperature has increased by around 2.2 degrees F over the past century. Without significant reductions in greenhouse gas emissions, Oregon's annual temperature is projected to increase by 5 degrees F by mid-century and by 8.2 degrees F by the 2080s.

According to the report, the total area of land burned by wildfire each year has increased in Oregon over the past 35 years, and wildfires have grown larger and spread into higher elevations during this period. We note that the number of days with extreme wildfire danger has more than doubled since 1979 throughout much of the Pacific Northwest. Drought, increased aridity, and reductions in relative humidity contribute to growing fire risks. As global temperatures increase, wildfires and fire seasons are expected to become larger and longer across the region.

Is There a Rigorous Methodology for Attributing Wildfires to ACC?

FACT 8. Recent advances in climate attribute studies (https://www.scientificamerican.com/article/attribution-science-linking-warming-to-disasters-is-rapidly-advancing/; accessed April 2, 2024), climate tracking satellites, long-term trend analyses, and computer simulation models demonstrate a statistically robust association between specific climate variables related to ACC and wildfire activity globally and across the West. Extreme climate attribution science (National Academy of Sciences, Engineering, Medicine, 2016) is one of the fastest developing climate assessment fields (hundreds of publications). Such attribution studies also have been used to estimate economic damages (Frame et al., 2020) from climate change—induced extreme events. That is, what would the damages be with and without ACC?

FACT 9. Rigorous peer-reviewed studies and metaanalyses (synthesis studies) show a consistent pattern of increased wildfire activity in the West linked to specific climate variables associated with ACC. The following studies apply to such conditions.

- Westerling et al. (2006) published a comprehensive time series of 1166 large (>8000 ha) forest wildfires for 1970 to 2003 and compared fire data to corresponding hydroclimatic and land surface variables noting that the incidence of wildfires increased in the mid-1980s in forested areas. Increases were strongly associated with rising temperatures during spring and summer. The length of the wildfire season also increased by 78 days, due to ACC.
- Denninson et al. (2014) reported increasing wildfire activity across the West was attributed in part to warmer and drier summer conditions (drought severity). For all ecoregions combined, large fires increased at a rate of seven per year, while total fire area increased at a rate of 355 km^2 per year.
- Westerling (2016) reaffirmed the tight association between wildfire activity and the relatively high cumulative warm-season actual evapotranspiration and early spring snow melt. The increase was attributed again to spring and summer temperature increases.

- Abatzoglou and Williams (2016) noted that anthropogenic increases in temperature and vapor pressure deficit significantly enhanced fuel aridity across western forests during 2000–2015, contributing to 75% more forested area experiencing high fire-season fuel aridity and an average of nine additional days per year of high fire potential. ACC accounted for ~55% of observed increases in fuel aridity and wildfire potential in recent decades.
- Holden et al. (2018) showed how declines in summer precipitation and rain days associated with GHG increases are the primary driver of increases in wildfire area in the West. Their findings are consistent with further decreases anticipated in summer precipitation and longer dry periods between rain events and this is very similar to the vapor pressure deficit as a key indicator of wildfire activity.
- Abatzoglou et al. (2021) reported that the 2020 Labor Day fires in Oregon exceeded the area burned in any single year for at least the past 120 years, contributing to hazardous air quality and massive smoke plumes. Unusually warm conditions with limited precipitation occurred in the 60 days prior to the fires. Exceptionally strong winds and dry air drove rapid rates of fire spread. The concurrence of these drivers created conditions unmatched in the observational record.
- Mass et al. (2021) reported that the Labor Day fires of 2020 were driven by strong easterly and northeasterly highly unusual winds. Wildfires produced dense smoke that initially moved westward over the Willamette Valley and eventually covered the entire region. Air quality rapidly degraded to hazardous levels, representing the worst levels in recent decades (see below).
- Reilly et al. (2022) found that the 2020 fires in western Oregon and Washington were not unprecedented, and that there were even larger fires historically, but also noted that ACC is making the fire weather conditions associated with the 2020 fires more common.
- Hawkins et al. (2022) noted that ACC factors (fuel aridity, warmer temperatures during dry wind events) increased fuel aridity and likelihood of extreme fire weather by 40% in northern California and Oregon.
- Dahl et al. (2023) linked increases in burned forest area across the West, including southwestern Canada, to the vapor pressure deficit, meaning drier atmospheric conditions produced drought-stressed plants and soils that readily burned. They used a robust global energy balance carbon-cycle model and a suite of downscaled climate models to attribute ACC to vapor pressure deficit from 1901 to 2021 and cumulative forest fire area from 1986 to 2021. ACC was responsible for 48% of long-term rise in vapor pressure deficit and, correspondingly, 37% of the cumulative area burned. Carbon emissions also contributed to nearly half the increase in drought- and fire-danger since 1901.

- MacDonald et al. (2023) synthesized the literature on climate-wildfire attribution studies, finding that there was a "striking increase" in annual area burned in the West related to increasing temperatures and the atmospheric vapor pressure deficit. ACC was the main driver behind wildfire activity, in addition to influencing other climate-related factors such as compression of the winter wet season. This trend is projected to increase without reductions in GHGs, the pathway the world is currently on.
- Turco et al. (2023) used the latest simulations for climate change attribution and detection studies, reporting that nearly all observed increases in burned area in California over the past half-century were attributed to ACC alone (summer temperature increases, dryness). Model simulations using ACC factors alone accounted for 172% (range 84%–310%) more area burned than simulations with natural processes only (no ACC in the model). Their results indicate that observed increases in burned area were primarily due to greater fuel aridity (from drying and summer temperatures).
- Higuera and Abatzoglou (2020) further demonstrated the connection between the record-setting heat wave in the western US and "extraordinary 2020 fire season." Accordingly, in just a few days, >760,000 ha burned in Oregon and Washington with millions of people enduring hazardous air, thousands of smoke-related deaths, over 10,000 structures damaged or destroyed, and dozens of lives directly lost. Extreme fire activity was attributed to the vapor pressure deficit as dry atmospheric air increased fuel aridity and dry fuels facilitated ignitions and rapid-fire spread, which is most problematic during fire containment. The authors' conclusion is directly relevant to the climate-fire interactions across the West:

Projected increases in fuel aridity in the coming decades make it unlikely that records from 2020 will stand for long. As a result, fire will increasingly become a driver of global change, catalyzing ecosystem shifts as landscapes adjust to a changing climate, and altering ecosystem services including carbon storage (Coop et al., 2020). Paramount for minimizing the negative human impacts of wildfires is addressing the root causes of anthropogenic climate change.

FACT 10. Timing, extent, and severity of wildfires in the West are strongly influenced by specific ACC factors related to vapor pressure deficit and hotter summer temperatures documented repeatedly in the above facts (e.g., Westerling, 2016) and supporting material. Other climate factors also contribute to increasing wildfire activity, including unusually strong winds (Higuera and Abatzoglou 2020), a higher incidence of lightning (Romps et al., 2014), longer fire seasons, and decreased snowpack.

Importantly, the vapor pressure deficit and summer temperatures are likely to further increase in the decades ahead based on projected carbon emissions scenarios, meaning even more extreme wildfire events are forecasted, as corroborated by global increases in wildfire activity as reported by the United Nations Environment Programme Report (2022).

How do Extreme Wildfires Impact the Built Environment?

FACT 11. According to Oregon Senate Bill 82, wildfires cost the state some $3 billion in structure losses in this decade alone. The 2020 Labor Day wildfires were the most destructive urban-wildland fires on record, killing 11 people, destroying 4300 homes, and triggering $422 million in Federal Emergency Management Agency (FEMA) aide. The cities of Phoenix, Talent, Detroit, and Gates were severely damaged by the Almeda and Santiam Fires, respectively (see Chapter 10). All told, there were 21 fires in Oregon in summer/fall of 2020, 12 of which started over the Labor Day weekend, affecting smoke and air quality levels with massive smoke plumes.

FACT 12. Excessive fine particulate matter (i.e., PM2.5 pollutants, Reid and Maestas, 2019) from wildfires can cause human health problems, including chronic obstructive pulmonary disease, acute lower respiratory illness, asthma, ischemic heart disease, and lung cancer. Health issues disproportionately impact vulnerable populations, including children with respiratory ailments, pregnant woman, the elderly, people of color, and economically disadvantaged as noted in the Oregon Health Authority (2021). Health problems are amplified in airsheds closest to a specific fire event, but impacts can extend over vast distances depending on wind direction, the Jetstream, and other factors. As an example, the figure below shows how smoke originating from fires in Oregon in 2017 not only impacted Oregon but was carried by the Jetstream across the continent (Figure 8.5).

FACT 13. Based on a regional climate and health monitoring report issued for Oregon (Clackamas, 2019), some of the damages caused by greenhouse gas emissions are as follows:

- Heat waves, extreme weather events, conditions that promote the spread and growth of disease-causing insect and bacteria populations, and poor air quality.
- Harm to some communities more than others, including those that work in physical and social environments, as well as individual factors that play important roles in determining vulnerability and resiliency to health impacts.
- Among the health conditions reported, those related to poor air quality, asthma-like symptoms and allergic disease affect the greatest number of people.
- Health conditions related to extreme heat are of growing concern.

(a)

(b)

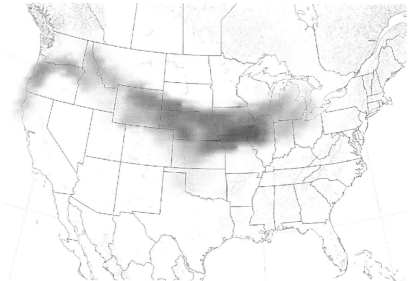

FIGURE 8.5 NASA Earth Observatory smoke (a) and (b) aerosol index (darkest colors highest aerosols) images using Suomi NPP OMPS data for September 4, 2017 (From: https://earthobservatory.nasa.gov/images/90899/smoke-pall-spans-the-united-states; accessed April 2, 2024).

- Projected impacts to the 3 Oregon counties include more extreme heat days, poorer air quality days, larger wildfires, and heavier rainfall increasing the risk of floods and landslides.
- Changes in the regional climate are already impacting health from deteriorated water and air quality, heat waves, and increased allergens.

The report also noted the following:

- Extreme weather events will lead to increases in floods, storms, and wildfires including more instances where people may become trapped and unable to escape. Changes in air quality are strongly linked to hotter, drier conditions as the region experiences more smoke. Warmer temperatures and less high-altitude snowpack create drier and longer summers and increase the risk of wildfires.
- Overwhelming evidence indicates that burning of fossil fuels is emitting unprecedented levels of carbon dioxide and other greenhouse gases that cause heat domes and extreme drought in places that are directly associated with the kind of conditions that cause more extreme fire seasons (larger fires, longer fire seasons, more loss of life, property, and worsened air quality).
- Since 2000 alone, there has been only 1 year (2006) that Oregon has not been locked into at least moderate drought conditions (https://www.drought.gov/states/oregon#:~:text=from%202000%20%2D%202020,affected%208.34%25%20of%20Oregon%20land; accessed April 2, 2024), and only in four of these years did it not have severe drought. Drought conditions are linked to rising temperatures along with winters warming faster than summers, declining snowpack, and earlier snow melt—the exact conditions that are associated with large wildfires.

FACT 14. Wildfire smoke projections from ACC increases indicate that more than 82 million people could experience a 57% and 31% increase in the frequency and intensity, respectively, of "smoke waves" (large smoke events) and those events will likely be greatest for Northern California, Western Oregon, and the Great Plains (Liu et al., 2016).

FACT 15. California fires also occurring around the same time (September 7–8, 2020) produced unhealthy smoke plumes carried aloft by the Jetstream over continental distances (https://videos.space.com/m/ELQ1EbRp/smoke-from-california-wildfires-seen-from-space?list=9wzCTV4g; accessed April 2, 2024).

FACT 16. Oregon's air quality index repeatedly broke records during the Labor Day 2020 fires (Table 8.1). Air quality index (AQI) monitoring showed several areas in Oregon breaking records (red) for at least 4 days during the Labor Day 2020 fires. AQI > 150 is considered "unhealthy;" >200 is considered "very unhealthy."

TABLE 8.1 Air Quality Index Monitoring as Reported by the Oregon Department of Environmental Quality for the Labor Day 2020 Fires in Various Locations in Oregon. AQI Values of Concern are Noted in Red Highlight

City	Previous Record	AQI 7-Sep	AQI 8-Sep	AQI 9-Sep	AQI 10-Sep	AQI 11-Sep	AQI 12-Sep	AQI 13-Sep
Portland	157	84	18	66	215	287	388	477
Bend	231	107	37	54	97	485	500	404
Medford	319	66	94	45	207	321	325	319
Klamath Falls	254	76	54	35	73	189	331	223
Eugene	291	106	342	239	387	447	438	457

Source: From https://deqblog.com/2020/09/16/wildfire-smoke-brings-record-poor-air-quality-to-oregon-new-data-shows/; accessed April 2, 2024

FACT 17. The State of Oregon Department of Environmental Quality (2023) reported Oregon alarming smoke-related unhealthy air across substantial portions of Oregon. The report noted:

- Across Oregon, smoke from wildfires is causing increases in AQI values that are "unhealthy for sensitive groups" (USG) or worse (as also noted above).
- More recently, there have been fires starting earlier, in mid-July, and lasting longer, until early October. Some recent impacts have come from British Columbia and Central Washington. In 2020, large fires also occurred in the Cascade Mountains and on the coast.
- Wildfire smoke emits a wide variety of pollutants as particulate matter (PM2.5 and PM10), black carbon, nitrogen dioxide, carbon monoxide, volatile organic compounds, polycyclic aromatic hydrocarbons, and metals. Of these pollutants, PM2.5 may represent the greatest health concern since it can be inhaled deeply into the lungs and a fraction may even reach the bloodstream. Volatile organic compounds can cause early symptoms such as watery eyes, respiratory tract irritation, and headaches. Higher levels of ozone (smog) can also be formed from an increase in the precursor pollutants: nitrogen dioxide and volatile organic compounds. As noted above in the Oregon Health Authority Report, the state and the county experienced smoke-related health problems in 2020, particularly an increase in the number of asthma related hospital admissions and comorbidity associated with COVID-19 exacerbated by smoke inhalation.
- Before 2015, Portland, Oregon, did not have a single day with air quality $\geq$ USG from wildfire smoke since air quality monitoring began in 1985.

From 2015 to 2022, however, Portland had 26 ≥ USG days or 3.3 ≥ USG days/year. In 2020, Portland had its first days over the unhealthy AQI level with 3 very unhealthy and 5 hazardous days. In 2022, Portland had 3 ≥ USG days.
- The Portland area was impacted by smoke from Nakia Creek Fire in Clark County Washington just north of the Oregon border. The fire was only around 800 ha but its proximity to the Portland−Vancouver area created smoky conditions for around a week.

Finally, the Oregon report concludes:

- AQI categories from wildfire smoke have been increasing since around 2012, with more frequent days at more "unhealthy" or worse levels, including the record-breaking events of September 2020. If these trends continue, Oregon should expect to see an increasing number of days with an AQI ≥ USG in summer across the state. This will include areas that have, in recent history, not seen serious smoke impacts until recently, including the North Coast, Willamette Valley, and Portland.

FACT 18. Large fires in Oregon were very costly from 2020−22 based on data obtained from the National Interagency Fire Center (Table 8.2). For the 3-year period, fire suppression totaled $512.5M, large fire duration averaged 55.6 days, and ~760,000 ha burned.

FACT 19. According to the IPCC (2023), human activities, principally through greenhouse gases (e.g., fossil fuels, land-uses), have unequivocally caused global overheating, with surface temperatures reaching 1.1°C above 1850−1900 in 2011−2020. Global emissions have continued to rise with ongoing contributions from unsustainable energy use, land use impacts, lifestyles, and patterns of consumption (high confidence).

FACT 20. Based on the IPCC (2023), global net anthropogenic emissions were about 12% higher than in 2010 and 54% higher than in 1990, with the largest share and growth in gross emissions in CO_2 from fossil fuels combustion and industrial processes (high confidence).

FACT 21. In 2021, CO_2 emissions came from transportation (35%), electricity (31%), industry (15%), residential and commercial (11%), and other nonfossil fuel combustion (8%) that together includes burning of coal, oil, and natural gas (EPA 2021).

Based on the above well-documented facts, we have the following recommendations for preparing fire-adapted ecosystems and communities for even more ACC-induced wildfire events, within a reasonable degree of certainty:

- ACC factors predispose regions to extreme wildfire activity from elevated temperatures (especially summer months), greater vapor pressure deficit, changes to air circulation patterns (high winds), declining rain days and snow levels, early timing of spring snowmelt, and lengthening of the fire season. An immediate transition out of fossil fuels is essential in keeping wildfire within HRV. We must also protect from logging far more forests

TABLE 8.2 Large (>40,000 acres; >16,000 ha) Fires in Oregon, 2020–22, Including Start and Containment Dates, Suppression Costs, and Duration Burning

	Start date	Contained or last report date	Acres	Suppression cost ($M)	Duration (days)
Large fires (2020)					
Lionshead	16-Aug	12-Nov	204,469	65.44	88
Beachie Creek	16-Aug	28-Oct	193,573	29.84	73
Holiday Farm	7-Sep	27-Oct	173,393	29.1	50
Riverside	8-Sep	26-Nov	138,054	20.48	79
Archie Creek	8-Sep	26-Nov	131,542	40	79
Brattain	7-Sep	5-Oct	50,951	9.9	28
Indian Creek	16-Aug	16-Sep	48,128	7	31
Star Mountain	8-Sep	11-Sep	48,000	6	3
Large fires (2021)					
Bootleg	6-Jul	13-Aug	413,717	100.9	38
Cougar Peak	7-Sep	20-Oct	91,810	26	43
Green Ridge	9-Jul	13-Oct	43,694	4.3	96
Large fires (2022)					
Double Creek	30-Aug	21-Oct	171,532	39.5	52
Cedar Creek	4-Aug	24-Nov	127,311	133.75	112
Willow Creek	28-Jun	4-Jul	40,274	0.285	7
Totals			1,876,448	512.495	779[a]

[a] Average fire duration = 55.6 days.

Sources: From https://www.predictiveservices.nifc.gov/intelligence/2020_statssumm/intro_summary20.pdf; https://www.predictiveservices.nifc.gov/intelligence/2021_

with high carbon concentrations (e.g., older forests, large, old trees, complex early seral) for their natural climate benefits.
- Extreme wildfire events have thus far impacted western communities, as an example, in terms of loss of life (11 died in the Oregon Labor Day, 2020 fires), destruction of property (~4300 structures in the Labor Day, 2020 fires), and economic damages in the billions of dollars (SB82). This is in addition to the federal government having spent $512.5 million in wildfire suppression in just the last 3 years (2020−2022), and FEMA spending another $422 million for just the Labor Day fires. Much of this money would be better spent in protecting communities from wildfires by working from the home-out, instead of logging in the wildlands (see Chapter 10). It also means working with wildfire under safe conditions for ecosystem benefits and natural fuel reduction (Calkin et al., 2023; Law et al., 2023; Baker et al., 2023b) and suppressing fires when there is a threat to towns.
- Smoke and the air quality index have broken records across the West, contributing to human health ailments and comorbidities during the COVID-19 pandemic. Community smoke shelters and improvements to air quality, by reducing air pollution from multiple point sources, would ameliorate this effect to people some extent.
- The impacts of wildfire are likely to increase without substantial reductions in GHGs, meaning throughout the world people will suffer even greater losses, affecting all levels of society. Impacts will escalate in the very near future as the world soon crosses the 1.5°C threshold due to unprecedented emissions.

8.6 CONCLUSIONS

Understanding the causes and effects of wildfire in forest ecosystems depends on the temporal and spatial scale of examination. In this regard, fire triangles are a common starting point for conceptualizing the suite of biophysical factors operating at particular scales as well as cross-scale interactions (Figure 8.6). Taken together, the fire triangle is defined by a hierarchy of temporal and spatial conditions that shape forest biomass burned over time and space. At the smallest scale, the fire triangle links oxygen, heat, and flammable vegetation in time scales of hours to years. At the next temporal and spatial scale, the fire event triangle links weather, fuels, and topography as factors that influence ignition probability, rate of fire spread, and fire intensity over seasons and years (Rothermel, 1972; Bowman et al., 2009). On decadal-to-millennial time scales, the fire triangle describes variables that determine the characteristic pattern, frequency, and intensity of fire at landscape and broader scales, reflecting the linkages among vegetation as a determinant of fuel, climate conditions as creators of fire weather, and ignition sources, be they human or natural (Parisien and Moritz, 2009; Krawchuk and Moritz, 2011). Our understanding of the paleofire record suggests that a larger and longer time scale should be considered in HRV,

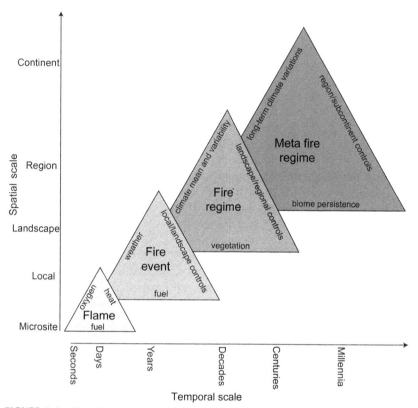

FIGURE 8.6 Controls of fire at multiple temporal and spatial scales conceptualized as fire behavior triangles. The side of each triangle indicates the dominant drivers at different temporal and spatial scales, and the overlap of triangles shows their nested nature. Paleoecological data suggest the need for a broader conceptualization of fire regimes that considers the variability of fire characteristics over the lifespan and spatial extent of a biome. *From Whitlock et al. (2010) and Modified from Parisien and Moritz (2009).*

spanning historical, contemporary, and potential future conditions. A meta-fire regime triangle describes insights gained from the range of conditions that govern fire over the duration of a vegetation type at time scales of centuries to millennia, as well as the fire–climate–vegetation–human linkages that are shaping current and future fire conditions.

Understanding past human, vegetation, climate linkages of fire regimes has gained wider attention and appreciation in the face of projected climate change. Although many definitions of a fire baseline implicitly consider time, baselines used to inform management decisions have been static and based on just the last few decades. That approach inherently misses a great deal of variability in fire activity, especially in fire regimes characterized by less-frequent, mixed- to high-severity fires.

What may seem like a stationary response on short time scales is often nonstationary when viewed on longer time scales and over a broader range of bioclimatic forces (Swetnam, 1993). In many parts of the western United States, for example, current levels of fire are considerably less than what climate would predict based on long-term linkages. This notion of a present-day fire deficit in many forest types implies that recent fire management has decoupled the natural relationship between area burned and climate (Marlon et al., 2012). We recommend that observed and projected fire conditions be understood in terms of: (1) the interactive effect of ACC, logging and roads, and human-caused ignitions in shifting fire regimes toward tipping points, especially on private lands where logging tends to be greatest and most fires spill over into the built environment (Downing et al., 2022); (2) the social and ecological benefits of managing wildfires for ecosystem benefits versus the potential deleterious consequences of increasing intervention (also see Chapter 10); and (3) a sufficiently long baseline of fire history information to capture the historical range of fire variability over a broad range of vegetation and climate conditions, while taking into account traditional Indigenous knowledge and the evolutionary history and habitat needs of native plant and animal species (see Chapters 1—4, and DellaSala et al., 2022).

Finally, preparing for a future with more fires in the coming decades is one of our greatest social—ecological challenges—one that will require expertise of multiple disciplines. Most importantly, ecological, climate, conservation ecologists, human health experts, social scientists, community planners, and fire historians working together. As part of this challenge, we need to rethink HRV strategies so that they provide information needed to address the current and future adaptive capacity of ecosystems. A comprehensive approach must be rooted in understanding fire's long-term role in different settings and climate conditions as well as the very real challenges facing fire-vulnerable communities, individuals, and the built environment.

ACKNOWLEDGEMENT

The authors would like to thank Cathy Whitlock, William Nanavati, and Shaye Wolfe for important insights, input, and contributions to the chapter through Section 8.3.

REFERENCES

Abatzoglou, J.T., Williams, A.P., 2016. Impact of anthropogenic climate change on wildfire across western US forests. Proc. Natl. Acad. Sci. USA 113, 11770—11775.

Abatzoglou, J.T., Rupp, D.E., O'Neill, L.W., Sadegh, M., 2021. Compound extremes drive the western Oregon wildfires of September 2020. Geophys. Res. Lett. 48. https://doi.org/10.1029/2021GL092520.

Alt, M., McWethy, D.B., Everett, R., Whitlock, C., 2018. Millennial scale climate-fire-vegetation interactions in a low-elevation mixed coniferous forest, Mission Range, northwestern Montana. Quat. Res. 90, 66—82. https://doi.org/10.1017/qua.2018.25.

Anderson, R.S., Allen, C.D., Toney, J.L., Jass, R.B., Bair, A.N., 2008. Holocene vegetation and fire regimes in subalpine and mixed conifer forests, southern Rocky Mountains, USA. Int. J. Wildland Fire 17, 96−114.

Archibald, S., Lehmann, C.E.R., Gómez-Dans, J.L., Bradstock, R.A., 2013. Defining pyromes and global syndromes of fire regimes. Proc. Natl. Acad. Sci. U.S.A. 110, 6442−6447.

Baker, B.C., Hanson, C.T., 2022. Cumulative tree mortality from commercial thinning and a large wildfire in the Sierra Nevada, California. Land 11 (7), 995. https://doi.org/10.3390/land11070995.

Baker, W.L., 2017. Restoring and managing lowseverity fire in dry-forest landscapes of the western USA. PLoS One 12 (2), e0172288. https://doi.org/10.1371/journal.pone.0172288.

Baker, W.L., Hanson, C.T., Williams, M.A., DellaSala, D.A., 2023a. Countering omitted evidence of variable historical forests and fire regime in western USA dry forests: the low-severity-fire model rejected. Fire 6, 146.

Baker, W.L., Hanson, C.T., DellaSala, D.A., 2023b. Harnessing natural disturbances: a nature-based solution for restoring and adapting dry forests in the western USA to climate change. Fire 6 (11), 428. https://doi.org/10.3390/fire6110428.

Bartlein, P.J., Anderson, K.H., Anderson, P.M., Edwards, M.E., Mock, C.M., Thompson, R.S., Webb, R.S., Webb III, T., Whitlock, C., 1998. Paleoclimate simulations for North America over the past 21,000 years: features of the simulated climate and comparisons with paleoenvironmental data. Quat. Sci. Rev. 17, 549−585.

Bowman, D.M.J.S., et al., 2009. Fire in the earth system. Science 324, 481−484.

Boyd, R., 1990. Demographic history, 1174−1874. In: Suttles, W. (Ed.), Handbook of North American Indians: Northwest Coast, vol. 7. Smithsonian Institution, Washington, DC, pp. 135−148.

Bradley, C.M., Hanson, C.T., DellaSala, D.A., 2016. Does increased forest protection correspond to higher fire severity in frequent-fire forests of the western United States? Ecosphere 7. https://doi.org/10.1002/ecs2.1492.

Briles, C.E., Whitlock, C., Bartlein, P.J., 2005. Postglacial vegetation, fire, and climate history of the Siskiyou Mountains, Oregon, USA. Quat. Res. 64, 44−56.

Briles, C.E., Whitlock, C., Skinner, C.N., Mohr, J., 2011. Holocene forest development and maintenance on different substrates in the Klamath Mountains, northern California, USA. Ecology 92, 590−601.

Brooks, M.L., D'Antonio, C.M., Richardson, D.M., Grace, J.B., Keeley, J.E., DiTomaso, J.M., Pyke, D., 2004. Effects of invasive alien plants on fire regimes. Bioscience 54, 677−688. https://doi.org/10.1641/0006-3568(2004)054[0677:EOIAPO]2.0.CO;2.

Brown, P.M., Kaufmann, M.R., Shepperd, W.D., 1999. Long-term, landscape patterns of past fire events in a montane ponderosa pine forest of central Colorado. Landsc. Ecol. 14, 513−532.

Brugger, S.O., Chellman, N.J., McConnell, J.R., 2022. High-latitude fire activity of recent decades derived from microscopic charcoal and black carbon in Greenland ice cores. Sage Journals 33 (2). https://doi.org/10.1177/09596836221131711.

Brunelle, A., 2022. Interactions among the fire, vegetation, the North American monsoon and the El Niño-southern oscillation in the North American desert southwest. Front. Ecol. Evol. 10. https://doi.org/10.3389/fevo.2022.656462.

Buechling, A., Baker, W.L., 2004. A fire history from tree rings in a high-elevation forest of Rocky Mountain National Park. Can. J. For. Res. 34, 1259−1273.

Calkin, D.E., Barrett, K., Cohen, J.D., Quarles, S.L., 2023. Wildland-urban fire disasters aren't actually a wildfire problem. PNAS 120 (51), e2315797120. https://doi.org/10.1073/pnas.2315797120.

Carcaillet, C., Thinon, M., 1996. Pedoanthracological contribution to the study of the evolution of the upper treeline in the Maurienne valley (North French Alps): methodology and preliminary data. Rev. Palaeobot. Palynol. 91, 399–416.

Chavardes, R.D., et al., 2022. Converging and diverging burn rates in North American boreal forests from the Little Ice Age to the present. Int. J. Wildland Fire 31 (12), 1184–1193. https://doi.org/10.1071/WF22090.

Cheng, L., et al., 2023. Another year of record heat for the oceans. Adv. Atmos. Sci. 40, 963–974. https://doi.org/10.1007/s00376-023-2385-2.

Clackamas, M., 2019. Regional climate and health monitoring report. Washington Counties. https://www.washingtoncountyor.gov/hhs/documents/regional-climate-and-health-monitoring-report/download?inline.

Clark-Wolf, K., Higuera, P.E., Shuman, B.N., McLauchlan, K.K., 2023. Wildfire activity in northern Rocky Mountain subalpine forests still within millennial-scale range of variability. Environ. Res. Lett. 18, 094029. https://doi.org/10.1088/1748-9326/acee16.

Colombaroli, D., Gavin, D.G., 2010. Highly episodic fire and erosion regime over the past 2,000 y in the Siskiyou Mountains, Oregon. Proc. Natl. Acad. Sci. U.S.A. 107, 18909–18914. https://doi.org/10.1073/pnas.1007692107.

Conedera, M., Tinner, W., Neff, C., Meurer, M., Dickens, A.F., Krebs, P., 2009. Reconstructing past fire regimes: methods, applications, and relevance to fire management and conservation. Quat. Sci. Rev. 28, 555–576.

Dahl, K.A., Abatzoglou, J.T., Phillips, C.A., Pablo Ortiz-Patrida, J., Licker, R., Merner, L.D., Ekwurzel, B., 2023. Quantifying the contribution of major carbon producers to increases in vapor pressure deficit and burned area in western US and southwestern Canadian forests. Environ. Res. Lett. 18, 064011. https://doi.org/10.1088/1748-9326/acbce8.

Daniau, A.-L., Sánchez Goñi, M.F., Martinez, P., Urrego, D.H., Bout-Roumazeilles, V., Desprat, S., Marlon, J.R., 2013. Orbital-scale climate forcing of grassland burning in southern Africa. Proc. Natl. Acad. Sci. U.S.A. 110, 5069–5073.

DellaSala, D.A., Hanson, C.T., 2019. Are wildland fires increasing large patches of complex early seral forest habitat? Diversity 11, 157. https://doi.org/10.3390/d11090157.

DellaSala, D.A., Baker, B.C., Hanson, C.T., Ruediger, L., Baker, W., 2022. Have western USA fire suppression and megafire active management approaches become a contemporary Sisyphus? Biol. Conserv. https://doi.org/10.1016/j.biocon.2022.109499.

Denninson, P.E., Brewer, C.C., Arnold, J.D., Moritz, M.A., 2014. Large wildfire trends in the western United States, 1984–2011. Geophys. Res. Lett. 41, 2928–2933. https://doi.org/10.1002/2014GL059576.

Downing, W.M., Dunn, C.J., Thompson, M.P., Caggiano, D., Short, K.C., 2022. Human ignitions on private lands drive USFS cross-boundary wildfire transmission and community impacts in the western US. Sci. Rep. 12, 2624.

Environmental Protection Agency, 2021. Overview of greenhyouse gases. https://www.epa.gov/ghgemissions/overview-greenhouse-gases.

Flannigan, M.D., Krawchuk, M.A., de Groot, W.J., Wotton, B.M., Gowman, L.M., 2009. Implications of changing climate for global wildland fire. Int. J. Wildland Fire 18, 483–507.

Frame, D.J., Wehner, M.F., Noy, I., Rosier, S.M., 2020. The economic costs of Hurricane Harvey attributable to climate change. Climatic Change 160, 271–281. https://doi.org/10.1007/s10584-020-02692-8.

Gavin, D.G., Hallett, D.J., Feng, S.H., Lertzman, K.P., Prichard, S.J., Brown, K.J., Lynch, J.A., Bartlein, P., Peterson, D.L., 2007. Forest fire and climate change in western North America: insights from sediment charcoal records. Front. Ecol. Environ. 5, 499–506.

Gavin, D.G., Brubaker, L.B., Greenwald, D.N., 2013. Postglacial climate and fire- mediated vegetation change on the western Olympic Peninsula. Washington. Ecol. Monogr. 83, 471–489.

Hanson, C.T., Odion, D.C., DellaSala, D.A., Baker, W.L., 2009. More-comprehensive recovery actions for northern spotted owls in dry forests: reply to Spies et al. Conserv. Biol. 24, 334–337. https://doi.org/10.1111/j.1523-1739.2009.01415.x.

Hanson, C.T., 2022. Cumulative severity of thinned and unthinned forests in a large California wildfire. Land 11 (3), 373. https://doi.org/10.3390/land11030373.

Harmon, M.E., Hanson, C.T., DellaSala, D.A., 2022. Combustion of aboveground wood from live trees in megafires, CA, USA. Forests 13 (3), 391. https://doi.org/10.3390/f13030391.

Harris, N.L., et al., 2016. Attribution of net carbon change by disturbance type across forest lands of the conterminous United States. Carbon Bal. Manag. 11, 24.

Hawkins, L.R., Abatzoglou, J.T., Li, S., Rupp, D.E., 2022. Anthropogenic influence on recent severe autumn fire weather in the west coast of the United States. Geophys. Res. Lett. 49, e2021GL095496.

Henne, P.D., Elkin, C.M., Reineking, B., Bugmann, H., Tinner, W., 2011. Did soil development limit spruce (Picea abies) expansion in the Central Alps during the Holocene? Testing a palaeobotanical hypothesis with a dynamic landscape model. J. Biogeogr. 38, 933–949.

Hessburg, P.F., Prichard, S.J., Hagmann, R.K., Povak, N.A., Lake, F.K., 2021. Wildfire and climate change adaptation of western North American forests: a case for intentional management. Ecol. Appl. https://doi.org/10.1002/eap.2432.

Heyerdahl, E.K., McKenzie, D., Daniels, L.D., Hessl, A.E., Littell, J.S., Mantua, N.J., 2008. Climate drivers of regionally synchronous fires in the inland Northwest (1650–1900). Int. J. Wildland Fire 17, 40–49.

Higuera, P.E., Brubaker, L.B., Anderson, P.M., Hu, F.S., Brown, T., 2009. Vegetation mediated the impacts of postglacial climate change on fire regimes in the south-central Brooks Range, Alaska. Ecol. Monogr. 79, 201–219.

Higuera, P.E., Whitlock, C., Gage, J., 2011. Linking tree-ring and sediment-charcoal records to reconstruct fire occurrence and area burned in subalpine forests of Yellowstone National Park, USA. Holocene 21, 327–341.

Higuera, P.E., Briles, C.E., Whitlock, C., 2014. Fire-regime complacency and sensitivity to centennial-through millennial-scale climate change in Rocky Mountain subalpine forests, Colorado, USA. J. Ecol. 102, 1429–1441.

Higuera, P.E., Abatzoglou, J.T., 2020. Recored-setting climate enabled the extraordinary 2020 fire season in the western United States. Global Clim. Change 27. https://doi.org/10.1111/gcb.15388.

Higuera, P.E., Shuman, B.N., Wolf, K.D., 2021. Rocky Mountain subalpine forests now burning more than any time in recent millennia. Proc. Natl. Acad. Sci. USA 118, e2103135118.

Holden, Z.A., Swanson, A., Luce, C.H., Jolly, W.M., Maneta, M., Oylen, J.W., Warren, D.A., Parsons, R., Affleck, D., 2018. Decreasing fire season precipitation increased recent western US forest wildfire activity. Proc. Natl. Acad. Sci. USA 115, E8349–E8357.

Iglesias, V., Yospin, G.I., Whitlock, C., 2015a. Reconstruction of fire regimes through integrated paleoecological proxy data and ecological modeling. Front. Plant Sci. 5, 785.

Iglesias, V., Krause, T.R., Whitlock, C., 2015b. Complex response of subalpine conifers to past environmental variability increases understanding of future vulnerability. PLoS One 10, e0124439.

Iglesias, V., Whitlock, C., Krause, T.R., Baker, R.G., 2018. Past vegetation dynamics in the Yellowstone region highlight the vulnerability of mountain systems to climate change. J. Biogeogr. 45, 1768–1780.

IPCC, 2023. AR6 Synthesis Report: Climate Change 2023. https://www.ipcc.ch/report/sixth-assessment-report-cycle/. (Accessed 22 April 2024).

Johnston, J.D., Schmidt, M.R., Merschel, A.G., Downing, W.M., Coughlan, M.R., Lewis, D.G., 2023. Exceptional variability in historical fire regimes across a western Cascades landscape, Oregon, USA. Ecosphere 14 (12), e4735.

Keane, R., Loehman, R.A., 2019. Historical range of variation. In: Encyclopedia of Wildfires and Wildland-Urban Interface Fires. https://doi.org/10.1007/978-3-319-51727-8_255-1.

Kehrwald, N.M., Whitlock, C., Barbante, C., Brovkin, V., Daniau, a.-L, Kaplan, J.O., Marlon, J.R., Power, M.J., Thonicke, K., van der Werf, G.R., 2013. Fire research: linking past, present, and future data. EOS Trans. Am. Geophys. Union 94, 421–422.

Kelly, R., Chipman, M.L., Higuera, P.E., Stefanova, I., Brubaker, L.B., Hu, F.S., 2013. Recent burning of boreal forests exceeds fire regime limits of the past 10,000 years. Proc. Natl. Acad. Sci. U.S.A. 110, 13055–13060.

Kitzberger, T., Brown, P.M., Heyerdahl, E.K., Swetnam, T.W., Veblen, T.T., 2007. Contingent Pacific-Atlantic Ocean influence on multi-century wildfire synchrony over western North America. Proc. Natl. Acad. Sci. U.S.A. 104, 543–548.

Krawchuk, M.A., Moritz, M.A., Parisien, M., Van Dorn, J., Hayhoe, K., 2009. Global pyrogeography: the current and future distribution of wildfire. PLoS One 4, e5102.

Krawchuk, M.A., Moritz, M.A., 2011. Constraints on global fire activity vary across a resource gradient. Ecology 92, 121–132.

Law, B.E., Hudiburg, T.W., Berner, L.T., Kent, J.J., Buotte, P.C., Harmon, M.E., 2018. Land use strategies to mitigate climate change in carbon dense temperate forests. Proc. Natl. Acad. Sci. USA 115 (14), 3663–3668. https://doi.org/10.1073/pnas.1720064115.

Law, B., Bloemers, R., Colleton, N., Allen, M., 2023. Redefining the wildfire problem and scaling solutions to meet the challenge. Bull. At. Sci. https://thebulletin.org/premium/2023-11/redefining-the-wildfire-problem-and-scaling-solutions-to-meet-the-challenge/. (Accessed 22 April 2024).

Lesmeister, D.B., Davis, R.J., Sovern, S.G., Yang, Z., 2021. Northern spotted owl nesting as fire refugia: a 30-year synthesis of large wildfires. Fire Ecology 17, 32.

Littell, J.S., McKenzie, D., Peterson, D.L., Westerling, A.L., 2009. Climate and wildfire area burned in western U.S. ecoprovinces, 1916–2003. Ecol. Appl. 19, 1003–1021.

Littell, J.S., McKenzie, D., Wan, H.Y., Cushman, S.A., 2018. Climate change and future wildfire in the western United States: an ecological approach to nonstationarity. Earth's Future 6, 1097–1111. https://doi.org/10.1029/2018EF000878.

Liu, J., Mickley, L.J., Sulpirizio, M.P., Dominici, F., Yue, X., Ebisu, K., Brooke Anderson, G., Khan, R.F.A., Bravo, M.A., Bell, M.L., 2016. Particular air pollution from wildfires in the western US under climate change. Clim. Change 138, 655–666. https://doi.org/10.1007/s10584-016-1762-6.

Long, C., Whitlock, C., Bartlein, P.J., 2007. Holocene vegetation and fire history of the Oregon Coast Range, USA. Holocene 17, 917–926.

MacDonald, G., et al., 2023. Drivers of California's changing wildfires: a state-of-the knowledge synthesis. Int. J. Wildland Fire 32, 1039–1058. https://doi.org/10.1071/WF22155.

Marguerie, D., Krause, T.R., Whitlock, C., 2021. What was burning? Charcoal identifications supplement an early-Holocene fire-history reconstruction in Yellowstone National Park, USA. Quat. Int. 593, 256–269. https://doi.org/10.1016/j.quaint.2020.09.033.

Marlon, J.R., Bartlein, P.J., Gavin, D.G., Long, C.J., Anderson, R.S., Briles, C.E., Brown, K.J., Colombaroli, D., Hallett, D.J., Power, M.J., Scharf, E.A., Walsh, M.K., 2012. Long-term perspective on wildfires in the western USA. Proc. Natl. Acad. Sci. U.S.A. 109, E535–E543.

Marlon, J.R., Barlein, P.J., Daniau, A.-L., Harrison, S.P., Power, M.J., Tinner, W., Maezumie, S., Vanniere, B., 2013. Global biomass burning: a synthesis and review of Holocene paleofire records and their controls. Quat. Sci. Rev. 65, 5–25.

Mass, C.F., Ovens, D., Conrick, R., Saltenberger, J., 2021. The September 2020 wildfires over the Pacific Northwest. Weather Forecast. 38, 1843–1865. https://doi.org/10.1175/WAF-D-21-0028.1.

McWethy, et al., 2013. A conceptual framework for predicting temperate ecosystem sensitivity to human impacts on fire regimes. Global Ecol. Biogeogr. 22, 900–912.

Meyer, G.A., Wells, S.G., Jull, A.J.T., 1995. Fire and alluvial chronology in Yellowstone National Park: climatic and intrinsic controls on Holocene geomorphic processes. Geol. Soc. Am. Bull. 107, 1211–1230.

Mildrexler, D.J., Berner, L.T., Law, B.E., Birdsey, R.A., Moomaw, W.R., 2023. Protect large trees for climate mitigation, biodiversity, and forest resilience. Conserv. Sci. Pract. 5, e12944. https://doi.org/10.1111/csp2.12944.

Millspaugh, S.H., Whitlock, C., Bartlein, P.J., 2000. Variations in fire frequency and climate over the past 17,000 yr in central Yellowstone National Park. Geology 28, 211–214.

Morgan, P., Heyerdahl, E.K., Gibson, C.E., 2008. Multi-season climate synchronized forest fires throughout the 20th century, northern Rockies, USA. Ecology 89, 717–728.

Mustaphi, C.J.C., Pisaric, M.F., 2014. A classification for macroscopic charcoal morphologies found in Holocene lacustrine sediments. Prog. Phys. Geogr. 38, 734–754.

National Academy of Sciences, Engineering, and Medicine, 2016. Attribution of Extreme Weather Events in the Context of Climate Change. The National Academies Press, Washington, DC. https://doi.org/10.17226/21852.

National Climate Assessment, 2023. Air quality. https://nca2023.globalchange.gov/chapter/14/.

Nanavati, W., Whitlock, C., de Porras, M.E., Gil, A., Navarro, D., Neme, G., 2022. Disentangling the last 1000 years of human-environment interactions along the eastern side of the southern Andes (34-52oS lat.). Proc. Natl. Acad. Sci. USA 119 (9), e2119813119. https://www.pnas.org/doi/10.1073/pnas.2119813119.

Odion, D.C., Moritz, M.A., DellaSala, D.A., 2010. Alternative community states maintained by fire in the Klamath Mountains, USA. J. Ecol. 98, 96–105.

Odion, D.C., Hanson, C.T., Arsenault, A., Baker, W.L., DellaSala, D.A., Hutto, R.L., Klenner, W., Moritz, M.A., Sherriff, R.L., Veblen, T.T., Williams, M.A., 2014. Examining historical and current mixed-severity fire regimes in ponderosa pine and mixed- conifer forests of Western North America. PLoS One 9, e87852.

Oregon Department of Energy, 2023. Sixth Climate Assessment. https://energyinfo.oregon.gov/blog/2023/1/11/occris-sixth-climate-assessment-outlines-climate-change-effects-on-oregon.

Oregon Global Warming Commission, 2023. Biennial Report to the Oregon Legislature 2023. https://static1.squarespace.com/static/59c554e0f09ca40655ea6eb0/t/64275b98de28d74ea4a9 6dc3/1680300956035/2023-Legislative-Report.pdf.

Oregon Health Authority, 2021. Climate and health in Oregon 2021–2022 Report. https://www.oregon.gov/oha/PH/HEALTHYENVIRONMENTS/CLIMATECHANGE/Documents/le-1052 51_23.pdf.

Parisien, M.-A., Moritz, M.A., 2009. Environmental controls on the distribution of wildfire at multiple spatial scales. Ecol. Monogr. 79, 127–154.

Parks, S.A., Abatzoglou, J.T., 2020. Warmer and drier fire seasons contribute to increases in area burned at high severity in western US forests from 1985 to 2017. Geophys. Res. Lett. 47, e2020GL089858.

Pfeiffer, M., Spessa, A., Kaplan, J.O., 2013. A model for global biomass burning in preindustrial time: LPJ-LMfire (v1.0). Geosci. Model Dev. (GMD) 6, 643–685.

Pierce, J.L., Meyer, G.A., Jull, A.J.T., 2004. Fire-induced erosion and millennial-scale climate change in northern ponderosa pine forests. Nature 432, 87–90. https://doi.org/10.1038/nature03058.

Reid, C.E., Maestas, M.M., 2019. Wildfire smoke exposure under climate change: impact on respiratory health of affected communities. Curr. Opin. Pulm. Med. 25 (2), 179–187. https://doi.org/10.1097/MCP.0000000000000552.

Reilly, M.J., Halofsky, J., Zuspan, A., Raymond, C., 2022. Cascadia burning: the historic, but not historically unprecedented, 2020 wildfires in the Pacific Northwest, USA. Ecosphere 13 (6). https://doi.org/10.1002/ecs2.4070.

Romero-Lankao, et al. (Eds.), 2014. Climate Change 2014: Impacts, Adaptation, and Vulnerability. Part B: Regional Aspects. Contribution of Working Group II to the Fifth Assessment Report of the Intergovernmental Panel on Climate Change. Cambridge University Press, Cambridge, United Kingdom and New York, NY, USA, pp. 1439–1498.

Romme, W.H., 1982. Fire and landscape diversity in subalpine forests of Yellowstone National Park. Ecol. Monogr. 52, 199–221.

Romps, D.M., Seeley, J.T., Vollaro, D., Molinari, J., 2014. Projected increase in lightning strikes in the United States due to global warming. Science 346 (6211), 851–854. https://doi.org/10.1126/science.1259100.

Rothermel, R.C., 1972. A mathematical model for predicting fire spread in wildland fuels. USDA Forest Service Research Paper INT USA, p. 40.

Schiller, C.M., Whitlock, C., Alt, M., Morgan, L.A., 2020. Vegetation responses to quaternary volcanic and hydrothermal disturbances in the northern Rocky mountains and greater Yellowstone ecosystem (USA). Palaeogeogr. Palaeoclimatol. Palaeoecol. 559, 109859.

Schoennagel, T., Sherriff, R.L., Veblen, T.T., 2011. Fire history and tree recruitment in the Colorado Front Range upper montane zone: implications for forest restoration. Ecol. Appl. 21, 2210–2222.

Sherriff, R.L., Veblen, T.T., Sibold, J.S., 2001. Fire history in high elevation subalpine forests in the Colorado Front Range. Ecoscience 8, 369–380.

Sibold, J.S., Veblen, T.T., 2006. Relationships of subalpine forest fires in the Colorado Front Range with interannual and multidecadal-scale climatic variation. J. Biogeogr. 33, 833–842.

State of Oregon Department of Environmental Quality, 2023. Wildfire Smoke Trends and the Air Quality Index. https://www.oregon.gov/deq/wildfires/Documents/WildfireSmokeTrendsReport.pdf.

Swetnam, T.W., 1993. Fire history and climate change in giant sequoia groves. Science 262, 885–889.

Swetnam, T.W., Betancourt, J.L., 1990. Fire-southern oscillation relations in the southwestern United States. Science 249, 1017–1020.

Taylor, A.H., Skinner, C.N., 2003. Spatial patterns and controls on historical fire regimes and forest structure in the Klamath Mountains. Ecol. Appl. 13, 704–719.

Tepley, A.J., Swanson, F.J., Spies, T.A., 2013. Fire-mediated pathways of stand development in Douglas-fir/western hemlock forests of the Pacific Northwest, USA. Ecology 94, 1729–1743.

Tepley, A.J., Swanson, F.J., Spies, T.A., 2014. Post-fire tree establishment and early cohort development in conifer forests of the western Cascades of Oregon, USA. Ecosphere 5, 80.

Trouet, V., Taylor, A.H., Wahl, E.R., Skinner, C.N., Stephens, S.L., 2010. Fire-climate interactions in the American west since 1400 CE. Geophys. Res. Lett. 37, L04702. https://doi.org/10.1029/2006GL027502.

Turco, M., Abatzoglou, J.T., Zhuang, Y., Jerez, S., Lucas, D.D., AghaKouchak, A., Cvijanovic, I., 2023. Anthropogenic climate change impacts exacerbate summer forest fires in California. Proc. Natl. Acad. Sci. USA 12 (25), e2213815120. https://doi.org/10.1073/pnas.2213815120.

United Nations Environment Programme, 2022. Spreading Like Wildfire: The Rising Threat of Extraordinary Landscape Fires. https://www.unep.org/resources/report/spreading-wildfire-rising-threat-extraordinary-landscape-fires.

Vachula, R.S., Russell, J.M., Huang, Y., Richter, N., 2018. Assessing the spatial fidelity of sedimentary charcoal size fractions as fire history proxies with a high-resolution sediment record and historical data. Palaeogeogr. Palaeoclimatol. Palaeoecol. 508, 166–175.

Walsh, M.E., Whitlock, C., Bartlein, P.J., 2010. 1200 years of fire and vegetation history in the Willamette Valley, Oregon and Washington. Palaeogeogr. Palaeoclimatol. Palaeoecol. 297, 273–289.

Westerling, A.L., Hidalgo, H.G., Cayan, D.R., Swetnam, T.W., 2006. Warming and earlier spring increases western US forest wildfire activity. Science 313, 940–943.

Westerling, A.L.R., 2016. Increasing western US forest wildfire activity: sensitivity to changes in the timing of spring. Phil. Trans. R. Soc. B 371, 20150178.

Whitlock, C., Larsen, C.P.S., 2001. Charcoal as a fire proxy. In: Smol, J.P., Birks, H.J.B., Last, W.M. (Eds.), Tracking Environmental Change Using Lake Sediments: Terrestrial, Algal, and Siliceous Indicators, Vol 3. Kluwer Academic, pp. 75–97. https://doi.org/10.1007/0-306-47668-1_5.

Whitlock, C., Marlon, J., Briles, C., Brunelle, A., Long, C., Bartlein, P., 2008. Long-term relations among fire, fuel, and climate in the north-western US based on lake-sediment studies. Int. J. Wildland Fire 17, 72–83.

Whitlock, C., Higuera, P.E., McWethy, D.B., Briles, C.E., 2010. Paleoecological perspectives on fire ecology: revisiting the fire-regime concept. Open Ecol. J. 3, 6–21.

Whitlock, C., Dean, W.E., Fritz, S.C., Stevens, L.R., Stone, J.R., Power, M.J., Rosenbaum, J.R., Pierce, K.L., Bracht-Flyr, B.B., 2012. Holocene seasonal variability inferred from multiple proxy records from Crevice Lake, Yellowstone National Park, USA. Palaeogeogr. Palaeoclimatol. Palaeoecol. 331–332, 90–103.

Whitlock, C., McWethy, D.B., Tepley, A.J., Veblen, T.T., Holz, A., McGlone, M.S., Perry, G.L.W., Wilmshurst, J.M., Wood, S.W., 2015. Past and present vulnerability of closed-canopy temperate forests to altered fire regimes: a comparison of the Pacific Northwest, New Zealand, and Patagonia. Bioscience 65, 151–163.

Whitlock, C., Colombaroli, D., Conedera, M., Tinner, W., 2018. Land-use history as a guide for forest conservation and management. Conserv. Biol. 32, 84–97. https://doi.org/10.1111/cobi.12960.

World Meteorological Organization, 2023. WMO Global Annual to Decadal Climate Update. https://library.wmo.int/records/item/66224-wmo-global-annual-to-decadal-climate-update.

Zald, H.S.J., Dunn, C.J., 2018. Severe fire weather and intensive forest management increase fire severity in a multi-ownership. Ecol. Appl. 28, 1068–1080. https://doi.org/10.1002/eap.1710.

FURTHER READING

Abatzoglou, J.T., Williams, A.P., Barbero, R., 2019. Global emergence of anthropogenic climate change in fire weather indices. Geophys. Res. Lett. 46, 326—336.

Littell, J.S., Peterson, D.L., Riley, K.L., Liu, Y., Luce, C.H., 2016. A review of the relationships between drought and forest fire in the United States. Global Change Biol. 22, 2353—2369.

Section III

Managing Mixed- and High-Severity Fires

Chapter 9

Postfire Logging Disrupts Nature's Phoenix

Dominick A. DellaSala[1], David B. Lindenmayer[2], Chad T. Hanson[3] and Jim Furnish[4]

[1]Wild Heritage, A Project of Earth Island Institute, Berkeley, CA, United States; [2]Fenner School of Environment and Society, The Australian National University, Canberra, ACT, Australia; [3]John Muir Project of Earth Island Institute, Berkeley, CA, United States; [4]Consulting Forester, Gila, NM, United States

9.1 POSTFIRE LOGGING AND THE KNEE-JERK RESPONSE TO FIRE

Three things are just about guaranteed every fire season: (1) forests will burn over large areas, occasionally reaching megafire proportions under extreme fire-weather (see Chapter 2); (2) land managers will proclaim burnt areas to be disasters in need of "restoration," proposing expansive postfire logging and tree planting to speed up "recovery;" and (3) decision-makers (government officials, politicians) will race to optimize the economic value of fire-killed and live trees before rot sets in, bypassing environmental safeguards to quickly get the cut out. The equivalent of a wash, rinse, and repeat cycle of misinformation and hyperbola used to justify logging and road building after fires.

Shortly after fires have been extinguished, so-called salvage logging of dead and frequently live trees happens and, in intensively managed areas, most often includes road building (e.g., "temporary roads"), replanting with commercial trees with genomes selected for site-specific conditions, seeding with nonnative plants, use of straw bales for erosion abatement (with mixed results), and spraying herbicide or using mechanical methods to suppress native vegetation—especially shrubs—that land managers think might compete with commercially valuable trees. Such forestry activities may make sense only if forests are viewed as commodities, but there are substantial tradeoffs given that they disproportionately target the most ecologically important areas where economic values are also highest, thereby setting up conflicts with increasing regularity, as big fires become more frequent in a changing climate.

As discussed throughout this book, fear of fire is coupled to socioeconomic drivers that result in command-and-control actions during (see Chapter 10) and after fires (this chapter) (also see DellaSala et al., 2022 for an update on command-and-control consequences). Misperceptions about postfire landscapes begin with the branding of the term "salvage" and postfire landscapes as "wastelands." Salvage is literally defined as "the act of saving goods or property that were in danger of damage or destruction; save from ruin, destruction, or harm, and collect discarded or refused material" (WordNet Dictionary). It also refers to "an amount estimated as expected to be realized or actually realized on sale of a fixed asset at the end of its useful life—used in calculating depreciation" (Merriam-Webster online. http://www.merriam-webster.com/dictionary/salvage).

This pretty much sums it up regarding postfire logging messaging and framing to build public support for this controversial set of logging actions. That is, the prevailing messaging view on fire is this: "disaster" (blackened forest) caused by fire is bad and "recovery" (green forest) via logging is good because burned areas are "destroyed" by fire (dead trees) so why not "restore" them with logging and planting live trees? The "discarded" materials in this case are fire-killed trees, "salvaged" mostly large trees before they "depreciate" in economic value or at the end of the fixed asset's useful life, to translate this into general terms/framing. But there is no ecological basis for postfire logging, and ecologists should refrain from using the term "salvage" as well. In reality, a fire-dependent ecosystem is not being salvaged from a disaster but, rather, is being degraded/set back by logging and related actions compared with the ecologically beneficial role that fire just performed.

In this chapter, we summarize how postfire logging and related activities can lead to compounded ecosystem disturbances (Paine et al., 1998) which, if implemented over large landscapes, may exceed disturbance thresholds, in flipping entire areas to altered ecosystem dynamics that trigger type conversions (also referred to as "landscape traps;" Lindenmayer et al., 2011, 2022). We also discuss how the postfire message framing is used as a driver for lifting environmental safeguards as proposed by decision makers wanting to replace fire-dependent, high-quality, complex early seral forests with highly altered ecosystem states characterized by biologically sterile tree plantations (i.e., essentially tree crops, planted with a few commercially valued tree species, often grown in dense rows and treated with herbicides and fertilizers). In many cases, tree plantations and intensively postfire logged areas with resulting slash debris and combustible invasive grasses have, ironically, burned in uncharacteristically intense fires (e.g., Odion et al., 2004; Thompson et al., 2007; Zald and Dunn, 2018; Lindenmayer et al., 2023).

Postfire logging proposals also tend to increase in proportion to the size of an individual fire (especially megafires), the accessibility of burned areas (e.g., high road densities), and the economic interests in expediting logging before trees degrade in economic value. To make matters worse, these activities are poised to scale up in places where climate change is expected to trigger more fires in the coming decades (see this chapter). Postfire logging may also interact with the effects of climate change that then accumulate over space and time, proving most impactful for rare and declining wildlife associated with complex early seral forests (see Chapters 3,4 and 8).

Four case studies illustrate the kinds of ecosystem degradation typically associated with postfire logging: (1) the Biscuit Fire of 2002 in southwest Oregon; (2) the Rim Fire of 2013 in the Sierra region of central California; (3) the Jasper Fire of 2000 in the Black Hills, South Dakota; and (4) montane ash-eucalypt forest fires of Victoria, Australia. For the case studies, we provide exemplary methods for reducing the ecological footprint of postfire logging where intervention occurs for economic reasons and recommendations for conserving ecologically valuable postfire landscapes where conservation is the priority. We stress that there is no ecological basis for postfire logging and, if forests are to be managed for ecological integrity, postfire logging—and its associated activities (chemical and mechanical removal of native shrubs and the establishment of artificial tree plantations)—are not a management practice that should continue. This particular chapter contains a mix of science, conservation, exemplary postfire logging issues, and our personal experiences in extreme postfire logging projects.

9.2 CUMULATIVE EFFECTS OF POSTFIRE LOGGING AND RELATED ACTIVITIES

Intensively managed postfire areas lack the pulse of legacy structures created by fire that are characteristic of complex early seral forests because most, if not all, of the ecologically valued dead and live trees (Box 9.1) are removed in logging operations (Appendix 9.1). These impacts occur when the post disturbance landscape is especially vulnerable to compaction of fragile postfire soils. Chronic management impacts can inhibit the development of complex postfire seral stages for decades to centuries, given slow rates of soil establishment in low soil productivity places (see McIver and Starr, 2000), and the removal of biological legacies over large areas (Lindenmayer et al., 2004, 2008; DellaSala et al., 2014). This affects a broad suite of postfire-dependent species, most notably, cavity-nesting (Figure 9.1) and shrub-nesting birds (Burnett et al., 2012; Hanson, 2014; Lindenmayer et al., 2018; Thorn et al., 2018).

Notably, in congressional testimony to the House Subcommittee on Resources (November 10, 2005, hearing on HR4200), University of Washington Professor Jerry Franklin stated, "Timber salvage is most appropriately viewed as a 'tax' on ecological recovery. The tax can be very large or relatively small depending upon the amount of material removed and the logging techniques that are used."

Response of fire-adapted species and communities to postfire logging depends on the scale, intensity (McIver and Starr, 2000; Lindenmayer and Noss, 2006), natural disturbance history (frequency, intensity) (Reeves et al., 2006; Hutto, 2006), and species-specific tolerance to logging on top of natural

BOX 9.1 Biological Legacies as The Building Blocks For Nature's Phoenix

Nothing in a forest is wasted, especially after a fire, as biological legacies link pre- and postdisturbance conditions, life and death in the forest, and aquatic and terrestrial ecosystems. Biological legacies such as large snags and downed logs typically have long "residence" times, persisting for decades to centuries and spanning successional stages. They include predisturbance elements (large live and dead trees, shrubs) that survive, persist, or regenerate in the burn area and are an important seed source for recolonization of plants in the new forest. They perform vital ecosystem functions such as anchoring soils (e.g., large root wads of live and dead trees); recycling nutrients (e.g., downed logs decomposed by detritovores); storing carbon long term (given slow rates of decomposition) and sequestering it, providing microsites for recolonizing plants and wildlife (e.g., so-called nurse logs that are substrate for conifer seedlings, large snags that provide shade for seedlings), and acting as refuges for numerous species (e.g., downed logs as moisture sites for salamanders, fungi, and invertebrates). Snags are used by hundreds of wildlife species for foraging (because they harbor numerous insects, particularly the larval stages), nesting, hiding, roosting, perching, and denning (examples include cavity-nesting birds, bats, and mammals, including many rare species). Many insectivorous species that use snags, in turn, perform vital trophic functions that help keep insects in check after fire. When large snags along streams eventually topple into the riverbed, they become hiding cover for fish, and pulses of postfire sedimentation (typically in the first winter after a fire) create spawning grounds for native fish, linking aquatic and terrestrial ecosystems. Despite their ecological importance, however, biological legacies are most often considered a "wasted resource" that will otherwise "rot" and need to be replaced by tree seedlings artificially grown in nurseries and planted in areas after burns, frequently in dense rows resembling corn fields, particularly in the western United States. The typical argument is that postfire logging and subsequent conifer plantings are needed to leap-frog over successional stages to a "forest," even though those actions degrade one of the most biologically diverse seral stages—complex early seral forest—and does not create a diverse forest ecosystem but, rather, creates a biologically diminished and simplified crop for lumber and wood fiber.

disturbances (i.e., cumulative effects). Documented impacts span a broad range of taxa, ecosystem processes, and forest functions (see Karr et al., 2004; Lindenmayer et al., 2004, 2008, 2018; Hutto, 2006; DellaSala et al., 2006, 2014; Hanson and North, 2008; Thorn et al., 2018; Bowd et al., 2019; see also Appendix 9.1) that can be summarized as follows:

- Extensive degradation of stand structure and function (legacies).
- Loss of soil nutrients, soil horizon damage from pile burning and ground-based logging machinery, damage to mycorrhizae and below ground chemical/nutrient exchange among plants.
- Chronic sedimentation and erosion.

FIGURE 9.1 (a) Forest fragment in 1991 before the 2003 Wedge Fire, ~60 km north of Columbia Falls, Montana. (b) Subset of avian survey plots ($n = 5000$ stations) distributed across >100 fire areas in western Montana since 1998. Exact center of the image had a nesting black-backed woodpecker (*Picoides arcticus*) after the Wedge Fire. (c) Postfire logging eliminated all biological legacies over a large landscape, including remaining nesting habitat for populations of black-backed woodpeckers not detected after the logging across the sample grids, including the center image area. *Courtesy R. L. Hutto, University of Montana.*

- Reduction in carbon storage and lost sequestration (wood products no longer sequester carbon).
- Increased fine fuel loads and reburn severity (e.g., Donato et al., 2006; Thompson et al., 2007).
- Habitat loss for threatened, endangered, and sensitive species.
- Reduced habitat and prey for apex predators and forest carnivores.
- Greatly reduced snag densities for cavity-nesting birds and mammals.
- Exotic species invasions, especially in highly disturbed areas such as along roads and fuel breaks.
- Reduced resilience and resistance of postfire landscapes to future disturbances (landscape trap).

Nearly unanimous results like those presented in Appendix 9.1 and illustrated in Figure 9.2 show a widespread and consistent pattern of postfire logging impacts across taxa and regions; that is, this type of logging has arguably more severe side effects (so to speak!) than logging in green forests. In addition, logging creates a damaging feedback loop with fires whereby areas burn in a fire, are then logged and planted with commercial species, only to burn even more intensely in the next fire, to be logged again and so on (Figure 9.3) (i.e., landscape traps; Lindenmayer et al., 2011, 2022).

FIGURE 9.2 Impacts of postfire logging on soil in two areas in southwest Oregon: the Quartz Creek Fire area, showing extensive soil ruts from dragging logs upslope on private lands (a and b); and the Biscuit Fire area, showing soil damage from burning logging slash on public lands (c). Onset of productive soil horizons spans human generations, and thus soil degradation is a chronic postfire disturbance. *(c) Photos by D. DellaSala.*

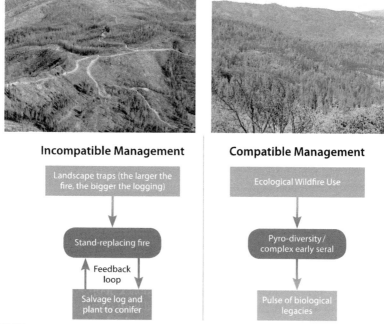

FIGURE 9.3 Incompatible versus ecologically compatible postfire management in large fire complexes. Most postfire management in the western United States follows the flow chart on the left.

Contributing to this feedback is the combination of postfire logging and removal of native shrubs through herbicides or shrub mastication—as is commonly practiced in the western United States—that dramatically increases the spread of invasive (and often highly combustible) weeds (McGinnis et al., 2010). This management-created feedback may accelerate in a changing climate in places where more fires are expected to trigger more logging responses, which already is occurring in much of the western United States and Australia.

9.3 CASE STUDY POSTFIRE LOGGING LESSONS

Biscuit Fire of 2002, Southwest Oregon

The Biscuit Fire of 2002 encompassed a fire perimeter of nearly 200,000 ha of southwest Oregon's Klamath Mountains, burning in a natural mosaic pattern of mixed severities (29% high-severity, 30% moderate-severity, and 41% low-severity fire [http://fsgeodata.net/MTBS_Uploads/data/2002/maps/OR42441 12390420020713_map.pdf]; Figure 9.4) the way nature has been doing for

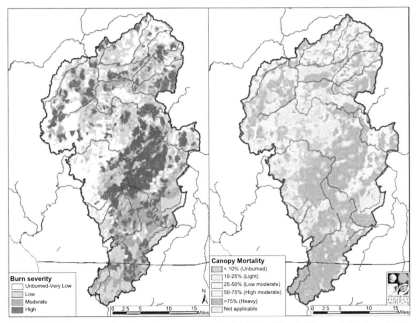

FIGURE 9.4 Burn severity (a) and canopy mortality (b) in the Biscuit Fire, as defined by the US Department of Agriculture postfire assessment team. It should be noted that, because the Forest Service's initial assessment was conducted very soon after the Biscuit fire, the canopy mortality map above is an overestimation (see http://fsgeodata.net/MTBS_Uploads/data/2002/maps/OR4244112390420020713_map.pdf), due to the fact that it was conducted too early to detect post-fire conifer responses, such as flushing (of pines) or epicormic branching (of Douglas-fir), as discussed in the next case study. *Courtesy of the Conservation Biology Institute.*

millennia (fires of this nature were common historically in this region during droughts).

At the time this was one of the nation's largest recorded fires in what is considered the most ecologically important (biodiverse) landscape in western North America that is largely unprotected (WWF Global 200 ecoregion; see DellaSala et al., 1999; Olson et al., 2012; https://www.oneearth.org/ecoregions/klamath-siskiyou-forests/). Using the Biscuit Fire as an example, we show (1) how context and scale matter in postfire management decisions; (2) how decisions by federal land managers (US Department of Agriculture Forest Service and US Department of the Interior Bureau of Land Management) are often at odds with postfire rejuvenation of areas with high conservation value; and (3) a prioritization process for minimizing ecological damage in large postfire landscapes where the pressure to log greatly outweighs conservation. This case study has broader implications in postfire management because the Biscuit logging project at the time was precedent setting (e.g., national legislation was proposed to expedite logging after fires in all the national forests and President George W. Bush visited the site to announce sweeping logging policies under the "Healthy Forest Restoration Initiative" at the time). We reiterate that there is no ecological justification for postfire logging. Given that land managers and decision makers already slated this area for massive and controversial logging, however, we present an approach that would have reduced some of the logging damage and perhaps some of the controversy.

To begin, context and scale matter in understanding patterns and processes in nature and are especially relevant in prioritization schemes. The use of "ecological screens" illustrates approaches that include recognition of context and scale in designating "go" and "no-go" zones that may be useful in reducing postfire management conflicts.

By "context," we mean knowledge of ecological condition, function, and management history that can be used to place a particular site or project area within its larger setting (Slosser et al., 2005; Perry et al., 2008). Along with context, planning at multiple scales is fundamental to understanding postfire processes and effects of management. Because ecological processes operate at multiple scales, the relative size of a management unit, the watersheds within which it lies, and the time frame over which natural processes operate all need to be factored into whether and how to treat landscapes following large fires.

Biscuit Project Scope

Unfortunately, in the planning stages for actions after the Biscuit Fire, the federal agencies completely downplayed one of the largest postfire logging proposals on public lands in history by focusing on the relatively smaller area proposed for logging over the much larger burn perimeter (USFS and BLM,

2004) and, in doing so, purposefully masked the impacts and importance of context and scale. For instance, federal agencies claimed that their activities would minimally impact the burn area because only ~4% (~8000 ha and 877,920 cubic meters of timber) of the 200,000-ha burn perimeter was to be logged. An additional 12,600 ha was to be either seeded or planted with nursery-grown conifer seedling stock, construction of 480 km of fuel management zones (FMZs) was proposed to remove 5600 ha (2360 cubic meters) of timber with the stated purpose of lowering fuel hazards, and another 8000-ha project-wide would be mechanically "thinned" under the rubric of fuels reduction, with 33,160 ha scheduled for prescribed burning even though the fire already lowered fuels. In actuality, the scale of postfire management was not 4% as claimed; rather ~51,360 ha (25%) would receive some form of postfire management activities. But this tells only part of the story as none of the planning involved issues of scale or context (see Box 9.2). Because postfire logging was heavily concentrated in high-severity fire areas, the effects on complex early seral forest from landscape fragmentation in particular were noticeably higher (Figure 9.5).

We note that the misuse of scale generally reflects the playbook of the Forest Service to routinely minimize ecosystem damages by inappropriately comparing proposed logging units to the entire burn area (thousands of hectares), resulting in the appearance of a nonsignificant impact area, while ignoring the location of logging units most often matches 1:1 with high conservation value areas as illustrated herein.

BOX 9.2 Land Use And Postfire Logging

While the Biscuit Fire project area is governed by several resource management laws and forest planning documents, two particular policies stand out the Northwest Forest Plan (USFS and BLM, 1994), which governs the management of nearly 10 million hectares of federal lands, and the Roadless Conservation Rule (USFS, 2001), which protected over 25 million hectares of inventoried roadless areas (IRAs) across the nation. The Northwest Forest Plan resulted in dramatic reductions in logging levels on federal lands that included, in part, late-successional reserves (LSRs) managed for late-seral species; however, some logging is permitted only if it is "conservative" or "prudent" and consistent with the development of late-seral conditions (USFS and BLM, 1994). The Roadless Conservation Rule prohibited logging in IRAs (which lack roads and are at least 2000 ha), with the exception of "primarily small tree thinning" where fire is a concern. Further, several watersheds in the Biscuit Fire area are managed for their wild and scenic character under the National Wild and Scenic Rivers Act (1968). Depending on specific categories, this encompassed an adjoining corridor of approximately 400 m on either side of the designated river.

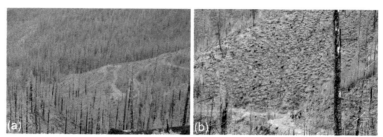

FIGURE 9.5 Biscuit postfire logged area in southwest Oregon (2002) showing landscape fragmentation from roads and clearcuts (a) and near complete removal of biological legacies (b). *Photos courtesy D. DellaSala.*

Context and Scale Matter

Before the Biscuit Fire of 2002, the area had been nominated for national monument protections by conservation groups because its regional context lies within the globally important Klamath-Siskiyou ecoregion as defined by WWF, in part because it has the highest concentration of rare plants of any national forest group in the United States, the largest complex of Inventoried Roadless Areas (IRAs) along the Pacific Coast from the Mexican to Canadian borders, and one of the best wild salmon fisheries in the region (DellaSala et al., 1999). This is clearly one of the last places that should be logged given its regional and global context, fires have been an ongoing source of natural landscape heterogeneity associated with the region's extraordinary beta-diversity (species turnover across environmental gradients) (Odion et al., 2010), and due to a number of microclimatic features may function as climate refugia for scores of rare plants, invertebrates, amphibians, mollusks, and the like (Olson et al., 2012).

In the landscape where the Biscuit Fire occurred, the greatest proportion of larger trees was concentrated within Late-Successional Reserves (LSRs) and IRAs—rather than distributed randomly throughout the project area—because of prior logging that had taken much of what was there and due to natural vegetation patterns associated with varied topography. Not only were the majority of proposed postfire management activities concentrated in those two conservation areas but also a disproportionate amount of the total expected 877,920 cubic meters of logging volume (Figure 9.6) would occur in some of the most sensitive areas. Notably, the largest LSR in the project area lies along the border of the Kalmiopsis Wilderness, a strategically important inland-coastal corridors for plant and wildlife movements (Olson et al., 2012). Furthermore, 90% of the proposed logging units lie within watersheds whose streams flow directly into wild and scenic rivers located within the burn perimeter (overall: scale and context matter).

When the appropriate scale is considered, rather than the percentage of the total burn perimeter to be logged, strikingly, 70% of the project-area

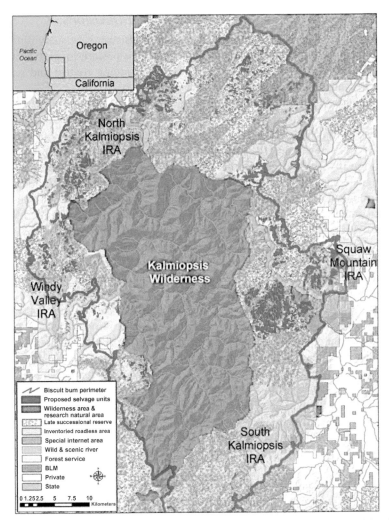

FIGURE 9.6 Map of the Biscuit Fire burn perimeter, showing agencies' proposed logging locations over major land management designations such as IRAs, LSRs, and other special interest areas.

volume would come from the collection of LSRs that represented only 42% of the total burn area. Similarly, a high percentage of proposed logging units (52%) would occur within IRAs. Finally, a total of 55 watersheds were proposed for logging to varying degrees, and therefore logging opportunities were not uniformly distributed but, rather, were clustered in areas of high ecological importance, making the project footprint much larger than claimed.

Integrating Context and Scale into Project Decisions

The lack of attention to context and scale in the Biscuit case study illustrates how land managers can grossly underestimate the postfire logging footprint. Thus, land managers would benefit from incorporating context and scale in decision-making to truly assess project impacts.

As an example, researchers (Beschta et al., 2004; Karr et al., 2004) proposed the application of "ecological screens" for minimizing postfire logging damage in ecologically sensitive areas. In the case of the Biscuit Fire, three types of screens have been proffered: administrative, operational, and ecological (Strittholt and Rustigian, 2004). Administrative screens are areas designated as off limits to logging by existing forest planning documents and environmental laws (e.g., congressionally and administratively withdrawn areas identified as no-logging areas in the Northwest Forest Plan, wilderness areas designated by Congress, IRAs). Operational screens are areas where steep terrain or lack of roads inhibit entry. Ecological screens are fine-scale filters related to specific retentions (e.g., large dead and live trees—biological legacies) to minimize impacts on site. Using multiple screens would yield a much different outcome.

After careful consideration of context and scale and applying the screens, a less ecologically damaging and more constrained response to the Biscuit project would yield a much-reduced logging "footprint" while producing significant timber volume if that is the objective. For instance, a total of 3950 ha and an estimated volume of 177,000–224,200 cubic meters of timber would be available for logging using ecological screens (Figure 9.7), compared with the agencies' alternative of 7877 ha yielding an estimated 877,920 cubic meters. Under the ecological screens approach, postfire logging would be permitted only under strict guidelines such as those recommended by Beschta et al. (2004).

Biscuit Fire Case Study Conclusions

In the absence of context and scale, postfire landscapes are treated as wastelands to be "restored" via logging, leading to devaluing of complex early seral forests and minimizing disproportionate impacts. Given the high risk of doing further damage to biodiverse postfire landscapes through large-scale logging and associated conifer planting and shrub removal, proceeding judiciously is most prudent, especially in areas of high ecological significance where context matters most. Incorporating ecological screens into project-level decisions allows for proper attention to context and scale. Even in go zones (logging units), however, managers must proceed cautiously because there is a preponderance of evidence that postfire logging disrupts natural processes and harms the development of complex early seral conditions. Within LSRs, where a "conservative" amount of logging is permitted under the Northwest Forest Plan, managers should maintain all biological

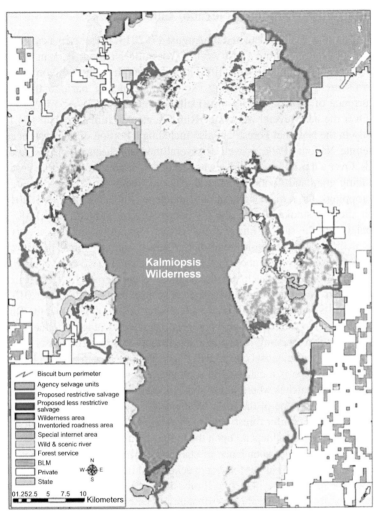

FIGURE 9.7 Comparison of agencies' logging units (*gray*) and units based on administrative, operational, and ecological screens highlighting two impact levels—restricted logging (*red*) and less restrictive (*purple*)—within the Biscuit Fire burn perimeter. *Courtesy of the Conservation Biology Institute.*

legacies. To do otherwise would place postfire landscapes with high conservation value at risk of significant ecological damage. Our findings are important for policy makers considering mandating logging, shrub removal, and tree planting following large-scale fire events, as is debated often by decision makers wanting to bypass environmental safeguards under the assumption that postfire landscapes are wastelands in need of recovery.

Rim Fire of 2013, Sierra Nevada, California

The Rim Fire was first detected on August 17, 2013, burning in a canyon in the Stanislaus National Forest just west of Yosemite National Park in the central Sierra Nevada Mountains of California. Like the Klamath-Siskiyou region, the Sierra's are globally outstanding ecologically due, in part, to the periodic occurrence of mixed-severity fire (DellaSala et al., 2017).

Over the next several weeks, the Rim fire would ultimately span 104,176 ha, mostly in the National Forest, but also including a portion of the western edge of Yosemite National Park, as well as several thousand hectares of private timberlands. Over a third of the fire area was comprised of nonconifer prefire vegetation, including grassland, foothill chaparral, oak woodlands, and numerous large rock outcroppings; the remainder comprised montane conifer forest (USFS, 2014a, b). Soon after the smoke cleared, the US Forest Service—which keeps 100% of the revenue from the sale of timber from postfire logging projects—was—as in the Biscuit fire—again proposing one of the largest national forest timber sales in history. Conservative members of Congress threatened to override environmental laws to mandate that such postfire logging occur across the Rim Fire area, including in Yosemite National Park. In response, in the autumn of 2013, some 250 scientists sent a letter to Congress opposing postfire logging in the Rim Fire area, urging lawmakers to instead appreciate the high ecological value of postfire habitat by not weakening or rolling back federal environmental laws (DellaSala et al., 2013). The scientists concluded: "Though it may seem at first glance that a post-fire landscape is a catastrophe ecologically, numerous scientific studies tell us that even in patches where forest fires burned most intensely the resulting post-fire community is one of the most ecologically important and biodiverse habitat types in western conifer forest. Post-fire conditions serve as a refuge for rare and imperiled wildlife that depend upon the unique habitat features created by intense fire. These include an abundance of standing dead trees or 'snags' that provide nesting and foraging habitat for woodpeckers and many other wildlife species, as well as patches of native flowering shrubs that replenish soil nitrogen and attract a diverse bounty of beneficial insects that aid in pollination after fire .... This post-fire habitat, known as 'complex early seral forest', is quite simply some of the best wildlife habitat in forests and is an essential stage of natural forest processes. Moreover, it is the least protected of all forest habitat types and is often as rare, or rarer, than old-growth forest, due to damaging forest practices encouraged by post-fire logging policies."

The scientists' letter carried the day with regard to the legislative threat, and the bill did not pass, but the Forest Service continued to move forward with its plan to log the Rim Fire area, as we will see next.

Overestimation of High Fire Severity

The US Forest Service's "rapid assessment" of the Rim Fire used satellite imagery obtained just weeks after the fire, reporting that ~40% of the area had experienced high-severity fire effects. They immediately released these results to the press stating that the Rim Fire created a "moonscape" that had been "nuked" (Cone, 2013). However, this was an exaggeration used to justify massive amounts of posture logging across thousands of hectares (USFS, 2014a). Not only was this effort to deny the ecological value of postfire habitat inaccurate and misleading, but the Forest Service's initial assessment also greatly exaggerated the fire severity by failing to account for postfire responses such as "flushing" in certain pine species (Hanson and North, 2009) and other rapid postfire vegetation regrowth. Through "flushing," conifers—including the most common species in the Rim Fire (ponderosa pine [*Pinus ponderosa*])—may initially seem to be dead because they have no remaining green needles after fire, but they can produce new green needles from surviving terminal buds at the ends of branches 1 year after fire, after the time period when satellite images are taken to assess fire effects (Figure 9.8).

Through this natural adaptation to mixed-severity fire, numerous areas that initially seem to have very high, or 100%, tree mortality ultimately have many or most trees survive, particularly larger overstory trees (Hanson and North, 2009). Within the Rim Fire area, flushing was common and pervasive among ponderosa pines, and some other species, by the spring and summer of 2014, resulting in many forested areas that looked quite different than they did when satellites were flying over (Figure 9.9).

However, when using satellite imagery 1 year after the fire, as accessed through the Monitoring Trends in Burn Severity (MTBS) system led by the US Geological Survey, high-severity fire comprised just under 20% of the Rim Fire (www.mtbs.gov), not 40% as initially claimed by the Forest Service. The effect of this change can be seen in the difference between the Forest Service's preliminary overestimate and the MTBS assessment 1 year after fire; the latter shows much less high-severity fire, much smaller high-severity fire patches, and far more internal heterogeneity within large, high-severity fire patches (i.e., low-/moderate-severity inclusions within high-severity fire patches; Figure 9.10). None of these changes, however,

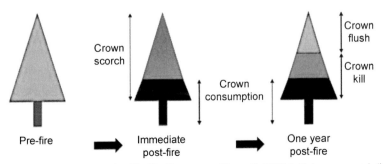

FIGURE 9.8 Process of postfire "flushing" among conifers with 100% initial crown scorch (i.e., no remaining green needles).

FIGURE 9.9 Early stages of postfire flushing of ponderosa pines (a, b), with 100% initial crown scorch, in May 2014. *Photo by Chad Hanson.*

FIGURE 9.10 A large difference in the amount of high-severity fire is seen between the Forest Service's preliminary Rapid Assessment of Vegetation (RAVG) condition and the Monitoring Trends in Burn Severity (MTBS) assessment 1 year after the Rim fire.

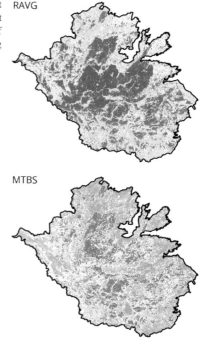

were taken into account when the Forest Service issued the final decision to conduct postfire logging (including both "salvage" logging and roadside logging along nonpublic roads) over 14,000 ha of the Rim Fire area (USFS, 2014a). Moreover, as with the Biscuit Fire, the agency also minimized the overall effects of the planned logging by noting that the logging would comprise <14% of the total area within the Rim Fire's perimeter (USFS, 2014a). Because less than two-thirds of the fire area comprised conifer forest, however, and because only about one-fifth of this conifer forest actually experienced high-severity fire, the logging planned by the Forest Service represented the removal of most of the Rim Fire area's "complex early seral forest." Complex early seral forest here as elsewhere was created by high-severity fire within mature/old conifer forest (DellaSala et al., 2014; see also Chapters 1—4), and once again the Forest Service misrepresented scale and context (and timing of flushing) to get the cut out.

Undisclosed Effects on California Spotted Owls (Strix occidentalis occidentalis)

California spotted owls use forests burned by low-/moderate-severity fire or unburned forests for nesting and roosting, and they preferentially select unlogged, high-severity fire patches occurring in older conifer forest for foraging (hunting) (Bond et al., 2009; Lee, 2018), given the rich small-mammal prey base in complex early seral forest habitat (Bond et al., 2013; also Chapter 4). Thus, the species seems quite adapted to fires if sufficient postfire habitat is available in an unlogged condition and in a mixture of patch severities.

Based on the Forest Service's preliminary Rapid Assessment of Vegetation of fire severity in the Rim Fire, the agency falsely concluded that the majority of spotted owl territories in the Rim Fire area would have little or no chance of being occupied by owls after the fire. This was based on the assumption that territories with substantial levels of high-severity fire would not likely retain owl occupancy (USFS, 2014a). The agency assessment was not based upon the MTBS fire severity data 1 year after the fire and it therefore substantially overestimated high-severity fire effects on owls in these areas, as discussed above (see Figure 9.10). Moreover, it included only effects to the ~120-ha cores of owl territories (USFS, 2014b), rather than effects on the biological territory used for foraging by owls—a much larger area with a radius of at least 1.5 km, in general (Bond et al., 2009). In addition, when complex early seral forest is removed by postfire logging and other postfire management (e.g., removal of native shrubs via mastication or herbicides), owl occupancy in the affected territories is usually eliminated (Lee et al., 2012; Clark et al., 2011, 2013; Lee, 2018); mixed-severity fire alone has not been found to reduce occupancy (Lee et al., 2012; Lee, 2018). Scientists have recommended that postfire logging be completely avoided within at least 1.5 km of spotted owl nest/roost sites (Bond et al., 2009, 2022; Clark et al., 2013; Lee, 2018).

When the results of the Forest Service's own surveys of California spotted owl 1 year after the Rim Fire became available—weeks before final decision documents were issued—independent owl scientists analyzed the data for the territories fully surveyed and made some remarkable discoveries. First, they found 39 occupied spotted owl territories, and occupancy was 92% of prefire "historical" territories (territories occupied in one or more years in the past), which is substantially higher than the average annual occupancy in unburned mature forest, which generally ranges from 60% to 76% (in any given year, not all spotted owl territories are occupied; the owls occasionally leave a territory and return one or more years later) (Lee and Bond, 2015). Second, they found that increasing high-severity fire did not reduce the occupancy of spotted owl pairs in the Rim Fire area, and even the territories with mostly high-severity fire had over 90% occupancy; some of the occupied territories were entirely within the boundaries of large high-severity fire patches with significant low-/moderate-severity inclusions (Lee and Bond, 2015). These findings were submitted to the Forest Service over a week before the decision to log the Rim Fire area was signed, but the agency did not disclose this information in its final decision. The independent analysis found that postfire logging units are located in every single occupied spotted owl territory in the Rim Fire area, and in numerous cases the majority of the entire territory would be postfire logged (Lee and Bond, 2015). In the Rim Fire area on national forest lands, this entails removal of all but 10 large snags per hectare where the prelogging snag patches typically have well over 125 large snags per hectare—generally over 90% removal (USFS, 2014a, b). Implementation of this logging project began in September 2014 (Figure 9.11).

FIGURE 9.11 Postfire clearcutting (a–c) in the Stanislaus National Forest in occupied California spotted owl territories in the autumn of 2014. *Photo by Chad Hanson.*

Natural Postfire Conifer Regeneration

One of the key rationales used by the Forest Service to justify the Rim Fire logging project was the argument that little or no postfire conifer regeneration would occur within the large, high-severity fire patches except within several dozen meters of the patch edges (USFS, 2014a, b). The agency also argued that it needed to clearcut thousands of hectares of complex early seral forest in the Rim Fire area ostensibly to cover the costs of artificially planting conifers where the Forest Service claimed conifers would not naturally regrow (USFS, 2014a, b).

Site visits by scientists in the spring and autumn of 2014, however, revealed abundant natural conifer regeneration (Figure 9.12), even deep within the interior of large, high-severity fire patches in the Rim Fire area (see also Chapter 2). Though these findings were conveyed to the Forest Service in the form of comments and photographs, the agency did not incorporate this information into the decision documents or provide any information on the amount of natural conifer regeneration in high-severity fire patches within the fire area to justify their claims (USFS, 2014a, b). Troublingly, postfire logging—especially ground-based tractor logging, which comprises nearly all of the planned logging in the Rim Fire area (USFS, 2014a)—kills most of the existing natural postfire conifer regeneration, literally crushing it under the treads of heavy logging machinery and as logs are skidded to landings, as was also the case in the Biscuit fire (Donato et al., 2006).

In reality, killing off the existing natural conifer regeneration in the Rim Fire area through logging will cost taxpayers many millions of dollars. The Forest Service estimates that it will generate about $500 per hectare in revenue

FIGURE 9.12 Natural postfire conifer regeneration (a, b), generally numbering hundreds of seedlings per hectare, within the interior of a large high-severity fire patch in the Rim Fire area in December 2014. *Photos by Chad Hanson.*

from the logging project (USFS, 2014a). However, the agency received only about one-third of that amount in the timber sales after implementing the Rim Fire logging project decision. Further, Forest Service documents show that artificial planting (including site preparation and planting expenses) costs over $1700 per hectare, and sometimes more (USFS, 2014c). Therefore, on any given hectare, when postfire logging kills natural conifer regeneration, the net cost to taxpayers for replanting the areas is at least $1200 per hectare. The real cost could be even higher given that most of the timber sale receipts are required to be allocated to future postfire logging projects, not replanting, under the "Salvage Sale Fund."

Moreover, after soil damage from postfire tractor logging on fragile soils and artificial conifer planting with nursery-grown seedlings that are not naturally adapted to the microsites where they are planted, plantings commonly fail—often extensively. When this occurs, the US Forest Service tends to conduct a second project involving intensive herbicide application to eliminate what is seen as competing vegetation, followed by additional attempts to establish conifer tree plantations. In some cases, the Forest Service is on the third iteration of this practice in a single fire area following initial postfire logging (e.g., USFS, 2014e). Notably, extensive surveys of the Forest Service's field plot locations by independent researchers 5 years postfire within unlogged large high-severity fire patches found abundant natural conifer regeneration at nearly every location, contrary to the claims and predictions of the agency (Hanson and Chi, 2021).

As the environmental assessment for another, much smaller, Forest Service postfire logging project—the Aspen postfire logging project in the Sierra National Forest—admitted, "Foregoing recovery and reforestation treatments would save taxpayers approximately $3,287,000 of appropriated funding needed to implement these activities" (USFS, 2014d). Because of the massive size of the Rim Fire, the net cost to taxpayers—just on this issue alone—could, by conservative estimates, be more than $15 million.

Rim Fire Case Study Conclusions

In the final decision documents for the Rim Fire area logging project, the Forest Service stated that the number one reason that the agency chose to propose and implement the project is that it would generate millions of dollars in revenue for the agency's budget (USFS, 2014a). The agency also noted that it was urgently interested in selling the timber to private logging companies and beginning logging as soon as possible to minimize natural postfire decay of merchantable timber and maximize the revenue to the Forest Service (USFS, 2014a). In this context of financial conflict of interest, the actual data regarding fire severity, flushing, spotted owl occupancy, natural postfire conifer regeneration, and the rarity and ecological value of complex early seral forest were largely ignored or subordinated by an agency eager to begin logging and maximize financial returns. To add insult to injury, in field trips led by the Forest Service and attended by two of us (DDS, CTH) in 2019, the agency proudly showcased postfire logging projects within a rare complex early seral forest that also included wetlands having very high bird diversity, including black-backed woodpecker (*Picoides arcticus*), northern goshawk (*Accipiter gentilis*), California spotted owl (*Strix occidentalis occidentalis*), and great gray owl (*Strix nebulosi*) nesting, and other areas where large trees were to be removed. The logging was supported by local timber companies and even a representative of The Nature Conservancy who stated his main concern was about getting more logs to the mills.

Under the current regional forest plan that governs all national forests in the Sierra Nevada Mountains, the Southern Cascades in California, and the Modoc Plateau in northeastern California, there are no protections for complex early seral forest created by high-severity fire in mature/old conifer stands (USFS, 2004), and thus those sites were unfortunately logged and logs sent to the mills as advocated for by timber companies and organizations like The Nature Conservancy.

Postfire logging in the Sierra Nevada has been found to significantly reduce overall avian biodiversity (Burnett et al., 2012) and harm rare and imperiled wildlife species like the California spotted owl (Lee et al., 2012; Hanson et al., 2018, 2021) and black-backed woodpecker (Hanson and North, 2008; Odion and Hanson, 2013; Siegel et al., 2013). Moreover, the avian species associated with the habitat created by high-severity fire, including several shrub-nesting species, are experiencing population declines, whereas birds associated with unburned forest are experiencing no such trend (Hanson, 2014). Ongoing fire suppression, postfire logging, subsequent eradication of native shrubs through mechanical or chemical means, establishment of artificial conifer plantations, and mechanical thinning have been identified as major threats to these declining species (DellaSala et al., 2014, 2017; Hanson, 2014). The ecological importance and rarity of complex early seral forest created by high-severity fire needs to be recognized and administrative and legislative protections put in place to maintain and recover this important habitat, which is currently in substantial deficit relative to its extent before fire suppression (DellaSala et al., 2014, 2017; Odion et al., 2014).

Jasper Fire of 2000, Black Hills, South Dakota

On August 24, 2000, a woman dropped a match along a highway west of Custer City, South Dakota. Within 2 h, a pyro-cumulus cloud appeared over the southwestern part of the Black Hills National Forest (BHNF). Three days later nearly 33,600 ha had burned, including almost all of Jewel Cave National Monument area and about 8% of Black Hills National Forest (BHNF) lands. Conditions in August 2000 were very dry and hot, making containment extremely difficult.

Predictably, a battle over whether and how much fire-killed timber should be logged ensued immediately. Yet, as is often the case when the smoke clears, initial fears about the gravity of the fire were lessened by a postfire assessment revealing a mosaic of fire severities (Figure 9.13). The burned area had been managed primarily for timber production and was dominated by previously logged ponderosa pine (excluding Jewel Cave National Monument), where stems per acre were reduced and a stand of homogenous, evenly spaced mature trees were left. Thus, the Jasper Fire ran through a highly managed landscape thought to be resistant to "catastrophic fire."

The Forest Service quickly produced a postfire assessment, estimating 542,740 cubic meters of tree mortality, and proposed to postfire log 141,584 cubic meters, concentrating on areas that had been severely burned with near total mortality.

Notably, the National Environmental Policy Act (NEPA) contains a provision to exempt certain emergency actions from environmental review, and the BHNF requested that the Jasper Fire postfire logging project be exempted. As Forest Service Deputy Chief at the time, one of us (JF) reviewed such requests before they were submitted by Forest Service Chief Mike Dombeck to the Council for Environmental Quality (CEQ) for approval. CEQ was known to be very stingy in granting exemptions, however.

We (JF) elected to limit a request to the narrower circumstance where the Jasper Fire had burned through areas already under commercial timber sale contract (several timber sales totaling 103,828 cubic meters), rather than the

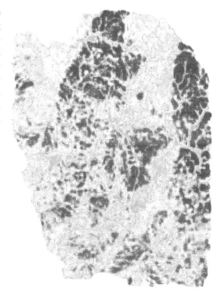

FIGURE 9.13 Burn intensities within the Jasper Fire burn perimeter, determined using ERDAS Imagery software to perform unsupervised classifications on the 3,4,7 band combination of Landsat 7 satellite imagery taken on September 3, 2000. Intensity included high (*red areas*, 12,718 ha, 38%), moderate (*yellow areas*, 10,421 ha, 31%), and low (*green areas*, 8047 ha, 24%) in Black Hills National Forest, as classified by the US Department of Agriculture Forest Service.

BHNF's comprehensive request to log as much of the timber in severely burned areas as possible. We acknowledged the time constraint—generally agreed to be about 1 year—until fire mortality begins to quickly lose commercial value. Annual timber harvest on the BHNF averaged about 165,182–188,789 cubic meters; thus, their plan to log 141,584 cubic meters equaled a "windfall substitution" of nearly a year's supply.

The chief's judgment was that the only postfire logging volume that merited emergency consideration from CEQ involved mortality in commercial timber sales where the government was involved in a contractual relationship with industry. Thus, we stipulated that only dead trees within existing timber sales could be logged under the exemption; any other postfire logging would have to follow normal NEPA procedures. Surviving trees could not be cut, even if they had been previously designated for logging. This resulted in far less postfire logging than the 141,584 cubic meters that the BHNF supervisor had requested.

In a political twist, this determination played out in the months subsequent to George W. Bush's election as president in November 2000 and before Chief Mike Dombeck's resignation in April 2001, shortly after Anne Veneman was appointed as the new Secretary of Agriculture. The election fundamentally altered the perspective of the administration from being cautious and environmentally sensitive to prologging. The result is that the final decision, made after completing the environmental impact statement, opted for postfire logging of the entire 141,584 cubic meters originally sought by the BHNF.

Jasper Case Study Conclusions

Consistent with the prevailing dogma of the time, local Forest Service officials viewed the Jasper Fire as an opportunity for unprecedented postfire logging because—timber production being of paramount concern—dead trees with commercial value should not be "wasted." The fire, the largest in recent history for the Black Hills, was termed a "disaster" even though most of the burn perimeter had low and moderate severity effects. Tree mortality was considered "lost" and of no value unless it was converted to wood products. The primary limitation for logging was time and thus constituted an emergency because time impinged on the capacity to maximize economics. Even though the BHNF produced an 80-page fire report within a few weeks, they claimed to lack the necessary staff and budget to comply with ordinary NEPA procedures.

By contrast, the more enlightened Forest Service Chief Mike Dombeck confined the NEPA "emergency exemption" to burned areas within commercial timber sales already in place before the fire and where the timber industry had a reasonable premise for economic loss. Beyond these contractual considerations, normal procedures would apply. Ultimately, we conclude that more postfire logging occurred because of political considerations than would have been authorized had the fire erupted in 1999 when a more environmentally supportive administration was in place.

Postfire conditions of the Jasper Fire area, particularly natural regeneration of pine seedlings, were affected by a severe drought from 2000 to 2008, resulting in ongoing reforestation efforts. Natural regeneration is slowly occurring throughout the Jasper burn area where adequate seed sources exist. There was widespread concern that fuel accumulations from snags would exacerbate ecosystem losses in the event of a reburn. However, no such reburn occurred and the Jasper Fire was ecologically beneficial other than in areas logged after fire.

2009 Wildfires, Victoria, Australia

This case study describes some aspects of the postfire logging operations following the February 2009 "Black Friday" wildfires in the wet montane ash forests of the Central Highlands of Victoria (Figure 9.14). Mountain ash (*Eucalyptus regnans*) and alpine ash (*Eucalyptus delegatensis*) are dominant in

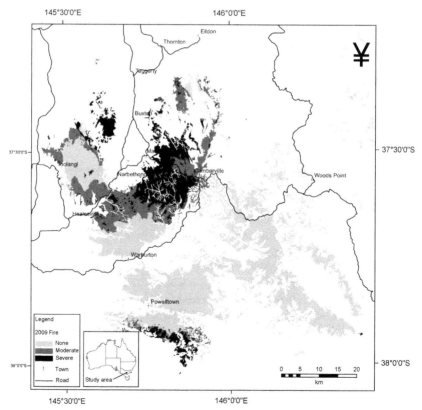

FIGURE 9.14 Mixed-severity fires of the 2009 "Black Friday" fires in wet montane ash forests of the Central Highlands of Victoria, Australia.

these forests. The 2009 wildfires burned more than an estimated 78,300 ha of montane ash forest (Burns et al., 2014). Many people think of the 2009 wildfires as a single conflagration; however, it was actually a number of different kinds of fires that varied markedly in severity over a 2-week period (Cruz et al., 2012). The most intense fires occurred in the afternoon and evening of Saturday, February 7, 2009 (Taylor et al., 2014), with some fire-affected areas reputed to have experienced among the most intense fires ever recorded, reaching 88,000 kW/m (Cruz et al., 2012). Indeed, these were the most destructive fires in Australian history in terms of loss of human life and property (Gibbons et al., 2012). By contrast, for 2 weeks after February 8, fires occurred in the forests around towns like Healesville at very low severity under a semisupervised "watching brief." Of course, such low-severity fires were markedly different in their effects on the forest and forest biodiversity relative to the fire in the afternoon and evening of February 7, 2009. These differences are critically important in terms of ecological and management understanding of montane ash forests, although some researchers have overlooked them and treated the 2009 fires as a single event, leading to flawed work.

Postfire logging is a prominent kind of logging in montane ash forests. It takes place after natural disturbances, especially wildfires, although it also occurs following windstorms in the region.

The steps involved in postfire logging operations are akin to conventional clearfelling (clearcutting) except that the removal of all merchantable trees on a site occurs in burned forest rather than green (unburned) forest. Following the completion of logging operations, slash such as tree heads and lateral branches are burned to produce a bed of ashes into which seeds of eucalypts are dropped to regenerate a new stand of trees. Hence, areas logged after fire are subject to three disturbances in rapid succession: wildfire, logging, and a regeneration burn.

As in western North America, postfire logging is conducted in an attempt to recover some of the economic value of the timber in burned stands. Significant areas of Central Highland ash forest were logged after the 1983 and 2009 wildfires. Of approximately 72,000 ha of montane ash forest that was burned in the fires that occurred in February 2009, about 3000 ha was logged, and logging (like in the western USA examples) was concentrated in areas of very high-severity fire following the 2009 fires (Figure 9.15). The area of forest logged after the 2009 fires was comparatively limited relative to the extensive and prolonged logging operations that followed the 1939 wildfires and that continued for more than 2 decades before finally being halted in the 1960s (Noble, 1977).

Much has been written about the *potential* impacts of postfire logging in montane ash forests (Lindenmayer and Ough, 2006; Lindenmayer et al., 2008). They include accelerating the loss of large, old, hollow-bearing trees (Lindenmayer and Ough, 2006) and damaging the fire-triggered

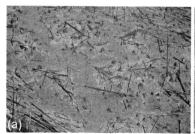 

FIGURE 9.15 Two views of post-fire logging operations at Paradize Plains, near Marysville in the Central Highlands of Victoria, showing extensive soil damage (a) and removal of biological legacies (b). *Photo by D. Lindenmayer.*

regeneration of understory and ground cover plants that are sensitive to mechanical disturbance by logging machinery. To quantify the actual impacts of postfire logging, we (D.L.) initiated formal experimental studies immediately following the 2009 wildfires. Two taxa—plants and birds—have been the focus of these studies, now in their 14th year. Results for birds in the postfire logging experiment indicate that very few species inhabit areas subject to postfire logging (Lindenmayer et al., 2018). One clear exception is the flame robin (*Petroica phoenicea*), which is closely associated with early succession (postfire and postlogging) environments in montane ash forests (Lindenmayer et al., 2014, 2022). No other species seem to be clear early successional specialists in this ecosystem, at least within the first 2 years after fire; however, several bird species returned to approximately prefire levels, or above prefire levels, by 3 or 4 years after fire (Lindenmayer et al., 2014; Appendix S3). Notably, subsequent studies of the 2009 fires revealed that the largest changes in avifauna occurrence were on severely burned sites but occupancy quickly returned postfire (Lindenmayer et al., 2022).

We note that the effects of postfire logging on plants are more severe than those of traditional clearfelling of "green" (unburned) forest. Postfire logging effects on plants are also more severe than those of wildfire. Some groups of plants, especially resprouting species like tree ferns and the musk daisy bush (*Olearia argophylla*), are particularly vulnerable to postfire logging. As an example, preliminary data indicate that overall plant species richness is reduced by almost 30% at sites that have been subject to postfire logging relative to clearfelled sites and sites subjected to high-severity wildfire (Bowd et al., 2018). In addition, our work to date has highlighted the negative impacts of postfire logging on soil nutrients and soil structure (Bowd et al., 2019); effects that remain evident for up to 80 years after logging.

The results of detailed experimental studies often become apparent only after many years of measurements, and this likely will be the case for our (DL) postfire logging work. Nevertheless, our work to date on plants in montane ash forests suggests the importance of conserving burned areas and exempting them from postfire logging operations (Bowd et al., 2021a). In addition, empirical evidence makes it clear that logging montane ash forests makes young stands regenerating after logging even more prone to subsequent crown fires (Taylor et al., 2014, 2020). Moreover, recent analyses using International Union for Conservation of Nature ecosystem assessment criteria suggest that the mountain ash ecosystem should be classified as critically endangered because of its risk of collapse in the next 30—50 years, particularly as a result of additional wildfires and ongoing logging (Burns et al., 2014). Given the relationship between logging and fire, together with the parlous state of this ecosystem, the amount of additional logging and possible future salvage logging needs to be carefully constrained.

9.4 CONCLUSIONS

Based on a review of the literature and the case studies presented from four regions herein, postfire logging severely impedes natural postfire processes by removing one of the rarest and most biodiverse wildlife habitat in forest ecosystems, compacting soils, causing chronic erosion, delaying natural succession, and introducing or spreading invasive species (effects are more severe for ground- and cable-based logging than helicopter logging), among other damages. Further, rather than jump-starting forests as claimed, postfire logging damages or removes complex early seral forests and inhibits the return of forest ecosystem conditions over time by killing naturally regenerating tree seedlings, removing the very components (large dead, dying, and downed trees—biological legacies; DellaSala, 2020) crucial to natural succession, and eradicating core components of forest biodiversity, such as native shrub patches (Bowd et al., 2023). Postfire logging can also elevate fine fuels by removing the least combustible portion of trees (trunks), leaving logging slash (in places where logging slash is treated with pile burning, damage to soils can have long-term consequences), and spreading combustible invasive grasses. Naturally regenerating landscapes following fire are biologically rich and need to be conserved for their unique ecological value.

Therefore, we recommend managers: (1) allow postdisturbance landscapes to regenerate on their own because evidence from numerous studies indicates postdisturbance processes can be surprisingly productive (Turner and Dale, 1998; Donato et al., 2006; Lindenmayer et al., 2008, 2021; DellaSala et al., 2017; Hanson and Chi, 2021; Bowd et al., 2021b); (2) avoid road building (including temporary roads) because it damages regenerative processes, channels sediments to streams, acts as a weed conduit along the disturbed road prism, and introduces the potential for accidental or arson-related subsequent burns; (3) prohibit postfire logging in dense, mature/old forest stands that experience intense fire because those areas tend to provide the highest quality, and spatially rarest, complex early seral forest habitat (Hutto, 2006; DellaSala et al., 2014, 2017; Lindenmayer et al., 2019); (4) protect biological legacies (especially large dead and dying trees) to aid in regenerative processes; (5) intervene only in ways that work with or promote natural processes (i.e., do no harm); and (6) avoid fragile areas using ecological screens to minimize impacts.

The case studies presented demonstrate that ecosystem-damaging feedback is created when a large fire strikes an area, which is logged over large landscapes after the fire, and especially where such logging occurs disproportionately in the most ecologically important areas as we have demonstrated. These degrading activities are likely to scale up in places where climate change produces more and larger fires. Finally, land managers should steer clear of making the wrong spatiotemporal assessment about ecosystem damages from proposed logging and instead avoid high conservation value areas to properly take scale and context into consideration during postfire planning decisions.

HAVE BEEN CONDUCTED

Location	Attribute	Effects on ecosystems	Sources
Broadly applicable	Process, structure, and function	Altered and diminished structural complexity of stands, ecosystem processes and functions, and populations of species and community composition	Karr et al. (2004), Lindenmayer and Noss (2006), Burnett et al. (2012), DellaSala et al. (2014)
Mostly western United States but broadly applicable	Chronic soil erosion	Erosion is greatest when logging is associated with road building, conducted with ground-based log retrieval systems, or undertaken in areas with steep slopes and sensitive soils; erosion occurs when logs are dragged across steep slopes, and damage to soil horizons occurs from burning of slash piles	McIver and Starr (2000), Beschta et al. (2004), Karr et al. (2004)
	Aquatic and hydrological processes	Substantial disturbance of hydrological systems especially from chronic sediments from roads. Removal of burned trees that provide shade may hamper tree regeneration, especially in high-elevation or dry sites; increased frequency and magnitude of erosive high flows and raising of sediment loads cause changes that alter the character of river channels, harming aquatic species; construction and reconstruction of roads and landings accelerate runoff and chronic erosion harmful to aquatic systems	Karr et al. (2004)
	Ecosystem restoration	Postfire logging inconsistent with comprehensive restoration goals	Beschta et al. (2004), Donato et al. (2006), Swanson et al. (2011), DellaSala et al. (2014), Hanson (2014)

Continued

Location	Attribute	Effects on ecosystems	Sources
Western United States	Riparian areas	Inhibit riparian functions	Reeves et al. (2006)
	Logging slash and fuels	Increased combustible fuels left on site	Weatherspoon and Skinner (1995), Duncan (2002), Donato et al. (2006)
	Herbicides	Used to kill shrubs viewed as competitors of commercially valuable trees, though such shrubs are important to nutrient cycling and mycorrhizae development often lacking in industrial settings (e.g., private lands), and data do not indicate that shrub cover precludes conifer regeneration; herbicide spraying strongly tends to increase invasive weeds	Beschta et al. (2004), Shatford et al. (2007), McGinnis et al. (2010)
Pacific Northwest and northern California, United States	Threatened, endangered, and sensitive species	Local extirpation of northern spotted owl (*Strix occidentalis caurina*) territories (Clark et al., 2011, 2013) and California spotted owl territories	Lee et al. (2012), Clark et al. (2013)
West-central Alberta, Canada	Acorn predators	Alters the guild of acorn predators and may reshape the pattern of seedling establishment	Puerta-Pinero et al. (2010)
Victoria, Australia		After a 1939 wildfire in Victoria, logging contributed to shortage of cavity trees for more than 40 vertebrate species, including some endangered ones	Lindenmayer et al. (2004)

Mostly western United States; Quebec, Canada	Plant richness and biomass, understory vegetation	Reduced vegetation biomass, increased graminoid (grass) cover, overall reduced plant species richness, and survival of planted seedlings relative to unlogged areas; reduced understory abundance, richness, and diversity	McIver and Starr (2000), Donato et al. (2006), Titus and Householder (2007), Purdon et al. (2004)
Western United States; Victoria, Australia	Biological legacies	Removal of a large percentage of large, dead, woody structure significantly alters postfire wildlife habitat (partial removal less so)	Lindenmayer et al. (2004), Beschta et al. (2004), Russell et al. (2006), Lindenmayer et al. (2018), (2021), DellaSala (2020)
	Cavity-nesting mammals and birds	Removal of hollows reduced the persistence of an array of cavity-using species, including Leadbeater's possum (*Gymnobelideus leadbeateri*), an endangered arboreal marsupial	Lindenmayer and Ough (2006)
		Reduced multiaged montane ash forests that typically support the highest diversity of arboreal marsupials and forest birds	Lindenmayer and Ough (2006)
		Reduced the abundance and nesting density of cavity-nesting birds	Caton (1996), Hitchcox (1996), Hejl et al. (1995), Saab and Dudley (1998), Smucker et al. (2005), Hutto (2006), Hutto and Gallo (2006), Cahall and Hayes (2009) (some open-nesting birds increased), Hanson and North (2008), Hutto (2008), Burnett et al. (2012)

Continued

Location	Attribute	Effects on ecosystems	Sources
New England, United States	Resistance/resilience to disturbance	Postdisturbance logging and silvicultural attempts after hurricanes and insect outbreaks failed to improve resistance or resilience of forests and were degrading overall	Foster and Orwig (2006)
Mostly western United States	Exotic species	Increases in invasions related to soil disturbance, livestock, road pathways, greater human (vector of spread) site access, and herbicide spraying	McIver and Starr (2000), Beschta et al. (2004), Karr et al. (2004), McGinnis et al. (2010)
Southwest Oregon, United States	Conifer seedlings	Natural conifer regeneration 2 years after the 2002 Biscuit Fire, although variable, was abundant even in high-severity burn areas where conifer seedling densities (>120/ha) exceeded regional standards for fully stocked stands Postfire logging reduced median conifer regeneration density by 71%, affected conifer seedlings by damaging soils and by physically burying seedlings by woody material as a result of logging; significantly increased fine and coarse, woody fuel loads	Donato et al. (2006)
Canadian Rockies, Alberta, Canada; eastern Oregon	Apex predators and forest carnivores	Avoidance of logged areas in wolf-ungulate systems; postfire logging, thinning, and conversion from fir to pine adversely affects fishers	Bull et al. (2001), Hebblewhite et al. (2008)
Mediterranean and Sierra Nevada conifer forests	Carbon storage	Reduced from removal of woody biomass	Powers et al. (2013), Serrano-Ortiz et al. (2011)

Pacific Northwest United States	Burn severity	Increased between successive fire events in logged areas	Thompson et al. (2007)
Northeastern Alberta, Canada	Bryophytes	Negative effect on species richness and species composition	Bradbury (2006)
Northwest Quebec, Canada	Soil nutrients	Loss of calcium, magnesium, and potassium for at least 110-year timber rotation	Brain et al. (2000)
Central Europe, Bohemian Forest	Species using dead wood (epixylic bryophytes, bark beetles, wood-inhabiting fungi)	All dead-wood associated taxa studied reduced	Thorn et al. (2018)
North America, Europe, Asia	Breeding bird assemblages	Persistent impacts to rare, common, and dominant species, evolutionary lineages for rare functional groups	Georgiev et al. (2020)

REFERENCES

Beschta, R.L., Rhodes, J.J., Kauffman, J.B., Gresswell, R.E., Minshall, G.W., Karr, J.R., Perry, D.A., Hauer, F.R., Frissell, C.A., 2004. Postifre management on forested public lands of the western United States. Conserv. Biol. 18, 957–967.

Bond, M.L., Lee, D.E., Siegel, R.B., Ward Jr., J.P., 2009. Habitat use and selection by California Spotted Owls in a postfire landscape. J. Wildl. Manag. 73, 1116–1124.

Bond, M.L., Lee, D.E., Siegel, R.B., Tingley, M.W., 2013. Diet and home-range size of California spotted owls in a burned forest. Western Birds 44, 114–126.

Bond, M.L., Chi, T.Y., Bradley, C.M., DellaSala, D.A., 2022. Forest management, barred owls, and wildfire in Northern Spotted Owl territories. Forests 13, 1730. https://doi.org/10.3390/f13101730.

Bowd, E.J., Lindenmayer, D.B., Banks, S.C., Blair, D.P., 2018. Logging and fire regimes alter plant communities. Ecol. Appl. 28, 826–841. https://doi.org/10.1002/eap.1693.

Bowd, E.J., Banks, S.C., Strong, C.L., Lindenmayer, D.B., 2019. Long-term impacts of wildfire and logging on forest soils. Nat. Geosci. 12, 113–118. https://doi.org/10.1038/s41561-018-0294-2.

Bowd, E., Blair, D.P., Lindenmayer, D.B., 2021a. Prior disturbance legacy effects on plant recovery post-high severity wildfire. Ecosphere 12 (5), e03480. https://doi.org/10.1002/ecs2.3480.

Bowd, E., McBurney, L., Lindenmayer, D.B., 2021b. Temporal patterns of vegetation recovery after wildfire in two obligate seeder ash forests. For. Ecol. Manag. 496. https://doi.org/10.1016/j.foreco.2021.119409.

Bowd, E., McBurney, L., Lindenmayer, D.B., 2023. The characteristics of regeneration failure and their potential to shift wet temperate forests into alternate stable states. For. Ecol. Manag. 529, 120673. https://doi.org/10.1016/j.foreco.2022.120673.

Bradbury, S.M., 2006. Response of the post-fire bryophyte community to salvage logging in boreal mixedwood forests of northeastern Alberta, Canada. For. Ecol. Manag. 234, 313–322.

Brain, S., David, P., Ouimet, R., 2000. Impacts of wild fire severity and salvage harvesting on the nutrient balance of jack pine and black spruce boreal stands. For. Ecol. Manag. 137, 231–243.

Bull, E.L., Aubry, K.B., Wales, B.C., 2001. Effects of disturbance on forest carnivores of conservation concern in eastern Oregon and Washington. Northwest Sci. 75, 180–184.

Burnett, R.D., Preston, M., Seavy, N., 2012. Plumas Lassen Study 2011 Annual Report. U. S. Forest Service, Pacific Southwest Region, Vallejo, CA.

Burns, E.L., Lindenmayer, D.B., Stein, J.A., Blanchard, W., McBurney, L., Blair, D., Banks, S.C., 2014. Ecosystem assessment of mountain ash forest in the Central Highlands of Victoria, south-eastern Australia. Austral Ecol. 39, 1–14. https://doi.org/10.1111/aec.12200.

Cahall, R.E., Hayes, J.P., 2009. Influences of postfire salvage logging on forest birds in the eastern Cascades. For. Ecol. Manag. 257, 1119–1128.

Caton, E.L., 1996. Effects of Fire and Salvage Logging on the Cavity-Nesting Bird Community in Northwestern Montana. University of Montana, Missoula, Montana.

Clark, D.A., Anthony, R.G., Andrews, L.S., 2011. Survival rates of northern spotted owls in postfire landscapes of southwest Oregon. J. Raptor Res. 45, 38–47.

Clark, D.A., Anthony, R.G., Andrews, L.S., 2013. Relationship between wildfire salvage logging, and occupancy of nesting territories by Northern Spotted Owls. J. Wildl. Manag. 77, 672–688.

Cone, T., September 19, 2013. Nearly 40 Percent of Rim Fire Land a Moonscape. Associated Press news story, Fresno, CA, USA.

Cruz, M.G., Sullivan, A.L., Gould, J.S., Sims, N.C., Bannister, A.J., Hollis, J.J., Hurley, R.J., 2012. Anatomy of a catastrophic wildfire: the Black Saturday Kilmore East fire in Victoria, Australia. For. Ecol. Manag. 284, 269–285.

DellaSala, D.A., Reid, S.B., Frest, T.J., Strittholt, J.R., Olson, D.M., 1999. A global perspective on the biodiversity of the Klamath-Siskiyou ecoregion. Nat. Area. J. 19, 300–319.

DellaSala, D.A., Karr, J.R., Schoennagel, T., Perry, D., Noss, R.F., Lindenmayer, D., Beschta, R., Hutto, R.L., Swanson, M.E., Evans, J., 2006. Postfire logging debate ignores many issues. Science 314, 51–52.

DellaSala, D.A., et al., 2013. In: Open Letter to Members of Congress from 250 Scientists Concerned about Post-fire Logging. October 30, 2014. http://www.geoinstitute.org/images/stories/pdfs/Publications/Fire/Scientist_Letter_Postfire_2013.pdf.

DellaSala, D.A., Bond, M.L., Hanson, C.T., Hutto, R.L., Odion, D.C., 2014. Complex early seral forests of the Sierra Nevada: what are they and how can they be managed for ecological integrity? Nat. Area. J. 34, 310–324.

DellaSala, D.A., et al., 2017. Accommodating mixed-severity fire to restore and maintain ecosystem integrity with a focus on the Sierra Nevada of California, USA. Fire Ecol. 13, 148–171.

DellaSala, D.A., 2020. Fire-mediated biological legacies in dry forested ecosystems of the Pacific Northwest, USA. In: Beaver, E.A., Prange, S., DellaSala, D.A. (Eds.), Disturbance Ecology and Biological Diversity. CRC Press Taylor and Francis Group, LLC, Boca Raton, FL, pp. 38–85.

DellaSala, D.A., Baker, B., Hanson, C.T., Ruediger, L., Baker, W., 2022. Have western USA fire suppression and active management approaches become a contemporary Sisyphus? Biol. Conserv. 268. https://doi.org/10.1016/j.biocon.2022.109499.

Donato, D.C., Fontaine, J.B., Campbell, J.L., Robinson, W.D., Kauffman, J.B., Law, B.E., 2006. Post-wildfire logging hinders regeneration and increases fire risk. Science 311, 352.

Duncan, S., 2002. Postfire Logging: Is it Beneficial to a Forest?. In: Science Findings Issue 47 USDA Forest Service, Pacific Northwest Research Station.

Foster, D.R., Orwig, D.A., 2006. Preemptive and salvage harvesting of New England forests: when doing nothing is a viable alternative. Conserv. Biol. 20, 959–970.

Georgiev, K.B., Chao, A., Castro, J., Chen, Y.-H., Choi, C., Fontaine, J.B., Hutto, R.L., Lee, E.-J., Müller, J., Rost, J., Żmihorski, M., Thorn, J., 2020. Salvage logging changes the taxonomic, phylogenetic and functional successional trajectories of forest bird communities. J. Appl. Ecol. 57, 1103–1112.

Gibbons, P., van Bommel, L., Gill, M.A., Cary, G.J., Driscoll, D.A., Bradstock, R.A., Knight, E., Moritz, M.A., Stephens, S.L., Lindenmayer, D.B., 2012. Land management practices associated with house loss in wildfires. PLoS One 7, e29212.

Hanson, C.T., Chi, T.Y., 2021. Impacts of postfire management are unjustified in spotted owl habitat. Front. Ecol. Evol. 9, 596282.

Hanson, C.T., Bond, M.L., Lee, D.E., 2018. Effects of post-fire logging on California spotted owl occupancy. Nat. Cons. 24, 93–105.

Hanson, C.T., Lee, D.E., Bond, M.L., 2021. Disentangling post-fire logging and high-severity fire effects for spotted owls. Birds 2, 147–157.

Hanson, C.T., North, M.P., 2009. Post-fire survival and flushing in three Sierra Nevada conifers with high initial crown scorch. Int. J. Wildland Fire 18, 857–864.

Hanson, C.T., North, M.P., 2008. Postfire woodpecker foraging in salvage-logged and unlogged forests of the Sierra Nevada. Condor 110, 777–782.

Hanson, C.T., 2014. Conservation concerns for Sierra Nevada birds associated with high- severity fire. Western Birds 45, 204—212.
Hebblewhite, M., Munro, R.H., Merrill, E.H., 2008. Trophic consequences of postfire logging in a wolf—ungulate system. For. Ecol. Manag. 257, 1053—1062.
Hejl, S.J., Hutto, R.L., Preston, C.R., Finch, D.M., 1995. Effects of silvicultural treatments in the Rocky Mountains. In: Martin, T.E., Finch, D.M. (Eds.), Ecology and Management of Neotropical Migratory Birds. Oxford University Press, New York, pp. 220—244.
Hitchcox, S.M., 1996. Abundance and Nesting Success of Cavity-Nesting Birds in Unlogged and Salvage-Logged Burned Forest in Northwestern Montana. M.S. Thesis. University of Montana, Missoula, Montana.
Hutto, R.L., 2006. Toward meaningful snag-management guidelines for postfire salvage logging in North America conifer forests. Conserv. Biol. 20, 984—993.
Hutto, R.L., 2008. The ecological importance of severe wildfires: some like it hot. Ecol. Appl. 18, 1827—1834.
Hutto, R.L., Gallo, S.M., 2006. The effects of postfire salvage logging on cavity-nesting birds. Condor 108, 817—831.
Karr, J.R., Rhodes, J.J., Minshall, G.W., Hauer, F.R., Beschta, R.L., Frissell, C.A., Perry, D.A., 2004. The effects of postifre salvage logging on aquatic ecosystems in the American West. Bioscience 54, 1029—1033.
Lee, D.E., Bond, M.L., Siegel, R.B., 2012. Dynamics of breeding-season site occupancy of the California spotted owl in burned forests. Condor 114, 792—802.
Lee, D.E., Bond, M.L., 2015. Occupancy of California spotted owl sites following a large fire in the Sierra Nevada, California. Condor 117, 228—236.
Lee, D.E., 2018. Spotted owls and forest fire: a systematic review and meta-analysis of the evidence. Ecosphere 9 (7), e02354.
Lindenmayer, D.B., Foster, D.R., Franklin, J.F., Hunter, M.L., Noss, R.F., Schmeigelow, F.A., Perry, D., 2004. Salvage harvesting policies after natural disturbance. Science 303, 1303.
Lindenmayer, D.B., Noss, R.F., 2006. Salvage logging, ecosystem processes, and biodiversity conservation. Conserv. Biol. 20, 949—958.
Lindenmayer, D.B., Ough, K., 2006. Salvage logging in the montane ash eucalypt forests of the central highlands of Victoria and its potential impacts on biodiversity. Conserv. Biol. 20, 1005—1015.
Lindenmayer, D.B., Burton, P.J., Franklin, J.F., 2008. Salvage Logging and Its Ecological Consequences. Island Press, Washington, DC.
Lindenmayer, D.B., Hobbs, R.J., Likens, G.E., Krebs, C.J., Banks, S.C., 2011. Newly discovered landscape traps produce regime shifts in wet forests. PNAS 108 (38), 15887—15891. www.pnas.org/cgi/doi/10.1073/pnas.1110245108.
Lindenmayer, D.B., Blanchard, W., McBurney, L., Blair, D., Banks, S.C., Driscoll, D.A., Smith, A.L., Gill, A.M., 2014. Complex responses of birds to landscape-level fire extent, fire severity and environmental drivers. Divers. Distrib. 20, 467—477.
Lindenmayer, D.B., McBurney, L., Blair, D., Wood, J., Banks, S.C., 2018. From unburnt to salvage logged: quantifying bird responses to different levels of disturbance severity. J. Appl. Ecol. 55, 1626—1636.
Lindenmayer, D.B., Westgate, M.J., Scheele, B.C., Foster, C.N., Blair, D.P., 2019. Key perspectives on early successional forests subject to stand-replacing disturbances. For. Ecol. Manag. 454, 117656. https://doi.org/10.1016/j.foreco.2019.117656.

Lindenmayer, D.B., Blanchard, W., Bowd, E., Scheele, B.C., Foster, C., Lavery, T., McBurney, L., Blair, D., 2021. Rapid bird species recovery following high-severity wildfire but in the absence of early successional specialists. Divers. Distrib. https://doi.org/10.1111/ddi.13611.

Lindenmayer, D.B., Taylor, C., Bowd, E., Likens, G.E., 2022. The interactions among fire, logging, and climate change has sprung a landscape trap in Victoria's montane ash forests. Plant Ecol. 223, 733–749. https://doi.org/10.1007/s11258-021-01217-2.

Lindenmayer, D.B., Yebra, M., Cary, G., 2023. Better managing fire in flammable tree plantations. For. Ecol. Manag. 528, 120641. https://doi.org/10.1016/j.foreco.2022.120641.

McGinnis, T.W., Keeley, J.E., Stephens, S.L., Roller, G.B., 2010. Fuel buildup and potential fire behavior after stand-replacing fires, logging fire-killed trees and herbicide shrub removal in Sierra Nevada forests. For. Ecol. Manag. 260, 22–35.

McIver, J.D., Starr, L., 2000. Environmental Effects of Postfire Logging: Literature Review and Annotated Bibliography. USDA Forest Service Pacific Northwest Research Station, Portland, OR. Gen. Tech. Report PNW-GTR-486.

Noble, W.S., 1977. Ordeal by Fire. The Week a State Burned Up. Hawthorn Press, Melbourne.

Odion, D.C., Hanson, C.T., 2013. Projecting impacts of fire management on a biodiversity indicator in the Sierra Nevada and Cascades, USA: the Black-backed Woodpecker. Open For. Sci. J. 6, 14–23.

Odion, D.C., Strittholt, J.R., Jiang, H., Frost, E., DellaSala, D.A., Moritz, M., 2004. Fire severity patterns and forest management in the Klamath National Forest, northwest California, USA. Conserv. Biol. 18, 927–936.

Odion, D.C., Moritz, M.A., DellaSala, D.A., 2010. Alternative community states maintained by fire in the Klamath Mountains, USA. J. Ecol. 98, 96–105.

Odion, D.C., Hanson, C.T., Arsenault, A., Baker, W.L., DellaSala, D.A., Hutto, R.L., Klenner, W., Moritz, M.A., Sherriff, R.L., Veblen, T.T., Williams, M.A., 2014. Examining historical and current mixed-severity fire regimes in ponderosa pine and mixed- conifer forests of western North America. PLoS One 9, e87852.

Olson, D., DellaSala, D.A., Noss, R.F., Strittholt, J.R., Kass, J., Koopman, M.E., Allnutt, T.F., 2012. Climate change refugia for biodiversity in the Klamath-Siskiyou ecoregion. Nat. Areas 32 (1), 65–74.

Paine, R.T., Tegener, M.J., Johnson, E.A., 1998. Compounded perturbations yield ecological surprises. Ecosystems 1, 535–545.

Perry, D.A., Oren, R., Hart, S.C., 2008. Forest Ecosystems, second ed. John Hopkins University Press, Baltimore, MD.

Powers, E.M., Marshall, J.D., Zhang, J., Wei, L., 2013. Post-fire management regimes affect carbon sequestration and storage in a Sierra Nevada mixed conifer forest. For. Ecol. Manag. 291, 268–277.

Purdon, M., Brais, S., Bergeron, Y., 2004. Initial response of understory vegetation to fire severity and salvage logging in the southern boreal forest of Quebec. Appl. Veg. Sci. 7, 49–60.

Puerta-Pinero, C., Sanchez, A., Leverkus, M.A., Catro, J., 2010. Management of burnt wood after fire affects post-dispersal acorn predation. For. Ecol. Manag. 260, 345–352.

Reeves, G.H., Bisson, P.A., Riema, B.E., Benda, L.E., 2006. Postfire logging in riparian areas. Conserv. Biol. 20, 994–1004.

Russell, R.E., Saab, V.A., Dudley, J.G., Rotella, J.J., 2006. N to wildfire and postfire salvage logging. For. Ecol. Manag. 232, 179–187.

Saab, V.A., Dudley, J.G., 1998. Responses of Cavity-Nesting Birds to Stand-Replacement Fire and Salvage Logging in Ponderosa pine/Douglas-fir Forests of Southwestern Idaho. USDA Forest Service, Ogden, Utah. Research paper RMRS-RP-11:1−17.

Serrano-Ortiz, P., Marañón-Jiménez, S., Reverter, B.R., Sánchez-Cañete, E.P., Castro, J., Zamora, R., Kowalski, A.S., 2011. Postfire salvage logging reduces carbon sequestration in Mediterranean coniferous forest. For. Ecol. Manage. 262, 2287−2296.

Shatford, J.P.A., Hibbs, D.E., Puettmann, K.J., 2007. Conifer regeneration after forest fire in the Klamath-Siskiyous: how much, how soon? J. For.

Siegel, R.B., Tingley, M.W., Wilkerson, R.L., Bond, M.L., Howell, C.A., 2013. Assessing Home Range Size and Habitat Needs of Black-Backed Woodpeckers in California: Report for the 2011 and 2012 Field Seasons. Institute for Bird Populations, Point Reyes Station, CA.

Slosser, N.C., Strittholt, J.R., DellaSala, D.A., Wilson, J., 2005. The landscape context in forest conservation: integrating protection, restoration, and certification. Ecol. Restor. 23, 15−23.

Smucker, K.M., Hutto, R.L., Steele, B.M., 2005. Changes in bird abundance after wildfire: importance of fire severity and time since fire. Ecol. Appl. 15, 1535−1549.

Strittholt, J.R., Rustigian, H., 2004. Ecological Issues Underlying Proposals to Conduct Salvage Logging in Areas Burned by the Biscuit Fire. Conservation Biology Institute, Corvallis, OR. Unpublished report.

Swanson, M.E., Franklin, J.F., Beschta, R.L., Crisafulli, C.M., DellaSala, D.A., Hutto, R.L., Lindenmayer, D., Swanson, F.J., 2011. The forgotten stage of forest succession: early- successional ecosystems on forest sites. Front. Ecol. Environ. 9, 117−125.

Taylor, C., McCarthy, M.A., Lindenmayer, D.B., 2014. Nonlinear effects of stand age on fire severity. Conserv. Lett. 7, 355−370.

Taylor, C., Blanchard, W., Lindenmayer, D.B., 2020. Does forest thinning reduce fire severity in Australian eucalypt forests? Conserv. Lett., e12766 https://doi.org/10.1111/conl.12766.

Thompson, J.R., Spies, T.A., Ganio, L.M., 2007. Reburn severity in managed and unmanaged vegetation in a large wildfire. PNAS 104, 10743−10748. https://doi.org/10.1073/pnas.0700229104.

Thorn, S., et al., 2018. Impacts of salvage logging on biodiversity: a meta-analysis. J. Appl. Ecol. https://doi.org/10.1111/1365-2664.12945.

Titus, J.H., Householder, E., 2007. Salvage logging and replanting reduce understory cover and richness compared to unsalvaged-unplanted sites at Mount St. Helens, Washington. Wes. N. Am. Naturalist 67, 219−231.

Turner, M.G., Dale, V.H., 1998. Comparing large, infrequent disturbances: what have we learned? Ecosystems 1, 493−496.

USFS, BLM, 1994. Record of Decision for Amendments to Forest Service and Bureau of Land Management Planning Documents Within the Range of the Northern Spotted Owl. USDA Forest Service and USDI Bureau of Land Management, Portland, OR. http://www.reo.gov/library/reports/newroda.pdf.

USFS, 2001. Forest Service Roadless Conservation Final Environmental Impact Statement. USDA Forest Service, Washington, DC.

USFS, BLM, 2004. Final Environmental Impact Statement: The Biscuit Fire Recovery Project. Rogue River-Siskiyou National Forest, Josephine and Curry Counties, OR.

USFS, 2004. Sierra Nevada Forest Plan Amendment, Final Supplemental Environmental Impact Statement and Record of Decision. U.S. Forest Service, Pacific Southwest Region, Vallejo, CA, USA.

USFS, 2014a. Rim Fire Recovery, Final Environmental Impact Statement and Record of Decision. U.S. Forest Service, Stanislaus National Forest, Sonora, CA, USA.

USFS, 2014b. Rim Fire Recovery, Vegetation Report. U.S. Forest Service, Stanislaus National Forest, Sonora, CA, USA.

USFS, 2014c. Big Hope Fire Salvage and Restoration Project. Environmental Assessment. U.S. Forest Service, Tahoe National Forest, Nevada City, CA, USA.

USFS, 2014d. Aspen Recovery and Reforestation Project. Environmental Assessment. U.S. Forest Service, Hi Sierra Ranger District, Sierra National Forest, Prather, CA, USA.

USFS, 2014e. Power Fire Reforestation Project, Proposed Action. U.S. Forest Service, Eldorado National Forest, Placerville, CA, USA.

Weatherspoon, C.P., Skinner, C.N., 1995. An assessment of factors associated with damage to tree crowns from the 1987 wildfire in northern California. For. Sci. 41, 430–451.

Zald, H.S.J., Dunn, C.J., 2018. Severe fire weather and intensive forest management increase fire severity in a multi-ownership landscape. Ecol. Appl. https://doi.org/10.1002/eap.1710.

FURTHER READING

USFS, 2002. Burned Area Emergency Rehabilitation Report for the Biscuit Fire. Rogue and Siskiyou National Forests, Medford, OR.

Chapter 10

Forest Managers Play the Backcountry Logging Fiddle as Towns Burn down

Dominick A. DellaSala
Wild Heritage, A Project of Earth Island Institute, Berkeley, CA, United States

10.1 THE DAY CLIMATE CHANGE CAME KNOCKING AT MY DOOR

The morning of September 8, 2020 started out routinely for me at 6 a.m. with a top off at the espresso machine, while I opined about an alarming climate situation that could trigger a really big fire consequential to unprepared western towns. A heat dome was camped over the region for weeks and a lengthy drought created parched, fire-prone landscapes that primed the fire pump, so I thought. Little did I realize that, later in the day, myself and thousands of fellow residents would be fleeing a fast-moving wildfire tsunami cresting over my home town and nearby communities, started by an arsonist during extreme fire weather.

Much of Oregon's wildland—urban interface was on fire that fatal day, as 30 fires raged across ~250,000 ha (1 million ac) of the state, causing 40,000 people to flee toward safer ground, while another half million were placed on evacuation alert (a lower evacuation level; https://en.wikipedia.org/wiki/2020_Oregon_wildfires). All told, the Oregon towns of Phoenix, Talent, Detroit, and Gates were damaged severely, with 11 lives lost and scores of families displaced by human-caused and climate-driven wildfires. Although thankfully the Almeda fire in southern Oregon's Rogue Valley stopped but a few blocks short of my house, I too was a fire evacuee for 10 days. When I returned home, smoke blanketed the sky for weeks, obscuring the Siskiyou and Cascade Mountains until October rains brought welcomed relief.

The so-called Almeda fire (named for its point of origin—Almeda Drive) was lit by an arsonist around 11 a.m. that day in a popular dog-walking park just 11 km (7 miles) south of Talent in the touristy town of Ashland, best known for its Shakespeare Festival. With easterly winds gusting to 64 km per hour (40 mph), fire quickly spread northward across weed-infested dried out fields, pitching embers into the parched canopy of the popular Bear Creek Greenway (an urban−wildland oasis of tall streamside cottonwoods that connects several towns). The fire skipped back and forth over two freeways (Interstate-5 to the east runs north-south; Highway 99 adjacent to I-5 also runs north-south bisecting Talent). No "fuel break" or suppression forces could possibly contain the fast-moving flame front as large portions of my community were soon engulfed in flames.

At 1 p.m. that day, I was hosing down my roof in case of passing embers, all the while making sure rain gutters were cleared of debris as the fire rapidly advanced. Shortly after, a neighbor's police band radio broadcasted evacuation orders (there was no central warning system!), prompting a rush to stuff musical instruments, computers, family photos, espresso machine, and some spare clothing into the back of my car. Streets were clogged with fleeing traffic having only one main artery along Highway 99 northbound. A closed I-5 rerouted traffic into this mass exodus, creating even more gridlock. Blackened by toxic fumes (structures burning), the sky turned afternoon daylight into an eerie Alaska-like summer evening twilight.

Having lived in my town for some 2 decades, and knowing it like the back of my hand, I quickly spied backcountry roads to plot an escape route away from the chaotic highway exodus, while I shockingly observed the fire heading in the same direction as evacuees toward the larger city of Medford, Oregon, where some 90,000 residents ($\sim$11 km to the north) were also on evacuation alert. Going northbound and away from the fire's origin seemed sensible at the time, with the ignition point in Ashland to the south, but it soon became unsafe with strong winds directing the fire in the path of the main evacuation corridor.

By early evening, the Almeda fire breached nearly all containment lines in downtown Talent despite persistent fire-retardant drops made by jumbo-jet airtankers and scores of brave fire-fighters working around-the-clock. First responders could not save hundreds of structures, and community water systems broke down, leaving fire crews with no water.

As the fire began taking out homes (including those of friends and businesses frequented for over 20 years), another one was soon started by a second arsonist just 3.2 km (2 miles) to the north in the equally small town of Phoenix, Oregon. Both fires (collectively named Almeda even though they never merged) left a destructive wake of apocalyptic looking, skeletonized structures (some completely erased from their building footprint) and bombed out cars and vans. Such devastating effects occurred along a 48 km (30 miles) stretch (Figure 10.1a−b), consuming entire neighborhoods ($\sim$2800 structures). Similar scenes like this were playing out in Gates, Detroit, and other portions of Oregon and have now unfortunately become a consequence of the rising costs of climate chaos and unprepared communities throughout the Pacific Northwest and greater western United States in general.

Forest Managers Play the Backcountry Logging Fiddle Chapter | 10 337

FIGURE 10.1

FIGURE 10.1 cont'd

Forest Managers Play the Backcountry Logging Fiddle Chapter | 10 339

FIGURE 10.1 Two views (a, b) of mobile homes destroyed by the Almeda fire; (c) a fire-prepared church untouched by flame; (d) unusual creek-side surviving house with dense vegetation; (e) interesting surviving treehouse (notice the burned-out structures right next to it); and (f) postfire reconstruction with flammable wood (no home hardening). *All photos, D. DellaSala.*

Later that evening, exploding propane tanks (used for heating mobile homes) lit up the night sky like Independence Day fireworks, casting embers sideways in gusting winds and causing adjacent homes to ignite like lined-up dominos. A nearby church survived unscathed as it was smartly constructed with metal roofing and fire-resistant siding, screened vents, and scant surrounding flammable vegetation (Figure 10.1c). A few hundred meters away, an auto-repair shop constructed of cinder block and metal roofing (which should have been fire resistant) burst into flames; apparently embers somehow came in contact with explosive chemicals inside, possibly entering a poorly screened vent. A creek-side home and nearby child's treehouse (Figure 10.1d–e) miraculously survived despite vegetation touching structures. Presumably either the winds shifted, sparing them, or relatively moist streamside vegetation created a "wet-blanket" to steer flames away. The treehouse survival is ironic as everything around it burned to the ground yet a child's playhouse nestled neatly within branches survived the hottest flames.

Devastating scenes like this in a growing number of communities are the new climate-fire "abnormality" as the parched and overheated West burns millions of homes built in unsafe places lacking land-use zoning to constrain with sprawl and effective home-hardening (see https://headwaterseconomics.org/natural-hazards/structures-destroyed-by-wildfire/accessed March 1, 2023) to protect existing structures. Some towns in northern California have even burned down more than once just decades apart as developers simply rebuilt over the same burn footprint with the same flammable structures lacking home hardening and defensible space (https://en.wikipedia.org/wiki/List_of_California_wildfires#Areas_of_repeated_ignition; accessed March 1, 2023). Just like the rebuild in Talent, this is clearly the definition of crazy!

The vast majority of wildland fires in the United States are started by people (Abatzoglou and Williams, 2016), while many are from fallen powerlines (energy infrastructure), arson, accidents, and by prescribed ignitions that sometimes go astray (lightning is also a major factor). But despite the escalating damages, nearly all the focus on fire "prevention" is misdirected at backcountry logging on public lands primarily via "thinning" that does nothing to help homeowners, communities, or businesses prepare. Most disastrous urban fires in Oregon, for instance, are actually fires spilling over from heavily logged private lands, and not public lands (Downing et al., 2022), made flammable by expansive clearcut logging that has transformed fire-resilient native forests into fire-prone plantations and slash piles (Zald and Dunn, 2018). Yet nearly all the emphasis of federal spending is on promoting at all costs "active management" on public lands, even though the vast majority of homes in the interface are actually on private lands (Schoenaggel et al., 2017; Baker, 2022).

Whether a home burns or not is all about the structure itself and if homeowners pruned vegetation within a 30-m ($\sim$100 ft) circular zone nearest the structure (Cohen, 2000; Gibbons et al., 2012; Syphard et al., 2014).

Unfortunately, an overwhelming amount of public money is spent on backcountry logging and prescribed fire that is failing to limit large fires from spilling over into urban areas. This is largely because under high to extreme fire weather, large fires can easily hop across 8-lane highways (as in the Almeda case), and, in at least one case, breached the kilometers-wide Columbia River Gorge east of Portland during the wind-driven 2017 Eagle Creek fire.

Wind-driven, intense fires are becoming more common, as the planet overheats and generates more active and explosive fire weather. With more fires either starting at or spreading into residential areas, a new approach is needed to address the root causes of the crisis at hand and not just the effects.

The most effective course of action to prepare communities for fire includes the following:

1. Solve for root causes by breaking the destructive climate-fire feedback loop via drastically cutting emissions across all sectors, including forestry.
2. Contain ex-urban sprawl into fire-prone areas through proper land zoning and insurance rate adjustments (higher insurance rates for unprepared building in unsafe areas vs. discounts for home hardening and defensible space).
3. Redirect fire spending toward urban planning by preparing existing structures from the home-out (home hardening, defensible space), while also planning evacuation corridors that have sufficient traffic redirection, real-time warning systems, community smoke and evacuation shelters, and other community fire prevention measures.
4. Seasonally close and obliterate roads as federal lands alone have enough roads to circumnavigate the global nearly 16 times with many of roads failing and not needed but that provide a conduit for unwanted ignitions (e.g., cigarette buts, sparks, illegal campfires).
5. Implement new powerline safety technologies to reduce accidental sparks (see https://www.nfpa.org/News-and-Research/Publications-and-media/Blogs-Landing-Page/Fire-Break/Blog-Posts/2019/12/06/new-technology-to-diagnose-power-grid-failure-holds-promise-for-wildfire-prevention; accessed July 3, 2023).
6. Surgically thin small trees in young plantations and treat flammable logging slash otherwise primed to burn hottest and fastest (Bradley et al., 2016), especially on private lands where most fires spill over into urban areas (Downing et al., 2022).
7. Protect mature and old-growth forests and large trees from logging as natural climate solutions and fire sanctuaries, recognizing that denser mature and old forests in places burn at lower severities in wildfires (e.g., Lesmeister et al., 2021).

This comprehensive strategy would begin the lengthy process of decoupling extreme fire weather from large fires and is complimentary to emissions reductions across all sectors by incorporating natural climate solutions and community resilience planning. It would solve for root causes rather than escalating the effects of logging and fires. And it is important to note that, even if we stopped all emissions today, given the centuries lag time (inertia) built into the biosphere, oceans, and atmosphere feedback systems, there is no quick-fix or magic bullet to reverse the trajectory we are on. Anyone proposing strategies that claim all we need to do is to massively thin and burn forests in the next decade or so is literally blowing smoke!

10.2 THE FIDDLE PLAYERS

> To do something trivial and irresponsible in the midst of an emergency; legend has it that while a fire destroyed the city of Rome, allegedly emperor Nero played his violin, revealing a total disregard for his people and empire (https://www.classicfm.com/music-news/nero-fiddled-rome-burned-trump-violin-meme/#:~:text=%E2%80%9CNero%20fiddles%20while%20Rome%20burns%E2%80%9D%20has%20become%20a%20phrase%20used,his%20people%20while%20they%20suffered; accessed March 7, 2023).

All Fire Management is Politics

Politicians are under intense pressure from constituents and donors that got them elected, who expect something in return (quid-pro-quo). In the shark-infested waters of wildfire politics, boatloads of money are thrown at fire suppression and logging ("active management") expecting a return on the investment via greatly scaled up and damaging treatments regardless of whether they work or not. Misinformation is routinely used to silence legitimate scientific concerns about limitations of expansive forest management and cumulative collateral damages to ecosystems from them. The federal land management agencies and timber industry have duped the public into accepting a sort of fiddling (logging) in the backcountry even while towns kilometers away from treated areas burn down. Message framing is the means for winning debates, and well-funded misinformation, and even disinformation campaigns, are often aimed at the public, politicians, and at scientists willing to speak out when the data and the issue warrant a major course correction (DellaSala, 2021).

Overly simplistic and false message framing used to back aggressive thinning and massive fire suppression goes something like this (think about this in Orwellian terms!):

- The only "good fire" is a low intensity fire; we need more good fire and less "bad fire."
- Thinning (euphemism for logging) and prescribed fire (good fire) prepares the area for more good fire.

- "Active management" with minimal environmental safeguards is needed in most places even in protected areas because they will burn up otherwise and the review process is overly cumbersome ("analysis paralysis;" see Hessburg et al., 2021 vs. DellaSala et al., 2022).
- Thinning is much better than letting trees die from fire and insects.
- Thinning prevents fire-related carbon losses by storing carbon in wood products.
- Thinning promotes "forest health and resilience" needed in a changing climate.
- Thinning takes only the small trees and shrubs to protect the large trees.

Such messaging not only often wins the public debate, but convinces politicians to pump large sums of money into largely ineffective approaches because they appear simple, sound logical at face value, and present what seems to be a quick-fix solution when in fact complexity is the rule in large fires governed mainly by top-down extreme fire weather interacting with bottom up industrial logging practices (e.g., see Bradley et al., 2016; Zald and Dunn, 2018; DellaSala et al., 2022). The deception of fire doublespeak messaging is making the situation far worse because ecosystem damages are trivialized in favor of quick fixes that do not work.

Some elected officials have told me privately that when it comes to legislating for fire prevention, simple messaging and easy to understand "optics" is the *raison d'etre* so long as it promotes a quick path to fire and smoke-free summers. They are generally willing to back research that claims fire-fixes are within grasp by scaling up more of the same failing measures (e.g., Hessburg et al., 2021). This politically motivated response creates a command-and-control attitude about nature that will cause cumulative ecosystem and climate damages (DellaSala et al., 2022). Small tree thinning and prescribed burning, while in some cases can slow fire spread in low-moderate fire weather if done right (see below), will not stop weather/climate-driven wildfires and thinning can actually increase fire severity by radically changing the forest microclimate if too much biomass is removed as is often the case (DellaSala et al., 2022).

I have also heard US Forest Service managers tell me flat out that they willingly throw everything they can at a fire suppression effort via massive fire crews and high-tech equipment even as they know they cannot possibly stop a fire running in extreme fire weather. They do this as a show of force to the public and to those that pay their salary, approve their budgets, evaluate their performance, and often pick up the tab for excessive costs. In other words, fire has become a blank check from Congress that must be fought at all costs even when it cannot be. Consequently, billions of dollars are spent addressing the effects (larger, more frequent burns) rather than the causes (e.g., climate change, ex-urban sprawl, lack of home hardening, logging, unmitigated suppression, human-caused ignitions). Meanwhile the excessive use of doublespeak messaging such as "fire remediation," "restoration," "resilience" is tipping the delicate balance from mixed severity fire with ecosystem benefits (pyrodiversity begets biodiversity) toward heavily logged and homogenized landscapes that blow up in climate-driven large fires. Anyone who stands in the way of the fire money train is attacked, and accused of spreading "misinformation" (DellaSala, 2021; DellaSala et al., 2022).

After my home town burned down, out of frustration and anger, I sent some of the same photos in Figure 10.1 to local politicians who I had briefed a year earlier on why communities need more assistance in fire prevention (home hardening, defensible space) instead of the millions of dollars wasted in the backcountry. Having their "optics" response and politics of fire on my mind, I mentioned that if this is really all about optics, then can you please explain to me what just happened to my community of 20 years?!

The homes and businesses of neighbors in Talent and Phoenix burned down while large sums of money were allocated to backcountry logging in the city of Ashland's municipal watershed, much of which is far removed from the nearest homes, including within remote roadless areas where helicopter-logging is offered as "fuels reduction." This logging has yet to be tested with backcountry wildfire, while large portions of the unprepared communities burned to the ground in the Almeda fire. Although Ashland escaped the devastating effects of the Almeda fire, this outcome had nothing to do with thinning in the backcountry but instead was because the winds sent flames racing toward Talent and safely away from Ashland. Had the winds been blowing the opposite direction, the consequences would have been devastating to Ashland instead, given the lack of home hardening there as well.

Despite significant investments in backcountry logging, today the communities of Oregon and the West writ-large are no safer as the home ignition problem is largely unaddressed, especially in the rebuild of fire-destroyed towns (see Figure 10.1f). Meanwhile, the city of Ashland, the Forest Service, the Bureau of Land Management (BLM), and The Nature Conservancy (TNC) are heavily invested in thinning, including removing large live and dead fire-resistant trees, away from residential areas. Although I have witnessed first-hand the devastating consequences of the current strategy, there has been no constructive dialog about the limited efficacy of this strategy, or a re-evaluation of planning priorities in light of the consequences. Instead, proponents have chosen to defend at all costs a losing and damaging proposition.

TNC and Collaboratives as Backcountry Fiddlers

TNC is the nation's largest international conservation organization with an annual operating budget of >$1 billion (more than the annual GDP of 26 countries) and net assets >$6.6 billion. While the organization is responsible for many important conservation achievements, such as some 50 million hectares of protected areas globally, and there are certainly conservation-dedicated people in their workforce, TNC's Fire Learning Network (FLN) operates in the shadows with most of the rest of the nongovernmental organization (NGO) community excluded. TNC's aggressive thinning projects stem from questionable fire science and carefully controlled stakeholder collaboratives that promote logging as fuels reduction with any local opposition dismissed, not invited to participate, not allowed to vote on decisions, or worse, reputations attacked.

TNC almost never publicly advocates for new protected forests in the United States, and most often teams with state and federal agencies, timber companies, and other extractive interests via its FLN and "stakeholder collaboratives" (https://firenetworks.org/fln/; accessed April 1, 2024). It has a special penchant for corporate board members with deep pockets (https://africanarguments.org/2023/08/revealed-big-conservation-ngos-majority-governed-finance-africa-carbon-markets/; accessed April 1, 2024). TNC collaboratives resemble membership only clubs that support TNC's core principals and scientific perspectives which, in turn, provide cover for federal agencies to claim community support around otherwise controversial logging projects. TNC's collaboratives—and collaboratives within which TNC is a major science contributor—are considered "community of practice" groups as noted by TNC itself (Bassett, 2018). Membership in the group or its governing board appears to be restricted to accepting "fuels reduction" via "active management" and TNC's narrow view of "good fire" (e.g., low intensity burns) (https://www.fs.usda.gov/rm/pubs_other/rmrs_2005_kaufmann001.pdf; accessed April 1, 2024). These "collaborative" organizations are often infused with timber industry representatives, federal land contractors, and organizations with direct financial interests in federal land logging projects.

Additionally, federal land managers often require their preferred collaborative partners, such as TNC, to remain silent when damaging logging projects are proposed or implemented. Those who oppose such activities are routinely pushed out of the collaborative, or will not get future contracts or Master Stewardship Agreements that allow collaborative organizations to operate on federal lands administered by federal agencies.

Such a management structure dramatically narrows the perspective of participating organizations and allows federal land managers and high value collaborative partners with grant funding capacity to dominate the land management decision space and the implementation of large-scale projects that are often harmful to the environment and local conservation priorities.

Here are five examples, four of which were originally presented in DellaSala et al. (2020), where TNC and their collaboratives rolled over local conservation groups by pushing aggressive "fuels reduction" and "restoration" logging projects, making claims that such projects would slow and reduce fire intensity and even eventually get rid of the smoke from large wildfires.

Rogue Basin, Southwest Oregon—TNC's role in this basin passes mainly through the Southern Oregon Forest Restoration Collaborative (SOFRC). The collaborative does not include a single conservation group on its governing board, and instead is populated by retired Forest Service, BLM, Oregon Department of Forestry officials, forest contractors with direct financial interests in logging and land management projects, as well as, for many years, the Vice President of a major regional timber industry lobby, the Southern Oregon Timber Industry Association. They are also members of the Rogue Forest Partners, who actively lobby for federal lands logging that their members then implement.

Despite the repeated opposition of seven local conservation groups, SOFRC prioritized a large-scale commercial logging project with no diameter limit and heavy canopy removals, under the rubric of "thinning," on 440,000 ha (1.1M acres) on *"all lands"* within the basin, including areas where most forms of logging are otherwise prohibited. This includes prioritizing logging within Late-Successional Reserves (LSRs) that are supposed to be managed for closed-canopy old forest species and critical habitat of the federally threatened northern spotted owl (*Strix occidentalis caurina*) that also requires closed canopies (Figure 10.2), Research Natural Areas, the Cascade-Siskiyou National Monument, Inventoried Roadless Areas, and other priority conservation areas identified by local conservation groups (DellaSala et al., 2020).

FIGURE 10.2 Thinning/logging project, promoted by TNC and the Southern Oregon Forest Restoration Collaborative within critical habitat of the federally threatened northern spotted owl on the Rogue-Siskiyou National Forest, Ashland Oregon. Pictured in this photo is a large gap in critical habitat created by logging that will reduce canopy closure below the 60% canopy threshold considered vital to owl site occupancy and nesting. This type of logging to create forest openings across significant portions of owl territories and critical habitat is considered a form of habitat degradation, not restoration. *Photo D. DellaSala.*

Based on federal documents obtained under the Freedom of Information Act, TNC supported aggressive forest canopy reductions that would degrade critical owl habitat under the assumption that it is a "short-term" impact despite independent research (Odion et al., 2014) and a government report (Raphael et al., 2013) showing the opposite. Spotted owls require dense canopy closure of at least 60% for nesting and roosting, yet such logging would lower the overstory canopy below those levels and, if implemented over large areas, would impact far more older forests than projected forest fires (Odion et al., 2014).

The collaborative's thinning/logging proposals are based on highly disputed claims that there is an overabundance of closed canopy, dense forest habitat due to fire suppression, which is the case in at least some areas, but not as much as TNC claims. According to TNC documents closed canopy stands must be "thinned" to open them up and "let the sunlight in!" (this same sunlight messaging has been used by land managers to promote clearcut logging projects!). Importantly, some Forest Service scientists have reported that reducing forest density and canopy can actually increase wildfire severity, based on a large-scale analysis by Lesmeister et al. (2021).

It should be noted that in this region, closed canopy mature forests are known to burn in lower fire severities compared to clearcuts and tree plantations possibly because of the presence of Pacific madrone (*Arbutus menziesii*) that may act as a sort of "wet blanket" effect on fires due to high moisture content of madrones (Odion et al., 2004).

TNC's highly refutable claims about too much closed canopy forests are made without cross validation of their limited fire-scar sampling with historical photo documentation and historical records that were made available to them by conservation groups to help TNC improve its estimates and approaches (DellaSala et al., 2020). Most disturbingly, TNC prioritizes logging of older forests like LSRs and owl critical habitat instead of previously logged and degraded forests where fire risks are much higher. Large tree logging and excessive canopy reductions in their treatment prescriptions would damage the last remaining older forests in the region, dry out forest stands, regenerate dense, highly flammable young growth, and reduce microclimate conditions that serve as fire and climate refugia. Interestingly, information made available through the Freedom of Information Act demonstrates that a TNC scientist in southern Oregon was actively editing official government documents for the Medford District BLM, providing citations to promote logging in old forests 180–240 years old as a "temporary" loss and attempting to spin it as wildfire management. The project was proposed with an alternative developed to implement TNC's "Rogue Basin Strategy" and was ultimately withdrawn over strong public opposition and unacceptable impacts to the Oregon red tree vole (*Arborimus longicaudus*), a major food source for the spotted owl, and another closed-canopy species.

At public meetings in the Applegate community of southwest Oregon, scientists and community activists requested that TNC withdraw their aggressive logging proposals, and presented photo documentation along with historical evidence that many of forests that TNC prioritized for open-canopy treatments were actually historically dense and closed forests and that excessive canopy removals were unwarranted. Instead of embracing this chance to calibrate models and build trust, TNC decided to go elsewhere where opposition was less organized and logging could proceed apace. By replacing real community involvement with narrowly tailored collaboratives, federal land managers are utilizing this approach throughout the West, like an unvetted management plan.

Sierra Nevada, California—Immediately following the largest fire on record at that time, the Rim fire of 2013, collaboratives consisting of those with a postfire logging agenda advocated for converting complex early seral forests (snag forests) on the Stanislaus National Forest into "feed stock" for woody biomass power plants. The local collaborative, which included TNC, advocated for clearcutting postfire wildlife habitats, bulldozing them into huge burn piles for chipping and hauling to carbon-polluting biomass energy facilities or incinerating the ecologically vital snags on site with accelerants, in order to plant "new forests" (tree plantations) on short-rotation logging cycles. Many participants had ties to the timber industry and extraction-oriented state and federal agencies, including those that champion well-funded, destructive postfire logging and biomass burning that result in high carbon pollution (e.g., the Sierra Nevada Conservancy's Stanislaus National Forest biomass proposals in the Rim fire; https://sierranevada.ca.gov/snc-launches-biomass-utilization-fund/; accessed March 7, 2023). Notably, at a 2014 Stanislaus National Forest field trip of the Rim fire area led by the Forest Service that we attended, a TNC representative who was part of the collaborative advocated for sending more logs from "salvage" operations to needy local mills as "restoration," sounding more like a logging company than a conservation organization. This included logs coming from exceptionally rare wetland inclusions in complex early seral forests (Figure 10.3).

The project advocate for postfire logging here was mainly the Sierra Nevada Conservancy, which received government "housing funds" that were supposed to support displaced residents but instead were used to clearcut the forest for biomass burning. TNC supported postfire logs going to mills despite legal challenges by NGOs and to my knowledge never publicly opposed biomass logging.

Greater Santa Fe, New Mexico—TNC's work in the Santa Fe region is premised on "landscape resilience" as the claimed goal for logging and burning the forests of the Sangre de Cristo Mountains via its involvement with the Greater Santa Fe Fireshed Coalition (GSFFC) and Santa Fe National Forest. As in other case studies, TNC is right on message promoting "good fire" (see PREFACT, 2018), following its "community of practice"

FIGURE 10.3 (a) Unlogged rare wetland inclusion within a complex early seral forest showing abundant biological legacies and prolific natural conifer establishment 5 years post-Rim fire (study plots showed abundant conifer establishment and not just this photo; taken in 2018). The area was important habitat for fire-dependent black-backed woodpeckers (*Picoides arcticus*), imperiled California spotted owls (*Strix occidentalis occidentalis*), wetland bird species, and numerous native plants and other wildlife. (b) The site was logged in 2019 because the Forest Service claimed it would be slow to regenerate without logging and planting and this type of "active management" was needed for "resilience" and "restoration" purposes despite evidence to the contrary. *Photos Chad Hanson, with permission.*

membership club with voting members consisting of forest industry groups, practitioners, watershed associations, and prothinning government agencies. As in other examples, local conservation groups are not part of the inner circle of voting partners, although some can participate as nonvoting "advisors."

The collaborative has been backing large-scale forest clearing projects of ~20,000 ha on the Santa Fe National Forest. Notably, the work of the Southwest Jemez Mountains Landscape Collaborative (which does include three conservation groups) is illustrated on a Forest Service roadside kiosk showcasing a project to remove a "mistletoe infestation" (mistletoe is a native species) that resulted in what looks very much like a clearcut with some whip-trees in the background and a weed infested savanna in the foreground (Figure 10.4). We have to ask—how can anyone with a straight face claim that this is "restoration?"

FIGURE 10.4 "Dwarf mistletoe control" project supported by the Southwest Jemez Mountains Landscape Collaborative. Note the burn-related soil damages and invasive weeds in the foreground (e.g., the *yellow* flowering invasive weed is Mullein *Verbascum thapsus*). *Photo D. DellaSala was taken right next to the Forest Service kiosk proclaiming "restoration" of a mistletoe infected stand. The doublespeak of "restoration" messaging lives on!*

Another example is the Forest Service's "Santa Fe Mountains Landscape Resiliency Project," supported by the GSFFC, of which TNC is a member.

Here, the Forest Service proposes massive thinning and prescribed fire mainly in closed canopy forests claimed to be well-above historic tree densities, despite some of the same fire-scar sampling problems noted in the Rogue Basin case study (i.e., extrapolating sampling beyond the study area). The project covers the Espanola and Pecos/Las Vegas Ranger Districts and includes landscape-scale thinning (8400 ha) and pile burning (17,200 ha). Note the widely scattered, scant trees (susceptible to blow down) and lack of an understory in heavily thinned and burned areas on the Santa Fe National Forest where collaboratives operate freely, despite significant opposition, and essentially as cover for the Forest Service (Figure 10.5). The GSFFC's mission and methods, and the aggressiveness of the Santa Fe Mountains Project, are opposed by many members of the public and some conservation groups, mostly community-based ones. Going forward with the project without a comprehensive environmental impact statement is opposed by some 98% of the project scoping comments received by the Forest Service, yet the agency stubbornly refuses to change direction, given it has the backing of a collaborative (pers. comm. S. Hyden; https://www.theforestadvocate.org/fm/sf/co/; accessed March 23, 2023).

FIGURE 10.5 Heavily thinned and burned stand resulting in type-conversion to savannah on the Santa Fe municipal watershed in a project supported by the Greater Santa Fe Fireshed Coalition. Note the wide spacing between tree clumps and the lack of a forest understory. Thinning/burning will likely dry out the understory, expose soils to flammable weeds, greater wind speeds and tree tipping, all of which can spread fire rapidly even if it does not reach the canopy (note: rapid fire spread is what kills fire fighters in uncontrollable blazes). *Photo D. DellaSala.*

Eastern Oregon Dry Pine and Mixed Conifer Forests—The Ochoco Collaborative (covering 340,000 ha), led by TNC, supported the Forest Service efforts to increase large-tree logging in so-called "fuels reduction" projects, and the Deschutes Collaborative (~103,000 ha), of which TNC is a member, pressed the agency to remove the "Eastside Screens" that were then functioning administratively as a prohibition on logging large, old trees above a certain diameter (>50 cm, >21 in diameter-at-breast height). Large tree logging up to 150 years old (>76 cm dbh) for large firs was then permitted in forests that otherwise still have an historic old growth deficit (Mildrexler et al., 2020; DellaSala and Baker, 2020; although at the date of this writing that decision was temporarily rescinded in federal court on scientific and other shortcomings). Meanwhile, the collaborative ignores surrounding private lands populated by logged over, flammable forests. In other words, older forests targeted for logging contained some of the last large tree remnants (economically valuable) and high-quality wildlife habitat (DellaSala and Baker, 2020). The implications of the collaboratives' position are profound, given the cover provided to the agency to act against the public wishes for an environmental impact statement instead of a less informed environmental assessment (lower standard), the need for full disclosure of likely impacts, and science presented in court hearings that refuted agency logging proposals.

Bootleg Fire, Southern Oregon—The Bootleg Fire near Beatty, Oregon affected nearly 160,000 ha during the hot July of 2021. Contrary to a single cherry-picked photo often used by TNC and shockingly even by some prominent scientists to promote thinning, claiming it caused the fire to lay down (low intensity), GIS mapping and Google Earth imagery actually revealed that not only did thousands of hectares of recent thinning, fuel breaks, and other forest management fail to stop or slow the fire's spread rate, but the fire moved fastest through those thinned areas. The nonprofit Los Padres ForestWatch (https://www.opb.org/article/2021/07/25/did-forest-management-play-role-in-bootleg-fire-growth/; accessed August 30, 2023) meticulously pored over state and federal logging and livestock grazing records as well as historical satellite and Google images to compile the most extensive forest management history dataset ever released on this fire that says a lot more than a single promo photo. Key findings included:

- The Bootleg fire was driven eastward by strong winds, and hot, dry conditions across nearly 60 km of ponderosa pine (*Pinus ponderosa*) and lodgepole pine (*Pinus contorta*) forest, grassland, and shrub steppe. About 64% of the burned area is managed by the Forest Service, and the rest is private and Tribal land.
- During the first 6 days, it burned nearly 10,000 ha of national forest subject to thinning, prescribed fire, fuel breaks, and other vegetation removals promoted as "fuel reduction" as part of the Black Hills and East Hills projects approved in 2012 and 2018, respectively.

- Thousands of additional hectares with "thinning" and other forest management from various logging projects burned during the first several days. The fire quickly grew to 80,000 ha and officials reported extreme fire weather and fire behavior so intense that firefighters were forced to disengage and retreat.
- Across this same period of significant fire growth, the Bootleg fire burned thousands of hectares of intensively managed private forest.
- On July 19, the fire made its largest single-day northerly run, spreading 8 km in about 13 h into the Sycan Marsh Preserve—across which TNC had been conducting commercial "thinning" operations and using prescribed fire for many years.
- According to Forest Service data, 75% of the national forest that burned was part of an active grazing allotment. Nearly 16,000 ha of the burn was grassland or a mix of grass and shrubs known as shrub steppe, much of which has been grazed as well.
- On average the wildfire advanced 5.4 km per day through public and private forest lands that were managed in the past 2 decades compared to an average of 3.4 km per day through the Wilderness and Roadless Area to the east according to an analysis of the fire progression data (Figure 10.6).

The Bootleg fire raced through much of the landscape that was logged in one way or another since the 1970s, including over the last few years. Neither news stories, the use of a single photo, nor TNC reported the full management history and its influence on fire behavior. The Bootleg fire serves as an example of how large fires are most often driven by wind and

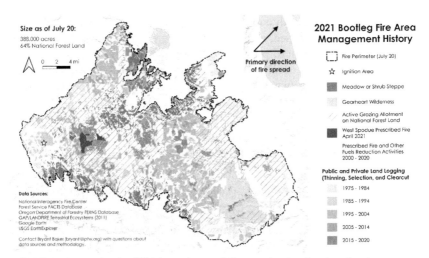

FIGURE 10.6 Bootleg fire 2021 in south-central Oregon, showing decades of active management and "fuels reduction" projects. *Courtesy of Los Padres ForestWatch.*

dry conditions exacerbated by climate change and misinformation. Consequently, these fires are inevitable across a variety of forest and nonforest ecosystems (Chapter 8) in the western United States, regardless of efforts to reduce vegetation via thinning and burning. And while detailed analysis and mapping are the tools of inquisitive truth-seeking scientists, anecdotal claims and single photos lacking supporting data should never be held up as proof of concept.

Collectively, the above five examples illustrate a growing rift between TNC's affiliations in collaboratives (running them, providing science to them, working against local NGOs) and the broader scientific and conservation community concerned about aggressive "fuels reduction" in TNC's FLN that is largely based on questionable fire science and a narrow view of "good" fire. As such, many conservation groups are now treating TNC's FLN as a leading threat to public lands conservation and legitimate ecological management efforts. TNC's image problem may stem from its cozy relationship with industry (see https://www.dogwoodalliance.org/2022/04/release-the-Nature-conservancy-exposed-for-promoting-industrial-logging-and-wood-products/; accessed March 7, 2023), its programmatic inflicted damages to local NGOs, its tone deafness of the impacts to faith-based and social justice groups downwind of biomass facilities (https://subscriber.politicopro.com/article/eenews/2023/08/23/logging-critics-go-after-the-Nature-conservancy-00112232; accessed August 30, 2023), and its board reliance on corporate board members with deep pockets (https://africanarguments.org/2023/08/revealed-big-conservation-ngos-majority-governed-finance-africa-carbon-markets/; accessed August 30, 2023). While they do not want to admit it, TNC has a major public relations problem tarnishing its brand as a "conservation organization." Even though TNC claims that is actively met with conservation groups over concerns raised by 158 conservation, faith-based, and social justice groups, TNC has stone-walled and ended any chance of productive dialog for over 2 years, with an terse response to move on despite the big differences remaining over approaches and concerns (https://www.dogwoodalliance.org/2022/04/release-the-Nature-conservancy-exposed-for-promoting-industrial-logging-and-wood-products/; accessed August 31, 2023).

10.3 COUNTERING FIRE HYPERBOLE AND DOUBLESPEAK

For TNC, the Forest Service, and collaboratives to improve credibility and build trust with the public and independent scientists trained in ecosystem and biodiversity assessments, they would need to stop promoting projects with questionable outcomes, exclude high conservation value areas identified by local NGOs, and severe ties with the biomass industry impacting social justice groups. TNC specifically would also need much greater transparency of its role in collaboratives and full disclosure of financial records linked to its FLN projects to avoid any conflict of interest (e.g., funds from federal agencies, state forestry, and timber companies with a financial stake in the outcome should be recognized in all TNC science documents as a potential conflict of interest). TNC and others could include local NGOs as equal status collaborative voting members, reach out to a broader base of scientists (and not just fuels scientists that support their position), and bolster fire-scar sampling with multiple lines of evidence (cross validation), including from historical photos and accounts, General Land Office surveys from the 1800s, and other reliable historical evidence (Odion et al., 2014; DellaSala et al., 2020). TNC and other land managers could also purge the overly simplistic good versus bad fire messaging by concentrating more on the ecological benefits of mixed-severity fires, including high-severity burn patches (large and small).

Finally, TNC's FLN needs to conduct a credible carbon life cycle analysis of its carbon footprint from logging projects advanced by collaboratives and their unwavering support for thinning, biomass burning, postfire logging, and highly suspect carbon offset projects that produce far more emissions than stated (https://www.bloomberg.com/news/features/2021-04-05/a-top-u-s-seller-of-carbon-offsets-starts-investigating-its-own-projects#xj4y7vzkg; accessed March 7, 2023). But will they do this or continue to promote the unabated "fuels reduction" narrative?

10.4 FIRE, FIRE, HOMES ON FIRE, AGAIN!

The same story about home losses and misdirected backcountry logging is rampant throughout fire-prone regions where towns have sprawled into harm's way due to lack of urban-growth boundaries and home hardening (https://headwaterseconomics.org/natural-hazards/structures-destroyed-by-wildfire/; accessed March 24, 2024). Fire fighters are routinely sent into dangerous places to stop fires, such as during extreme fire weather and in steep terrain, that they cannot possibly put out, while logging/thinning is held up with quasi-religious zeal as a way to convince donors, decision makers, land managers, NGOs, and the public to do even more of the same (DellaSala et al., 2022). And there is even a strong vocal push by some decision makers to bypass wilderness and other environmental protection measures to allow more bulldozers and heavy-handed treatments during fires that cannot possibly be put out or that are burning safely in the backcountry away from towns.

Well before the advent of militarized fire suppression, fires often spilled over into, or were exclusively within, urban areas, causing massive loss of life and property (Wikipedia lists many historical accounts—https://en.wikipedia.org/wiki/List_of_town_and_city_fires; accessed March 24, 2024). Some of the more infamous burns (human caused or otherwise) include the library and city of Alexandria (48 BCE), destruction of the Temple and City of Jerusalem (587 BCE), Great Fire of Rome (64—was Nero really playing the fiddle?), London (798, 1135, 1212, 1666), Moscow (1547, 1571, 1812), Boston (1760, 1872),

New Orleans (1788, 1794), New York (1835), St. Louis (1849), Portland, Maine (1866), Great Chicago Fire (1871, 1874), Wisconsin (Peshtigo fire, 1871), San Francisco (1906) and many, many more. Even more recently and with massive fire suppression forces, cities like Santa Rosa were destroyed twice in repeat fires of 1960 (Hanly Fire) and 2017 (Tubbs Fire). Notably, the Tubbs Fire took a much bigger toll of life and property despite nearly identical fire footprints mainly because more homes were built in harm's way in the time between fires. So naturally, there is a long history of fire impacting where people live, work, and play. Despite gains on fire suppression made when the Forest Service went to militarized firefighting operations post WW-II, and the advent of Smoky Bear messaging, towns continue to burn down, and the losses are mounting because of climate change, lack of land use planning, and lack of home hardening.

Notably in the mid-20th century, ex-urban sprawl was pushing deeper into the wildlands when the climate was much cooler (less area was burning). This insatiable drive to live in nature created a false sense of security as the Forest Service's mechanized fire-attack forces got really good at putting out nearly all fires (the 10 a.m. fire policy was created whereby the goal was to put out all fires by 10 a.m. the next day). However, hotter, drier conditions that started around the 1980s from a ramp up in climate change then set in motion the perfect climate—fire—urban storm. This includes expansion of towns like Ashland, Phoenix, Talent, and millions of homes built right in fire's path while effectiveness in putting out fires in extreme fire weather went way down. With this kind of clear and present danger it is troubling how so many politicians and land managers refuse to see the climate writing on the wall and instead think that they can somehow put the fire genie back in the bottle via command-and-control warfare with nature and massive fuel reduction spending.

10.5 WHAT IS "ACTIVE MANAGEMENT" AND WILL IT WORK?

Active management in this case encompasses a wide range of actions before, during, and after fire that often starts with thinning/logging and burning of sorts with the intent of reducing "hazardous fuels," includes massive fire suppression (e.g., bulldozers, airtankers, fire retardants, fire crews, back burns) when fires do start, and often ends with post-fire ("salvage") logging (Della-Sala et al., 2022). This combination of impacts accumulates over vast expanses (including suppression in wilderness areas and other protected areas) and is arguably the biggest current public lands threat and funding scheme in US public lands history, backed by government-funded research claims that large-scale active management can tame big fires (Hessburg et al., 2021). Intensive logging used in active management often is misdirected at large fire-resistant trees (contrary to claims otherwise), involves extensive road access and sometimes road building, along with pile burning that is damaging to soil horizons and mycorrhizae essential in forest growth and development (DellaSala et al., 2022).

Thinning (a form of active management) involves the killing of some number and type of trees for various claimed purposes, but it is most commonly deployed to "reduce fuels" in dry forests and has support in the scientific community, particularly among scientists funded by agencies, and some NGOs wedded to the premise that by treating the effects (preventing "bad fires") via logging and prescribed fire, one can largely downplay or claim that this is solving for the causes (climate change, industrial logging, ex-urban sprawl, poor urban planning, etc.).

When done properly, thinning of small trees followed by prescribed burning, or prescribed burning alone, can, in fact, reduce fire intensity under certain conditions (Zachmann et al., 2018). However, it remains controversial when mixed with large tree removals, and has significant downsides and limitations of scale (DellaSala et al., 2022) and effectiveness (Cary et al., 2017), especially when extreme fire weather is the main driver. All things being equal (winds, slope, aspect, fire weather), thinning small trees and prescribed burning can reduce fire intensity when fires are not raging and if everything is done right. But all things are seldom equal or done right, and the odds of co-occurrence of a fire hitting a to treated spot on the ground are slim (Schoenaggel et al., 2017). What has to be in place for it to work despite the slim odds is sort of a collection of "what ifs:"

- Fire hits the site during low-moderate fire weather—i.e., relatively calm winds, temperatures are moderate, and relatively humidity levels are high (no drought or heat dome).
- Thinning is from below the forest canopy (i.e., small trees) to limit wind penetrance, reduce understory and soil drying, and prevent quick vegetation regrowth by opening up the canopy too much and too fast.
- All large trees—those that occupy the forest canopy—are retained.
- Under certain conditions where too many large fire-intolerant trees have come in due to fire suppression, some of those trees could be killed (e.g., girdled) and left on site as snags or tipped into streams as habitat given their disproportionate value as carbon reservoirs (Mildrexler et al., 2020; 2023) and wildlife habitat because as trees age and grow big they naturally develop fire insulating properties (DellaSala and Baker, 2020).
- Small trees are thinned or treated via prescribed burning or lop and scatter (also chipping) of residues; in the case of burning, this involves the prescribed use of a drip torch over a defined area for ecosystem benefits and not "pile burning," which is otherwise very damaging to soils and mycorrhizae due to intensive, localized heating. Cultural burning practices and traditional ecological knowledge can also be most helpful.

Under extreme fire weather (excessive heat, drought, high winds) fires blow right through treated sites. Thus, billions of dollars are wastefully spent in a futile race to get to scale, which will never happen because of limitations on access, workforce, dollars, safety, low odds of co-occurrence, and especially, the escalating influence of climate change on fire behavior (DellaSala et al., 2022). Instead, focusing on areas closest to homes where funding can be redirected to achieve scalable outcomes is common sense fire risk reduction (Schoenaggel et al., 2017; Ager et al., 2021).

The reader should note that thinning is not some benign activity that removes just the small trees and shrubs but instead

- Can degrade closed-canopy habitat for imperiled species like the spotted owl (Odion et al., 2014) as in the TNC Rogue Basin project.
- Requires an extensive road network damaging to aquatics through sediment runoff, and that can act as conduits for weed infestations (Ibisch et al., 2016) and fire ignitions (Balch et al., 2017).
- Is often coupled with other cumulative stressors such as pile burning, livestock grazing, invasive species, and off-highway vehicles in the same space.
- Releases more carbon than fires, even severe ones (Campbell et al., 2012; Law et al., 2018), at an equivalent scale of treatment.
- Is seldom cost effective without massive federal subsidies or logging large fire-resistant trees to pay for the operations.

10.6 HAS ACTIVE MANAGEMENT BECOME A RELIGION OF SORTS? (CONCLUDING THOUGHTS)

Despite the extensive limitations and substantial collateral damages of many forms of active management, as a practice it has risen to a level resembling a quasi-religion, with the disciples of thinning refusing to question whether and when it works (on faith and narrowly focused research) and instead assuming collateral damages are merely short-term penance for some perceived good fire-Nirvana that awaits fire restoration redemption. Anyone questioning the doublespeak gospel is attacked as a fire heretic. Is there fuel science supporting their view—yes, of course. But there is also extensive ecological and climate science showing it is way overblown (limitations), often damaging, and ineffective at scale in a changing climate. Rather than recognize the uncertainties and limitations, their dialog is directing the messaging war on fire but at substantial costs to ecosystems, communities, and the climate.

Until land managers, TNC, and decision makers recognize that there are much bigger top down drivers of large fires at play (i.e., the root causes) more homes will be lost, more fire fighters will be placed in harm's way, more forests will be degraded by logging, and unprecedented levels of active management will contribute to the increasingly harmful climate-fire feedback loop. First and foremost, we must treat the root causes to address the effects of the climate-fire crisis.

Instead of doing it over and in the same way in community rebuilds—as in the case of fire damaged western towns like Talent—communities and decision makers should learn from the past by redirecting wildfire funds to proven community protection efforts (e.g., home hardening, defensible space, urban growth containment, smoke shelters, air filtration systems, emergency warning systems and evacuation plans) rather than backcountry logging. The only proven science for reducing fire risks to homes begins with the homeowner and that is where the money is best spent (an ounce of prevention is worth a pound of cure!). All of this could be coupled with better growth management to limit expansion of homes into fire-danger areas and changes to homeowner insurance policy rates that provide a cost-break to prepared homeowners.

Foresters can also make improvements by ending clearcut logging, focusing treatments on flammable plantations, not overselling thinning, closing, and obliterating roads to reduce unwanted ignitions, and protecting natural areas as natural climate solutions for their ability to sequester and store massive amounts of atmospheric carbon. And of course, none of this gets any better the longer the world remains addicted to fossil fuels. Clean renewable energy needs to prosper from roof top to roof top and scale up across all sectors and communities now.

Are there other forms of management that can be a solution? Yes, depending on whether they are rooted in ecosystem science and not just fuel treatments. Prescribed and cultural burning show promise but likewise have scale limitations, and the safe use of natural ignitions for ecosystem benefits at scale is widely supported in many science circles. However, communities and land managers are increasingly wary of the smoke and chance of an escaped fire. Under certain conditions, small tree removals can also be helpful especially closest to homes and within flammable tree plantations. Additionally, holistic restoration (and not just "fuels" management) should be directed at removing cumulative anthropogenic stressors (roads, livestock, ORVs, mines, biomass utilization, weeds, extractive uses on public lands) to give ecosystems a chance to adapt. Reintroducing beavers (*Castor canadensis*) in floodplains for flood control and water storage ("green infrastructure"), bringing back large carnivores and other imperiled species via reintroductions and recovery actions, along with rewilding landscapes would reconnect humanity with nature for the benefit of all, including replacing command-and-control attitudes toward fire and nature with working with nature (including fire) for ecosystem benefits.

In the long run, we need wildland fire—of all types—in dry forest ecosystems. If we cast aside inherent biases in how some see fire (destructive, needing to be tamed, always suppressed, only good fire, etc.), and look at it from the standpoint of what's best for fire-dependent ecosystems, then there really is no such thing as good or bad fire ecologically. Fire is a self-willed force of nature that has been the architect of pyrodiversity and a primer for the evolution of new species and adaptations since the dawn of plants some

400M years ago. It has worked in concert with pre-contact Indigenous peoples for millennia and still can. And, in all its forms, it is as right for dry forests as rain is for rainforests. We must learn to coexist as its neighbor and end the false narrative of command-and-control that will not end well for nature or us if we continue on this destructive path ecologically, socially, and climatically.

Acknowledgments

Reviews of this chapter were conducted by local activists Luke Ruediger (Klamath Forest Alliance), and Sarah Hyden (Forest Advocate).

REFERENCES

Abatzoglou, J.T., Williams, A.P., 2016. Does impact of anthropogenic climate change on wildfire across western US forests. Proc. Natl. Acad. Sci. 113, 11770−11775.

Ager, A.A., Evers, C.R., Day, M.A., Alcasena, F.J., Houtman, R., 2021. Planning for future fire: scenario analysis of an accelerated fuel reduction plan for the western United States. Landsc. Urban Plann. 215, 104212. https://doi.org/10.1016/j.landurbplan.2021.104212.

Baker, W.L., 2022. Defensible-space treatment of <114,000 ha 40 m from high-risk buildings near wildland vegetation could reduce loss in WUI wildfire disasters across Colorado's 27 million ha. Landsc. Ecol. 37, 2967−2976. https://doi.org/10.1007/s10980-022-01539-0.

Balch, J.K., et al., 2017. Human-started wildfire expand the fire niche across the United States. Proc. Natl. Acad. Sci. 114, 2946−2951.

Bradley, C.M., Hanson, C.T., DellaSala, D.A., 2016. Does increased forest protection correspond to higher fire severity in frequent-fire forests of the western United States? Ecosphere 7, 1−13.

Bassett, S., 2018. Greater Santa Fe Fireshed Wildfire Risk Assessment. TNC New Mexico Field Office. http://www.santafefireshed.org/blog2/2018/6/11/greater-santa-fe-fireshed-wildfire-riskassessment;. (Accessed 29 March 2023).

Campbell, J.L., Harmon, M.E., Mitchell, S.R., 2012. Can fuel-reduction treatments really increase forest carbon storage in western US by reducing future fire emissions? Front. Ecol. Environ. 10, 83−90. https://doi.org/10.1890/110057.

Cary, G.J., Davies, I.D., Bradstock, R.A., Keane, R.E., Flannigan, M.D., 2017. Importance of fuel treatment for limiting moderate-to-high intensity fire: findings from comparative fire modelling. Landscape Ecol. 32, 1473−1483. https://doi.org/10.1007/s10980-016-0420-8.

Cohen, J.D., 2000. Preventing disaster: home ignitability in the wildland-urban interface. J. For. 98, 15−21.

DellaSala, D.A., Ruediger, L., Hanson, C., 2020. A Science-Based Critique of the Nature Conservancy's Forest and Fire Management Programs With a Focus on Case-Study Areas in Western Fire-dependent Forests. Wild Heritage. https://wild-heritage.org/wp-content/uploads/2020/12/TNCfirereport-DellaSalaetal-2020.pdf;. (Accessed 29 March 2023).

DellaSala, D.A., Baker, W.L., 2020. Large trees: Oregon's bio-cultural legacy essential to wildlife, clean water, and carbon storage. Wild Heritage. https://wild-heritage.org/wp-content/uploads/2020/12/Large-Trees-Report-12.2020.pdf;. (Accessed 29 March 2023).

DellaSala, D.A. (Ed.), 2021. Conservation Science and Advocacy for a Planet in Crisis: Speaking Truth to Power. Elsevier, Oxford. https://doi.org/10.1016/C2016-0-03650-1.

DellaSala, D.A., Baker, B., Hanson, C.T., Ruediger, L., Baker, W., 2022. Have western USA fire suppression and active management approaches become a contemporary Sisyphus? Biol. Conserv. 268, 109499. https://doi.org/10.1016/j.biocon.2022.109499.

Downing, W.M., Dunn, C.J., Thompson, M.P., Caggiano, M.D., Shor, K.C., 2022. Human ignitions on private lands drive USFS cross-boundary wildfire transmission and community impacts in the western US. Sci. Rep. 12, 2624. https://doi.org/10.1038/s41598-022-06002-3.

Gibbons, P., van Bommel, L., Gill, M.A., Cary, G.J., Driscoll, D.A., Bradstock, R.A., Knight, E., Moritz, M.A., Stephens, S.L., Lindenmayer, D.B., 2012. Land management practices associated with house loss in wildfires. PLoS One 7, e29212.

Hessburg, P.F., Prichard, S.J., Hagmann, R.K., Povak, N.A., Lake, F.K., 2021. Wildfire and climate change adaptation for intentional management. Ecol. Appl. 31, e02432. https://doi.org/10.1002/eap.2432.

Ibisch, P.L., Hoffmann, M.T., Kreft, S., Pe'eer, G., Kati, V., Biber-Freudenberger, L., DellaSala, D.A., Vale, M.V., Hobson, R.R., Selva, N., 2016. A global map of roadless areas and their conservation status. Science 354, 1423−1427. https://doi.org/10.1126/science.aaf7166.

Law, B.E., et al., 2018. Land use strategies to mitigate climate change in carbon dense temperate forests. Proc. Natl. Acad. Sci. 115, 3663−3668. www.pnas.org/cgi/doi/10.1073/pnas.1720064115.

Lesmeister, D.B., Davis, R.J., Sovern, S.G., Yang, Z., 2021. Northern spotted owl nesting forests as fire refugia: a 30-year synthesis of large wildfires. Fire Ecol. 17, 32. https://doi.org/10.1186/s42408-021-00118-z.

Mildrexler, D.J., Berner, L.T., Law, B.E., Birdsey, R.A., Moomaw, W.R., 2020. Large trees dominate carbon storage in forests east of the cascade crest in the United States Pacific Northwest. Front. For. Glob. Chang. 3, 594274. https://doi.org/10.3389/ffgc.2020.594274.

Mildrexler, D.J., Berner, L.T., Law, B.E., Birdsey, R.A., Moomaw, W.R., 2023. Protect large trees for climate mitigation, biodiversity, and forest resilience. Conserv. Sci. Pract., e12944 https://doi.org/10.1111/csp2.12944.

Odion, D.C., Frost, E.J., Strittholt, J.R., Jiang, H., DellaSala, D.A., Moritz, M.A., 2004. Fire severity patterns and forest management in the Klamath National Forest, northwest California, USA. Conserv. Biol. 18, 927−936.

Odion, D.C., Hanson, C.T., DellaSala, D.A., Baker, W.L., Bond, M.L., 2014. Effects of fire and commercial thinning on future habitat of the northern spotted owl. Open Ecol. J. 7, 37−51.

PERFACT, 2018. Promoting Ecosystem Resilience and Fire Adapted Communities Together (PERFACT): Collaborative Engagement, Collective Action, and Co-Ownership of Fire. In: Semi-Annual Report Submitted by TNC to USDA Forest Service.

Raphael, M.G., Hessburg, P., Kennedy, R, Lehmkuhl, J., 2013. Assessing the Compatibility of Fuel Treatments, Wildfire Risk, and Conversation of Northern Spotted Owl Habitats and Populations in the Eastern Cascades: A Multi-Scale Analysis. https://www.researchgate.net/publication/280662277_Assessing_the_Compatibility_of_Fuel_Treatments_Wildfire_Risk_and_Conservation_of_Northern_Spotted_Owl_Habitats_and_Populations_in_the_Eastern_Cascades_A_Multi-scale_Analysis. (Accessed 29 March 2023).

Schoennagel, T., et al., 2017. Adapt to more wildfire in western North American forests as climate changes. Proc. Natl. Acad. Sci. 114, 4582−4590. https://doi.org/10.1073/pnas.1617464114.

Syphard, A.D., Brennan, T.J., Keeley, J.E., 2014. The role of defensible space for residential structure protection during wildfires. Intl. J. Wildl. Fire 23, 1165−1175.

Zachmann, L.J., Shaw, D.W.H., Dickson, B.G., 2018. Prescribed fire and natural recovery produce similar long-term patterns of change in forest structure in the Lake Tahoe basin, California. For. Ecol. Manag. 409, 276–287.

Zald, H.S.J., Dunn, C.J., 2018. Severe fire weather and intensive forest management increase fire severity in a multi-ownership landscape. Ecol. Appl. 28, 1068–1080. https://doi.org/10.1002/eap.1710.

FURTHER READING

Noss, R.F., et al., 2012. Bolder thinking for conservation. Conserv. Biol. 26, 1–4.

Westerling, A.R., 2016. Increasing western US forest wildfire activity: sensitivity to changes in the timing of spring. Phil. Trans. R. Soc. B 371, 20150178. https://doi.org/10.1098/rstb.2015.0178.

Chapter 11

Misinformation About Historical and Contemporary Forests Leads to Policy Failures: A Critical Assessment of the "Overgrown Forests" Narrative

Chad T. Hanson[1] and Bryant C. Baker[2]
[1]*John Muir Project of Earth Island Institute, Berkeley, CA, United States;* [2]*Wildland Mapping Institute, Ventura, CA, United States*

11.1 THE POPULAR NARRATIVE OF "OVERGROWN FORESTS"

Anyone paying attention to media coverage of forests and wildfires in recent years has read and heard the "overgrown forests" narrative many times. We are told that current dry forests of the western United States are "overgrown" due to decades of fire suppression and that these overly dense forests are causing wildfires to be too big and too severe, incinerating trees and sending their carbon into the atmosphere and converting many forests to shrublands, grasslands, or even bare ground. We are told this is because dense forests have too much biomass, too many trees, and therefore an excess accumulation of "fuel." It seems to make intuitive sense to many, on the surface; and, if it were true, it certainly sounds dire. Given the consistency and relentlessness of this messaging in the public dialog, many people do not even question the veracity of this narrative. But what does the science have to say about it? Is there a sound evidentiary basis for the overgrown forest narrative, or is there a disconnect between the public discussion and the science? That is the question we will explore in this chapter.

11.2 ARE CONTEMPORARY WESTERN US DRY FORESTS "OVERGROWN?"

The first significant problem with the overgrown forest narrative is that it overlooks the enormous amounts of carbon that have been removed from western US dry forests since the mid/late-1800s due to logging. These forested regions currently have carbon storage levels that are far lower than their biological potential (Erb et al., 2018; Strassburg et al., 2020; Schmitz et al., 2023). The overgrown forest narrative would have us continue, and increase, the removal of trees from these forests, ostensibly to "restore" them, yet this would only further exacerbate the carbon storage deficit that presently exists due to many decades of logging.

This narrative has been built upon a body of scientific studies—generally authored by scientists affiliated with agencies and entities that are financially involved in logging—that have been discredited scientifically for a pattern of evidentiary omissions and misrepresentations that created a "falsification of the scientific record," as documented in great detail in Baker et al. (2023a). These studies omitted readily available historical evidence regarding the density of smaller trees as well as oaks and other nonconifers. They also omitted historical evidence, including written descriptions and maps, of small and large high-severity fire patches. In the Sierra Nevada, for example, when these omitted tree data were included, it was revealed that historical tree density (stems/hectare) was 7 times greater than previously reported in mixed-conifer forests, and 17 times greater in ponderosa pine (*Pinus ponderosa*) forests—a finding that remains uncontested scientifically (Baker et al., 2018).

11.3 DO DENSER, MATURE, AND OLD FORESTS BURN MORE SEVERELY?

Though it may seem to make common sense from a layperson's standpoint that denser forests with greater biomass would have more "fuel" and, therefore, burn more severely in wildfires, a large body of science on this subject point in a very different direction. Importantly, some of this research is now being published by courageous scientists affiliated with agencies involved in commercial logging, like the US Forest Service, but these scientists are following the evidence where it leads, even when it contradicts the forest management policies of their own agency. For example, Lesmeister et al. (2021) recently published the findings of a massive, 30-year analysis of forest structure and wildfire severity using 472 wildfires spanning 2.1 million hectares of dry and mesic forests in the Pacific Northwestern United States. They reported that "open" forests often have "hotter, drier, and windier microclimates, and those conditions decrease dramatically over relatively short distances into the interior of older forests with multi-layer canopies and high tree density." In addition, the authors found higher wildfire severity in more open forests, especially in dry forest ecosystems. Notably, while research supports the conclusion that dense, mature/old forests tend to burn less severely in wildfires, young, tightly-packed and homogenous tree plantations tend to burn somewhat more severely (e.g., Odion et al., 2004).

Similarly, in an analysis of the 104,000-hectare Rim fire of 2013, a group of US Forest Service and university scientists (Lydersen et al., 2014) concluded: "Density of small to intermediate size trees ... were also related to Rim Fire severity, with plots with a greater small tree density tending to burn with lower severity."

These dense, mature forests have more shade due to higher canopy cover, and a cooler, moister microclimate that often makes such stands less likely to experience high-severity fire (Figure 11.1).

11.4 DOES "THINNING" REDUCE OVERALL SEVERITY IN WILDFIRES?

Denser, multilayered forests with greater biomass and canopy cover tend to experience lower fire severity due to their shadier, cooler microclimate, and the windbreak effect afforded by higher tree density, as discussed above (though extreme weather conditions can override these effects). Conversely, when trees are removed from such forests, a less-shaded, hotter, drier, and more wind-prone microclimate results, which can be exacerbated by the spread of highly flammable, nonnative grasses and weeds due to soil disturbance by heavy equipment.

FIGURE 11.1 A very dense, old stand that experienced lower-severity fire in the Creek fire of 2020, Sierra National Forest, southern Sierra Nevada, California, USA. *Photo by Chad Hanson.*

In the largest study of its kind ever conducted, analyzing 3 decades of fire data and 1500 wildfires across 9.5 million hectares of ponderosa pine/Jeffrey pine (*P. ponderosa* and *P. jeffreyi*) and mixed-conifer forests, Bradley et al. (2016) found that while weather/climate factors primarily drive wildfire severity, past logging is a key secondary factor. Specifically, forests protected from tree removal had the lowest fire severity, forests with some environmental protections but also some logging and other tree removal allowed had higher fire severity, and private forestlands with the fewest environmental protections and the most intensive tree removal had the highest fire severity, within the same forest types. In other words, the more tree removal that occurs, the more intensely wildfires burn (Figure 11.2). The authors concluded the following:

We found forests with higher levels of protection [from tree removal] had lower severity values even though they are generally identified as having the highest overall levels of biomass and fuel loading. Our results suggest a need to reconsider current overly simplistic assumptions about the relationship between forest protection and fire severity in fire management and policy.

Similar findings emerged from an analysis of dry forests in the >150,000-hectare Creek fire of 2020 in the southern Sierra Nevada. Based on the Forest Service's own data, forests with previous logging under the rubric of "fuel reduction"—specifically, mechanical thinning and postfire logging—had overall higher fire severity than unmanaged forests (Hanson, 2021).

More recent analyses have begun looking at another key question regarding mechanical thinning and wildfire severity in dry forests, related to overall combined tree mortality from thinning itself and subsequent wildfire. These

FIGURE 11.2 An open, low-density mixed-conifer forest resulting from previous commercial thinning, and which burned at high-severity in the Creek fire of 2020. *Photo by Chad Hanson.*

studies have consistently found that mechanical thinning kills more trees than it prevents from being killed in mature and old dry forests, including Hanson (2022) (pertaining to the Antelope fire of 2021 in northern California), Baker and Hanson (2022) (pertaining to the Caldor fire of 2021 in the northern Sierra Nevada), and DellaSala et al. (2022) (pertaining to the Wallow fire of 2011 in Arizona). Baker and Hanson (2022) explained why some studies have erroneously reported that mechanical thinning is effective as a wildfire management approach:

> *Despite controversy regarding thinning, there is a body of scientific literature that suggests commercial thinning should be scaled up across western US forest landscapes as a wildfire management strategy. This raises an important question: what accounts for the discrepancy on this issue in the scientific literature? We believe several factors are likely to largely explain this discrepancy. First and foremost, because most previous research has not accounted for tree mortality from thinning itself, prior to the wildfire-related mortality, such research has underreported tree mortality in commercial thinning areas relative to unthinned forests. Second, some prior studies have not controlled for vegetation type, which can lead to a mismatch when comparing severity in thinned areas to the rest of the fire area given that thinning necessarily occurs in conifer forests but unthinned areas can include large expanses of non-conifer vegetation types that burn almost exclusively at high severity, such as grasslands and chaparral. Third, some research reporting effectiveness of commercial thinning in terms of reducing fire severity has been based on the subjective location of comparison sample points between thinned and adjacent unthinned forests. Fourth, reported results have often been based on theoretical models, which subsequent research has found to overestimate the effectiveness of thinning. Last, several case studies draw conclusions about the effectiveness of thinning as a wildfire management strategy when the results of those studies do not support such a conclusion, as reviewed in DellaSala et al. (2022). (internal citations omitted)*

11.5 IS HIGH-SEVERITY FIRE CONVERTING DENSE DRY FORESTS TO NONFOREST?

The narrative that our "overgrown" forests are causing large fires that are burning too hot, converting forests to bare ground or shrubs, has become alarmingly common in the public dialog. This narrative is largely being driven by a series of hypothetical modeling studies authored in substantial part by scientists funded by entities that are financially involved in logging, and that actively use the overgrown forests narrative to promote more logging both before and immediately after wildfires occur.

One recent example of this is Davis et al. (2023). There are several serious methodological issues that immediately become apparent upon reading the article. The three most obvious study design flaws are the failure to (1) restrict plots to prefire conifer forest; (2) exclude postfire logged areas; and (3) account for the massive influence of very small plot size (they used a 0.01-ha plot size for their analyses) on the reporting of their central metric, i.e., the proportion of plots with/without postfire conifer regeneration. These three failures alone create a profound underreporting bias with regard to the proportion of plots with conifer regeneration, also expressed as the probability of postfire recruitment, as was established in Hanson and Chi (2021), a study that was not referenced by Davis et al. (2023).

Hanson and Chi (2021) found that, even after excluding postfire logged sites, 40% of the plot locations on a 200-meter grid established by the US Forest Service in the Rim fire of 2013 must be disqualified as potential data points for natural postfire conifer regeneration research. This was because they were naturally nonforested (plots located on solid granite outcroppings, or in the middle of creeks or wet meadows with no natural tree growth) or because of a total lack of any prefire mature conifers due to earlier (prefire) logging operations (most often, plots in the middle of logging roads and landings, as well as some very young tree plantations). Failure to omit such locations creates a major bias in the results, causing studies to falsely report that substantial proportions of high-severity fire areas are devoid of postfire conifer reproduction.

Study plot sizes that are far too small, especially when the proportion of plots without conifer seedlings is a primary metric being reported, as in Davis et al. (2023), will grossly underestimate conifer establishment. For example, even >300 m from the nearest live, surviving conifer, deep into the interior of the largest high-severity fire patches, Hanson and Chi (2021) found an average of 256 naturally regenerating conifer seedlings per hectare at 5 years postfire. However, if the proportion of plots without conifer seedlings was used as a metric, plot size profoundly influenced the results in such areas: 50% of such plots lacked conifer seedlings with a mere 5-meter-radius plot, while only 7% lacked conifer seedlings with a 20-meter-radius plot. Therefore, using a 5-meter-radius plot, and reporting the proportion of plots lacking conifer seedlings, would lead to wildly misleading results and the false implication that half of the area had been converted to nonforest (while Hanson and Chi, 2021 did not report quantitative data for a 40-meter radius, 100% of plots at this radius had natural conifer regeneration) (Figure 11.3).

In addition to these three design flaws, an even more extraordinary source of methodological bias in Davis et al. (2023) is how their "results" with regard to high-severity fire do not reflect actual, on-the-ground, postfire conifer regeneration but instead represent a fictionalized world into which the authors imported their field plot data, even for the present and recent-past timeframes (e.g., 2001–20). Specifically, the "high-severity fire scenario" in Davis et al. (2023) represents vast landscapes characterized by 10-year-old large high-severity fire patches that are over 150 m from the nearest live, surviving tree, and where there is only 10% forest cover within a 300-meter radius (i.e.,

FIGURE 11.3 Abundant natural conifer regeneration over 300 m into the interior of a large high-severity fire patch in the Rim fire, at 8 years postfire. *Photo by Chad Hanson.*

no live trees within 150 m, and very few live trees 150–300 m away). Yet, such areas comprise <1% of the forests (Hanson, 2018; DellaSala and Hanson, 2019). By extrapolating small areas to make sweeping conclusions based on hypothetical scenarios, Davis et al. (2023) manufactured a context that apparently forms the basis for a funding request to Congress for real-world management "interventions" and increased appropriations to the US Forest Service—the agency that funds many of the authors of Davis et al. (2023)—for such "interventions" which they make clear includes logging.

Similarly, another recent study by US Forest Service scientists, Steel et al. (2022), claimed that about 30% of conifer forests in the central and southern Sierra Nevada were converted to nonforest between 2011 and 2020 by high-severity fire and/or drought plus native bark beetles. The authors utilized a relatively new method that incorporates data from various remote sensing and field plot datasets into a model, but our analysis (Baker et al. *in prep*) of the data they made available indicates major inaccuracies in their methodologies. There are two main issues with their methods: (1) categorizing a "forest" as having >25% canopy cover (typically any tree-dominated ecosystem with >10% canopy cover is considered forested as reported by the Forest Service to the Food and Agriculture Organization); and (2) the underlying post-disturbance data they used were clearly inaccurate when compared to other readily available datasets such as LANDFIRE. As an example, we compared Steel et al.'s data to data from LANDFIRE for an area that burned in two adjacent wildfires in the Sierra Nevada. We found that between 34% of the area classified as "nonforest" by Steel et al. (due to their finding that canopy cover was <25%) has ≥30% canopy cover according to LANDFIRE. And an additional 17% of the area classified as "nonforest" by Steel et al. has between 20% and 30% canopy cover according to LANDFIRE, as this dataset only classifies canopy cover in 10% bins (see Figure 11.4). This was further corroborated by postfire aerial imagery of discrepancy areas, which clearly show that the landscape is still well-forested. Perhaps even more importantly, if the authors had gathered any field plot validation data, they would have found that, within the relatively minor portion of the landscape that their model correctly classified as high-severity, there is consistently abundant natural conifer regeneration, such as in the field plot study sites in Hanson (2018) and Hanson and Chi (2021), which were falsely classified by the Steel et al. (2022) hypothetical model as having been converted to nonforest.

In a particularly telling mishap, two aspects of the "overgrown forests" narrative recently came into irreconcilable conflict, as the US Forest Service proposed approximately 20,000 ha of commercial thinning in mature conifer forests that the agency said burned much too lightly in the Creek fire of 2020 and, supposedly, remained far too dense and overgrown. The agency seemed unaware of the illogic of this statement for, if being very dense and "overgrown" truly caused stands to burn at high-severity, why did these vast areas burn so lightly? Simultaneously, Steel et al. (2022) classified most of this area as having been converted to nonforest by high-severity fire. When this awkward discrepancy was pointed out in an administrative objection on the proposed logging project, the agency had to reluctantly admit that they had promoted false information.

Recently, the US Forest Service and the National Park Service have begun applying this particular version of the "overgrown forests" narrative to groves of giant sequoias (*Sequoiadendron giganteum*). They made numerous statements to the press, claiming that the minor portion of recent wildfires comprised of high-severity fire effects had converted those portions of the groves to nonforest, where giant sequoia reproduction is supposedly absent or sorely lacking. In each case, the agencies have issued "closure orders," prohibiting public access to these groves under penalty of substantial fines or potential arrest and prosecution if a member of the public dared to hike into these public lands to take some photos. Meanwhile, based on these claims, which the public was not allowed to verify, most of these groves were being

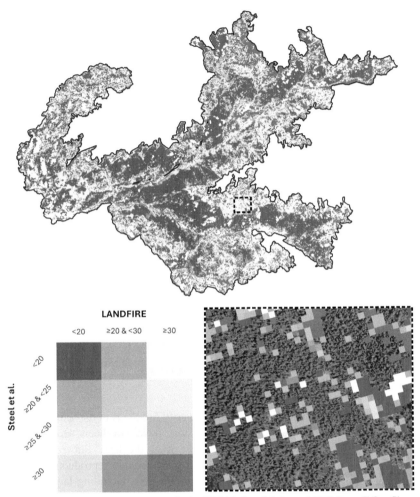

FIGURE 11.4 Map of the area burned in the 2017 McCormick fire and 2018 Donnell fire, Sierra Nevada (top). The color of each 30 × 30 m pixel corresponds to the bottom left matrix of % canopy cover from Steel et al. and LANDFIRE (note that LANDFIRE canopy cover values are only provided as 10% bins), and white areas were classified as "nonforest" prior to the wildfires by Steel et al. Two to 3-year postfire aerial imagery (NAIP) of the dashed box area shown in the top map clearly show areas that are still dominated by mature trees and which were classified as having <25% canopy cover by Steel et al. and ≥30% canopy cover by LANDFIRE (bottom right).

subjected to intensive management, including logging, herbicide spraying, and creation of tree plantations, without public notice or comment, and without environmental analysis. After it was pointed out that the legal footing of these closure orders was tenuous at best, the Forest Service and Park Service reluctantly allowed some very limited public access. This limited access created the opportunity to take the following representative photos

FIGURE 11.5 Abundant, vigorous natural giant sequoia reproduction in large high-severity fire patches within (a) Redwood Mountain Grove, Sequoia and Kings Canyon National Parks, at 2 years postfire (photo by Bryant Baker), (b) Alder Creek Grove, Giant Sequoia National Monument, at 3 years postfire (photo by Chad Hanson), (c) Nelder Grove, Sierra National Forest, at 6 years postfire (photo Bryant Baker), and (d) Nelder Grove, Sierra National Forest, at 6 years postfire. *Photo by Bryant Baker.*

(Figure 11.5) showing what these high-severity fire areas in the sequoia groves really look like—at least the portions that have not been surreptitiously managed and degraded, without oversight, after the fires. Subsequent peer-reviewed research confirmed that post-fire giant sequoia reproduction is most abundant and vigorous in high-severity fire patches, including large patches, compared to low/moderate-severity areas (Hanson et al., 2024a,b).

11.6 IMPLICATIONS OF PROLOGGING MISINFORMATION

Recently, a series of highly misleading studies by some US Forest Service scientists has led to forest policies and proposals that threaten to substantially increase tree mortality and climate change impacts.

For example, US Forest Service managers are now attempting to implement a misleadingly named logging project more than 80,000-hectares called the "Community Protection Project" on the Plumas National Forest in the northern Sierra Nevada (the vast majority of the area proposed for logging, under the rubric of "thinning," is in backcountry forests far from any homes or towns), and several other similar logging projects on nearby national forests range from 20,000 to 75,000 ha. These massive logging projects are predicated on a Forest Service study, North et al. (2022), which claims that over two-thirds of the trees in Sierra Nevada forests must be logged and removed supposedly to restore natural forest density and promote resilience by reducing potential for tree mortality. However, this study was found to represent a "falsification of the scientific record" for several reasons (Baker et al., 2023a).

Specifically, North et al. (2022) relies on previous studies by Collins and Stephens, which reported that there were only about 50—75 trees per hectare in historical Sierra Nevada forests, based on c.1911 Forest Service field surveys. However, as found in Baker et al. (2018), the Collins and Stephens work omitted the small-tree data in those historical datasets and failed to use correction factors that the Forest Service itself, a century ago, repeatedly stated were needed to avoid severe underestimations of forest density. The surveys were based on visually estimated distance from the transect line, but surveyors consistently overestimated distance (e.g., they would see 30 or 40 feet to their left and right but would assume they were seeing 66 feet left and right), causing a major underestimation of forest density. The findings in Baker et al. (2018) are uncontested by the Forest Service.

Additionally, North et al. (2022) misleadingly claimed that "current" forests have approximately 350—500 trees per hectare but inexplicably used data from 2011 to represent supposed "current" conditions, and they failed to mention that over 90% of their study areas have burned in mixed-intensity wildfires since 2011 and/or experienced drought-induced tree mortality over the past decade such that a large portion of the live trees that existed a decade ago are now snags and downed logs.

The bottom line is that North et al. (2022) severely underreported historical forest density by using previous historical density estimates that have been discredited and superseded, and they overreported existing live tree forest density by using 2011 as their "current" data, even though that fire and drought since 2011 have dramatically reduced live tree density in their study areas.

Studies that have claimed success of logging projects on reducing tree mortality from native bark beetles (many taxa) generally do not consider the thinning-caused mortality when considering the concept of a successful project. For instance, Fettig et al. (2012) examined the effect on bark beetle-induced tree mortality of various levels of thinning in comparison to unthinned areas in mixed-conifer forests in the Sierra Nevada. While they demonstrated lower relative bark-beetle induced tree mortality in thinned units, the authors did not consider the initial mortality event caused by the thinning itself. Their measure of success was whether the level of tree mortality in thinned stands was less than in the unthinned stands, but apparently mortality was only significant to determine success if caused by bark beetles. When reanalyzing their data here, it is actually quite simple to glean that the overall mortality (i.e., mortality from thinning plus mortality from subsequent bark beetles) in the three thinning prescriptions was substantial (109—289 trees killed per hectare

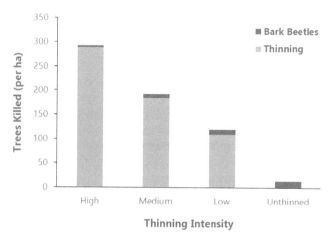

FIGURE 11.6 Data from Fettig et al. (2012) showing tree mortality from thinning versus bark beetles in a thinning project implemented ostensibly to reduce and prevent tree mortality. Fettig et al. only showed mortality from bark beetles in their original graphs.

on average) compared to the overall tree mortality from bark beetles in the unthinned stands (approximately 13 trees killed per hectare on average). Granted, the number of trees killed by bark beetles was lower in the thinning units (3—11 trees killed per hectare on average) compared to the unthinned stand (13 trees killed per hectare on average), but this pales in comparison to overall number of trees killed due to the thinning itself (see Figure 11.6). Another way to view this is that approximately 289 trees per hectare were killed in the most intensive treatment by the thinning itself in order to prevent 10 trees from being killed in the future by bark beetles.

Six et al. (2014) notes a similar pattern:

Although more trees were killed overall in control units during the outbreak, all controls still retained a greater number of residual mature trees than did thinned stands as they entered the post-outbreak phase.

And a separate study in ponderosa pine forests in the Black Hills demonstrated that far more trees were killed through the actual thinning process than through a subsequent bark beetle outbreak that was more severe than that experienced in the study by Fettig et al. (2012). Negron et al. (2017) examined stands in which the overall mortality (again, mortality caused by thinning plus mortality caused by bark beetles) was 242.6 trees killed per acre on average in thinned stands compared to 87.7 trees killed per acre in unthinned stands (results were expressed on a per-acre basis in the study, so we present them in the same way here). As with other similar studies, the thinning was the primary source of mortality in the stand rather than bark beetles. By the end of the outbreak, there were more trees in the unthinned stands (203.2 trees per acre on average) compared to the thinned stands (55 trees per acre on average) as well as more basal area (which could be considered a proxy for both biomass and carbon storage; 67.8 square feet per acre compared to 32.3 square feet per acre).

In Sierra Nevada mixed-conifer and ponderosa pine forests after the major drought occurring 2012−17, Restaino et al. (2019: Figures 3 and 4) reported mixed effects of increasing forest basal area on tree mortality from drought and native bark beetles, with no clear relationship. Restaino et al. (2019), Figure 5 reported that thinned forests had approximately the same or higher tree mortality from drought/beetles compared to unthinned forests for three of four conifer species studied. Only ponderosa pine had a slightly lower probability of mortality in thinned forests than in unthinned forests, but the difference was only 15% on average, while Figure 2a of their study showed that thinning itself killed about 35% of the forest basal area before the drought occurred. Thus, thinning once again killed more trees than it prevented from being killed, even for the one conifer species that had somewhat lower probability of tree mortality in thinned areas.

Cumulative tree mortality in thinned stands and subsequent natural disturbances should be more explicitly reported by researchers who are actively testing hypotheses about the effectiveness of mechanical thinning. Otherwise, studies will again fail to divulge or disclose how mechanical thinning, conducted ostensibly to reduce stand densities and reduce competition-related tree mortality, kills far more trees than it prevents from being killed by natural disturbances. The adverse climate change impacts of this sort of misrepresentation, in terms of increased carbon emissions, would be profound if the North et al. (2022) recommendation to log most of the trees in Sierra Nevada forests, based on their falsification of the scientific record (see Baker et al., 2023a), was implemented. Even the largest wildfires consume less than 2% of overall tree biomass on average (Harmon et al., 2022)—far less than even a very light-touch thinning operation. And Bartowitz et al. (2022) found that logging conducted as commercial thinning, which involves removal of some mature trees, substantially increases carbon emissions relative to wildfire alone—by about threefold or more, in fact—and commercial thinning "causes a higher rate of tree mortality than wildfire." The authors concluded:

> Our results and the majority of full-carbon accounting studies conclude that any type of harvest (logging or commercial thinning) decreases forest carbon storage ... and this research shows harvest emits more carbon per unit area than fire at all scales.

The difference in carbon emissions between thinned and unthinned forests would be even larger if thinning plus wildfire is compared to wildfire alone (no thinning). This is arguably a more relevant comparison, since thinned forests are of course expected to burn in wildfires.

11.7 CONCLUSIONS

While the "overgrown forests" narrative persists in public discourse and among many policy-makers, land managers, and even some conservation groups, there is significant evidence that calls the entire concept into question. Mature/old forests with higher tree densities, and greater biomass and carbon storage levels, tend to burn at low- to moderate-severity due to microclimate effects, while thinned or otherwise logged forests generally experience greater tree mortality from the combined effects of the logging itself and subsequent wildfire. Similarly, more trees tend to be removed from forests to prevent drought and bark beetle mortality events than either of those natural disturbances would typically kill to begin with. And thinning as a wildfire management approach tends to increase overall carbon emissions. That contemporary forests are much more resilient than is portrayed is good news from a biodiversity and natural climate solutions, or "proforestation," standpoint as dense, mature forests draw down and store far more carbon than very young or open, low-density forests (Moomaw et al., 2019). However, ongoing misinformation about these issues must be addressed in order to implement policy solutions that will better protect forests from misguided management practices.

This raises an important question: what are better policy solutions? We recommend prioritizing fire over tree removal. This includes wildland fire use (also called managed wildfire) and Native American cultural burning in forest wildlands, along with prescribed fire by land management agencies where low-severity fire is in deficit (Baker et al., 2023b). We note the wealth of scientific literature establishing that fire alone can be effectively allowed or introduced on to the forested landscape, without prior thinning or other tree removal, conducted during early or late fire season in mild to moderate fire weather (e.g., Stephens and Finney, 2002; Knapp and Keeley, 2006; Knapp et al., 2005, 2007; van Mantgem et al., 2011, 2013, 2016; Stephens et al., 2021). We agree with the conclusion of North et al. (2015) that "fire is usually more efficient, cost-effective, and ecologically beneficial than mechanical treatments." There may of course be exceptions in some circumstances for various reasons. In site-specific locations where thinning is warranted either instead of or prior to fire, such as to create defensible space adjacent to homes for protection against wildfires, we recommend non-commercial thinning of small trees beneath the forest canopy and limbing-up of mature trees.

REFERENCES

Baker, B.C., Hanson, C.T., 2022. Cumulative tree mortality from commercial thinning and a large wildfire in the Sierra Nevada, California. Land 11, 995.

Baker, W.L., Hanson, C.T., Williams, M.A., 2018. Improving the use of early timber inventories in reconstructing historical dry forests and fire in the western United States: reply. Ecosphere 9, e02325.

Baker, W.L., Hanson, C.T., Williams, M.A., DellaSala, D.A., 2023a. Countering omitted evidence of variable historical forests and fire regime in western USA dry forests: the Low-Severity-Fire Model rejected. Fire 6, 146.

Baker, W.L., Hanson, C.T., DellaSala, D.A., 2023b. Harnessing natural disturbances: a nature-based solution for restoring and adapting dry forests in the western USA to climate change. Fire 6, Article 428.

Bartowitz, K.J., Walsh, E.S., Stenzel, J.E., Kolden, C.A., Hudiberg, T.W., 2022. Forest carbon emission sources are not equal: putting fire, harvest, and fossil fuel emissions in context. Front. Forests Global Change 5, 867112.

Bradley, C.M., Hanson, C.T., DellaSala, D.A., 2016. Does increased forest protection correspond to higher fire severity in frequent-fire forests of the western USA? Ecosphere 7, e01492.

Davis, K.T., et al., 2023. Reduced fire severity offers near-term buffer to climate-driven declines in conifer resilience across the western United States. Proc. Natl. Acad. Sci. U.S.A. 120, e2208120120.

DellaSala, D.A., Baker, B.C., Hanson, C.T., Ruediger, L., Baker, W.L., 2022. Have western USA fire suppression and megafire active management approaches become a contemporary Sisyphus? Biol. Conserv. 268, 109499.

DellaSala, D.A., Hanson, C.T., 2019. Are wildland fires increasing large patches of complex early seral forest habitat? Diversity 11, 157.

Erb, K.H., et al., 2018. Unexpectedly large impact of forest management and grazing on global vegetation biomass. Nature 553, 73–76.

Fettig, C.J., Hayes, C.J., Jones, K.J., McKelvey, S.R., Mori, S.L., Smith, S.L., 2012. Thinning Jeffrey pine stands to reduce susceptibility to bark beetle infestations in California. U.S.A. Agric. Forest Entomol. 14, 111–117.

Hanson, C.T., 2018. Landscape heterogeneity following high-severity fire in California's forests. Wildl. Soc. Bull. 42, 264–271.

Hanson, C.T., 2021. Is "fuel reduction" justified as fire management in spotted owl habitat? Birds 2, 395–403.

Hanson, C.T., 2022. Cumulative severity of thinned and unthinned forests in a large California wildfire. Land 11, 373.

Hanson, C.T., Chi, T.Y., 2021. Impacts of postfire management are unjustified in spotted owl habitat. Front. Ecol. Evol. 9, 596282.

Hanson, C.T., Chi, T.Y., Baker, B.C., Khosla, M., Dorsey, M.K., 2024a. Post-fire reproduction of a serotinous conifer, the giant sequoia, in the Nelder Grove, California. Ecol. Evol. (in press).

Hanson, C.T., Chi, T.Y., Khosla, M., Baker, B.C., Swolgaard, C., 2024b. Reproduction of a serotinous conifer, the giant sequoia, in a large high-severity fire area. Fire 7, Article 44.

Harmon, M.E., Hanson, C.T., DellaSala, D.A., 2022. Combustion of aboveground wood from live trees in megafires, CA, USA. Forests 13, 391.

Knapp, E.E., Keeley, J.E., 2006. Heterogeneity in fire severity within early season and late season prescribed burns in a mixed-conifer forest. Int. J. Wildland Fire 15, 37–45.

Knapp, E.E., Keeley, J.E., Ballenger, E.A., Brennan, T.J., 2005. Fuel reduction and coarse woody debris dynamics with early season and late season prescribed fire in a Sierra Nevada mixed conifer forest. For. Ecol. Manag. 208, 383–397.

Knapp, E.E., Schwilk, D.W., Kane, J.M., Keeley, J.E., 2007. Role of burning on initial understory vegetation response to prescribed fire in a mixed conifer forest. Can. J, For. Res. 37, 11–22.

Lesmeister, D.B., Davis, R.J., Sovern, S.G., Yang, Z., 2021. Northern spotted owl nesting forests and fire refugia: a 30-year synthesis of large wildfires. Fire Ecol. 17, 32.

Lydersen, J.M., North, M.P., Collins, B.M., 2014. Severity of an uncharacteristically large wildfire, the Rim Fire, in forests with relatively restored frequent fire regimes. Forest Ecol. Manag. 328, 326–334.

van Mantgem, P.J., Caprio, A.C., Stephenson, N.L., Das, A.J., 2016. Does prescribed fire promote resistance to drought in low elevation forests of the Sierra Nevada, California, USA? Fire Ecol. 12, 13–25.

van Mantgem, P.J., Nesmith, J.C.B., Keifer, M., Brooks, M., 2013. Tree mortality patterns following prescribed fire for Pinus and Abies across the southwestern United States. Forest Ecol. Manag. 289, 463–469.

van Mantgem, P.J., Stephenson, N.L., Battles, J.J., Knapp, E.K., Keeley, J.E., 2011. Long-term effects of prescribed fire on mixed conifer forest structure in the Sierra Nevada, California. Forest Ecol. Manag. 261, 989–994.

Moomaw, W.R., Masino, S.A., Faison, E.K., 2019. Intact forests in the United States: proforestation mitigates climate change and serves the greatest good. Front. Forests Global Change 2, 27.

Negron, J.F., Allen, K.K., Ambourn, A., Cook, B., Marchand, K., 2017. Large-scale thinnings, ponderosa pine, and mountain pine beetle in the Black Hills, USA. For. Sci. 63, 529–536.

North, M.P., Stephens, S.L., Collins, B.M., Agee, J.K., Aplet, G., Franklin, J.F., Fule, P.Z.; (co-authored by U.S. Forest Service), 2015. Reform forest fire management. Science 349, 1280–1281.

North, M.P., Tompkins, R.E., Bernal, A.A., Collins, B.M., Stephens, S.L., York, R.A., 2022. Operational resilience in western US frequent-fire forests. For. Ecol. Manag. 507, 120004.

Odion, D.C., Frost, E.J., Strittholt, J.R., Jiang, H., DellaSala, D.A., Moritz, M.A., 2004. Patterns of fire severity and forest conditions in the western Klamath Mountains, California. Conserv. Biol. 18, 927–936.

Restaino, C., Young, D., Estes, B., Gross, S., Wuenschel, A., Meyer, M., Safford, H., 2019. Forest structure and climate mediate drought-induced tree mortality in forests of the Sierra Nevada, USA. Ecol. Appl. 29, e01902.

Schmitz, O.J., et al., 2023. Trophic rewilding can expand natural climate solutions. Nat. Clim. Change 13, 324–333.

Six, D.L., Biber, E., Long, E., 2014. Management for mountain pine beetle outbreak suppression: does relevant science support current policy? Forests 5, 103–133.

Steel, Z.L., et al., 2022. Mega-disturbances cause rapid decline of mature conifer forest habitat in California. Ecol. Appl. 33, e2763.

Stephens, S.L., Finney, M.A., 2002. Prescribed fire mortality of Sierra Nevada mixed conifer tree species: effects of crown damage and forest floor combustion. Forest Ecol. Manag. 162, 261–271.

Stephens, S.L., Thompson, S., Boisramé, G., Collins, B.M., Ponisio, L.C., Rakhmatulina, E., Steel, Z.L., Stevens, J.T., van Wagtendonk, J.W., Wilkin, K., 2021. Fire, water, and biodiversity in the Sierra Nevada: a possible triple win. Environ. Res. Commun. 3 (8), 081004.

Strassburg, B.B.N., et al., 2020. Global priority areas for ecosystem restoration. Nature 586, 724–729.

Chapter 12

Out of the Ashes, Nature's Phoenix Rises

Dominick A. DellaSala[1] and Chad T. Hanson[2]

[1]Wild Heritage, A Project of Earth Island Institute, Berkeley, CA, United States; [2]John Muir Project of Earth Island Institute, Berkeley, CA, United States

12.1 PYRODIVERSITY BEGETS BIODIVERSITY REAFFIRMED

Duncan Canyon on the Tahoe National Forest, California. Photo taken July 29, 2008 within the 2001 Star Fire area. *With permission from Doug Bevington.*

In both editions of this book, we presented compelling evidence of the ecologically beneficial role of mixed- and high-severity fires in western North America, Africa, Australia, and Europe. To cut to the chase, our main biodiversity findings are built on the prior edition, as follows:

- Large fires in fire-adapted forests are not currently ecological disasters; rather, they are catalyzing events for "pyrodiversity begets biodiversity," reaffirmed at least for the regions in this book. We recommend maintaining the patch complexity of large wildfires by working with fire for ecosystem benefits under safe conditions (when not threatening homes/towns). In doing so, unburned-low severity patches provide fire refugia for fire-avoiders while moderate-severe burned patches are habitat for fire-seekers (Della-Sala et al., 2017).
- Time-since-fire, site productivity, distance to seed source, type, amount, and distribution of biological legacies at multiple spatial scales (canopy gap, burn patch, landscape) dictate how much biodiversity is present postfire and the speed at which the renewal process unfolds. We recommend long-term field plots with appropriate plot sizes and adequate sampling to assess the regenerative timeline (Hanson and Chi, 2021). This should also include determining the impacts of active management and climate change on essential processes.
- The critical pulse of biological legacies (e.g., live and dead standing and fallen trees, reemerging shrubs, mycorrhizae, dormant seed banks, surviving and colonizing bird and mammal populations and pollinators) uniquely generated by high-severity flare ups (small and large patches) sustains postfire processes, ecosystem functions, and complex structures (e.g., large snags, logs) (Swanson et al., 2010; Donato et al., 2012; DellaSala, 2020). We recommend retention of biological legacies to aid in renewal processes, provide habitat for dependent species, and store carbon long-term (e.g., nearly all of the tree carbon after severe fires is transferred from live to dead pools and soils while new vegetation jump starts sequestration; Harmon et al., 2022).
- Mixed-severity fires cannot be mimicked by logging or by prescribed fire that relies on low intensity burns because such practices do not generate a pulse of biological legacies (unless they escape containment) and logging in particular removes them. In addition to working with natural ignitions, burn prescriptions should allow for flare ups to generate legacies at scale when safe to do so.
- The decade or two before conifer crowns close in on complex early seral forests is when Nature's Phoenix is at its grandest (Swanson et al., 2010). As crowns close, biodiversity declines, and competition for limited resources (light levels, nutrients, moisture) triggers tree mortality that eventually opens up the forest canopy (tree-fall gaps, vertical and horizontal layering)

and produces complex structures (snags, logs) associated with high levels of biodiversity through the old-growth stage and back again after the next disturbance (successional circularity at gap, patch, and landscape scales) (also—forests are "born complex;" Donato et al., 2012).
- Researchers should take note of time since fire before reacting prematurely, and even if conifers are delayed at first, the exceptional biodiversity characterized by complex early seral forests is lengthened (Swanson et al., 2010). We recommend designing studies on longer-timelines and/or using chronosequences (plots located in forests at different times since fire) to track conifer establishment and processes across spatial and temporal gradients and allowing this successional process to play out.
- The connection between young and old forests is broken when forests are logged especially at large scales as this can flip an entire landscape into a degraded, novel dynamism that severely compromises biodiversity (Figure 12.1).

12.2 A NATURE-BASED CORRECTION IS NEEDED IN ATTITUDES AND APPROACHES

In sum, what you see is what you get when it comes to fire perceptions and management approaches. Land managers, certain researchers, decision makers, and the media often see a forest, including protected areas, only for the live trees, overlooking the extraordinary wildlife habitat, and carbon storage, in complex early seral forests. By contrast, many ecologists and conservation biologists seek a rudimentary understanding of the causal mechanisms in successional circularity and how best to maintain their ecological integrity. This duality is implicit in how one sees nature generally (a commodity producer to be placed in command-and-control management for human benefit vs. an interconnected system intrinsically valued and self-willed). It has become so polarized that those seeing nature and burned areas for their intrinsic value, with questions and concerns about intensive management, are often personally attacked (DellaSala, 2021; DellaSala et al., 2022) even when the evidence shows intensively managed areas—not protected areas—burn at uncharacteristically severe levels (Bradley et al., 2016).

Science is built on rigorous testing and presentation of evidence designed to reject null hypotheses (questioning each other is a good thing and is supposed to be respectful, not personal). The entire field benefits from disagreements handled in the spirit of fact-finding expeditions, but it is collectively damaged when attacks are personalized (DellaSala, 2021). Criticizing the idea, not the person, is the high road most often missed by our opponents. We speculate that this overstep may stem from researchers dialed into federal research dollars designed to demonstrate the efficacy of active management without questioning limitations or collateral damages. Those of us that question are then mis-labeled as

FIGURE 12.1 (a) Circular succession showing old-growth forest (left) in the Duncan Canyon Roadless Area on the Tahoe National Forest, California affected by a severe fire (right) showing a complex early seral forest 18-years postfire. For example, biological legacies created by severe fire are the live and dead trees (especially large ones), shrubs, and seedbanks remaining postfire that "lifeboat" forests through succession to old growth and so on. (b) Two views 10 years after the Biscuit fire on the Rogue-Siskiyou National Forest, Oregon. Upper panel shows a site logged right after the fire and 10 years later with little regeneration. Lower panel just upslope shows an unlogged botanical area that has developed complex early seral features. Logging disrupts circular succession. *(a) Photo: C. Hanson. (b) Photo: D. DellaSala.*

"outside the main stream," not "consensus scientists," and "minority opinion scientists" (see Hessburg et al., 2021; vs. DellaSala et al., 2022) even though we typically have the support of conservation scientists, ecologists, naturalists and climate change scientists. Testing approaches, however, is not some kind of a popularity contest but rather is the search for truth as supported or refuted by evidence. Not accounting for or downplaying damages from active management is acting irresponsibly and complicit with the root causes of the climate-fire feedback system (DellaSala et al., 2022).

Throughout this book, we see fire through the lens of Nature's Phoenix, relying on conservation biology perspectives. In other words, we see the forests for more than just the live trees and Nature's natural processes as rejuvenation, not death-and-destruction. Some assumptions that need a Nature-based correction are as follows:

- While total area burning has been increasing since at least the 1980s (with annual variation creating "noise" around trend lines), whether high severity fire itself is increasing in extent or proportion depends on the region, study methods, data sources, scale of the analysis, and time period examined among other factors (summarized in Odion et al., 2014, 2016; DellaSala et al., 2022). A more comprehensive review of the literature shows that there are many questions around whether high severity is increasing (this is anything but settled) and that the low-moderate fire regime model in general is rejected as a hypothesis for most dry western forests (Baker et al., 2023). However, we note that climate change may alter the picture on fire severity especially under a worse case emissions scenario with more logging and suppression compounding the climate effects, which, of course, is the current pathway we find ourselves on.
- Much of the differences in high fire severity area and proportion of high severity in burn complexes stems from inherent biases in fire-scar sampling, the omission of multiple lines of evidence in fire review papers, including ignoring historical accounts and assessments, General Land Office surveys, and paleo-charcoal assessments (Baker et al., 2023). Ways to correct for bias—a standard operating procedure in science—are available in the literature (Williams and Baker, 2012; Odion et al., 2016) but are routinely ignored by fuel-centric researchers (Williams and Baker, 2012; Baker et al., 2023).
- Using a recent timeline as the "historical baseline" and then comparing that with contemporary changes in fire extent and severity, often done by land managers, is indicative of a "shifting baseline syndrome" (Papworth et al., 2009), whereby the baseline is shifted to a more recent time period while longer timelines involving ecosystem changes are ignored.
- In the late 1880s to early 1900s, wildfire activity was much greater in the US Rockies and portions of the Pacific Northwest (Egan, 2009) primarily due to prolonged drought. Small and large high severity patches (flare ups) did occur even in frequent-fire forests during droughts (DellaSala and Hanson,

2019). Yet, land managers and decision makers often incorrectly proclaim that current fire sizes are "unprecedented" and not "natural," and assume that large high severity patches historically were rare from frequent-fire forests. We agree that fire extent has been increasing since around the 1980s; however, we have repeatedly shown high severity fire is an important component of mixed-severity fires and high severity patch sizes currently do not appear to be increasing (DellaSala and Hanson, 2019).

- The size, location, and selection of sampling plots can bias postfire conifer establishment estimates, particularly such documented problems as sample plots not properly located within forests, or that were too small to pick up seedlings, leading to false claims of regeneration failures (Hanson and Chi, 2021).
- Thinning itself can cause extensive tree mortality not accounted for in mortality comparisons of areas burned with and without thinning (Hanson, 2022). Failing to account for tree mortality from thinning misleadingly inflates the efficacy of such management.

12.3 RESPOND TO THE ROOT CAUSES AND NOT JUST THE EFFECTS

Nearly all active management is narrowly focused on the effects of the climate-fire feedback rather than treating the root causes, and this failure to act on root causes is making things a lot worse. Bottom-up fuel-centric approaches, while having some merit under a narrow set of conditions, cannot solve for top-down drivers of wildfire activity already overwhelmed by the climate-fire feedback signal triggering large fires (Figure 12.2).

- The 2% or so of fires that escape containment annually do so under extreme fire weather as they rapidly become large (>40,000 ha) fires despite unprecedented suppression efforts (Congressional Research Service, 2023). Retrofitted airline tankers with massive fire-retardant drops are mainly a show of force that cannot possibly contain fire under extreme fire-weather, and are a waste of taxpayer dollars because what puts these fires out is eventual weather changes (e.g., fall rains).
- Extreme fire-weather (hot, dry, high winds) is exacerbated by anthropogenic climate change (ACC) that is overwhelming on-the-ground and in-the-air (air tankers) suppression. This influence will only increase (rapidly) under anticipated emissions scenarios, if we do not move away from fossil fuel and other carbon fuel consumption and dramatically increase protection of natural carbon sinks.
- ACC factors mostly attributed to increased fire activity include early onset of spring conditions, especially snowmelt; longer fire seasons; heat domes and drought (e.g., also expressed as the vapor pressure deficit and the Palmer Drought Index) that elevate evapotranspiration rates resulting in

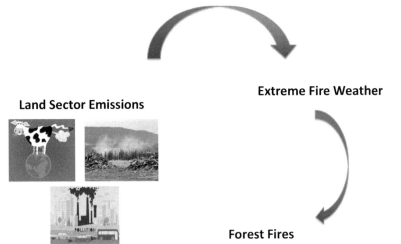

FIGURE 12.2 Illustrating the land sector emissions, extreme fire weather, forest fire feedback system. *Images from Istock (royalty free) and DellaSala (logging). Note: emissions from logging exceed that from natural disturbances combined (Harris et al., 2016), thus, while fires will contribute to emissions they do so on a much lower level.*

parched vegetation and soils; more lightning from greater energy in the atmosphere; and unusual katabatic winds (downdraft from smoke plumes that can cause hurricane force winds and pyro-tornadoes) (Dennison et al., 2014; Westerling, 2016; Abatzoglou and Williams, 2016; Holden et al., 2018; Hawkins et al., 2022; Dahl et al., 2023, also see Chapter 8). In general, ACC has greatly increased the odds of extreme fire weather now influencing fire behavior.

- Based on the above, the current path that fire managers, society, decision makers, and some conservation groups are on with respect to intensive management will fail because it is akin to the mythical Sisyphus pushing a giant boulder up an increasingly steep hill (DellaSala et al., 2022). It is just not possible to meaningfully change this trajectory by working on bottom-up factors while ignoring top-down drivers.

We close this section with this concern: too few legislators and land managers recognize that the only way out of this mess is to solve first and foremost for the root causes. Billions of taxpayer dollars are wastefully spent on back-country logging, supported by some conservation groups ostensibly more interested in photo-ops with misinformed legislators rather than solving for root causes, all the while "Rome is burning" while supporters fiddle away defiantly.

12.4 DOES ACTIVE MANAGEMENT WORK?

The short answer is—it depends on what, why, when, where, and how an area is being treated. Let us first explore the problems of most active management approaches and then what we support in the way of what might work instead, or at least have fewer collateral damages.

The What and Why of Active Management—Active management is shorthand (jargon) for just about anything a land manager wants to do to a pre- and postfire area, including, but not limited to, massive wildfire suppression (e.g., air tankers, ground crews, bulldozers even in wilderness areas) to stop fires even when they cannot possibly be stopped; clearcut logging and extensive roadside "hazard tree removal" to capture economic value under minimal environmental safeguards; postfire logging ostensibly to remove "fuels" and speed up forest succession even though this creates excessive amounts of fine fuels, spreads combustible invasive grasses, and kills natural conifer establishment; road building—permanent or "temporary"—to access areas for logging, causing massive sediment flow to streams; pile burning to reduce logging slash that then damages soils, seed banks, and mycorrhizae; prescribed fire in an attempt to suppress mixed-severity wildfires; livestock grazing to reduce fuels even though that causes weed invasion and soil compaction; and herbicide spraying to reduce conifer competition including killing many shrubs involved in nitrogen fixation and mycorrhizae connections, among other impacts.

The Where of Active Management—Active management actions are aimed at large burn complexes deemed "destroyed by fire" and in need of "restoration" to achieve more "climate resilient" landscapes and sometimes branded as "climate smart forestry." Impacts often accumulate spatially and temporally across large areas (cumulative effects), and in most cases, threaten fire-mediated biodiversity while setting up false expectations that "bad fire" can somehow be replaced by "good fire" via command-and-control of nature (DellaSala et al., 2022). Active management is also a buzz phrase for actions assumed to be benign, and is no doubt informed by public attitudes on fire. Unfortunately, in many cases, it is anything but benign and the buzz phrase (active management) is often used in double-speak misinformation marketing and branding to promote logging. Meanwhile, the public and media often do not bother to check the fine print.

The How of Active Management—Here are some examples covered in this book related to how active management is misused in large fire complexes.

- Despite massive branding, marketing, and communications spending by proponents, nothing they do in the way of "fuels management" can prevent or stop a pyro-tornado or any other extreme fire event from advancing on a community. Nonetheless, thinning and burning have become a sort of pseudo-religion, promoted as a panacea for the fire problem, with anyone questioning it and/or its impacts labeled an active management heretic. In reality, many forms of active management can cause unmitigated harm to ecosystems and the climate (DellaSala et al., 2022) but are discounted or ignored by supporters.

- Calls to scale up active management are being met by unprecedented amounts of wishful-thinking and massive amounts of funding to support even more of the same damaging approaches that can never achieve big reductions in wildfire activity, compared to working with wildfire for ecosystem benefits that does much better for biodiversity, carbon storage, and community safety with far less costs (DellaSala et al., 2017, 2022).
- Proponents expect that the billions of dollars being spent on active management will somehow cover areas big enough to make a difference in fire behavior but this too is unrealistic due to access problems, workforce limitations, climate effects, rapid growth of grasses, shrubs, and tree seedlings following management, and increased wind penetrance in areas where active management has reduced the forest canopy. Despite false claims to the contrary, expansive collateral damages will occur from cumulative management actions, including emissions from logging and other vegetation removal that often exceed those of natural disturbances combined (Harris et al., 2016; Harmon et al., 2022; DellaSala et al., 2022). These damages are almost never counted in active management evaluations, especially by legislators, managers, and fuel-centric researchers.
- Fuel breaks are championed as community protection yet are largely ineffective, costly, and damaging to ecosystems (Hanson, 2021). Clearing swaths of vegetation fragments ecosystems, introduces weeds and soil desiccation, must be frequently maintained, and does little to nothing in high wind events that can carry embers over large distances (including in at least one situation across the Columbia River Gorge in Oregon/Washington, and another across >15 km in the northern Sierra Nevada during the Dixie fire).
- Purposefully setting fires to reduce vegetation ahead of flame fronts is often used in fire-fighting operations to reduce fuels before the fire arrives. However, this is almost never monitored for effects on burn area and fire severity, especially impacts to fire refugia (unburned areas).
- While claims are made about the efficacy of active management (thinning, burning), the evidence is undermined by potential confounding factors on fire behavior such as time of day, wind speed and direction, humidity levels and temperature that may have had a much bigger influence on fire behavior than the actions themselves. These can shift hourly during a fire.

There is a Better Way—Despite the many limitations of active management, there are some approaches (active and passive) that we support in restoring wildfire-dependent ecosystems with some tweaking to limit damages as follows.

- Some forms of active management can be effective in temporarily reducing fire intensity if done right (see below), and the treated site happens to get lucky enough to be hit by a lightning bolt under low-moderate fire weather (very low odds, Schoennagel et al., 2017). However, with climate change,

this too will not alter fire activity much as more fires are spun out in the dangerous fire feedback connection to ACC factors.
- Increasing the scale and pace of natural ignitions for ecosystem benefits works. There is simply no better way to address wildfires at scale when conditions are safe to do so and it should be done in conjunction with community safety measures (Moritz et al., 2014; Schoennagel et al., 2017). Importantly, managing for wildfire use does not involve sitting back and letting everything burn. To the contrary, it means attempting to suppress the edges of wildfires that are near residential areas, monitoring spread rates, and employing the use of hand fire-lines instead of bulldozers in sensitive areas like wilderness. All of this can be done in the same fire complex by compartmentalization of suppression tactics, and it is without question a form of active management, but one that is highly cost-effective and provides clear benefits for ecosystems and communities. Unfortunately, while the Forest Service claims to be implementing its National Cohesive Wildland Fire Management Strategy (https://www.forestsandrangelands.gov/strategy/thestrategy.shtml; accessed September 13, 2023) that allows for wildfire use, pressure is mounting from the public and decisions makers not to do so for fear of smoke and escaped fires.
- The best way to reduce unwanted human-caused ignitions is to close and obliterate roads, limit public access during red flag (high risk) wildfire conditions, and insulate powerlines and implement safer technologies to prevent sparks from fallen lines. Notably, some 84% of wildfires are human-caused (Balch et al., 2017). Land managers in particular seldom consider anthropogenic ignitions in wildfire risk reduction yet they must and it is much easier to solve for this than the fuels issue.
- Get rid of the cows! Livestock tend to concentrate on native plants, thereby contributing to understory type conversions populated by flammable, invasive weeds (Beschta et al., 2012). The use of livestock exclosures, permanent retirement of public lands grazing leases with financial compensation (public "buyouts"), and weed eradication are essential to avoid understory type conversions. However, this is a futile effort if the vectors (cows, roads, logging machinery, off-highway vehicles) are not also curbed.
- Reduce or remove longstanding administrative barriers to Native American cultural burning for ecosystem and cultural benefits but do not expect this to appreciably change fire behavior at large scales due to limitations. It can, however, have beneficial local effects.
- When and where relying upon natural ignitions is infeasible, prioritize prescribed fire, recognizing that thinning or other tree removal activities may be unnecessary prior to burning, even in the densest forests, and if the goal is generally low fire severity, such burning can simply be conducted early in the fire season when fire weather is relatively mild (Stephens and Finney, 2002; van Mantgem et al., 2011). Where thinning is conducted,

limit it to genuinely small trees, without reducing dominant overstory trees and while maintaining representative (intact) understories. This can be followed with cultural or prescribed fire.
- Managers should use multiple lines of evidence (Odion et al., 2014; Baker et al., 2023) rather than relying strictly on uncorrected fire-scar sampling to determine how much and how often to intervene (Williams and Baker, 2012). This means in frequent fire systems, the time between actions may need to be extended to allow ecosystems to function properly. For example, many dry pine, dry mixed conifer forests historically had varied tree densities and understories yet are most often overly thinned and type converted to open, park-like savannas (e.g., Williams and Baker, 2012).
- Protect mature and old-growth forests from all forms of logging, including when they are affected by natural disturbances. There is absolutely no ecological reason for postdisturbance logging and thinning should be prioritized to the more flammable logged areas.
- Conduct prescribed and cultural burning practices under safe conditions and not during the shoulder season, during droughts, or high winds. As mentioned, pile burning is not the same as prescribed burning as it damages soils and mycorrhizae.

12.5 PUBLIC ATTITUDES ARE SHIFTING BUT NATURE'S PHOENIX REMAINS UNDERVALUED

While the public generally supports prescribed burning and some fires in the backcountry, attitudes have not changed enough about mixed severity fires that have a significant high-severity component, largely because of the prevalence of "good" versus "bad" fire myopia and the public's genuine fear of fire. The "good fire" messaging persists largely because of the public's limited understanding of fire and oversimplified messaging from groups like The Nature Conservancy, decision makers, and lands managers. Nonetheless, public polling shows the following mixed but somewhat encouraging signs especially in relation to climate change.

- In a large sample (n = 807) of adult residents in southern California, respondents were asked 40 questions on wildfire perception and knowledge, including fire risk reduction measures (Masri et al., 2023). Researchers reported that female gender, knowing a wildfire victim, and having a general interest in nature were main factors associated with perceptions about wildfires, and contributed to willingness to evacuate if threatened, and willingness to support a wildfire-related tax increase. Importantly, the majority (>57%) got the connection between climate change and wildfire increases.
- According to a poll conducted by YouGov for the Economist, as reported by FiveThirtyEight (https://fivethirtyeight.com/features/maui-wildfires-climate-change-public-opinion/; accessed September 10, 2023), right after the

deadly fire in Lahaina (Hawaii) (August 12−17, 2023), only 37% of Americans believed that fire was primarily the result of climate change with a similar share (36%) believing fires like this one happen from time to time! Not surprising, >60% that voted for Joe Biden as president got the climate connection, showing partisan views persist in spite of the rise in climate-related extreme events. The trend in extreme climate events destroying towns like Lahaina (2023; hurricane force winds), Paradise (CA, 2018; winds >96 kph; >60 mph), and the Labor Day fires (OR, 2020; winds >64 kph; >40 mph) is at least partially attributed to ACC and alarming to say the least.

- A large share (42%) of Americans say they are concerned with air quality (June−July 2023) that was compromised by Canadian wildfires of 2023, with 59% attributing these fires to climate change (https://today.yougov.com/topics/health/articles-reports/2023/07/07/four-10-americans-say-their-air-quality-wildfires; accessed September 10, 2023). Likewise, the split is astonishingly partisan with 84% of Democrats and only 32% of Republicans making the climate connection.
- With the increase in wildfire spillovers in urban areas at least since the first edition of this book, homeowners still are dangerously unprepared (though local examples of wildfire preparedness have increased markedly). For instance, in a 2021 poll conducted by Oregon State University of 458 residents in the so called wildland−urban interface (WUI) of Deschutes County, Oregon (where fires are frequent), only 38% had taken precautions within 5 years, including fire-resistant plants, nonflammable construction materials, pruning branches, pruning of combustible vegetation nearest the home (∼30 m), and fire-hardening of the structure itself (https://liberalarts.oregonstate.edu/sites/liberalarts.oregonstate.edu/files/deschutes_survey_report.pdf; accessed September 10, 2023).
- In a June 2−11, 2022 statewide poll by Oregon Values and Beliefs Center, Oregonians were mostly concerned about smoke, loss of fish and wildlife habitat, and severity of wildfires (https://www.opb.org/article/2022/07/04/poll-oregonians-united-in-concerns-about-wildfires/; accessed September 10, 2023). Less than half approved of how fires are currently managed, 78% said there should be periodic "controlled burns," 89% supported defensible space adjacent to homes, and 85% supported home hardening. However, there is still a large portion of society that believes wildfires with high severity components are destructive to fish and wildlife habitat.
- The public polling connection between "extreme weather" events (long hot periods, droughts, intense storms, floods, wildfires, sea level rise) and climate change is somewhat encouraging as a May 2−8, 2022 poll of 10,282 US adults across the United States conducted by Pew Research Center showed >80% of those sampled indicated that there was at least a "*little*" connection to climate change in the extreme weather events they

were experiencing (https://www.pewresearch.org/short-reads/2022/08/12/most-americans-who-have-faced-extreme-weather-see-a-link-to-climate-change-republicans-included/; accessed September 10, 2023). Some 95% of Democrats and Democratic-leaning Independents made the climate connection compared to 65% of Republican and Republican-leaning Independents (at least that is encouraging!). Not surprising, the wildfire-climate connection was strongest in the western United States.

- During a July 28-August 9, 2023 poll of 1071 adults conducted by EKOS Politics in Canada, climate change, by a wide margin (64%), stands out as the primary reason for the recent string of wildfires with 31% also assigning culpability to the forestry industry (https://www.ekospolitics.com/index.php/2023/08/public-attitudes-to-wildfires-and-the-logging-industry/; accessed, September 10, 2023).
- In a January 2−13, 2021 poll of 400 registered voters in 8 Western states, ~70% of respondents said wildfires are "more of a problem" than a decade ago, 42% of whom attributed the increase to climate change and 40% to drought (https://www.washingtonexaminer.com/politics/western-voters-concerned-wildfires-poll; accessed, September 10, 2023).

12.6 THE DISCONNECT BETWEEN INDEPENDENT RESEARCH AND WILDFIRE ATTITUDES

In the first edition of this book, we and others reported on a wildfire deficit for all fire severities across North America and globally. With this second edition, we reaffirm the fire deficit still exists but the gap between contemporary and historical fires could close soon with more large wildfires tied to extreme fire-weather caused by ACC. A study by Marlon et al. (2012), however, remains instructive regarding long timelines needed to fully assess fire cycles and define historical baselines (i.e., it is the antidote to the shifting baseline syndrome). In this case, the authors used a 3000-year sedimentary charcoal record in the western United States that showed how fires track climate over multi-century and millennial time scales. For instance, a lengthy cool down (Little Ice Age) dampened fires around 1400 to 1700 with prominent peaks in the Medieval Climate Warming Anomaly (950−1250 CE) and the late 1800s (warmer, drier). As we also discussed in Chapter 8 of this edition, it is important to understand the long arc of fire cycles before jumping to conclusions about "unprecedented" burning, particularly as fire activity relates to temperature and precipitation factors controlled by global processes like the interplay between El Niño and La Niña events. Nonetheless, the upward trend in area burning is likely to continue for decades until the climate stabilizes; however, rates of high severity fire remain proportional to fire sizes at least for the time being (DellaSala and Hanson, 2019). That is to say, nature will modify fire extent regardless of how much we try to suppress or alter fire behavior by ineffective bottom-up active management. Simply put, we can never put the fire "Genie" back in the bottle or even expect to lower fire intensity and fire spread rates when climate is the main driver. Claiming to do so is a form of climate denialism with champions complicit in the intertwined planetary biodiversity and climate crises.

In general, we do see some appreciable shifts in public attitudes about large fires having ecosystem benefits, including some changes with regard to the connection between wildfires and climate change and great attention to the importance of home hardening and defensible space to protect communities from wildfires. Nevertheless, the broader fire dialog too often continues to be greatly misinformed by the "good" versus "bad" fire messaging promoted by groups like The Nature Conservancy and public agencies involved in logging-fuels projects like the US Forest Service. This is underscored by the politics of fire hyperbole and spread of fire misinformation, routinely broadcasted by the media duped by false solutions claims, and supported by biased research and fuels-oriented limited research. The public generally remains supportive of wildfires as an essential ecosystem process but mostly if fires are "controlled" or "prescribed," with substantive misunderstandings regarding large fires and severe burns. The emphasis by too many decision makers thus far remains completely misplaced, focused on throwing even more money at the effects of fire rather than solving for the root causes. Treating bottom-up factors without solving for top-down drivers is risky business in North America, Europe, and Australia.

12.7 LESSONS LEARNED AND CLOSING REMARKS

It has been nearly a decade since the first edition of Nature's Phoenix and there are many lessons learned to build off, as follows:

- The fire deficit in forests is the result of decades of mechanized fire suppression that annually stamps out all but some 2% of fire starts (mostly since US Forest Service policies to extinguish all fire starts by 10 a.m. the next day, and accelerating after the advent of Smokey Bear following WWII). Despite lip service paid to fire suppression being a root cause of the fire problem and some calls to work with natural fire ignitions under safe conditions, the US Forest Service seldom implements its own natural wildfire ignition policy due largely to the fear of fire by the public aided by misinformation about good versus bad fires and fire hyperbole from decision makers with financial or political stakes in the outcomes. Notably, we are aware of certain Forest Service regional supervisors that continue to throw everything they can at a natural fire burning safely in remote areas simply because they can (blank check from Congress) and as a show of public force. Additionally, the Chief of the Forest Service has repeatedly put a temporary hold on prescribed fire during fire season to direct everything the

agency has at suppression, but this will only magnify the fire deficit and will not protect vulnerable communities.
- While there has been a shift in the fuel-centric community toward general acceptance of mixed severity fire, the active management approach remains centered on maintaining low-moderate fire even though this hypothesis was summarily rejected by a more complete evidence review of fire regimes (Baker et al., 2023). The low-moderate severity approach undermines the importance of high-severity patches that produce a pulse of biological legacies critical to sustaining biodiversity needs for decades to centuries (DellaSala and Hanson, 2019). This most biodiverse severe fire component is not recreated by prescribed fire or thinning, unless potentially via escaped burns. However, it may be possible to adjust burn prescriptions to accommodate more natural patterns and biological legacy formulations, perhaps in roadless areas and wilderness removed from towns, while wildland fire use guarantees natural mixed-severity habitat heterogeneity.
- The emphasis on intensive, mechanized fire management misses the mark on the circularity of succession that uniquely binds the dynamism of forests—ever changing—across spatial and temporal scales. Hence, Nature's Phoenix is much more complex than we can ever imagine or manage through logging, suppression, and other forms of active management, aside from wildland fire use.
- The good versus bad fire messaging is damaging to ecosystems that need the appropriate mixture of burn patches, including high severity ones. We urgently need a new eco-centric lexicon that purges the catastrophe-speak (good vs. bad fire), replaces burn scar with fire perimeter, eliminates "destroyed" or "consumed" by fire in favor of "rejuvenated" and "restored," and even fire severity itself needs to be replaced with fire effects. We recognize this is a tall order given cognitive dissonance and engrained terminology but nothing changes unless we lead by example.
- Whether high severity fire is increasing or not is informed by differences in studies and confounding factors affecting fire behavior; nevertheless, even if it is increasing in some areas, we cannot control this via active management given the increasing climate signal.
- Collateral ecosystem damages and emissions from intensive management will continue to accumulate over large areas until policies change.
- We need to have objective monitoring of the spatiotemporal extent of high severity fire and circular succession at ecologically meaningful postfire timescales and with scientifically defensible sampling (Hanson and Chi, 2021), before conclusions about "megafires" leading to ecosystem type shifts are given credence (see Chapter 2). Comprehensive monitoring of active management impacts also needs to be addressed rather than trivialized.
- It is certainly appropriate to grieve one's favorite campground or hiking trail that appears at face value "destroyed" by an intense burn, but this view misses the soon to emerge bounty of Nature's Phoenix. The only way to

shift from grief to nature celebration is to have people and the media see for themselves the magnificent process of renewal over time (1, 3, 5, 10 years postfire). An example is the so-called "destructive" 1988 fires in Yellowstone National Park that was soon flooded by photo-snapping tourists in search of colorful wildflower blooms with majestic elk foraging on the new growth amidst naturally regenerating forests. Some excellent time elapsed videography is available to marvel at nature's capacity to self-rejuvenate (https://vimeo.com/showcase/8470537; https://www.youtube.com/@thrivingwithfire1701; https://vimeo.com/greenoregon; accessed September 26, 2023; courtesy of Ralph Bloemers).

- There appears to be greater public awareness of the role of ACC in driving large fires. However, this connection remains highly partisan and polarized within political circles that continue to deny the causal mechanisms of large fires linked to carbon emissions from burning fossil fuels and land-use actions like forestry. The only way to interrupt the climate-fire feedback is to address emissions across all sectors, especially logging/thinning and road building.
- Simply put, when it comes to homes, if it was built in harm's way and not hardened to fire, and without appropriate defensible space pruning, it may very well burn if/when a fire passes through the area. We have witnessed this repeatedly in tragic urban wildfires that destroyed towns like Lahaina, Paradise, Talent, and Phoenix, and many more communities in North America, Europe, and Australia. The only way to resolve this is more effective land-use planning (we do not build on top of inactive volcanoes so why build in fire's path?), and by working from the home-out in fire risk reduction. Nothing managers do in the backcountry will solve for home loss; it all comes down to that 30-m zone around the home and the home itself as to whether it burns or not.

We pulled no punches in this book, and some may find our views offensive to organizations like The Nature Conservancy and federal researchers and land managers. However, we attack regressive, misinformed, and harmful ideas and policies, not people, and we seek to expose some of the shortcomings of biased research, and the approaches of conservation groups that appear to have drifted from their biodiversity mission. We also take note of the dedicated work of activists all over the world to solve for the mounting problems of climate chaos and its connection not only to fossil fuels but to the way forests and chaparral systems are being degraded by inappropriate active management. And we provide this edition as a valuable source for scientists incorrectly branded as being in the "minority," knowing instead that when a comprehensive review of the evidence is presented, their concerns are anything but the minority opinion (DellaSala, 2021; Baker et al., 2023). We stand with those willing to question what rightfully needs to be questioned, and for greatly improving fire management for the benefit of safe communities and fire-mediated biodiversity.

We also offer this book to those fighting back against what is routinely perceived as "agency deference" in the courts by providing activists with a scientific resource that may help in project-level appeals. After all, the only way to shift public attitudes is through sound science, activism, the courts, and seeing for yourself the remarkable miracle of Nature's Phoenix rising from the ashes by taking the time to witness a burned area spring to life. When it comes to a severely burned forest, the beauty or the destruction of the event is in the eyes of the beholder, but nature is the ultimate teacher if we are patient enough to listen, watch, and experience the rebirthing processes unfolding.

REFERENCES

Abatzoglou, J.T., Williams, A.P., 2016. Impact of anthropogenic climate change on wildfire across western US forests. Proc. Natl. Acad. Sci. USA. www.pnas.org/cgi/doi/10.1073/pnas.1607171113.

Baker, W.L., Hanson, C.T., Williams, M.A., DellaSala, D.A., 2023. Countering omitted evidence of variable historical forests and fire regime in Western USA dry forests: the low-severity fire model rejected. Fire 6, 146. https://doi.org/10.3390/fire6040146.

Balch, J.K., Bradley, B.A., Abatzoglou, J.T., Nagy, R.C., Fusco, E.J., Mahood, A.L., 2017. Human-started wildfires expand the fire niche across the United States. Proc. Natl. Acad. Sci. USA. www.pnas.org/cgi/doi/10.1073/pnas.1617394114.

Beschta, R.L., multiple authors, 2012. Adapting to climate change on western public lands: addressing the impacts of domestic, wild and feral ungulates. Environ. Manag. https://doi.org/10.1007/s00267-012-9964-9.

Bradley, C.M., Hanson, C.T., DellaSala, D.A., 2016. Does increased forest protection correspond to higher fire severity in frequent-fire forests of the western United States? Ecosphere 7, 1−13.

Congressional Research Service, 2023. Wildfire Statistics. https://sgp.fas.org/crs/misc/IF10244.pdf. (Accessed 14 September 2023).

Dahl, K.A., Abatzoglou, J.T., Phillips, C.A., Ortiz-Partida, J.P., Licker, R., Merner, L.D., Ekwurzel, B., 2023. Quantifying the contribution of major carbon producers to increases in vapor pressure deficit and burned area in western US and southwestern Canadian forests. Environ. Res. Lett. 18 (6).

DellaSala, D.A., Hutto, R.L., Hanson, C.T., Bond, M.L., Ingalsbee, T., Odion, D., Baker, W.L., 2017. Accommodating mixed-severity fire to restore and maintain ecosystem integrity with a focus on the Sierra Nevada of California, USA. Fire Ecol. 13, 148−171.

DellaSala, D.A., Hanson, C.T., 2019. Are wildland fires increasing large patches of complex early seral forest habitat? Diversity 11, 157. https://doi.org/10.3390/d11090157.

DellaSala, D.A., 2020. Fire-mediated biological legacies in dry forested ecosystems of the Pacific Northwest, USA. In: Beaver, E.A., Prange, S., DellaSala, D.A. (Eds.), Disturbance Ecology and Biological Diversity. CRC Press Taylor and Francis Group, LLC, Boca Raton, FL, pp. 38−85.

DellaSala, D.A. (Ed.), 2021. Conservation Science and Advocacy for a Planet in Crisis: Speaking Truth to Power. Elsevier, Oxford. https://doi.org/10.1016/C2016-0-03650-1.

DellaSala, D.A., Baker, B., Hanson, C.T., Ruediger, L., Baker, W., 2022. Have western USA fire suppression and active management approaches become a contemporary Sisyphus? Biol. Conserv. 268. https://doi.org/10.1016/j.biocon.2022.109499.

Dennison, P.E., Brewer, S.C., Arnold, J.D., Moritz, M.A., 2014. Large Wildfire Trends in the Western United States, 1984-2011. Geophysical Research Letters. https://doi.org/10.1002/2014GL059576.
Donato, D.C., Campbell, J.L., Franklin, J.F., 2012. Multiple successional pathways and precocity in forest development: can some forests be born complex? J. Veg. Sci. 23, 576—584.
Egan, T., 2009. The Big Burn. Houghton Mifflin Harcourt, Boston, NY.
Hanson, C.T., 2021. Smokescreen: Debunking Wildfire Myths to Save Our Forests and Our Climate. University Press Kentucky, Lexington, KY.
Hanson, C.T., 2022. Cumulative severity of thinned and unthinned forests in a large California wildfire. Land 11, 373.
Hanson, C.T., Chi, T.Y., 2021. Impacts of postfire management are unjustified in spotted owl habitat. Front. Ecol. Evol. 9, 596282.
Harmon, M.E., Hanson, C.T., DellaSala, D.A., 2022. Combustion of aboveground wood from live trees in megafires, CA, USA. Forests. Forests 13 (3), 391. https://doi.org/10.3390/fl3030391.
Harris, N.L., multiple authors, 2016. Attribution of net carbon change by disturbance type across forest lands of the conterminous United States. Carbon Bal. Manag. 11, 24. https://doi.org/10.1186/s13021-016-0066-5.
Hawkins, L.R., Abatzoglou, J.T., Li, S., Rupp, D.E., 2022. Anthropogenic influence on recent sever autumn fire weather in the West Coast of the United States. Geophys. Res. Lett. 49, e2021GL095496. https://doi.org/10.1029/2021GL095496.
Hessburg, P.F., Prichard, S.J., Hagmann, R.K., Povak, N.A., Lake, F.K., 2021. Wildfire and climate change adaptation of western North American forests: a case for international management. Ecol. Appl. 21, e02432.
Holden, Z.A., Swanson, A., Lucke, C.H., Affleck, D., 2018. Decreasing fire season precipitation increased recent western US forest wildfire activity. Proc. Natl. Acad. Sci. USA 115 (36). https://doi.org/10.1073/pnas.1802316115.
Marlon, J.R., multiple authors, 2012. Long-term perspective on wildfires in the western USA. Proc. Natl. Acad. Sci. USA. www.pnas.org/cgi/doi/10.1073/pnas.1112839109.
Masri, S., Shenoi, E.A., Garfin, D.R., Wu, J., January 2023. Assessing perception of wildfires and related impacts among adult residents of southern California. Int. J. Environ. Res. Publ. Health 20 (1), 815. https://doi.org/10.3390/ijerph20010815.
Moritz, M.A., multiple authors, 2014. Learning to coexist with fire. Nature 515, 58—66. https://doi.org/10.1038/nature13946.
Odion, D.C., multiple authors, February 2014. Examining historical and current mixed-severity fire regimes in ponderosa pine and mixed-conifer forests of western North America. PLoS One 9, 1—14.
Odion, D.C., Hanson, C.T., Baker, W.L., DellaSala, D.A., Williams, M.A., May 19, 2016. Areas of agreement and disagreement regarding ponderosa pine and mixed conifer forest fire regimes: a dialogue with Stevens et al. PLoS One. https://doi.org/10.1371/journal.pone.0154579.
Papworth, S.K., Rist, J., Coad, L., Milner-Gulland, E.J., 2009. Evidence for shifting baseline syndrome in conservation. Conserv. Lett. https://doi.org/10.1111/j.1755-263X.2009.00049.x.
Schoennagel, T., multiple authors, 2017. Adapt to more wildfire in western north american forests as climate changes. Proc. Natl. Acad. Sci. USA 114, 4582—4590. https://doi.org/10.1073/pnas.1617464114.
Stephens, S.L., Finney, M.A., 2002. Prescribed fire mortality of mixed-conifer tree species: effects of crown damage and forest floor combustion. For. Ecol. Manag. 162, 261—271.
Swanson, M.E., multiple authors, 2010. The forgotten stage of forest succession: early-successional ecosystems on forested sites. Front. Ecol. Environ. 9, 117—125. https://doi.org/10.1890/090157.

van Mantgem, P.J., Stephenson, N.L., Battles, J.J., Knapp, E.K., Keeley, J.E., 2011. Long- term effects of prescribed fire on mixed conifer forest structure in the Sierra Nevada, California. For. Ecol. Manag. 261, 989–994.

Westerling, A.L., 2016. Increasing western US forest wildfire activity: sensitivity to changes in the timing of spring. Philos. Trans. Royal Soc. B. https://doi.org/10.1098/rstb.2015.0178.

Williams, M.A., Baker, W.L., 2012. Variability of historical forest structure and fire across ponderosa pine landscapes of the Coconino Plateau and south rim of Grand Canyon National Park, Arizona, USA. Landsc. Ecol. 28, 297–310.

Index

'*Note:* Page numbers followed by "f" indicate figures, "t" indicate tables and "b" indicate boxes.'

A

Acer spp., 234
Active management, 345, 356−358, 381−383
  approaches, 30
  religion of sorts, 358−360
  public lands, 340
  work, 385−389
Africa
  savannah habitats, 217−218
  vegetated habitat, 211−212
Adaptations, 63−64
Administrative screens, 306
Aeolian sands, 232−233
Aerial photos, 9−10
African buffalo (*Syncerus caffer*), 216
Agile antechinus (*Antechinus agilis*), 107
Air quality index (AQI), 277, 281
Aleppo pine (*Pinus halepensis*), 111−112
Allelopathy, 174
Almeda fire, 335−336, 344
Alpine ash (*Eucalyptus delegatensis*), 318−319
American martens (*Martes americanus*), 111−112, 135b−136b
American three-toed woodpecker (*Picoides dorsalis*), 70−73
Andrena lapponica, 235
Angeles National Forest, 190
Animal fire adaptations, 13−14
Anthropogenic burning, 225
Anthropogenic climate change (ACC), 257−258, 384
  climate tipping point affect wildfires and people, 269−272
  extreme wildfires impact built environment, 275−281
  linking wildfire to, 269−281
  rigorous methodology for attributing wildfires to, 272−275
Anthropogenic disturbances, 33b
Anthropogenic megafires, 30
Aplodontidae (mountain beavers), 104−105
Apostlebird (*Struthidea cinerea*), 82−84
Arizona Game and Fish Department, 181
Aspen (*Populus tremuloides*), 41−42, 114, 116−117
Australia, Wildfires (2009), 220−223

B

Backcountry fiddlers, 344−354
  dwarf mistletoe control project, 350f
  unlogged rare wetland inclusion within complex early seral forest, 349f
Balsam fir (*Abies balsamea*), 240−242
Balsam poplar (*Populus balsamifera*), 240−242
Bandicoots (*Isoodon obesulus* and *Perameles nasuta*), 107−108
Bank voles (*Myodes glareolus*), 106−107, 235−236
Bark beetle
  affecting forest resilience, 150−153
  outbreaks, 137, 151b
  and biodiversity, 135b−136b
Barred owl (*Strix varia*), 32−33
Basal hollows, 101−102
Bats, 101−104, 203−204
  adult aquatic insects, 102−103
  adult mosquitoes, 101
  basal hollows, 101−102, 102f
  Coleoptera species, 103−104
  dead trees, 104
  echolocation frequencies, 101
  ecosystem function, 104
  high-quality habitats, 103−104
  median capture rate, 103
  nectar-feeding bats, 101
  nitrogen, 101
  postfire logging, 100−101
  pyrophilous, 103−104
  roosting sites, 103
  snag recruitment, 103

Bats (*Continued*)
  stream benthic macroinvertebrates, 102−103
  thermoregulation, 101−102
Bears (*Ursus* spp.), 99
Beaver (*Castor canadensis*), 99
Beetles, 133−137
Bembix rostrata, 233
Betula pendula, 234
Big Burn (1910), 3, 35−36
Big-cone douglas-fir (*Pseudotsuga macrocarpa*), 167−168
Bighorn sheep (*Ovis canadensis*), 99
Biodiversity, 15−17, 376, 380
  Canadian boreal ecosystem, 248−250
  sub-Saharan Africa, 211−212
Biomass, 363
Bird ecology, 64−84, 203−204
  fire effects, context dependent
    old-growth forests, 66−68
    postfire vegetation conditions, 68−79
    time since fire, 65−66
  fire risk reduction should be focused on human population centers, 84
  fire suppression should be focused on human communities, 84
  high-severity fires beget mixed-severity results, 85
  insights from bird studies, 64−84
  mimic nature, 87−89
  mitigate fire severity through thinning, 85
  postfire "salvage" logging, 86−87, 86f
  severe fire, bird species in other regions seem to require, 79−84
    buff-breasted flycatcher, 80
    conservation efforts, 80−82
    ecological integrity, 84
    eucalyptus forest species, 82−84
    fire-killed timber areas, 73−74
    foraging locations, 78
    habitat distribution patterns, 79−80
    logging techniques, 80−82
    mixed-conifer forests, 82−84
    montane ash forests, 82−84
    postfire management implications, 84
    vegetation types, 84
    wood-boring beetle larvae, 77
  temporal and vegetation conditions
    burned forests, 70−73
    burned-out root wads or uprooted trees, 70−73
    intact tops, broken-top snags, 75f
    mature-to old-growth forests, 70−73
    secondary cavity-nesting and snag-nesting species, 70−73
    seed dispersers, 76f
    severe fire, 68−70
Biscuit Fire (2002)
  megafires, 29−30
  Southwest Oregon, 301−307
    administrative screens, 306
    Biscuit Project Scope, 302−303
    burn perimeter, 307f
    burn severities, 301f, 304f
    canopy mortality, 304f
    case study conclusions, 306−307
    context and scale matter, 304−305
    ecological processes, 302
    ecological screens, 306
    implications, 302
    integrating context and scale into project decisions, 306
    operational screens, 306
Biscuit Fire, 302−303
Black ash (*Fraxinus nigra*), 240−242
Black-backed woodpecker (*Picoides arcticus*), 32, 70−74, 78−79, 85, 88, 249−250, 314−315
Black bears (*Ursus americanus*), 32, 114
Black-faced woodswallow (*Artamus cinereus*), 82−84
Black Friday Bushfires (1939), 34−35
Black Hills, 38−39, 315−318, 374−375
Black Hills National Forest (BHNF), 315
Black locust (*Robinia pseudoacacia*), 232
Black Mountain fire, 65
Black oak (*Quercus kelloggii*), 41−42
Black Saturday wildfires, 204
Black spruce (*Picea marina*), 240−242
Black-tailed jackrabbit (*Lepus californicus*), 106
Black wattle (*Acacia mearnsii*), 221−222
Blanket leaf (*Bedfordia salicina*), 197−198
Board of Forestry, 185
Bobcat (*Lynx rufus*), 113
Bolan Lake, 265
Bombus cryptarum, 233
Bontebok (*Damaliscus pygargus pygarus*), 225
Bootleg Fire, 352−354, 353f
Boreal black spruce (*Picea mariana*), 104
Boreal forest, 239
Bottom-up fuel-centric approaches, 384
Bracken fern (*Pteridium* spp.), 264

Brazilian Amazon, 30
Brown creeper (*Certhia americana*), 65
Brown thornbill (*Acanthiza pusilla*), 82—84
Brush mice (*Peromyscus boylii*), 106
Buff-breasted flycatcher (*Moucherolle beige*), 80
Buff-rumped thornbill (*Acanthiza reguloides*), 204
Building blocks for Nature's Phoenix, biological legacies as, 298b
Bureau of Land Management (BLM), 344
Bush rats (*Rattus fuscipes*), 107

C

Cactus mice (*Peromyscus eremicus*), 106
Calibration studies and process-based models, 261
California Board of Forestry and Fire Protection, 184
California Chaparral Institute (CCI), 63, 164, 184, 189
California Department of Fish and Wildlife (CDFW), 184
California desert Bighorn sheep (*Ovis canadensis nelsoni*)), 117
California Environmental Quality Act, 183, 185
California grizzly bear (*Ursus arctos californicus*), 163
California mice (*Peromyscus californicus*), 106
California mountains, 117
California spotted owl (*Strix occidentalis occidentalis*), 32—33, 69b—70b, 311—312, 313f, 314—315
California Statewide Fire History Database, 174
California voles (*Microtus californicus*), 106
California's Fire Resource and Assessment Program's Fire History Database, 179
Calliope hummingbird (*Selasphorus calliope*), 70—73
Canada lynx (*Lynx canadensis*), 112
Canadian boreal ecosystem, large fires in annual area burned in, 248f
area burned by decade in, 249f
biodiversity, 248—250
boreal forest and woodlands, 240f
boreal landscape in Terra Nova National Park, 243f
Canadian boreal forest and Alaska boreal interior, 241f
ecozones of Canada, 241f
fire cycle estimates for, 245t
fire regime of, 242—246
freshly killed trees, 251f
green halo, 238—239
land of extremes, 239
large fires, 244f, 247f
mixed-severity fires, 242—243
old-growth forests, 243
plants coping with fire, 242
residual structure and remnants left after large fire, 251f
Saproxylic insects, 249—250
temporal patterns of fire and changes, 246—248
vegetation, 240—242
Canadian lynx (*Lynx canadensis*), 99
Canopy bulk density (CBD), 142—143
Cape floristic region, 225
Carbon
cycle, 151b
storage, 364
3-carbon molecules (C3), 214
4-carbon molecules (C4), 214
Caribou (*Rangifer tarandus*), 120
Carnivores, 111—115
bears, 113—115
mesocarnivores and large cats, 111—113, 112f
seed dispersal by, 114b
Cassin's finch (*Haemorhous cassinii*), 70—73
Castoridae (beaver), 104—105
Catastrophic fire, 316
Ceanothus, 119—120, 163, 175—176, 265
Ceanothus megacarpus seedlings, 175—176
Cedar (*Calocedrus decurrens*), 265
Cedar Fire (2003), 189
Cedar Fire in San Diego County (2003), 186
Central Europe
Aeolian sands, 232—233
burned areas, 231
burned forest stands, 234—235
forest fires, 231
invertebrates and postfire habitat, 235—236
Natura 2000, 231
postfire succession, 233—236
setting, 231
Central Highland ash forest, 319
Central Highlands of Victoria, 318—319
Cerambycid *Arhopalus tristis* (F.), 103—104
Ceratodon purpureus, 234
Cerro Grande Fire (2000), 189

Chamaesphecia leocopsiformis, 233
Chamise (*Adenostoma fasciculatum*), 106, 176
Chaparral, 106, 171–172, 175
  analysis, 261
  crown fires, 166
  fire intensity, 166
  fire misconceptions, 168–182
    allelopathy, 174
    combustible resins and hydrophobia, 180–182
    confusing fire regimes, 169–170
    decadence, productivity, 172–173
    fire suppression myth, 174–175
    Native American burning practices, 171
    postfire recovery, 179
    succession, 171–172
    too much fire degradation, 175–179
    type conversion and prescribed fire, 179–180
  fire-severity, 167
  fire suppression paradigm, 163–166
    biodiversity, 163
    cognitive dissonance, 163
    fire-adapted ecosystem, 164
    oozing combustible resins, 165
    plant community, 164f
    surface fires, 166
    wildfires, 164–165
  paradigm change revisitation, 186–189
    extensive fuel treatments, 188
    fuels, 186
    horrific fires, 187–188
    straw man fallacy, 187
    terrestrial environments, 187
    wildfire, 186
  paradigm shift, 189–190
  plant communities, 168
  reconstructions, 13
  records, 260–261
  reducing cognitive dissonance, 182–186
    land management plans, 182
    local agency, 183–184
    media, 185–186
    state agency, 184–185
  resprouting species, 167f
  shrubs, 167f
  source area, 261
Charcoal accumulation rates, 261
Chipmunk species, 108–109
Chronic management impacts, 297

Clark's nutcracker (*Nucifraga columbiana*), 70–73
Clean renewable energy, 359
Cleveland National Forest, 38
Climate change, 201, 271, 296–297, 375
  anthropogenic climate change, 269–281
    climate tipping point affect wildfires and people, 269–272
    extreme wildfires impact built environment, 275–281
    rigorous methodology for attributing wildfires to, 272–275
  came knocking at my door, 335–342, 337f
  fire history across moisture gradient, 262–263, 262f
  fire-history analysis from lake sediments, 245
  fires and bark beetles affecting forest resilience in context of, 150–153
  fuel-limited fire regimes depicting interaction of, 259f
  historical range of variation (HRV), 267–268
    high severity fire, 268
    land managers and policymakers, 267
  looking back over paleo-record, 258–263
  models, 257
  reconstructing past fire regimes
    burned and unburned vegetation patterns, 258–260
    data and models types, 260f
    fire-scar tree-ring records, 258–260
    tree rings, 258–260
  sedimentary charcoal analysis, 260–261
  subalpine forests, 266f
  Western USA fire history case studies, 263–267
    Colorado Rocky Mountains, 265–267
    early Holocene, 263
    Greater Yellowstone Region, 264
    ocean atmosphere interactions, 263
    summer insolation, 263
    tree-ring and charcoal data, 263
    Western Cascades, 264
    winter insolation, 263
Climate-fire "abnormality", 340
Climate variability, 258–260
Coastal sage scrub, 106
Cognitive dissonance reduction, 182–186
Coleoptera species, 103–104
Colorado Front Range, 9–12
Colorado Rocky Mountains, 265–267

Combustible resins, 180–182
Command-and control approaches (CACA), 268
Community of practice groups, 345, 348–350
Community Protection Project, 372–373
Complex early seral forest, 14–15, 31–34, 37f, 308–311
Congo Basin, 212–213
Congressional testimony, 168–169
Conifer seedlings, 368
Consensus scientists, 381–383
Conservation Biology Institute, 172
Conservation science approaches (CSA), 268
Corynephorus canescens, 233
Council for Environmental Quality (CEQ), 316
Coyote (*Canis latrans*), 113
Cricetidae, 109
Crown fires, 166
Cumulative process, 179
Cumulative tree mortality, 375

D

Dark-eyed junco (*Junco hyemalis*), 70–73
Dead trees, 33–34
Deer (*Odocoileus* spp.), 99
Deer mice, 110
Deer mouse (*Peromyscus maniculatus*), 105
Dendroctonus bark beetles, 133–134, 149–150
Denser, 365
Denser hummock grass spinifex (*Triodia* spp.), 110
Deserts, 109–110
Dipodidae (jumping mice), 104–105
Diptera (flies), 103
Disaster, 296
Discarded materials, 296
Disturbances, 134–136
    dependent systems, 88–89
    systems, 65
Donner fire (1960), 31
Douglas-fir (*Pseudotsuga menziesii*), 35–36, 102–103, 133–134, 264
Douglas-fir bark beetle (*Dendroctonus pseudotsugae*), 133–134
Douglas-fir forests, 152
Douglas-fir/ninebark (*Physocarpus* sp.), 118–119
Douglas-fir-Western larch forest in Montana, 110

Douglas squirrels (*Tamiasciurus douglasii*), 108–109
Draft 2010 California Fire Plan, 176–178
Drier climate, 201
Drier microclimates, 364–365
Drought conditions, 277
Dry forests, 169, 357, 359–360
Dry ponderosa pine (*Pinus ponderosa*), 165
Dusky flycatcher (*Empidonax oberholseri*), 80
Dusky robin (*Melanodryas vittata*), 82–84
Dusky woodswallow (*Artamus cyanopterus*), 82–84

E

Eastern Cascades, 36
Eastern gray kangaroo (*Macropus giganteus*), 107–108
Eastern Oregon Dry Pine, 352
Eastern white pine (*Pinus strobus*), 240–242
Ecological processes, 239
Ecological resilience, 14
Ecological screens, 302, 306
Ecological trap, 86–87
Ecosystems, 116
    resilience and mixed-and high-severity fire, 14–15
El Niño-Southern Oscillation, 30–31
Elephant (*Loxodonta africana*), 215
Elk (*Cervus elaphus*), 32, 99
Elytropappus rhinocerotis (Renosterbos), 222b
Empirical field observations, 143
Empirical modeling systems, 138–139
Endangered Habitats League, 189
Endangered Species Act (ESA), 88
Engelmann spruce (*Picea engelmannii*), 133–134, 264
Environmental impact report (EIR), 184
Episodic disturbance agents, 116–117
Erethizontidae (porcupine), 104–105
Eucalyptus, 107
Eurasian pygmy shrew (*Sorex minutus*), 235–236
Eurasian shrew (*Sorex araneus*), 235–236
European badger (*Meles meles*), 114b
European beech (*Fagus sylvatica*), 233–234
European Forest Fire Information System, 231
Extreme fire-weather, 384

F

Fagus sylviatica, 234
Fauna
  and fire-affected habitat structures, 203–204
  of Mixed-severity Fire, 202–203
Feed stock, 348
Ferns (*Pteridium aquilinum*), 234
Festuca vaginata subsp., 233
Field-based research, 144–145
Field data, 10–11
Fiddle players, 342–354
  collaborative as backcountry fiddlers, 344–354
    dwarf mistletoe control project, 350f
    unlogged rare wetland inclusion within complex early seral forest, 349f
  fire management politics, 342–344
  The Nature Conservancy (TNC), 344–354
Financial pressure, 188
Fir (*Abies* spp.), 36, 264
Fire, 3, 133–137, 148, 215–218, 296, 359–360
  adapted ecosystems, 34–35, 164
  adapted forests, 380
  adapted species and communities, 297–299
  affecting forest resilience in context of climate change, 150–153
  behavior models, 140, 142, 147–148
  brands, 386
  countering fire hyperbola and doublespeak, 354–355
  degradation, 175–179
  faunal response to spatial outcomes of, 205
  fire, homes on fire, 355–356
  history, 258
    across moisture gradient, 262–263
    studies, 15
    tree ring reconstructions of, 11–13
  intensity, 166
  loving beetle communities, 104
  misconceptions, 168–182
  money train, 343–344
  prevention, 340
  refugia, 218
  regimes, 136–137, 169–170
  remediation, 343–344
  resilience, 343–344
  restoration, 343–344
  risk reduction should be focused on human population centers, 84
  scar tree-ring records, 258–260
  suppression
    human communities, 84
    myth, 174–175
    paradigm, 163–166, 186
    policies, 4
    shrublands, 175
    weather, 147–148
Fire activity, 31
Fire effects, context dependent
  postfire vegetation conditions, 68–79
  time since fire, 65–66
    biological diversity, 66
    Black Mountain fire, 65
    brown creeper (*Certhia americana*), 65
    disturbance-based systems, 65
    house wren (*Troglodytes aedon*), 65
    mixed-severity fires, 66
    old growth, 66–68
    plant and animal species, 65
    postfire period, 65
    sapsucker (*Sphyrapicus thyroideus*), 65
Fire exclusion/fuels buildup model, 6–7
Fire Learning Network (FLN), 344–345
Fire moss (*Ceratodon purpureus*), 242
Fire Resource and Assessment Program's Fire History Database (FRAP), 179
Fire severity, 167
  stand age influence on, 201–202
  thinning, 85
  time since fire influence, 6–8
    chaparral, 8
    forest understory, 8
    fuel load accumulation, 6–7
    long-unburned forests, 7–8
    missed fire cycle, 7
    severe stand-replacing fires, 6–7
Fireweed (*Chamerion angustifolium*), 242
Fisher (*Martes pennanti* or *Pekania pennant*), 112
Flame robin (*Petroica phoenicea*), 82–84, 204, 319–320
Flammulated owl (*Psiloscops flammeolus*), 70–73
Flushing, 308–309
Foliar moisture content (FMC), 142
Forage, 109
Foraging studies, 119–120
Forest mustelids, 99
Forest Service management indicator species, 69b–70b
Forest (s), 4–5, 7–8, 106–109

disturbances, 154
fires, 45–46, 231
resilience in context of climate change, 150–153
restoration, 87–88
service, 344
    to Forest and Agriculture Organization, 369–370
    succession process, 14
    thinning, 87–88
    and burning, 343
Foxes, 113
Freedom of Information Act, 347
Fuel accumulation, 6–7
"Fuel break", 335–336, 387
Fuel combustion, 279
Fuel management zones (FMZs), 302–303
Fuels
    centric community, 393
    management, 386
    reduction, 188, 345, 352, 354, 366
Funaria hygrometrica, 234
Fungal pathogens, 149–150

G

General Land Office records, 44–45
General Land Office survey data, 11–12, 38
General Land Surveys, 383
Geomyidae (gophers), 104–105
Geopyxis carbonaria, 234
Giant sequoias (*Sequoiadendron giganteum*), 370–372
Global climate change models, 257
Good fire, 345, 348–350, 389–391
Grampians National Park, 107
Grand Teton National Parks, 73–74
Grass Valley Fire, 188
Gray-collared chipmunks (*Tamias cinereicollis*), 108–109
Gray fantail (*Rhipidura albiscapa*), 82–84
Gray foxes (*Urocyon cinereoargenteus*), 113
Gray owls (*Strix nebulosa*), 32
Gray shrike-thrush (*Colluricincla harmonica*), 82–84
Gray wolves, 99
Great Fire, 34–35
Great gray owl (*Strix nebulosa*), 70–73, 314–315
Greater gliders (*Petauroides volans*), 107–108
Greater Santa Fe Fireshed Coalition (GSFFC), 348–350
Greater Yellowstone Region, 264
Greenbark (*Ceanothus spinosus*), 176
Greenhouse gas emissions (GHGs), 257
Greenhouse gases (GHGs), 271
Griffith Park Fire, 171
Grizzly bears (*Ursus arctos*), 99, 113–114
Gypsophila paniculata, 233

H

Habitat
    alteration, 215
    management, 220–223
    type, 110
Hairy woodpecker (*Picoides villosus*), 70–73
Hardwood forests, 82
Hare (*Lepus* spp.), 110
Harvest mice (*Reithrodontomys megalotis*), 106
Hazardous fuels, 356–357
Healthy Forest Restoration Initiative, 302
Healthy Forests Restoration Act [2003], 188
Heat dome, 335
Helichrysum arenarium, 233
Herbivores, 215–218
Heteromyidae (pocket mice and kangaroo rats), 104–105
High fire severity, overestimation of, 308–311
High-intensity fires, 164, 166, 168, 181
High-resolution charcoal investigations, 261
High severity fires, 117, 137–148, 152, 202, 380–381
    bark beetle infestation, 149–150
    beetle-killed fuels, 147–148
    beget mixed-severity results, 85
    Central Europe
        Aeolian sands, 232–233
        burned areas, 231
        burned forest stands, 234–235
        forest fires, 231
        invertebrates and postfire habitat, 235–236
        Natura 2000, 231
        postfire succession, 233–236
        setting, 231
    conflict between modeling and observational results, 147–148
    converting dense dry forests to nonforest, 367–372
    earlier hypotheses and current research, 3–14

High severity fires (*Continued*)
  ecosystem resilience and mixed-and high-severity fire, 14–15
  evidence for, 8–14
    aerial photos, 9–10
    charcoal and sediment reconstructions, 13
    direct records and reconstructions from early land surveys, 10–11
    historical higher-severity fire proportions, 9t
    historical reports, 10
    plant and animal fire adaptations, 13–14
    tree ring reconstructions of stand densities and fire history, 11–13
  fire frequencies, 15
  fire management policy, 4–5
  fire weather, 147–148
  forest management policies, 15–17
  high-severity fire patches in montane forests, 16f
  land managers, 3, 12–13
  lodgepole pine forests, 142–145
    active crown fire, 142
    beetle-affected *vs.* beetle-unaffected stands, 144
    burning conditions, 144–145
    canopy bulk density (CBD), 142–143
    fire behavior models, 142
    fire ignitions, 145
    lightning strikes, 145
    mountain pine beetle (MPB), 142
    prefire stand conditions, 144–145
    stand and landscape scales, 142
    stand-scale FMC, 142–143
  low-and low/moderate-severity fire, 3
  methodological considerations, 137–148
    active crown fire, 138–139
    beetle-affected forests, 140
    experimental burns, 137–138
    field experiments, 137–138
    field/remote sensing methods, 140
    fire behavior modeling, 137–138
    operational fire behavior models, 138–139
    outbreaks and fires, 140
    physics-based fire models, 138–139
    retrospective case studies, 137–138
    retrospective studies, 140, 141t
    wildfire behavior modeling, 138
  patches, 36
  scenario, 368–369
  spruce-fir forests, 145–147
    crown flammability, 150–151
    Engelmann spruce, 146
    foliar moisture and chemistry, 146
    spruce beetle outbreaks, 146–147
    subalpine landscape, 145–146
  Sub-Saharan Africa
    Africa's savannah habitat, 218
    big picture, 211–212
    coevolution of savannah, herbivores, and fire, 214–215
    fire important in, 212–213
    habitat changes forest loss/gain and considerations, 218–220
    habitat management through controlled burns, 220–223
    herbivores and fire, 215–218
    people and fire, 213–214
    southwestern cape renosterveld management, 223–225
  subsequent outbreaks, 148–150
    lodgepole pine forests, 148–150
    mortality, nonbeetle causes of, 149–150
    spruce-fir forests, 149
  time since fire influence fire severity, 6–8
  unnatural levels, 15–17
  Western US Forests, 6
    historical montane forest landscape, 6
    native vegetation, natural regeneration of, 5f
    regional climate and biophysical setting, 6
    time since fire influence fire severity, 6–8
High-severity fire patches, 311
High-severity reburned areas, 31
Hipparchia statilinus, 233
Historic range of variability (HRV), 257
Historical forests, 4
Hollow-bearing trees, 203–204
Home-ignition zone, 48
Hotter microclimates, 364–365
House wren (*Troglodytes aedon*), 65
Housing funds, 348
Human-caused fires, 88–89
Human communities, fire suppression should be focused on, 84
Human population centers, 84
Hydrophobia, 180–182
Hydrophobic soils, 180–181

I

Insects, 104, 134–136
Intense fires, 106, 341
Inventoried roadless areas (IRAs), 303b, 304
Invertebrates, 235–236

J

Jack pine (*Pinus banksiana*), 80–82, 143–144, 240–242
Jasper Fire (2000) Black Hills, 315–318
   Black Hills National Forest (BHNF), 315
   burn intensities, 318f
   case study conclusions, 317–318
   commercial timber sale, 316–317
   Secretary of Agriculture, 317
   tree mortality, 316
Jeffrey pine (*Pinus jeffreyi*), 366
Jewel Cave National Monument, 315
Joshua tree (*Yucca bevifolia*), 109

K

Kangaroo rats (*Dipodomys* spp.), 106
Kiggelaria africana (wild peach), 222b
Kirtland's warbler (*Setophaga kirtlandii*), 80–82
Klamath Mountains, 301–302
Klamath-Siskiyou ecoregion, 304
Knee-jerk response to fire, postfire logging and, 295–297
Kruger National Park, 221

L

Labor Day fires, 269
Lagomorphs (pika, hares, and rabbits), 104–106
Lake Arrowhead, 188
Lake sediments, 245
Land survey records, 11
LANDFIRE dataset, 369–370
Landscape
   megafires and landscape heterogeneity, 39–43
   resilience, 348–350
   traps, 296
Large carnivores, 111
Large cats, 111–113
Large fires, 242–243
Large high-severity fire patches, 13–14, 30
Large severe fire patches, 31–34
Lasioglossum nitidiusculum, 235

Late successional reserves (LSRs), 303b, 304, 346
Lazuli bunting (*Passerina amoena*), 70–73
Leadbeater's possum (*Gymnobelideus leadbeateri*), 203–204
Least chipmunks (*Tamias minimus*), 108–109
Lepidoptera (moths), 103
Lessons learned and closing remarks, 392–395
Lewis's woodpecker (*Melanerpes lewis*), 70–73
Lichens, 120
Lightning fire approach, 216
Limber pine (*Pinus flexilis*), 149–150
Lions (*Panthera leo*), 216
Little brown bat (*Myotis lucifugus*), 103
Little Ice Age (1600–1900 AD), 264
Livestock exclosures, 388
Living organisms, 13
Lizards, 203–204
Local agency, 183–184
Lodgepole pine (*Pinus contorta*), 35–36, 108, 133–134, 142–145, 148–150, 240–242, 264, 352
Long-eared myotis (*Myotis evotis*), 103
Long-term studies, 107–108
Los Angeles Times, 185
Los Padres National Forest, 170
Low-mixed severity approach, 393
Low-moderate fire regime model, 383
Low-severity fire, 258–260
Lynx (Lynx canadensis), 135b–136b

M

MacGillivray's warbler (*Geothlypis tolmiei*), 70–73
Madrean evergreen woodland, 115
Mammals
   bats, 101–104
   carnivores, 111–115
   ecological tolerance, 99
   flagship mammal species, 99
   management and conservation relevance, 120–121
   mixed-and high-severity wildfire, 99
   severe wildfire, 121–125
   small mammals, 104–110
   ungulates, 116–120
Management indicator species, 88
Manzanita (*Arctostaphylos* sp.), 163
Marble Cone Fire, 185–186
Marchantia polymorpha, 234

Marsupials, 203–204
Martens (*Martes* spp.), 111–112
Masked shrews (*Sorex cinereus*), 106–107
McNally fire (2002), 31
Mean Annual Rainfall (MAR), 217
Media, 185–186
Medieval Climate Anomaly, 264
Mediterranean hackberry (*Celtis australis*), 114b
Mediterranean-type climate shrublands, 168
Megafires, 27–30, 43–45
  beta and alpha diversity, 27–28
  biological legacies, 47
  catastrophes, 27
  climate change, 30
  complex early seral forests, 31–34, 37f
  ecologically beneficial, 27–28
  fire deficits, 44–45
  fire plumes, 27
  General Land Office records, 44–45
  global change agents, 30–31
  greenhouse gas emissions, 48
  historical evidence of, 34–39
    Black Friday Bushfires (1939), 34–35
    Black Hills, 38–39
    Eastern cascades and Southern cascades, 36
    Oregon Coast Range and Klamath Region, 36
    Rocky Mountain Region, 35–36
    Sierra Nevada, 38
    Southwestern United States and Pacific Southwest, 38
  keystone, 33–34
  and landscape heterogeneity, 39–43
    high-severity fire patches, 39, 39t, 42f
    natural postfire conifer regeneration, 40
    relative delta normalized burn ratio, 40t
  landscape-scale fires, 44
  language, 45–46
  large severe fire patches, 31–34
  postfire landscapes, 46–47
  pulse *vs.* chronic disturbance, 33b
  regional fire deficits, 29
  smoke plume, 28f
  spatiotemporal characteristics, 29
  stand-age analysis, 44–45
  top-down ecosystem process, 47–48
  in western United States, 34f
Melolontha hippocastani, 232–233
Merimna, 103–104
Merriam's kangaroo rat (*Dipodomys merriami*), 109
Mesocarnivores and large cats, 111–113, 112f
  fire affected sites, 111–112
  fisher, 112
  forested habitats, 111–112
  gray fox and coyote scat, 113
  marten foraging intensity, 111–112
  postfire logging, 113
  scat sampling, 112
  stone marten, 111–112
Message framing, 342
Mexican gray wolves (*Canis lupus baileyi*), 113
Mexican spotted owl (*Strix occidentalis lucida*), 32–33, 69b–70b
Microclimates, 364–365
Mid-elevation forests, 103
Midtropospheric surface-blocking events, 30–31
Miramichi fire, 34–35
Mistletoe infestation, 350
Mixed-conifer bird species, 80
Mixed-conifer forests, 11–12, 36, 352
Mixed responders, 64–65
Mixed severity, 211–212
  fire regimes, 258–260
Mixed-severity fire, 258–260, 380–381
  Central Europe
    setting, 231
  earlier hypotheses and current research, 3–14
  ecosystem resilience and mixed-and high-severity fire, 14–15
  evidence for, 8–14
    aerial photos, 9–10
    charcoal and sediment reconstructions, 13
    direct records and reconstructions, 10–11
    historical higher-severity fire proportions, 9t
    historical reports, 10
    plant and animal fire adaptations, 13–14
    tree ring reconstructions of stand densities and fire history, 11–13
  fire frequencies, 15
  fire management policy, 4–5
  forest management policies, 15–17
  high-severity fire patches in montane forests, 16f

land managers, 3, 12−13
large fires in Canadian boreal ecosystem
   biodiversity, 248−250
   fire regime of, 242−246
   green halo, 238−239
   land of extremes, 239
   plants coping with fire, 242
   temporal patterns of fire and changes in boreal, 246−248
   vegetation, 240−242
low-and low/moderate-severity fire, 3
mountain ash forests, southeast Australia
   conservation challenges and future fire, 205−207
   distribution of old-growth forests, 202
   fauna and fire-affected habitat structures, 203−204
   fauna of, 202−203
   faunal response to spatial outcomes, 205
   influence of stand age on fire severity, 201−202
   mountain ash life cycle, 199−201
   setting, 197−198
regimes, 14−15
Sub-Saharan Africa
   Africa's savannah habitat, 218
   big picture, 211−212
   coevolution of savannah, herbivores, and fire, 214−215
   fire important in, 212−213
   habitat changes forest loss/gain and considerations, 218−220
   habitat management through controlled burns, 220−223
   herbivores and fire, 215−218
   people and fire, 213−214
   southwestern cape renosterveld management, 223−225
time since fire influence fire severity, 6−8
unnatural levels, 15−17
Western US Forests, 6
   historical montane forest landscape, 6
   native vegetation, natural regeneration of, 5f
   regional climate and biophysical setting, 6
   time since fire influence fire severity, 6−8
Moderate-/high-severity fire, 31
MODIS Rapid Response System Global Fire Maps, 211f
Moisture gradient, fire history across, 262−263, 262f
Mojave Desert, 109
Monitoring Trends in Burn Severity (MTBS) system, 309−311
Montane-Eucalyptus forests, 34−35
Montane forests, 4−5
Moose (*Alces alces*), 114
Mortality, nonbeetle causes of, 149−150
Mount Graham red squirrels (*Tamiasciurus fremonti grahamensis*), 108−109
Mountain ash (*Eucalyptus regnans*), 107, 197, 318−319
Mountain ash forests, southeast Australia, 201
   Black Saturday, 204
   conservation challenges and future fire, 205−207
   distribution of old-growth forests, 202
   fauna and fire-affected habitat structures, 203−204
   fauna of, 202−203
   faunal response to spatial outcomes of fire, 205
   fire severity
      influence, 201−202
      patterns, 198
   high-severity wildfires, 204
   hollow-bearing trees, 203−204
   life cycle, 199−201
   logged forest areas, 206−207
   regeneration, 198
   risk assessment, 206
   seedlings, 198
   setting, 197−198
   shrub layers of, 197−198
   spatiotemporal mosaic, 206
   stags, 203f
   topographically sheltered, 200f
   Victorian Central Highlands (VCH), 198, 199f
Mountain bluebird (*Sialia currucoides*), 70−73
Mountain brushtail possum (*Trichosurus cunninghami*), 204
Mountain hemlock (*Tsuga mertensiana*), 265
Mountain lions (*Puma concolor*), 32, 113
Mountain pine beetle (MPB) (*Dendroctonus ponderosae*), 133−134, 142, 144
Mule deer (*Odocoileus hemionus*), 32, 118
Muridae (voles, mice, rats, and woodrats), 104−105

Musk daisy bush (*Olearia argophylla*), 197–198, 320
Myrtle beach (*Nothofagus cunninghamii*), 197–198

N

National Cohesive Wild-land Fire Management Strategy, 388
National Environmental Policy Act (NEPA), 316
National Interagency Fire Center, 35, 49
National Park Service, 189
National Wild and Scenic Rivers Act [1968], 303b
Native American burning practices, 171
Native biodiversity, 14
Natura 2000 network, 231
Natural climate solutions, 376
Natural disturbance forces, 3
Natural fire regime, 183
Natural postfire conifer regeneration, 42–43, 313–314, 316f
Natural regeneration, 318
"Natural" fire, 63–64
Naturally occurring megafires, 30
Nature-based correction, 381–384
Nature's natural processes, 383–384
Nature's Phoenix, 298b, 380–381, 383–384
Nature's Phoenix Rises
  active management work, 385–389
  disconnect between independent research and wildfire attitudes, 391–392
  lessons learned and closing remarks, 392–395
  nature-based correction, 381–384
  public attitudes, 389–391
  pyrodiversity begets biodiversity reaffirmed, 379–381
  respond to root causes and not just effects, 384–385
Nectar-feeding bats, 101
"Negative responders", 64–65
Nestucca fires, 36
NEXUS, 138–139
Nonbeetle causes of mortality, 149–150
Nongovernmental organization (NGO), 344–345
Nonnative plants, 295
North American fire science, 6–7
Northern flicker (*Colaptes auratus*), 70–73
Northern flying squirrels (*Glaucomys sabrinus*), 108–109
Northern goshawk (*Accipiter gentilis*), 314–315
Northern hawk owl (*Surnia ulula*), 70–73
Northern red-backed voles (*Myodes rutilus*), 106–107
Northern Sierra Nevada, 31
Northern spotted owl (*Strix occidentalis caurina*), 32–33, 346
Northwest Forest Plan, 306–307
Norway spruce (*Picea abies*), 233–234

O

Oak (*Quercus* spp.), 41–42, 265
Ocean atmosphere interactions, 263
Old-growth chaparral, 172–173
Old-growth forests, 66–68
  distribution, 202
  old-growth species and severe disturbance events, 69b–70b
Old-growth species and severe disturbance events, 69b–70b
Olea europaea (wild olive), 222b
Olive-sided flycatcher (*Contopus cooperi*), 70–73, 74f
Operational fire behavior models, 138–140
Operational screens, 306
Optics, 343
Oregon
  disastrous urban fires in, 340
  wildland–urban interface, 335
Oregon 2020 Labor Day fires, 269
Oregon Health Authority Report, 278
Oregon red tree vole (*Arborimus longicaudus*), 347
Overgrown forests, 370
  contemporary western US dry forests overgrown, 364
  denser, mature, and old forests burn, 364–365
    lower-severity fire in Creek fire, 365f
  high-severity fire converting dense dry forests to nonforest, 367–372
  implications of prologging misinformation, 372–375
  popular narrative of, 363
  thinning reduce overall severity in wildfires, 365–367
    low-density mixed-conifer forest, 366f
Oxypteris, 103–104

P

Pacific Decadal Oscillation, 30–31
Pacific fishers (*Pekania pennanti*), 32–33
Pacific madrone (*Arbutus menziesii*), 347
Pacific Southwest Region of the US Forest Service, 189
Pacific subtropical high-pressure system, 263
Pale rats (*Rattus tunneyi*), 107
Paleoecologists, 13
Patch-mosaic burning strategy, 218
Peak detection, 261
Pholiota highlandensis, 234
Pine forests, 232
Pine siskin (*Spinus pinus*), 70–73
Pinnacles National Park, 180
Pinus spp., 234
Pinyon mice (*Peromyscus truei*), 106–107
Plant
  coping with fire, 242
  fire adaptations, 13–14
Pocket mice (*Chaetodipus* spp.), 106
Politicians, 342
Polyphylla fullo, 233
Pomaderris (*Pomaderris aspera*), 197–198
Ponderosa pine (*Pinus ponderosa*), 6, 11–12, 32–33, 102–103, 149–150, 264, 308–309, 352, 364, 374–375
Ponderosa pine-dominated forests, 38–39
Ponderosa pine/bluebunch wheatgrass (*Pseudoroegneria spicata*), 118–119
Ponderosa pine/Douglas-fir (*Pseudotsuga menziesii*), 35–36
"Positive responders", 64–65
Postfire "salvage" logging, 86–87, 86f
Postfire landscapes, 46–47, 308
Postfire logging, 205–206, 319
  2009 Wildfires, Victoria, Australia, 318–321
    effects, 318–319
    intense fires, 318–319
    mixed-severity fires of, 320f
    nature ecosystem assessment criteria, 320–321
    operations at Paradize Plains, 320f
    potential impacts, 319–320
  Biscuit Fire of 2002, Southwest Oregon, 301–307
    administrative screens, 306
    Biscuit Project Scope, 302–303
    burn perimeter, 307f
    burn severities, 301f, 304f
    canopy mortality, 304f
    case study conclusions, 306–307
    context and scale matter, 304–305
    ecological processes, 302
    ecological screens, 306
    implications, 302
    integrating context and scale into project decisions, 306
    operational screens, 306
  case study, 301–321
  Jasper Fire of 2000, Black Hills, South Dakota, 315–318
    Black Hills National Forest (BHNF), 315
    burn intensities, 318f
    case study conclusions, 317–318
    commercial timber sale, 316–317
    Secretary of Agriculture, 317
    tree mortality, 316
  and knee-jerk response to fire, 295–297
  land use and, 303b
  proposals, 296–297
  and related activities cumulative effects, 297–301
    biological legacies as building blocks for Nature's Phoenix, 298b
    incompatible *vs.* ecologically compatible postfire management, 300f
    on soil in two areas in southwest Oregon, 300f
  Rim Fire (2013), Sierra Nevada, California, 308–315
    case study conclusions, 314–315
    Forest Service, 308
    high-severity fire, 312f
    Monitoring Trends in Burn Severity (MTBS), 309–311
    natural postfire conifer regeneration, 313–314, 316f
    overestimation of high fire severity, 308–311
    ponderosa pines, 310f
    postfire "flushing" among conifers, 310f
    rapid assessment, 308–309
    undisclosed effects on California spotted owls, 311–312, 313f
Postfire management implications, 47, 84, 323
  fire risk reduction on human population centers, 84
  high-severity fires beget mixed-severity results, 85

Postfire management implications (*Continued*)
  mimic nature, 87–89
  mitigate fire severity, 85
  postfire salvage logging, 86
Postfire recovery, 179
Postfire rehabilitation, 87
Postfire vegetation conditions, 68–79
Potoroo (*Potorous tridactylus*), 107–108
Prefire forest patches, 69b–70b
Prescribed burning, 87–88, 180
Prescribed fire, 179–180
Proforestation, 376
Prologging misinformation, implications of, 372–375
Pruning plants, 169
Public attitudes, 389–391
Pyrodiversity begets biodiversity reaffirmed, 379–381
  circular succession, 382f
  Duncan Canyon on Tahoe National Forest, 379f
Pyronema omphalodes, 234
Pyrophilous Buprestids, 103–104

Q
Quaking aspen (*Populus tremuloides*), 152–153

R
Rabbits, 108
Raccoon (*Procyon lotor*), 113
Radio-collared black bears, 115
Rapid Assessment of Vegetation (RAVG), 312f
Recovery via logging, 296
Red alder (*Alnus rubra*), 264
Red crossbill (*Loxia curvirostra*), 70–73
Red fox (*Vulpes vulpes*), 113
Red maple (*Acer rubrum*), 240–242
Red-necked wallaby (*Macropus rufogriseus*), 107–108
Red phase of outbreaks, 142–143
Red pine (*Pinus resinosa*), 240–242
Red spruce (*Pinus rubens*), 240–242
Reduce fuels, 357
Reintroducing beavers (*Castor canadensis*), 359
Remote sensing methods, 140, 178
Renewal processes, 380
Resprouters, 168
Resprouting eucalyptus trees, 67f
Resprouting process, 176

Rhinoceros (*Diceros bicornis* or *Ceratotherium simum simum*), 225
Rhizina undulata, 234
Rhyolite plateaus, 264
Rim Fire (2013)
  California, 308–315
    case studies, 314–315
    Forest Service, 308
    high-severity fire, 312f
    Monitoring Trends in Burn Severity (MTBS), 309–311
    natural postfire conifer regeneration, 313–314, 316f
    overestimation of high fire severity, 308–311
    ponderosa pines, 310f
    postfire flushing among conifers, 310f
    rapid assessment, 308–309
    undisclosed effects on California spotted owls, 311–312, 313f
Rim fire, 32, 46–47
Ringtail (*Bassariscus astutus*), 113
Robinson Fire, 144
Rocky Mountain fires (1988), 73–74
Rocky Mountain Region, 35–36
Rocky Mountain Subalpine forests
  climate change, 150–153
    bark beetle outbreaks and carbon cycle, 151b
    compounded disturbances, 150–151
    regeneration, 152
    warming climate, 150–151
  fire, beetles, and interactions, 133–137
    bark beetle outbreaks and biodiversity, 135b–136b
    disturbance interactions, 136–137
    ecosystems, 133–134
    fire regimes, 136–137
    forest landscapes, 134–136
    modern fire suppression, 133–134
    postdisturbance regeneration, 134–136
    predisturbance conditions, 134–136
    spatial heterogeneity, 136–137
    upper montane forests, 133–134
    wildfires and insect outbreaks, 134f
  high severity fires, 137–148
Rocky Mountains, 145–146, 149–150
Rodents, 99, 104–105
Rodeo-Chedeski fire, 29–30
Rogue Basin, 345–347
Root causes, 384–385
Rutstroemia carbonicola, 234

S

Sage scrub habitats, 175
Salvage, 296
  logging, 295
    operations, 205–206
San Bernardino Mountains, 113
San Diego County, 183–184
San Felipe Valley, 180
San Francisco Bay-area publication, 185–186
San Gabriel Mountains, 117
San Jacinto Mountains, 118
San Joaquin Valleys, 176
Sandy desert habitats, 110
Santa Barbara County Fish and Game Commission, 170
Santa Clara Valley, 176
Santa Fe Mountains Landscape Resiliency Project, 350–351
Santa Monica Mountains National Recreation Area (SMMNRA), 189
Santiago Canyon fire, 38
Sapsucker (*Sphyrapicus thyroideus*), 65
Saskatoon (*Amelanchier alnifolia*), 242
Scarlet robin (*Petroica boodang*), 204
Sciuridae (squirrels, chipmunks, and marmots), 104–105
Scots pine (*Pinus sylvestris*), 232
Sea surface temperature, 30–31
Sediment reconstructions, 13
Sedimentary charcoal analysis, 260–261
Seed dispersal by carnivores, 114b
Seedling abundance, 201
Serengeti ecosystem, 216
Sessile oak (*Quercus petraea*), 233–234
Severe fire, 99
  bird species, 79–84
Severe wildfire, number of studies by taxa showing directional response to, 121–125
Shifting baseline syndrome, 383
Shrub steppe, 353
Shrubland ecosystems, 171
Shrublands, 117, 163, 165–166
Sierra Nevada, 38, 308–315, 348, 364
  forests, 372–373
  mixed-conifer, 374–375
Sierra Nevada bighorn sheep (*Ovis canadensis sierrae*), 117
Sierra Nevada Mountains, 31
Silver birch (*Betula pendula*), 233–234
Silver-eye (*Zosterops lateralis*), 82–84
Silver wattle (*Acacia dealbata*), 197–198, 204
"Single-species management", 88
Sitka spruce (*Picea sitchensis*), 264
Small mammals, 100–101, 104–110
  assemblages, 104–105
  chaparral and coastal sage scrub, 106
  deer mice, 110
  deer mice increase after severe fire in a variety of habitats, 105f
  deserts, 109–110
  fire effects, 105
  forests and woodlands, 106–109
SMMNRA's approach, 189
Smoke, 281
Smoke plume, 28f
Smoldering logs, 168
Snowshoe hares (*Lepus americanus*), 108
Social–ecological challenges, 283
Soil seed bank, 173
Solar radiation, 263
Soricidae (shrews), 104–105
South Africa's Kruger National Park, 211–212
Southern California's largest fire, 186
Southern Cascades, 36
Southern greater glider (*Petauroides volans*), 204
Southern Oregon, 352–353
Southern Oregon Forest Restoration Collaborative (SOFRC), 345–346
Southern Oregon Timber Industry Association, 345–346
Southern red-backed voles (*Myodes gapperi*), 106–107
Southern Rocky Mountains, 15–17, 150–151
Southern sassafras (*Atherosperma moschatum*), 197–198
Southwest Oregon, 301–307, 345–346
Southwestern cape renosterveld management, 223–225
Species, 65
Spergula morisonii, 233
Spotted owl (*Strix occidentalis*), 68
Spotted pardalote (*Pardalotus punctatus*), 82–84
Spruce (*Picea* spp.), 114
Spruce beetle (*Dendroctonus rufipennis*), 133–134, 152
Spruce-Fir Forests, 145–147, 149
Spruce sawyer (*Monochamus scutellatus*), 249–250

Stakeholder collaborative, 345
Stand densities, tree ring reconstructions of, 11–13
Standard fire behavior models, 138–139
Stanislaus National Forest, 308
Stipa borysthenica, 233
Stone marten (*Martes foina*), 111–112
Stony desert habitats, 110
Storrie fire (2000), 31
Striped skunk (*Mephitis mephitis*), 113
Sub-Saharan Africa, Mixed-and High-severity Fires in
  adaptive management strategy, 218
  Africa's savannah habitat, 218
  big picture, 211–212
  biodiversity, 211–212
  bird and mammal populations, 218
  coevolution of savannah, herbivores, and fire, 214–215
  Congo Basin, 219
  cycad's coralloid root adaptation, 214
  fire important in, 212–213
  habitat changes forest loss/gain and considerations, 218–220
  habitat management through controlled burns, 220–223
  herbivores and fire, 215–218
  MODIS Rapid Response System Global Fire Maps, 211f
  patch burn, 224f
  patch mosaic recovery, 238f
  people and fire, 213–214
  southwestern cape renosterveld management, 222b, 223–225
  successful conservation strategies, 213–214
  vegetation dynamics, 223f
  vegetation dynamics mediated by fire and rainfall, 223f
Subalpine fir (*Abies lasiocarpa*), 133–134, 240–242, 265–267
Subalpine forests, 147–148
Succession rather than destruction, 171–172
Sugar maple (*Acer saccharum*), 240–242
Sugar pine (*Pinus lambertiana*), 40
Summer insolation, 263
Superb fairywren (*Malurus cyaneus*), 82–84
Suppression forces, 335–336
Surface fire spread model, 138–139, 166
Swamp wallaby (*Wallabia biocolor*), 107–108
Synansphecia muscaeformis, 233

T

Talpidae (moles), 104–105
Tamarack (*Larix laricina*), 240–242
Tamias
  T. amoenus, 108–109
  T. ruficaudus, 108–109
Testing approaches, 381–383
The Nature Conservancy (TNC), 344–354
  thinning/logging project, 346f
Thimbleberry (*Rubus parviflorus*), 119–120
Thinning, 353, 357
  reduce overall severity in wildfires, 365–367
Thresholds of potential concern (TPCs), 218
Timber salvage, 297
Time-since-fire, 380
Tipping points, 269–270
Topography, 146–147
Towns Burndown
  active management, 356–358
  active management become religion of sorts, 358–360
  climate change came knocking at my door, 335–342
  countering fire hyperbola and doublespeak, 354–355
  fiddle players, 342–354
  fire, fire, homes on fire, 355–356
Townsend's chipmunks (*Neotamias townsendii*), 108–109
Townsend's solitaire (*Myadestes townsendi*), 70–73
Treating bottom-up factors, 392
Tree ferns (*Dicksonia antarctica* and *Cyathea australis*), 197–198
Tree ring reconstructions of stand densities, 11–13
Tree species, 152–153
Tree squirrels, 108–109
Tree swallow (*Tachycineta bicolor*), 70–73
Trembling aspen (*Populus tremuloides*), 240–242
Tubbs Fire (2017), 186
Turkish red pine (*Pinus brutia*), 111–112
Tygerberg Preserve
  controlled burns at, 225
  patch mosaic recovery at, 238f
  Western Cape, 223–224
Type conversion, 175, 179–180

U

Undisturbed climax chaparral, 172
Ungulates, 116—120
  burned sites, 102—103
  caribou, 120
  Ceanothus, 119—120
  ecosystem process, 116
  fawn-to-doe ratios, 119
  fire size, 116—117
  green forage availability, 117
  high-severity fire, 117
  mule deer, 118
  papillary morphology, 119—120
  postfire plant regeneration, 116—117
  thimbleberry, 119—120
  vegetation development and productivity, 116
Unhealthy for sensitive groups (USG), 278
US Fish and Wildlife Service, 69b—70b
US Forest Service, 188—190, 314, 364—365, 369—370, 392—393
US General Land Office, 10—11
USDA Forest Service, 4—5

V

Varied sittella (*Daphoenositta chrysoptera*), 82—84
Vegetation fragments ecosystems, 387
Vegetation Treatment Plan (VTP), 184
Ventana Wilderness, 185—186
Verbascum phoeniceum, 233
Victorian Central Highlands (VCH), 197—198
Viejas Mountain, 182

W

Wallow fire (2011), 29—30, 32
Water repellency, 181
Welcome swallow (*Hirundo neoxena*), 82—84
Western balsam bark beetle (*Dryocoetes confusus*), 149—150
Western bluebird (*Sialia mexicana*), 70—73
Western Cascades, 264
Western hemlock (*Tsuga heterophylla*), 264
Western larch forest in Montana, 110
Western red cedar (*Thuja plicata*), 264
Western US Dry forests overgrown, contemporary, 364
Western US Forests, 6
Western wood-pewee (*Contopus sordidulus*), 80

Wet blanket effect on fires, 340, 347
Wet temperate forests, 265
Whitebark pine (*Pinus albicaulis*), 113—114, 240—242, 264
White birch (*Betula papyrifera*), 240—242
White-browed scrubwren (*Sericornis frontalis*), 82—84
White-fronted honeyeater (*Purnella albifrons*), 82—84
White pine (*Pinus strobus*), 233—234
White spruce (*Picea glauca*), 240
Wildebeest (*Connochaetes taurinus*), 216
Wildfires (2009)
  Victoria, Australia, 318—321
    effects, 318—319
    intense fires, 318—319
    mixed-severity fires of, 320f
    nature ecosystem assessment criteria, 320—321
    operations at Paradize Plains, 320f
    potential impacts, 319—320
Wildfires, 134—136, 154, 202—203
  attitudes and independent research, disconnect between, 391—392
  behavior modeling, 138
  historical range of variation (HRV), 267—268
  linking wildfire to anthropogenic climate change, 269—281
  looking back over paleo-record, 258—263
  smoke, 278
  thinning reduce overall severity in, 365—367
  understanding past, present, and future of, 257—258
  Western USA fire history case studies, 263—267
Wildland fires, 84, 340
Wildland—urban interface (WUI), 335, 390
Wildlife Society and Society for Conservation Biology, The, 69b—70b
Williamson's sapsucker (*Sphyrapicus thyroideus*), 65, 70—73, 74f
Willie wagtail (*Rhipidura leucophrys*), 82—84
Windier microclimates, 364—365
Winter insolation, 263
Wolves (*Canis lupus*), 32, 99
Wombats (*Vombatus ursinus*), 107—108
Woodland caribou (*Rangifer tarandus caribou*), 120

Woodlands, 106–109
Woodrats (*Neotoma* spp.), 106
Woody species, 116–117
World Meteorological Organization, 270

Y

Yellow birch (*Betula alleghaniensis*), 240–242
Yellow-faced honeyeater (*Lichenostomus chrysops*), 82–84, 204
Yellow-necked mouse (*Apodemus flavicollis*), 235–236
Yellowstone National Park, 143–144, 393–394
Yosemite National Park, 308

Z

Zaca Fire (2007), 170f
Zerynthia polyxena, 233

9780443137907